Comprehensive Virology 3

Comprehensive Virology

Edited by Heinz Fraenkel-Conrat
University of California at Berkeley

and Robert R. Wagner
University of Virginia

Volume 1: *Descriptive Catalogue of Viruses* – by Heinz Fraenkel-Conrat

Reproduction

Volume 2: *Small and Intermediate RNA Viruses* – Contributors: J.T. August, L. Eoyang, A. Siegel, V. Hariharasubramanian, L. Levintow, E. R. Pfefferkorn, D. Shapiro, and W. K. Joklik

Volume 3: *DNA Animal Viruses* – Contributors: L. Philipson, U. Lindberg, J. A. Rose, N. P. Salzman, G. Khoury, B. Moss, B. Roizman, and D. Furlong

Volume 4: *Large RNA Viruses* – Contributors: P. W. Choppin, R. W. Compans, R. R. Wagner, and J. P. Bader

Volume 5: *Bacterial DNA Viruses* – Contributors: R. Sinsheimer, D. S. Ray, E. P. Geiduschek, M. E. Gottesman, R. Weisberg, and L. M. Kozloff

Structure and Assembly

Contributors include: H. Vasken Aposhian, W. Bauer, D. L. D. Caspar, P. W. Choppin, R. W. Compans, L. V. Crawford, F. A. Eiserling, W. Fiers, John Finch, R. M. Franklin, H. S. Ginsberg, L. Hirth, P. Hung, J. King, Y. Okada, R. R. Rueckert, T. I. Tikchonenko, T. W. Tinsley, J. Vinograd, R. C. Williams, W. B. Wood

Regulation and Genetics

Contributors include: J. Atabekov, D. Baltimore, D. Bishop, G. Bruening, R. Calendar, A. Campbell, P. Cooper, W. Doerfler, W. Eckhart, B. Fields, H. Ginsberg, H. Hanafusa, A. Huang, E. M. J. Jaspars, J. Kates, E. D. Kilbourne, A. Lewis, B. McAuslan, D. Nathans, E. R .Pfefferkorn, C. Pringle, J. R. Roberts, H. L. Sänger, F .W. Stahl, J. H. Subak-Sharpe, W. Szybalski, L. Van Vloten-Doting, and P. Vogt

Interaction of Viruses and Their Hosts

Effects of Physical and Chemical Agents

Comprehensive

Edited by

Heinz Fraenkel-Conrat
Department of Molecular Biology and Virus Laboratory
University of California, Berkeley, California

and

Robert R. Wagner
Department of Microbiology
University of Virginia, Charlottesville, Virginia

Virology

3
Reproduction
DNA Animal Viruses

PLENUM PRESS • NEW YORK AND LONDON

Library of Congress Cataloging in Publication Data

Fraenkel-Conrat, Heinz, 1910-
Reproduction: DNA animal viruses.

(Their Comprehensive virology, v. 3)
Includes bibliographies.
1. Viruses—Reproduction. I. Wagner, Robert R., 1923- joint author. II. Title.
III. Series: Fraenkel-Conrat, Heinz, 1910- Comprehensive virology, v. 3.
[DNLM: 1. Virus diseases. 2. Viruses. QW160 F799ca]
QR357.F72 vol. 3 [QR470] 576'.6484 74-17457
ISBN 978-1-4684-2705-9 ISBN 978-1-4684-2703-5 (eBook)
DOI 10.1007/978-1-4684-2703-5

Softcover reprint of the hardcover 1st edition 1974
A Division of Plenum Publishing Corporation
227 West 17th Street, New York, N.Y. 10011

United Kingdom edition published by Plenum Press, London
A Division of Plenum Publishing Company, Ltd.
4a Lower John Street, London W1R 3PD, England

Foreword

The time seems ripe for a critical compendium of that segment of the biological universe we call viruses. Virology, as a science, having only recently passed through its descriptive phase of naming and numbering, has probably reached that stage at which relatively few new—truly new—viruses will be discovered. Triggered by the intellectual probes and techniques of molecular biology, genetics, biochemical cytology, and high-resolution microscopy and spectroscopy, the field has experienced a genuine information explosion.

Few serious attempts have so far been made to chronicle these events. This comprehensive series, which will comprise some 6000 pages in a total of about 22 volumes, represents a commitment by a large group of active investigators to analyze, digest, and expostulate on the great mass of data relating to viruses, much of which is now amorphous and disjointed and scattered throughout a wide literature. In this way, we hope to place the entire field in perspective as well as to develop an invaluable reference and sourcebook for researchers and students at all levels. This series is designed as a continuum that can be entered anywhere but which also provides a logical progression of developing facts and integrated concepts.

The first volume contains an alphabetical catalogue of almost all viruses of vertebrates, insects, plants, and protists, describing them in general terms. Volumes 2–5 deal primarily, though not exclusively, with the processes of infection and reproduction of the major groups of viruses in their hosts. Volume 2 deals with the simple RNA viruses of bacteria, plants, and animals; the togaviruses (formerly called arboviruses), which share with these only the feature that the virion's RNA is able to act as messenger RNA in the host cell; and the reoviruses of animals and plants, which all share several structurally singular features, the most important being the double-strandedness of their multiple RNA molecules. This grouping, of course, has only slightly more in its favor than others that could have been or indeed were considered.

Volume 3 addresses itself to the reproduction of all DNA-containing viruses of vertebrates, a seemingly simple act of classification, even though the field encompasses the smallest and the largest viruses known.

The reproduction of the larger and more complex RNA viruses represents the subject matter of Volume 4. These share the property of lipid-rich envelopes with the togaviruses included in Volume 2. They share as a group, and with the reoviruses, the presence of enzymes in their virions and the need for their RNA to become transcribed before it can serve messenger functions.

Volume 5 attends to the reproduction of DNA viruses in bacteria, again ranging from small and simple to large and complex.

Aspects of virion structure and assembly of many of these viruses will be dealt with in the following series of volumes, while their genetics, the regulation of their development, viroids, and coviruses will be discussed in subsequently published series. The last volumes will concentrate on host–virus interactions, and on the effects of chemicals and radiation on viruses and their components. At this juncture in the planning of *Comprehensive Virology*, we cannot foresee whether certain topics will become important aspects of the field by the time the final volumes go to press. We envisage the possibility of including volumes on such topics if the need arises.

It is hoped to keep the series at all times up to date by prompt and rapid publication of all contributions, and by encouraging the authors to update their chapters by additions or corrections whenever a volume is reprinted.

Contents

Chapter 1

Parvovirus Reproduction

James A. Rose

Chapter 2

Reproduction of Papovaviruses

Norman P. Salzman and George Khoury

Chapter 3

Reproduction of Adenoviruses

Lennart Philipson and Uno Lindberg

Chapter 4

The Replication of Herpesviruses

Bernard Roizman and Deirdre Furlong

Chapter 5

Reproduction of Poxviruses

Bernard Moss

CHAPTER 1

Parvovirus Reproduction

James A. Rose

Laboratory of Biology of Viruses
National Institute of Allergy and Infectious Diseases
National Institutes of Health
Bethesda, Maryland 20014

1. INTRODUCTION

1.1. Definitions

Parvoviruses are the smallest DNA-containing vertebrate viruses. The generic designation, parvovirus (*parvus* = small), was first proposed in 1965 (Lwoff and Tournier, 1966) and finally accepted in 1970 (Andrewes, 1970). These agents are assembled in the cell nucleus and are icosahedral particles 18–26 nm in diameter, about the size of animal cell ribosomes. They possess considerable heat and acid stability and are not inactivated by lipid solvents. Their densities in CsCl solution are relatively high (about 1.40 g/cm^3) owing to the high DNA content of the particle (20–25%). The capsid proteins of the group members thus far studied can be resolved into three polypeptide components, and all parvoviruses appear to contain a single-stranded DNA genome. Although certain insect and bacterial viruses resemble parvoviruses in many respects (see Sect. 1.2), these have been classified separately (Lwoff and Tournier, 1966) and are not given detailed consideration in this review.

1.2. Classification

Since the first characterization of a parvovirus, rat virus or RV, by Kilham and Oliver (1959), the number of viruses found to have similar general properties has steadily increased. Mayor and Melnick (1966) initially called attention to the fact that these small DNA-containing viruses comprised a distinct group, which they tentatively called "picodnaviruses." Some viruses in this group have been shown to be related by serological or biochemical techniques or both, but genetic relationships among most members remain to be more clearly defined. Existing relatedness data, however, can be coupled with host specificities to provide a useful classification (Table 1) which reveals two prominent features of the parvoviruses: (1) they consist of a number of distinct species which naturally infect a wide variety of animal hosts including man, monkeys, cats, pigs, cattle, dogs, rodents,

TABLE 1

Classification of Parvoviruses

Nondefective Parvoviruses
Rodent viruses
Rat virus (RV), H-3 virus, X14 virus, L-S virus, hemorrhagic encephalopathy virus of rats (HER), Kirk virus
H-1 virus, HT virus (tentative)
HB virus (tentative)
minute virus of mice (MVM)
Feline and related viruses
Feline panleukopenia virus (FPV),[a] mink enteritis virus (MEV)
Porcine viruses
Porcine parvovirus (PPV), KBSH virus (identical to PPV?)
Bovine virus
Bovine hemadsorbing enteric virus (Haden virus)
Canine virus
Minute virus of canines (MVC)
Unclassified viruses
TVX virus
LuIII virus
RT virus
Defective Parvoviruses[b]
Human and simian AAV[c]
AAV-1
AAV-2 (H, M strains)
AAV-3 (H, K, T strains)
AAV-4
Bovine AAV
AAV X_7
Avian AAV (AAAV)
Canine AAV (CAAV)?

[a] LV (= leopard virus) is the prototype strain (Johnson, 1965*a*).
[b] Adenovirus-associated viruses (AAV) or adeno-satellite viruses (ASV).
[c] AAV-1H, AAV-2H, AAV-3H, and AAV-4M are the prototype strains (Hoggan, 1971).

TABLE 2

Chronology of Discovery of Parvoviruses

Virus[a]	Source from which first recovered	Primary natural host	Reference
MEV	Mink liver and spleen	Mink	Wills (1952)
RV	Rat tumor	Rat	Kilham and Oliver (1959)
H-1	HEp 1 cells[b]	Rat?	Toolan *et al.* (1960)
Haden	Calf feces	Cattle	Abianti and Warfield (1961)
H-3	HEp 3 cells[b]	Rat	Dalldorf (1960)
L-S	Rat tumor	Rat	Lum and Schreiner (1963)
X14	Rat mammary tissue	Rat	Payne *et al.* (1963)
HT	Human placenta[b]	Originally rat?	Toolan (1964)
HB	Human placenta[b]	?	Toolan (1964)
AAV-1	SV15 stock	Rhesus monkey?	Atchison *et al.* (1965)
FPV	Leopard spleen	Cat	Johnson (1965*a*)
MVM	Mouse adenovirus stock	Mouse	Crawford (1966)
AAV-2	Ad 12 stock	Man	Hoggan *et al.* (1966)
AAV-3	Ad 7 stock	Man	Hoggan *et al.* (1966)
PPV	Hog cholera virus stock	Pig	Mayr and Mahnel (1966)
AAV-4	SV15 stock	African green monkey	Parks *et al.* (1967*a*)
AAAV	Quail bronchitis virus stock	Bird	Dutta and Pomeroy (1967)
HER	Rat CNS tissue	Rat	El Dadah *et al.* (1967)
MVC	Dog feces	Dog	Binn *et al.* (1968)
CAAV	ICHV stock	Dog	Domoto and Yanagawa (1969)
Kirk	Detroit-6 cells? human serum?	Originally rat?	Boggs (1970)
AAV X_7	Bovine adenovirus type I stock	Cattle	Luchsinger *et al.* (1970)
KBSH	KB cells[c]	Pig	Hallauer *et al.* (1971)
TVX	Amnion cells[c]	?	Hallauer *et al.* (1971)
Lu III	Lu 106 cells[c]	?	Hallauer *et al.* (1971)
RT	Rat fibroblasts[d]	?	Hallauer *et al.* (1971)

[a] Abbreviations defined in Table 1.
[b] Detected after 3–4 blind passages in newborn hamsters.
[c] Continuous human cell lines.
[d] Continuous rat cell line.

and birds, and (2) they can be divided into two major groups, (a) nondefective and (b) defective parvoviruses (Melnick, 1971), based on whether or not they are capable of autonomous reproduction. In general, the viruses listed in Table 1 represent those which are best characterized at present. To provide additional perspective, their chronological sequence of discovery, original source, and primary natural host are noted in Table 2.

1.2.1. Nondefective Parvoviruses

Rodent viruses comprise the largest group of these viruses. The origins of all except Kirk virus (Boggs *et al.*, 1970; Mirkovic *et al.*, 1971) have been discussed in detail in recent reviews (Toolan, 1968, 1972). Kirk virus was found in a line of Detroit-6 cells which had been inoculated with plasma from an individual who had ingested MS-1 infectious hepatitis serum. A close antigenic relationship between Kirk and H-3 virus (Mirkovic *et al.*, 1971) places it in the rodent group. Based on serological tests rodent viruses can be divided into four subgroups: (1) RV, H-3, X14, L-S, HER, and Kirk virus (Moore, 1962; Payne *et al.*, 1964; El Dadah *et al.*, 1967; Nathanson *et al.*, 1970; Lum, 1970; Mirkovic *et al.*, 1971), (2) H-1 and HT virus (Toolan, 1964), (3) HB virus (Toolan, 1964), and (4) MVM (Crawford, 1966). In addition, viruses within subgroups (1) and (2) can be distinguished from each other by red blood cell hemagglutination (HA) patterns (Moore, 1962; Toolan, 1968; Mirkovic *et al.*, 1971). Serological cross-reactions between members from different subgroups have been found among MVM, RV, and H-1 (Hoggan, 1971; Cross and Parker, 1972), but they could be demonstrated only by fluorescent-antibody (FA) staining tests. This evidence suggests a relatively distant antigenic relatedness among the three viruses and justifies their present separate grouping. Whether H-1, HT, and HB are basically rodent or human viruses is still open to question. All three were apparently recovered from human tissues (Table 2). However, these viruses have been tentatively included within the rodent group because human antibody is rare (Toolan, 1968), while H-1 antibody is found frequently in rats (Kilham, 1966; Cross and Parker, 1972), and because all three viruses are pathogenic in newborn hamsters (Toolan, 1968), a feature not yet observed with viruses outside the rodent group.

Johnson and Cruickshank (1966) were the first to conclude that FPV and MEV might be parvoviruses. These viruses are closely related by serum neutralization (Gorham *et al.*, 1966; Johnson *et al.*, 1967). Whether or not they are essentially identical remains to be settled. Hemagglutination-inhibition (HI) tests have not revealed any antigenic relationship between FPV and rodent, porcine, bovine, or unclassified parvoriruses (Hallauer *et al.*, 1971).

Both Haden and PPV were also identified as parvoviruses subsequent to their initial recoveries (Horzinek *et al.*, 1967; Storz and Warren, 1970). The relationship between PPV and KBSH has been investigated in some detail (Hallauer *et al.*, 1971, 1972). KBSH is one of many similar isolates recovered by Hallauer *et al.* (1971) from a number of continuous human cell lines. Owing to a lack of clear distinction between KBSH and PPV by HI tests, it was suggested that KBSH and PPV may be identical and that KBSH might represent a PPV contamination introduced when cells are dispersed with hog trypsin (Hallauer *et al.*, 1971). Hallauer *et al.* (1971) also recovered three other serologically distinct parvoviruses from continuous cell lines, two from human cells and one from rat fibroblasts. These were designated TVX, LuIII, and RT viruses, respectively. While TVX was recovered from several different cell lines, LuIII and RT could be obtained from only single cell lines. Based on HI testing, little, if any, antigenic relatedness was demonstrable among these viruses or between them and rodent viruses, FPV, PPV, or Haden. TVX, LuIII, and RT have therefore been left unclassified. As in the case of KBSH, Hallauer *et al.* (1971) have suggested that TVX, LuIII, and RT probably arose as contaminants. One might suppose that adaptation to new host cells could have resulted in antigenic alterations which presently mask the origins of these viruses. Nucleic acid-hybridization tests should prove useful in assessing possible genetic relationships between these and other parvoviruses.

1.2.2. Defective Parvoviruses

Viruses in this group are capable of reproducing only when the cells they infect are also infected with an adenovirus (Atchison *et al.*, 1965; Hoggan *et al.*, 1966; Smith *et al.*, 1966; Parks *et al.*, 1967*a*). Because it was initially observed that they only multiplied together with a helper adenovirus, they were called adenovirus-associated viruses, or AAV (Atchison *et al.*, 1965). The name adeno-satellite virus, or ASV, has also been proposed (Mayor and Melnick, 1966) owing to analogies between these viruses and the defective satellite tobacco necrosis virus (STNV) (Kassanis, 1962; Reichmann, 1964).* The AAV designation has been widely used, however, and will be employed in this review.

The human and simian AAV group have been best studied, and four serotypes are well defined (Hoggan *et al.*, 1966; Parks *et al.*, 1967*a;* Rose *et al.*, 1968). Only types 2 and 3 (AAV-2 and AAV-3) can

* STNV is a small RNA-containing plant virus whose multiplication also depends on a larger helper virus, tobacco necrosis virus (TNV). Like AAV, STNV strains are serologically unrelated to their helpers, appear to code for their own coat protein, and can interfere with the multiplication of the helper virus itself (see Sects. 1.2.2, 3.2.6, and 3.3.1; Kassanis, 1962; Reichmann, 1964). STNV and the AAV may thus be considered as parasites of their respective helper viruses, representing the smallest entities so far known to enter into such a relationship.

be shown to be serologically related by complement fixation (CF), FA, and serum neutralization (Hoggan, 1971), whereas all four serotypes are genetically related by nucleic acid-hybridization tests (Rose *et al.*, 1968; Koczot and Rose, unpublished). In addition, there are three AAV-3 strains definable by serum neutralization (Hoggan, 1971) and nucleic acid hybridization (Rose and Hoggan, unpublished). The same techniques, however, do not clearly reveal differences between the H and M strains of AAV-2 (Rose *et al.*, 1968; Hoggan, 1971). Hoggan (1971) has investigated antigenic relatedness between each of the four AAV serotypes and several nondefective parvoviruses (RV, H-1, MVM, and Haden). Although this study did not reveal any serological relationships between these viruses and the AAV, nucleic acid-hybridization tests will be required to rule out the presence of possible genomic homologies. In this regard it should be noted that whereas antigenic relatedness can only be demonstrated between AAV types 2 and 3, the genomes of all four AAV serotypes seem to share about the same fraction of sequences in common (see Sect. 2.2.6; Rose *et al.*, 1968; Koczot and Rose, in preparation.) In no instance thus far has either antigenic relatedness or nucleic acid-sequence homology been shown to exist between an AAV and an adenovirus (Atchison *et al.*, 1965; Hoggan *et al.*, 1966, Smith *et al.*, 1966, Parks *et al.*, 1967*a*; Rose *et al.*, 1968).

A striking feature of the human and simian AAV is that two discrete populations of progeny particles are synthesized: One type of particle contains single "plus" DNA strands; the other contains single "minus" DNA strands (Sects. 2.2.1 and 2.2.2; Mayor *et al.*, 1969*b*; Rose *et al.*, 1969). It is expected that this is a property common to all AAV, but confirmatory studies with AAV outside the human and simian group remain to be done.

Only within the past few years have other species-related AAV been identified. Luchsinger *et al.* (1970) have reported serological and physical properties of what appears to be a bovine AAV. This virus, called AAV X_7, was found to be antigenically unrelated to the four human and simian serotypes on the basis of CF and HI tests. Interestingly, AAV X_7 can hemagglutinate certain red blood cells, a property common to all nondefective parvoviruses, but so far observed with only one other AAV, i.e., AAV-4 (Ito and Mayor, 1968). In addition, canine (Domoto and Yanagawa, 1969) and avian (Dutta and Pomeroy, 1967; El Mishad *et al.*, 1973) AAV have been described. The latter virus also appears to be antigenically unrelated to the human and simian serotypes (El Mishad *et al.*, 1973). There is some uncertainty over the canine AAV since it has been found to cross antigenically with

AAV-3 (Onuma, 1971). In other studies, however, serological cross-reactions could not be detected between CAAV and any of the AAV-3 strains (Hoggan, personal communication).

It should be noted that certain other viruses bear a close resemblance to the parvoviruses. Included are the *Escherichia coli* bacteriophage ϕX174 (Sinsheimer, 1959*a,b*) and densonucleosis virus (DNV) of *Galleria mellonella* (Meynadier *et al.*, 1964; Kurstak, 1971).* These viruses are listed in Volume 1 and discussed in detail elsewhere in this series.

2. BIOLOGICAL PROPERTIES

Knowledge of the prominent biological properties of the various parvoviruses is essential to understanding their mode of reproduction. Additional information may be obtained from several other reviews (Kilham, 1966; Toolan, 1968; Rapp, 1969; Hoggan, 1970). Two general features of nondefective parvoviruses have been emphasized repeatedly by many investigators: (1) They multiply best in rapidly dividing cells, and (2) they often produce only subtle and transient cytopathic effects (CPE), making virus detection frequently difficult and uncertain. Aside from searches for most-sensitive cells, measures successfully used to uncover parvoviruses have included blind passages both *in vitro* and *in vivo* (Kilham and Oliver, 1959; Toolan *et al.*, 1960; Toolan, 1964; Kilham and Maloney, 1964; Robey *et al.*, 1968) and repeated extraction of cell sheets with a glycine buffer (Hallauer *et al.*, 1971). Viral assays have been carried out by a tissue culture infective dose ($TCID_{50}$) procedure (Johnson, 1967; Mayr *et al.*, 1968; Binn *et al.*, 1970; Hallauer *et al.*, 1972), HA titration (Cole and Nathanson, 1969), or by plaquing (Ledinko, 1967; Tennant *et al.*, 1969; Tattersall, 1972*b*; Bates and Storz, 1973).

CPE produced by the defective parvoviruses would be obscured by

* Both ϕX174 and DNV are similar to the parvoviruses in size and morphology (Tromans and Horne, 1961; Kurstak, 1971). In addition, ϕX174 and probably DNV also contain single-stranded DNA genomes (Sinsheimer, 1959*b*; Barwise and Walker, 1970; Kurstak, 1971). There is a reasonable expectation that reproduction of the single-stranded DNA progeny of all these small viruses proceeds by at least some common mechanisms [e.g., a double-stranded replicative form of DNA (Sinsheimer *et al.*, 1962)]. Evidence has been put forth that DNV particles, like those of the human and simian AAV, may consist of virions which contain either plus or minus strands of DNA (Barwise and Walker, 1970; Kurstak, 1971). Antigenic relatedness has not been demonstrable between DNV and the four human and simian AAV serotypes or several nondefective parvoviruses including RV, H-1, MVM, and Haden (Hoggan, 1971).

that of their helper adenoviruses. Optimal AAV reproduction is not determined solely by AAV–helper multiplicity relationships since the efficiency of adenoviruses as helpers may depend both on the serotype and on the particular cell line used (Smith and Thiel, 1967; Boucher *et al.*, 1969; Atchison, 1970). The AAV have been assayed by CF antigen titration (Hoggan *et al.*, 1966), particle counting in the electron microscope (Atchison *et al.*, 1965; Parks *et al.*, 1967*a*), HA titration (AAV-4 and AAV X_7 only; Ito and Mayor, 1968; Luchsinger *et al.*, 1971), and FA titer (Ito *et al.*, 1967; Blacklow *et al.*, 1967*a*; Smith and Thiel, 1967).

2.1. Stability

Parvoviruses are among the most stable of all vertebrate viruses. They withstand heating at 56°C for 1 or more hours, and some have been shown to survive temperatures of 75–80°C for 30 minutes or longer (Kilham and Oliver, 1959; Greene, 1965; Johnson and Cruickshank, 1966; Binn, *et al.*, 1970; Siegl *et al.*, 1971). This resistance to heating is one factor which mitigates much of the difficulty in recovering parvoviruses which are free of infectivity from extraneous or helper viruses since it is often possible to selectively heat-inactivate such contaminants. All parvoviruses resist inactivation by ether or chloroform (Kilham and Oliver, 1959; Atchison, 1965; Johnson and Cruickshank, 1966; Binn *et al.*, 1968; Storz and Warren 1970; Siegl *et al.*, 1971). To the extent studied, they have resisted inactivation by trypsin, pepsin, papain (Johnson and Cruickshank, 1966; Rose *et al.*, 1966; Mayr *et al.*, 1968; Siegl *et al.*, 1971), and deoxyribonuclease (Vasquez and Brailovsky, 1965; Rose *et al.*, 1966; Siegl *et al.*, 1971), and they have withstood incubation at *p*H 3 (Greene, 1965; Mayr *et al.*, 1968; Storz and Warren, 1970). In addition, they can remain viable for years on laboratory bench tops or when stored in the cold (Hoggan, personal communication).

2.2. Host Range and Pathogenicity

2.2.1. Nondefective Parvoviruses

Data regarding both *in vitro* and *in vivo* host range capabilities of the parvoviruses are incomplete. In general, the host range of most nondefective parvoviruses is rather restricted. All rodent viruses have been grown in either rat cells or hamster cells or both (Toolan, 1968; Mirkovic *et al.*, 1971; Hallauer *et al.*, 1972). Only MVM grows well in mouse cells (Crawford, 1966; Hallauer *et al.*, 1972). Kirk virus appears to propagate best in Detroit-6 cells, but attempts at producing it in other human cells or simian cells have been unsuccessful (Mirkovic *et al.*, 1971). H-1 and H-3 are additionally capable of replicating in several human and simian cell lines (Toolan and Ledinko, 1965; Hallauer *et al.*, 1972). Of interest is the finding that although H-1 would not grow in secondary human embryonic lung cells, Ledinko and Toolan (1968) found that it multiplied in these cells when adenovirus type 12 (Ad 12) was present as a helper. In this regard it should be noted that Mirkovic *et al.* (1971) observed a hundredfold increase in HA titers of Kirk virus when Detroit-6 cells were coinfected with Ad 7. These investigators also reported a fourfold stimulation of X14 HA titer when rat embryo cells were coinfected with the same adenovirus. In *in vivo* infections by rodent viruses, naturally occurring RV, H-1, and L-S antibody have been found in rats (Kilham, 1966; Robey *et al.*, 1968; Lum, 1970), MVM antibody in rats and mice (Kilham and Margolis, 1970; Parker *et al.*, 1970*b*), and, rarely, H-1 antibody in man (Toolan, 1964, 1968). Rodent viruses are not known to cause disease under natural circumstances, but, experimentally, RV and the H viruses can produce either fatal infection in newborn hamsters or osteolytic lesions which result in a "mongoloidlike" appearance (Toolan *et al.*, 1960; Kilham, 1961*a,b*). Also, RV and closely related strains have been reported to cause CNS lesions in newborn hamsters (Kilham and Margolis, 1964; Nathanson *et al.*, 1970), kittens (Kilham and Margolis, 1965), and rats (Nathanson *et al.*, 1970). Inoculation of H-1 into both man (Toolan *et al.*, 1965) and monkey (Toolan, 1966) has apparently resulted in viremia without detectable illness.

The several strains of FPV have been grown in feline kidney cells (Johnson, 1965*a,b*; Johnson *et al.*, 1967). Natural susceptibility to these viruses includes animals belonging to three different families: (1) cats (domestic cats, lions, panthers, leopards, tigers), (2) mink, and (3) coatimundi and raccoons (Johnson and Halliwell, 1968). Ferrets, which are closely related to mink, have been experimentally infected with FPV (Johnson *et al.*, 1967). FPV strains are the only parvoviruses strongly linked to naturally occurring disease, causing cerebellar hypoplasia in kittens, or anorexia, diarrhea, and occasionally death in all natural hosts (Johnson, 1965*a*; Johnson *et al.*, 1967; Johnson and Halliwell, 1968).

In studies with cells from a variety of species, PPV was found to grow in pig cells and several continuous lines of human cells (Mayr *et al.*, 1968; Cartwright *et al.*, 1969; Hallauer *et al.*, 1972). The possibility that PPV may be a frequent contaminant of continuous human cell lines has been noted (Hallauer *et al.*, 1971). Serological studies reveal that PPV frequently infects swine (Cartwright *et al.*, 1969), and it has been suggested that PPV infection might be related to infertility and abortion in pigs (Cartwright and Huck, 1967; Cartwright *et al.*, 1969; Johnson, 1969).

Haden virus replicates best in cells derived from its host species (Bates and Storz, 1973). There is a high incidence of natural infection among cattle (Abinanti and Warfield, 1961; Spahn *et al.*, 1966; Storz *et al.*, 1972). Storz *et al.* (1972) have also reported significant HI titers against Haden virus in sera from cynomolgus monkeys, guinea pigs, dogs, goats, and horses. Experimentally, virus inoculation in calves has been associated with diarrhea, respiratory symptoms, and a rise in specific antibody (Spahn *et al.*, 1966).

Binn *et al.* (1970) tested the growth of MVC in a variety of primary and serial cell cultures, but only detected CPE in a continuous line of dog epithelial cells. They also observed a high incidence of neutralizing antibody in German shepherds and beagles.

Replication of TVX and LuIII seems restricted to continuous human cell lines (Hallauer *et al.*, 1972). RT virus could be grown only in the continuous line of rat fibroblast cells from which it was initially recovered.

2.2.2. Defective Parvoviruses

Among the defective parvoviruses, AAV-1, -2, and -3 seem able to replicate in nearly any cell line which can also be productively infected with an adenovirus (Hoggan *et al.*, 1966; Casto *et al.*, 1967*a*; Smith and Gehle, 1967; Blacklow *et al.*, 1968*b*; Ishibashi and Ito, 1971; Boucher *et al.*, 1969). However, inefficient or possibly incomplete replication of helper adenoviruses may still support good AAV production as indicated by potentiation of AAV-1 in cells where helpers multiply poorly if at all (Smith and Gehle, 1967; Mayor and Ratner, 1972) and by the reported stimulation of AAV replication by DNA-minus temperature-sensitive adenovirus mutants (Ito and Suzuki, 1970; Ishibashi and Ito, 1971; Mayor and Ratner, 1973). On the other hand, neither Ad 12-transformed cells (Hoggan *et al.*, 1966) nor Ad 2-transformed rat embryo cells (Rose *et al.*, unpublished results), which express an early event in the adenovirus cycle (i.e., T-antigen synthesis), have been found to support AAV reproduction. Moreover, AAV-1 did not replicate in African green monkey kidney (AGMK) cells which were coinfected with the E46⁻ strain of Ad 7 (Blacklow *et al.*, 1967*a*). E46⁻ grows poorly in AGMK cells (Rowe and Baum, 1964; Feldman *et al.*, 1965), although both its DNA and RNA are extensively synthesized (Reich *et al.*, 1966). These latter findings suggest that cells must permit expression of a relatively late event(s) in the adenovirus cycle if AAV replication is to occur. A simian serotype, AAV-4, is apparently more host-restricted than the other AAV serotypes since serial passage of this virus seems to require African green monkey cells (Hoggan, 1970).

Antibodies to all four AAV serotypes have been found in both man and monkeys (Blacklow *et al.*, 1968*a*). However, serological data plus the occasional recovery of AAV-4 from AGMK cells indicate that AAV-2 and -3 are human viruses which occasionally infect monkeys, whereas AAV-4 is an African green monkey virus which rarely infects man (Blacklow *et al.*, 1968*a*; Parks *et al.*, 1970). It has been suggested that AAV-1 is of rhesus origin (Rapoza and Atchison, 1967; Parks *et al.*, 1970). Only AAV-2 and -3 have been isolated from man (Blacklow *et al.*, 1967*b*), and, as is the case in cell culture, it seems that AAV replication *in vivo* requires a helper adenovirus (Blacklow *et al.*, 1968*b*, 1971*b*).

Bovine AAV has been propagated in bovine kidney cells with bovine adenoviruses as helpers (Luchsinger *et al.*, 1970), and avian AAV in chicken embryos with avian adenoviruses (El Mishad *et al.*, 1973). Serological findings by Luchsinger *et al.* (1970) and Luchsinger and Wellermans (1971) suggest that AAV X_7 antibody may occur with moderate frequency in both human and cattle sera. The status of canine AAV (Domoto and Yanagawa, 1969) as a distinct entity has not been established because antigenic comparisons with other AAV are still uncertain (see Sect. 1.2.2).

2.3. Persistent Infection and Latency

As already noted, the presence of parvoviruses is often inapparent. Because of the stability of these agents, cross-contaminating infections can readily occur in the laboratory, both in animals and in cell cultures (Johnson, 1965*b*). This fact may compound the difficulty of determining the origin of viruses which suddenly appear. Several studies, however, indicate that parvoviruses can persist in cells, becoming apparent only after their replication is enhanced (or possibly induced) during passage (Kilham and Oliver, 1959; Toolan *et al.*, 1960; Robey *et al.*, 1968) or by certain manipulations such as shipping or freeze-storage of cells (Hallauer *et al.*, 1971). Nathanson *et al.* (1970) suggested the possibility that cyclophosphamide may activate the replication of latent HER virus in rats, and Payne *et al.* (1963, 1964) noted a possible enhancement of X14 virus replication following X irradiation of rats. Whether parvovirus latency simply reflects an inability to detect very low levels of persistent infection or actually results from the carriage by cells of a subviral component(s) is not yet clear. Smith *et al.* (1968) found that after infection of human amnion cells with AAV alone, virus could be recovered from a fraction of subcultured or cloned cells, but that disruption of these cells resulted in more than a 99% decrease in recoverable AAV, consistent with the existence of a latent subviral form. Uncommonly, AAV-2 has been recovered from human embryonic kidney cells, whereas AAV-4 has been found more frequently in various lots of AGMK cells (Hoggan, 1970). This persistence of AAV in kidney cells is interesting in view of the rapid loss of ability to recover AAV following helper-free infection in the laboratory in these and other cells (Atchison *et al.*, 1965; Casto *et al.*, 1967*a*; Mayor *et al.*, 1967; Hoggan, 1973). Two possibilities that could explain AAV persistence are (1) the coincident presence of an undetected helper virus or (2) a nonhelper-dependent carriage of the AAV genome by the cell. Evidence for the latter possibility comes from a recent study (Hoggan, 1973) in which AAV-carrier lines of Detroit-6 cells were established. Infectious AAV could be recovered from one line following infection with helper adenovirus, whereas another line only yielded AAV antigen after helper challenge, thus suggesting that this line may only contain a portion of the AAV genome. A nonhelper-dependent carriage of the AAV genome by cells in an integrated state would help to explain how the AAV are maintained in nature, but AAV stability and the widespread occurrence of adenovirus infection in man and animals seem sufficient to account for AAV survival. An animal reservoir in which AAV is not defective must also be considered as a possible sustaining factor (Blacklow *et al.*, 1968*a*). In any event, both nondefective and defective parvoviruses should be recognized as potential, tenacious contaminants of cell lines, virus stocks, and certain vaccines.

2.4. Defectiveness

Because of their absolute requirement for a helper adenovirus, the AAV have been referred to as "nonconditional defective viruses" (Melnick and Parks, 1966). There is, of course, a qualification concerning the nonconditional defectiveness of the AAV since it can be argued that an undiscovered, permissive host cell might exist. Furthermore, in addition to adenoviruses, there might yet be another virus which is capable of assisting AAV multiplication. Although extensive tests with a wide range of both RNA and DNA viruses have not revealed other viruses which are able to support the reproduction of infectious AAV (Atchison *et al.*, 1965; Hoggan *et al.*, 1966; Atchison, 1970), it was discovered that AAV replication could be partially helped by herpesviruses (see Sect. 5; Atchison, 1970; Blacklow *et al.*, 1970, 1971*a*; Boucher *et al.*, 1971; Rose and Koczot, 1972; Dolin and Rabson, 1973).

Within the group of nondefective parvoviruses, Ledinko and Toolan (1968) and Ledinko *et al.* (1969) found that H-1 exhibited what might be called "conditional defectiveness" (Melnick and Parks, 1966). H-1 infection of human lung cells yielded viral antigen in the absence of infectious virus synthesis, but infectious virus was produced when cells were coinfected with Ad 12. The basis for H-1 restriction in these cells is not known, and the helper function(s) provided by adenovirus may or may not correspond to a requirement(s) for AAV multiplication. Adenovirus enhancement of Kirk and X14 virus replication have been mentioned (Sect. 2.2.1). The mechanism of this adenovirus helper activity is also not known.

2.5. Interference

Both homologous and heterologous interference have been observed with certain defective parvoviruses. In infections with specific human or simian AAV serotypes, virus production is decreased at high AAV multiplicities. For example, dual infections of KB cells with helper Ad 2 and AAV-1, -2, or -3 provided highest AAV yields (averaging 1–3 $\times$ 10^4 $TCID_{50}$/cell) with AAV multiplicities in the range of 10–20 $TCID_{50}$/cell (Rose and Koczot, 1972; Rose *et al.*, unpublished results). Increases in AAV multiplicity to 100 or more $TCID_{50}$/cell usually resulted in a tenth or less of this optimal yield. Reduction in AAV synthesis associated with undiluted (high multiplicity) passage has also been found with AAV-4 and helper SV15 in BSC-1 cells (Torikai, *et al.*, 1970). "Light" AAV-4 particles were not augmented during three serial undiluted passages which resulted in approximately a hundredfold decrease in complete virus particles as measured by HA titer. Thus, in contrast to defective interfering particles produced during vesicular stomatitis virus infections (Huang and Wagner, 1966), the decrease in synthesis of complete AAV is not associated with an absolute increase in incomplete particles. It is possible, though, that a relative increase of particles capable of interference did occur. The specific mechanism of homologous AAV interference is not clear.

Heterologous interference has been found between AAV serotypes, between AAV and their helper adenoviruses, and between AAV and SV40, a virus without discernible AAV helper activity. In mixed infections with two AAV serotypes, one serotype may dominate (Torikai and Mayor, 1969) and even suppress detectable growth of the other serotype (Hoggan, 1971). AAV interference with replication of helper adenoviruses has been frequently observed (Hoggan *et al.*, 1966; Smith *et al.*, 1966; Casto *et al.*, 1967*a,b*; Parks *et al.*, 1967*a*, 1968). This interference is apparently not mediated by an interferonlike inhibitor, requires infectious AAV, and is directly related to AAV dosage (Casto *et al.*, 1967*a,b*; Parks *et al.*, 1968). Cell type may also influence AAV interfering activity. For example, the replication of Ad 7 or SV15 was more restricted in KB cells than in human kidney cells when coinfections were carried out with purified AAV-4 (Parks *et al.*, 1968). AAV interference with helper multiplication does not appear to be exerted at the level of adsorption, penetration, or uncoating, and it is abolished when AAV infection is delayed for 8 hours following adenovirus infection (Parks *et al.*, 1968). Furthermore, AAV yield is unaffected by the presence or absence of AAV interference with adenovirus multiplication, at least indicating that interference and helper activity can be dissociated (Parks *et al.*, 1968). Parks *et al.* (1968) have pointed out that the acquisition of adenovirus resistance to AAV interference is temporally associated with the onset of adenovirus DNA synthesis.

At the cellular level Casto *et al.* (1967*a*) found that AAV interference decreased adenovirus infectious centers, and they suggested that most of the decrease in adenovirus yield might be due to a failure of AAV-infected cells to produce adenovirus. Parks *et al.* (1968), however, reported that AAV did not diminish the number of adenovirus infectious centers, and hence the number of cells yielding virus, but caused a 90% reduction in single-cell yields of infectious adenovirus. Since the adenovirus particle/PFU ratio was the same with or without AAV inhibition, Parks *et al.* (1967*a*) concluded that the decrease in infectious yield resulted from reduced adenovirus production and not a change in the quality of the virus.

Examination of thin sections of doubly infected cells in the electron microscope clearly reveals that nuclei contain predominantly *either*

AAV or adenovirus particles (Archetti *et al.*, 1965; Atchison *et al.*, 1966; Mayor *et al.*, 1967; Torpier *et al.*, 1971; W. Hall, B. Hobbs, and J. Rose, unpublished results). This evidence suggests that, in a given cell, growth of one virus is accompanied by suppression of growth of the other virus. Figures 1B and 2 demonstrate the crystalline character of replicating AAV-2 in the cell nucleus; negatively stained, purified AAV particles are also shown for comparison (Fig. 1A). AAV particles could be detected in KB cell nuclei within 20 hours after coinfection with helper Ad 2, which is consistent with reported titration data (Rose and Koczot, 1972). At 43 hours post-infection (Fig. 1B) numerous rodlike crystalline aggregates of AAV particles can be seen together with a few scattered adenovirus particles, indicating that both AAV and adenovirus can be synthesized in the same cell. Figure 2 shows a portion of a cell nucleus with ordered arrays of AAV but no adenovirus. Of interest is the association of AAV particles with a nucleolus, a feature often observed in these preparations. After dual infections with AAV-1 and herpes simplex virus type 1 (HSV-1), a partial helper of AAV replication, Henry *et al.* (1972) observed similar arrays of *empty* AAV particles in the nuclei of human embryonic lung cells and called attention to nucleolar replacement by large numbers of these empty particles. The authors did not say whether or not the presence of AAV particles was associated with exclusion of herpesvirus or *vice versa*.

In tests with several RNA and DNA viruses Casto *et al.* (1967*b*) found that AAV interfered only with adenovirus and SV40 plaque formation. Because the range of viruses inhibited by AAV is narrow, it should be considered that adenoviruses and SV40 might share some close biochemical relationship. It is, perhaps, relevant that these viruses can interact biologically (Rabson *et al.*, 1964) and genetically (Baum *et al.*, 1966; Kelly and Rose, 1971).

No parvovirus has yet been shown to possess oncogenic capability (Kilham and Oliver, 1959; Kilham and Maloney, 1964; Toolan, 1967; Hallauer *et al.*, 1972). On the other hand, certain parvoviruses have been reported to interfere with natural or virus-induced tumorigenesis. Toolan (1967) found that the incidence of tumors in control hamsters was twenty times greater than that in hamsters which had acquired mongoloidlike deformities as a result of injection with H-1 virus. These mongoloid hamsters retain high titers of H-1 antibody throughout their lives (Toolan, 1964). Toolan and Ledinko (1968) also found that when H-1 was injected together with Ad 12, the incidence of Ad 12 tumors in newborn hamsters was reduced more than 50%. In other studies with hamsters, Kirschstein *et al.* (1968) showed that coinfection with AAV-1

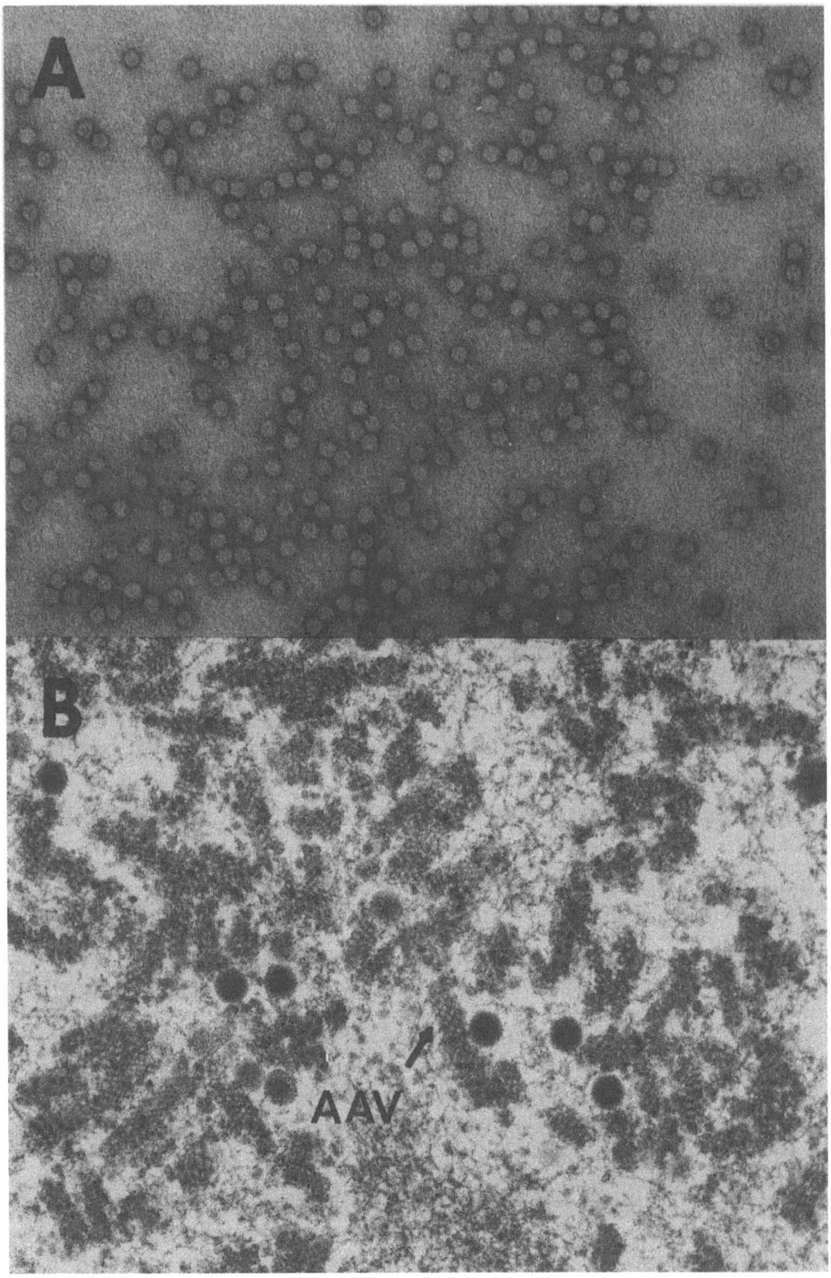

Fig. 1. (A) CsCl-purified AAV-2 particles negatively stained. ×105,000. (B) Thin section through cell pellet fixed with glutaraldehyde and osmium tetroxide 43 hours after virus infection. KB cells in suspension culture were infected with multiplicities of 10 $TCID_{50}$/cell of AAV-2 and 100 $TCID_{50}$/cell of Ad 2. A single nucleus containing crystalline bundles of AAV particles (arrow) and occasional adenovirus virions is shown. ×66,500. Courtesy of W. Hall, B. Hobbs, and J. Rose (unpublished results).

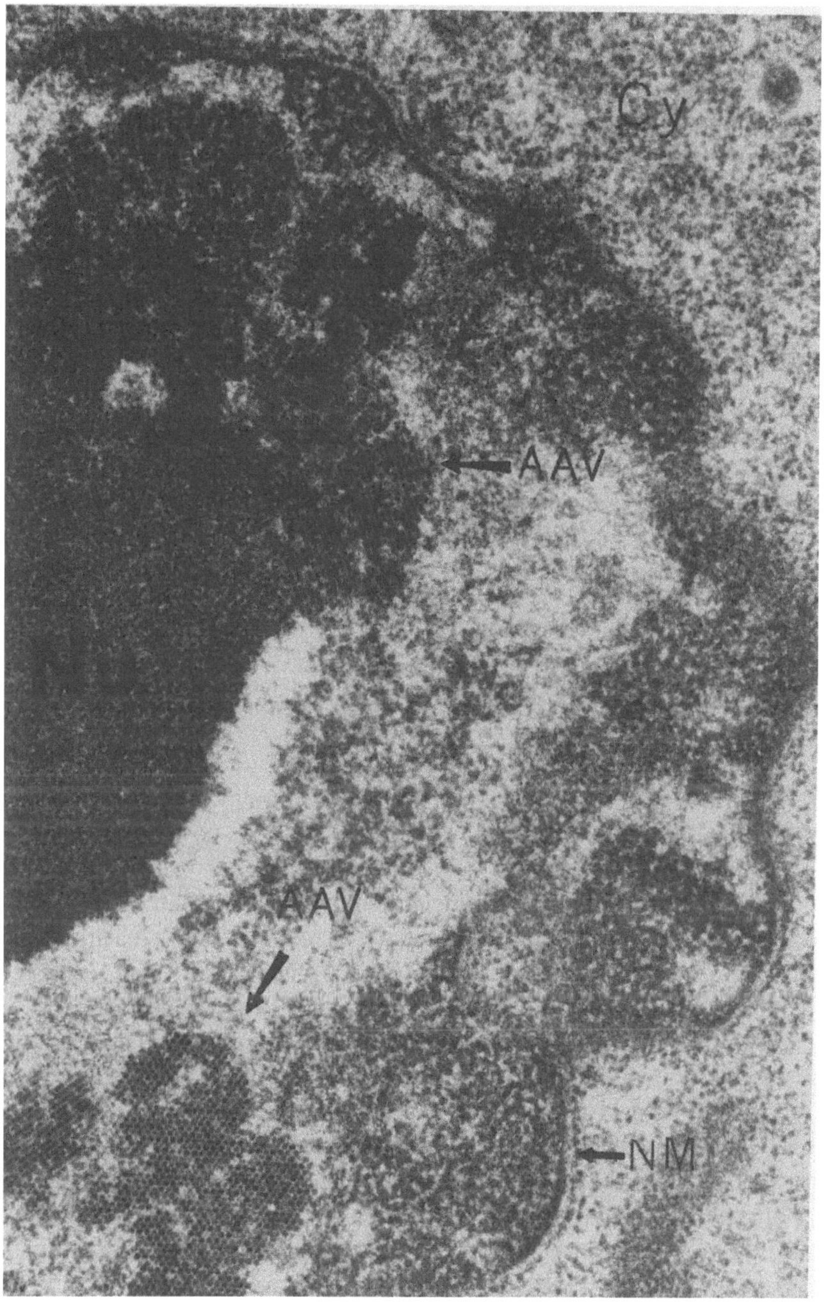

Fig. 2. Thin section through the same cell pellet described in Fig. 1B. Crystalline arrays of AAV particles are present in the nucleus of a cell, but adenovirus particles are not seen. Nu, nucleolus; NM, nuclear membrane; Cy, cytoplasm. ×47,500. Courtesy of W. Hall, B. Hobbs, and J. Rose (unpublished results).

decreased by two-thirds the incidence of Ad 12-induced tumors, and Mayor *et al.* (1973) demonstrated even greater interference by AAV-1 with Ad 31 oncogenesis. No interference between AAV and SV40 oncogenicity was found, suggesting that AAV protection may be adenovirus-specific (Mayor *et al.*, 1973).

In hamster embryo cells *in vitro*, transformation by three adenoviruses (SA7, SV11 or Ad 12) was inhibited by AAV-1 (Casto and Goodheart, 1972). An 80% reduction in SV11- or Ad 12-transformed cell foci was achieved with 10 adenovirus plaque-inhibiting units (Casto *et al.*, 1967*b*) per cell of AAV. Like plaque inhibition, transformation inhibition appears to be one-hit with respect to AAV, and the two processes seem equally efficient (Casto *et al.* 1967*b*; Casto and Goodheart, 1972). It is not known whether AAV interference with adenovirus replication, *in vitro* adenovirus transformation, or *in vivo* adenovirus tumorigenesis involves the same or different mechanisms. In view of AAV interference with SV40 replication (Casto *et al.*, 1967*b*), it is interesting that AAV does not inhibit SV40 oncogenesis (Mayor *et al.*, 1973).

3. PROPERTIES AND COMPONENTS OF VIRIONS

3.1. Physicochemical Properties of Virions

3.1.1. Purification

Parvoviruses adhere to nuclear material, and current methods used to release virus include sonication followed by treatment with combinations of nucleases, proteolytic enzymes, and detergent (Vasquez and Brailovsky, 1965; Rose *et al.*, 1968; Robinson and Hetrick, 1969; McGeoch *et al.*, 1970; Rose and Koczot, 1971). Banding of viruses in CsCl density gradients achieves a final state of high purity (Crawford, 1966; Hoggan *et al.*, 1966; Rose *et al.*, 1966; Parks *et al.*, 1967*b*). Other procedures such as differential filtration through Millipore membranes (Atchison *et al.*, 1965; Hoggan *et al.*, 1966), banding in potassium tartrate gradients (Breese *et al.*, 1964; Vasquez and Brailovsky, 1965), and zonal rotor centrifugation (Luchsinger *et al.*, 1971) have also been successful. Because complete AAV particles are about 0.05 g/cm^3 more dense in CsCl than their helper adenoviruses, separation of the two viruses is possible, and twice rebanding of the AAV component yields AAV essentially free of adenovirus. The particles shown in Fig. 1A are from a preparation of this type. There are few empty capsids, and adenovirus particles could not be found. However, contaminating adenovirus at concentrations up to 10^6–10^7 particles/ml could escape detection by the electron microscope (Parks *et al.*, 1967*a*). Other criteria of AAV purity include hybridization specificity of extracted DNA (Rose *et al.*, 1968) and inability to detect AAV or helper replication on subsequent passages (Atchison *et al.*, 1965).

Parvovirus particle/PFU (or $TCID_{50}$) ratios have been determined only in a few instances. The ratio for MVM virus is 300–400:1 (Tattersall, 1972*b*); that for AAV-4 is 10–100:1 (Ito *et al.*, 1967; Parks *et al.*, 1967*a*).

3.1.2. Virion Morphology

Almost all of the viruses listed in Table 1 have been visualized and measured in the electron microscope. A detailed account of the bulk of this work is given in a recent review (Hoggan, 1971). In brief, all parvoviruses are icosahedral in symmetry and most are 20–22 nm in diameter (Fig. 1A). Although some apparent size variation may be due to technical factors, small differences in diameter appear to exist between certain viruses, e.g., H-3, RV, and HB have diameters of approximately 19–20 nm, but H-1, HT, Haden, MVM, and AAV 1, -2, and -3 have diameters in the range of 21–22 nm (Karasaki, 1966; Hoggan, 1971). Owing to the small size of parvoviruses, detailed analysis of virion substructure has been difficult. Some investigators have surmised that the capsids of several viruses in the rodent group are composed of 32 capsomers (Vasquez and Brailovsky, 1965; Karasaki, 1966; Mayor and Jordan, 1966). However, Mayor *et al.* (1965) have concluded that the AAV-4 particle consists of 12 capsomers, whereas Smith *et al.* (1966) believe that AAV-3 capsids consist of a netlike reticulum, similar to that described for reovirus (Vasquez and Tournier, 1964).

3.1.3. Density and Sedimentation Constant

Reported buoyannt densities in CsCl of the different parvoviruses are in the range of 1.38–1.47 g/cm³ (McGeoch *et al.*, 1970; Siegl *et al.*, 1971; Hoggan, 1971). Values for the same virus may show modest differences between different laboratories, e.g., values of 1.39, 1.422, and 1.47 g/cm³ have been published for H-1. Most parvoviruses, however, have densities near 1.40 g/cm³. In addition, lighter virus bands are frequently observed at densities between 1.38–1.34 g/cm³ and 1.32–1.30 g/cm³ (Payne *et al.*, 1964; Crawford, 1966; Robinson and Hetrick, 1969; Torikai *et al.*, 1970; Siegl, 1972; Usategui-Gomez *et al.*, 1969). DNA can be extracted from particles in intermediate density bands and has been found to be considerably smaller than genome size (i.e., 20–30% of strand length); its origin, whether viral or host, is uncertain (Torikai *et al.*, 1970; Siegl, 1972). Compared to the primary band, infectivity is greatly reduced in the lighter bands which contain large proportions of empty particles (Payne *et al.*, 1964; Greene and Karasaki, 1965; Crawford, 1966; Robinson and Hetrick, 1969; Usategui-Gomez *et al.*, 1969; Johnson *et al.*, 1971; Tattersall, 1972*b*). Another band, more dense than the primary band, is often seen at 1.44–1.47 g/cm³. The particles in this band seem smaller than those in the primary band, and, in the case of AAV, they are known to be infectious (Johnson *et al.*, 1971; Hoggan, 1971). With AAV, DNA molecules extracted from either primary- or dense-band particles appear similar in size, and dense particles can be generated when primary-band particles are treated with sodium dodecyl sulfate (Rose and Koczot, unpublished results). The dense particles probably represent virions whose capsids are incomplete or altered, and not particles with a larger genomic species. Because AAV light-band parti-

TABLE 3

Sedimentation Constants and Estimated Particle Weights of Parvovirus Virions

Virus	Sedimentation constant ($s_{20,w}$), S	Particle weight $\times 10^{-6}$	Reference
MVM	110	—	McGeoch *et al.* (1970)
RV	110, 122	6.6	McGeoch *et al.* (1970), Salzman and Jori (1970)
H-1	110	—	McGeoch *et al.* (1970)
KBSH	105	5.3	Siegl *et al.* (1971)
AAV-1	104, 125	5.4	Crawford *et al.* (1969), Rose *et al.* (1971)
AAV-2	125, 120	5.4	Rose *et al.* (1971), Rose and Koczot (1972)
AAV-4	137	5.4	Mayor *et al.* (1969*b*)
ϕX174[a]	114	6.2	Sinsheimer (1959*a*)

[a] Given for comparison.

cles and complete adenovirus particles have similar densities, it is not possible to free adenovirus from contaminating AAV by density equilibrium centrifugation in CsCl (Rose *et al.*, 1966; Smith *et al.*, 1966).

Sedimentation coefficients and estimates of particle weights of parvovirus virions are given in Table 3. Differences among reported values may not be significant.

3.1.4. Chemical Composition

Parvoviruses have been found resistant to ether or chloroform inactivation, consistent with an absence of essential lipid in the virion (Kilham and Oliver, 1959; Abinanti and Warfield, 1961; Atchison *et al.*, 1965; Johnson and Cruickshank, 1966; Horzinek *et al.*, 1967; Binn *et al.*, 1968). Inhibition of virus growth by agents which block DNA synthesis has provided evidence that parvoviruses contain a DNA genome (Payne *et al.*, 1963; Ledinko, 1967; Mayr *et al.*, 1968; Storz and Warren, 1970; Binn *et al.*, 1970), and this has been extensively confirmed by direct recovery of DNA from purified virus preparations (Table 4). Chemical analyses of the DNA content of parvoviruses have been reported for H-1 (25%; Cheong *et al.*, 1965), RV (25.6%; Salzman and Jori, 1970), and AAV-1 and AAV-4 (18.9% and 26.5%; Parks *et al.*, 1967*b*). These values are in the same range as that determined for ϕX174 (25%; Sinsheimer, 1959*a*).

3.2. DNA

The DNA of only a few parvoviruses has been extensively characterized, mostly due to difficulty in obtaining sufficient quantities of purified virus. A large portion of this work has been done with the human and simian AAV because of interest in the biochemistry of their defectiveness and the relative ease with which these agents can be grown and purified from their helper adenoviruses. Efficient extraction of parvovirus DNA requires somewhat vigorous treatment with proteolytic enzymes or hot detergent or both (Crawford, 1966; Rose *et al.*,1966). Combined treatment produces DNA recoveries of 60–80% (Rose *et al.*, 1966). Alkaline extraction of DNA is simpler, yielding a higher proportion of unbroken molecules than after enzyme treatment (Koczot *et al.*, 1973). DNA can be released in 0.1 N NaOH and then purified by pelleting through a CsCl solution of density 1.5 g/cm^3 (McGeoch *et al.*, 1970) or released by sedimenting purified virus into a

TABLE 4

Physical Properties of Parvovirus DNA Molecules[a]

Virus	ρ CsCl,[b] g/cm³	T_m,[c] °C	Sedimentation constant,[d] S	Length[e] μm	Genome mol. wt.[f] $\times 10^{-6}$	Comments	Reference
MVM	1.722	—	—	1.2	1.5	Single-stranded, linear	Crawford (1966), Crawford *et al.* (1969)
RV	1.726	—	27	1.5 ± 0.2	1.6	Single-stranded, linear	May and May (1970), Salzman *et al.* (1971)
H-1	1.720	—	27.8	—	1.7	Single-stranded	Usategui-Gomez *et al.* (1969), McGeoch *et al.* (1970)
KBSH	1.724	—	24	—	1.4	Single-stranded	Siegl (1972)
AAV-1	1.717; 1.729	91.5 (54.2)	15.5	1.38 ± 0.05	1.35	Duplex DNA is formed by annealing of linear plus and minus strands which are separately encapsidated; plus and	Parks *et al.* (1967*b*), Rose *et al.* (1966, 1968, 1969), Rose and Koczot (1971), Koczot *et al.*, 1973, Koczot and
AAV-2	1.714; 1.726	90.4 (51.5)	15; 24	1.38 ± 0.06	1.35		
AAV-3	1.715; 1.727	90.8 (52.4)	15	1.39 ± 0.06	1.35		
AAV-4	1.720; 1.728	93.0 (58)	15.7	1.5 ± 0.2	1.5		

						minus strands have self-complementary ends (inverted terminal repetition) which can anneal to form single-stranded circles	Rose, unpublished results
Ad 2	1.716	92.5 (57)	31.1	12.6 ± 0.8	24.4	Double-stranded, linear with inverted terminal repetition	Pina and Green (1965), Green *et al.* (1967), Garon *et al.*, 1972, Wolfson and Dressler, 1972
φX174	1.725	—	27.6	1.77 ± 0.13	1.7	Single-stranded, covalently closed circles	Sinsheimer (1959*b*), Freifelder *et al.* (1964), Studier (1965), Szybalski (1968)

[a] Ad 2 and φX174 DNA data are given for comparison.
[b] Bouyant densities of both duplex and denatured AAV DNA are given. All buoyant densities relative to *E. coli* DNA at 1.710 g/cm³.
[c] Melting temperature in SSC (0.15 M NaCl plus 0.015 M Na citrate) of the double-stranded form of AAV DNA and Ad 2 DNA. Mole % G+C in duplex molecules calculated from T_m is given in brackets.
[d] The sedimentation coefficients of both double-stranded and single-stranded molecules are given for AAV-2; all other AAV values are for double-stranded molecules.
[e] Obtained by electron microscopy. AAV molecular lengths are those of duplex molecules.
[f] Best estimate from available data.

TABLE 5

Base Composition of Parvovirus DNA and *In Vivo* Synthesized AAV RNA

Virus	Nucleotides, moles/100 moles A	T(U)	G	C	G+C content, %	Reference
RV	26.8	29.6	20.6	22.9	43.5	McGeoch *et al.* (1970)
MVM	26.5	32.7	19.5	21.4	40.9	McGeoch *et al.* (1970)
H-1	25.5	29.3	22.6	22.6	45.2	McGeoch *et al.* (1970)
AAV-2						
(H)[a]	20.5	26.5	26.7	26.3	53.0	Rose and Koczot (1971)
(L)[b]	25.2	21.7	26.6	26.5	53.1	Rose and Koczot (1971)
(RNA)[c]	26.2	20.7	26.4	26.7	53.1	Rose and Koczot (1971)
φX174[d]	24.6	32.8	24.2	18.4	42.6	Sinsheimer (1959*b*)

[a] Minus strand of AAV DNA.
[b] Plus strand of AAV DNA.
[c] AAV RNA selected by hybridization with double-stranded (unfractionated) AAV DNA.
[d] Data given for comparison.

sucrose gradient containing 0.3 N NaOH (Koczot *et al.*, 1973). The latter method also provides information concerning the homogeneity of DNA preparations.

Physical and chemical properties of DNA from several parvoviruses are given in Tables 4 and 5, along with those of Ad 2 and φX174 DNA which are provided for comparison.

3.2.1. Strandedness

Excluding AAV DNA, physical and chemical properties of DNA extracted from parvoviruses clearly indicate that these molecules are single-stranded (Tables 4 and 5). Moreover, for MVM, RV, and H-1, reaction with formaldehyde is consistent with the presence of single-stranded DNA in the virion (Crawford, 1966; Salzman and Jori, 1970; Usategui-Gomez, 1969). Based on acridine orange staining, X14 particles also appear to contain single-stranded DNA (Mayor and Melnick, 1966). Initially, conflicting results with acridine orange led to some confusion whether AAV particles contained single- or double-stranded DNA (Mayor *et al.*, 1965; Atchison *et al.*, 1966; Mayor and Melnick, 1966). Staining reactions indicative of single-stranded DNA could be obtained but only when freshly prepared, purified preparations of AAV were used (Mayor and Melnick, 1966). In addition, the virus reaction with formaldehyde also suggested that DNA was single-stranded in the virion (Mayor *et al.*, 1969*a*). These findings, however, were hard to explain in view of physical and chemical data which showed that extracted AAV DNA was double-stranded (Rose *et al.*, 1966, 1968; Parks *et al.*, 1967*b*). The problem was eventually resolved when Crawford *et al.* (1969) found a close similarity in physical properties of MVM, ϕX174, and AAV-1 virions and suggested that the molecular weight of AAV DNA in the particle should not be greater than that of MVM or ϕX174 (1.7×10^6), a value equivalent to half the molecular weight of extracted AAV DNA. They speculated that this discrepancy could be explained if single, complementary strands of DNA were separately encapsidated in different AAV particles. Double-stranded DNA molecules, each composed of strands donated by two particles, could then be generated *after* DNA was released from virions. Rose *et al.* (1969) tested this hypothesis by analyzing the density of DNA extracted from a mixture of AAV-3 particles which contained either bromodeoxyuridine-substituted (heavy) or -unsubstituted (light) DNA. If double-stranded DNA was formed as a result of the annealing of single strands deriving from different virions, a DNA density hybrid should be produced (i.e., DNA duplexes having a buoyant density intermediate between the densities of duplexes composed of either light or heavy strands). This was found to be the case, and density distributions of sheared and unit-length duplex molecules confirmed that density hybrids were constructed by a lateral pairing of heavy and light strands, not an end-to-end joining of heavy and light duplexes by short lateral linkages. Evidence for AAV-4 encapsidation of single, complementary DNA strands was also independently reported by Mayor *et al.* (1969*b*). In these experiments it was shown that single-stranded DNA could be extracted from virus at low ionic strength and that this DNA could subsequently hybridize to form duplex molecules. It is not surprising that annealing of AAV DNA strands is highly efficient under usual conditions of extraction (SSC, 50°C), since the $C_0t_{1/2}$ value for intact AAV strands (as determined in 0.14 M sodium phosphate buffer at 60°C) is 7.4×10^{-4} (Carter *et al.*, 1972). Preparations of AAV DNA similarly extracted from the other human and simian serotypes are double-stranded, and a single-stranded component is not evident (Rose *et al.*, 1966, 1968; Parks *et al.*, 1967*b*). Furthermore, when fragmented and denatured, this DNA completely reassociates (Carter *et al.*, 1972). These observations indicate that there is little or no excess of one strand over the other in progeny virus, and thus suggest an equivalent synthesis of both strand species in infected cells. With the possible exception of DNV (Kurstak, 1971), the presence of single, complementary strands in different AAV particles represents a circumstance not yet found with any other virus. The biochemical basis for the production and "segregation" of complementary strands as well as the possible relationship of these features to dependence on adenovirus is not clearly understood.

3.2.2. Composition

It has been possible to preparatively separate the complementary AAV DNA strands, thus permitting determination of the base composition of each and analysis of strand-specific transcription *in vivo* (Berns and Rose, 1970; Rose and Koczot, 1971). Separation can be accomplished because the strands contain different amounts of thymidine which, in turn, results in a differential incorporation of the density label, 5-bromodeoxyuridine. Under labeling conditions a 70–80% substitution of thymidine occurs, and when DNA is denatured and banded in neutral CsCl, two single-stranded species (heavy strand $\cong$ 1.830 g/cm^3; light strand $\cong$ 1.798 g/cm^3) and a renatured duplex component ($\cong$ 1.790 g/cm^3) are seen. Double-stranded DNA can be formed only when heavy and light strands are mixed (Berns and Rose, 1970). Further purification of each strand species is achieved by sucrose sedimentation (Rose and Koczot, 1971). Base composition of purified AAV-2 heavy and light strands is shown in Table 5, along with the composition of DNA from other parvoviruses. There is considerable similarity among values for MVM, RV, and H-1 DNA, but the composition of AAV DNA is clearly different, especially with regard to thymidine content in the light strand. The thymidine content of heavy strands is 22% greater than that of light strands, and complementarity between the strand species is reflected by equivalence of the paired bases. The base composition of AAV-2 RNA synthesized in cells coinfected with Ad 2 is similar to that of light strands (Table 1). In agreement with these data, AAV-2 RNA synthesized throughout the infectious cycle has been found to hybridize only with heavy strands (Rose and Koczot, 1971; Carter *et al.*, 1972). The same is true for AAV RNA synthesized with herpes simplex virus, a partial helper of AAV replication (Carter and Rose, 1972; Carter *et al.*, 1972). The conclusion that the heavy-strand species alone serves as the transcriptional template *in vivo* is therefore strongly supported, although transcription from the light strand might go undetected if synthesis was very limited, or yielded RNA molecules only transiently stable. Since transcriptional function can be assigned only to heavy strands, they have been designated "minus strands," whereas light strands are "plus strands"

(Rose and Koczot, 1971). In the case of other parvoviruses, the transcriptional specificity of viral DNA has not been reported.

3.2.3 Size and Shape

Based on sedimentation and electron microscopy, estimated molecular weights of DNA from different parvoviruses are fairly comparable (Table 4). Values are in the range of $1.35–1.7 \times 10^6$ daltons, but differences may be more apparent than real, perhaps mainly relating to methodology or technical factors. For example, estimates of AAV DNA size based on sedimentation coefficients are 10–20% greater than those based on measurements in the electron microscope (Table 4; Rose *et al.*, 1966; Parks *et al.*, 1967*b*). Direct visualization has not revealed circular structures in MVM and RV DNA preparations (Crawford *et al.*, 1969; Salzman *et al.*, 1971). However, rare circular forms were seen when DNA was extruded from AAV-1 particles as a result of osmotic shock (Vernon *et al.*, 1971). Recent studies have demonstrated that extracted AAV DNA strands are capable of forming single- or double-stranded circular molecules closed by hydrogen-bonded duplex segments (see Sect. 3.2. 5; Koczot *et al.*, 1973). There is thus far no evidence that any parvovirus contains a covalently closed single- or double-stranded genome. A comparison of physical and chemical properties of parvovirus and ϕX174 virions strongly suggests that only a single strand of DNA is present within parvovirus particles (Crawford *et al.*, 1969). The finding that parvoviruses apparently contain linear DNA represents a major distinction between them and ϕX174, possibly reflecting a basic difference(s) in their mechanisms of DNA replication.

3.2.4. Infectivity

AAV-1 DNA is the only parvovirus DNA that has been shown to be infectious (Hoggan *et al.*, 1968; Boucher *et al.*, 1971). Infectivity is demonstrable with extracted (double-stranded) DNA and is dependent upon the presence of a helper adenovirus. Since infection with DNA circumvents the processes of adsorption, penetration, and uncoating, the continued requirement for helper virus indicates that a defective step(s) in AAV replication must exist at some stage after the intracellular release of its DNA (uncoating). It is not known whether infection can be evoked with single-stranded DNA, i.e., purified preparations of plus or minus strands. Attempts at determining the infectious capability of encapsidated plus or minus strands have been made. Particles containing either plus or minus strands can be separated in CsCl because of a density difference resulting from the differential incorporation of bromodeoxyuridine by the two strand species (Berns and Rose, 1970; Berns and Adler, 1972). Although the fractionated particles appear to be equally infectious, definitive purities of only 90% could be demonstrated for either type of particle. This, coupled with the insensitivity of infectivity assays (± 0.5 log), does not allow a firm conclusion (Rose and Dolin, unpublished results; Berns and Hoggan, personal communication).

3.2.5 Terminal Structure

The single-stranded DNA extracted from MVM and RV appears to be linear (Crawford *et al.*, 1969, Salzman *et al.*, 1971). Single-stranded, linear DNA is also released from AAV virions (Rose *et al.*, 1966, 1969; Parks *et al.*, 1967*b*; Koczot *et al.*, 1973), but these molecules consist of plus and minus strands which can anneal to form double-stranded structures (Mayor *et al.*, 1969*b*; Rose *et al.*, 1969; Berns and Rose, 1970). The absence of covalently closed, circular molecules in preparations of extracted parvovirus DNA argues that their genomes are linear in the virion. It is remotely possible, however, that a closed, circular genome might be efficiently opened during the extraction process. If so, for AAV at least, evidence strongly suggests that such cleavage would have to occur predominantly within a specific region. This conclusion arises from the finding by Koczot *et al.* (1973) that purified plus or minus strands of AAV-2 DNA contain self-complementary terminal sequences which can anneal to produce single-stranded circular molecules closed by relatively short duplex segments. Up to 75% of strands were capable of circularizing, an indication that most strands begin and end at or near the same specific points. Recently, Gerry *et al.* (1973) have shown conclusively that AAV DNA is not randomly permuted. There is no evidence of terminal complementarity with DNA from nondefective parvoviruses, but only MVM and RV DNA strands have been examined (Crawford *et al.*, 1969; Salzman *et al.*, 1971). It is not yet known whether self-complementary terminal sequences are present in DNA strands from all the AAV species, but they are present in the DNA of all the human and simian serotypes (AAV-1, -2, and -3) so far studied (Koczot *et al.*, 1973; Koczot and Rose, unpublished results).

Terminal-sequence complementarity is not unique to AAV DNA. The single strands of adenovirus DNA also have self-complementary ends, but they are the only other viral DNA molecules known to possess this structural feature (Garon *et al.*, 1972; Wolfson and Dressler, 1972). Double-stranded adenovirus DNA molecules, therefore, contain an inverted terminal repetition (Garon *et al.*, 1972), a type of repetition that would also be present in linear duplex AAV DNA molecules (Koczot *et al.*, 1973). In view of the fact that AAV reproduction is unconditionally dependent on a helper adenovirus, it is especially notable that the genomes of both viruses contain a similar and, perhaps exclusive, form of terminal repetition. For either virus the specific function(s) of these sequences is presently undefined, but they could play a role in DNA synthesis (Koczot *et al.*, 1973; Sect. 5.2.2). Furthermore, since herpesviruses can partially assist AAV replication, it also must be considered that their genomes might contain a similar type of repetition.

Physical and chemical properties of the self-complementary AAV terminal sequences have been explored in some detail (Carter *et al.*, 1972; Koczot *et al.*, 1973). The discovery of these sequences in AAV DNA resulted from the observation that about 50% of purified minus strands chromatographed as double-stranded DNA on hydroxyapatite, and that this DNA was partially resistant to nuclease S_1, an enzyme which specifically degrades single-stranded DNA (Carter *et al.*, 1972). The nuclease data suggested that self-annealed sequences represented not more than 20–25% of total DNA strand, equivalent to 400–500 base pairs. Subsequent electron microscopic examination revealed that 70–75% of molecules in plus- or minus-strand preparations were circular (Fig. 3A), indicating that self-annealed sequences were terminally located (Koczot *et al.*, 1973). Most molecules were of unit length, but, as expected, occasional linear and circular oligomers could be found. The closure of circles by duplex, hydrogen-bonded deoxyribonucleotide segments was substantiated by their sensitivity to exonuclease III and their quantitative regeneration after alkali denaturation (Koczot *et al.*, 1973). In addition, the self-annealed segments had a thermal stability (T_m) comparable to that of double-stranded DNA molecules, indicating the presence of highly ordered base pairing. A problem existed, however, concerning the length and configuration of these duplex regions. The most likely order of terminal bases, shown diagrammatically in Fig. 3A, would produce a projection or "panhandle" on circular molecules. Based on the length predicted from nuclease data, such panhandles should be visible in the electron microscope. Possible panhandles were observed (Koczot *et al.*, 1973) but were smaller than expected, equivalent to only 4–8% of total DNA strand (100–200 base pairs; Fig. 3A). Their existence, which would confirm the mode of terminal base arrangement, was uncertain because (1) projections were seen on relatively few circles (10–20%), (2) occasional circles had more than one projection, and (3) similar projections were also observed on single-stranded SV40 circular molecules. However, when heteroduplex molecules, prepared with any combination of AAV-1, -2, or -3 DNA, are spread in high concentrations of formamide, self-annealed strand ends can be specifically located and panhandle structures readily identified (Fig. 3B; Koczot and Rose, in preparation). These are about the same length as those seen on single-stranded circles (Fig. 3A). Thus, the model in Fig. 3A appears to depict the correct terminal base order. The same conclusion has been also reached by Berns and Kelly (personal communication), who have found that projections on minus strand oligomers are separated by distances equal to the length of monomeric strands. The reason for greater nuclease S_1 resistance than can be accounted for by panhandle length is not clear. Perhaps small, self-annealed hairpin regions are also present in the DNA strand.

Examination of double-stranded AAV DNA in the electron microscope has revealed that 5–15% of the duplex molecules can form circles closed by hydrogen-bonded regions (Koczot *et al.*, 1973). The double-stranded circles could result from annealing of intact strands with either (1) strands missing some or all of their terminal sequences or (2) with a subpopulation of permuted strands (Koczot *et al.*, 1973). Evidence has recently been obtained that duplex AAV circles are closed by overlaps equal to 1.5–6% of the genome's length (Gerry *et al.*, 1973). Approximately half the circles could be opened by exonuclease III and the remaining half by T5 exonuclease, indicating that the nucleotide sequence of AAV DNA contains a limited number of permutations, the start points of all being confined to a region equivalent to less than 6% of the genome. Furthermore, these investigators also observed that up to 34% of duplex linear monomers could be converted to circles or oligomers after a 1% digestion of DNA with exonuclease III, suggesting that AAV DNA molecules may also contain a type of terminal repetition similar to that present in linear bacteriophage DNA (natural terminal repetition; Thomas and MacHattie, 1967). It is possible, however, that duplex circles and oligomers resulted from annealing of terminal self-complementary sequences. Considering that a natural terminal repetition might exist in addition to the terminal self-complementary sequences (inverted terminal repetition), models can be constructed which contain both types

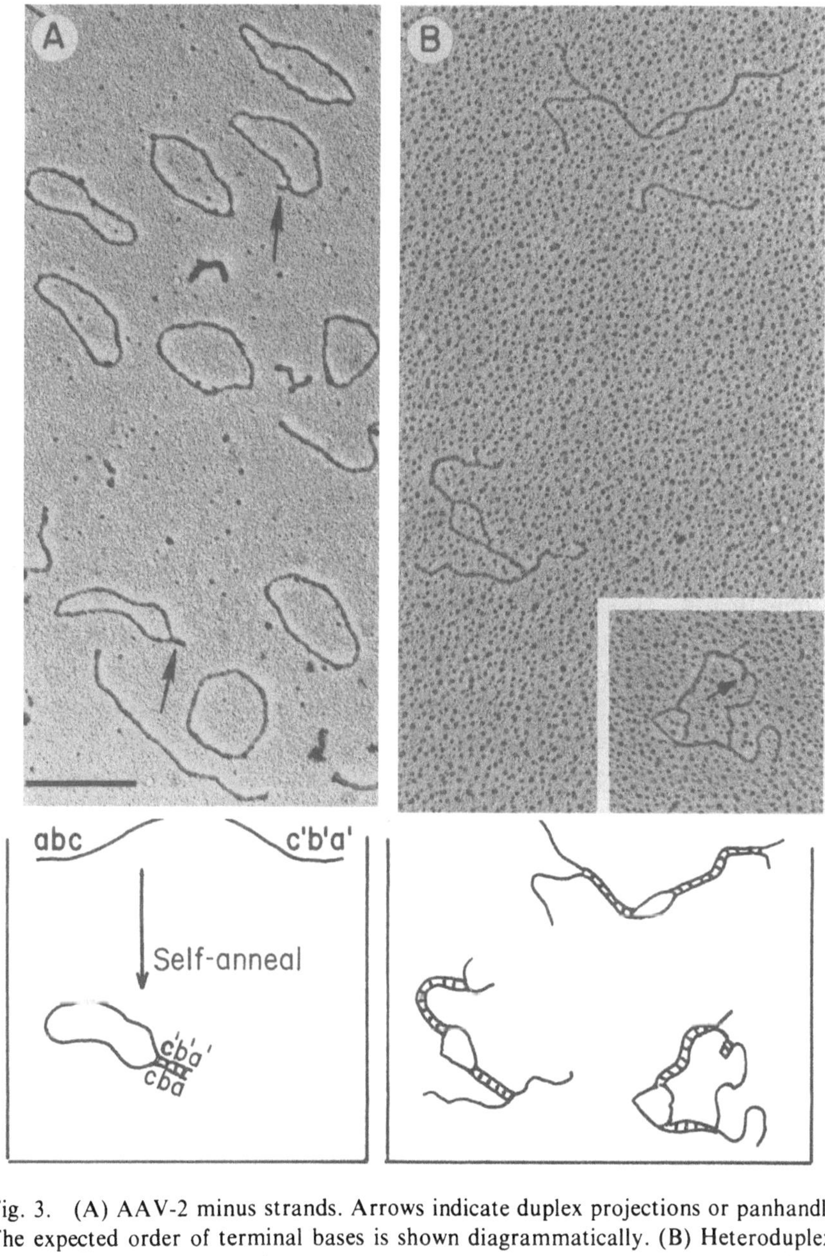

Fig. 3. (A) AAV-2 minus strands. Arrows indicate duplex projections or panhandles. The expected order of terminal bases is shown diagrammatically. (B) Heteroduplexes formed between AAV-1 and AAV-3 DNA strands and spread in 85% formamide. The ends of one strand have self-annealed to produce a circular heteroduplex molecule (inset), and the resulting duplex panhandle can be seen (arrow). The structure of heteroduplex molecules is also represented diagrammatically. Duplex regions are indicated by crossbars. The line superimposed on the electron micrograph in A is equivalent to 0.5 μm. From Koczot and Rose (in preparation).

of terminal repetition while still being compatible with all the observed findings (Gerry *et al.*, 1973).

3.2.6. Genetic Relatedness

Assessments of nucleic acid homology have been made only among genomes from the four human and simian AAV serotypes. With a technique employing synthetic RNA, interserotypic homologies ranged from 27–37% (Rose *et al.*, 1968). Recent DNA–DNA competition-hybridization studies have indicated, however, that values obtained with synthetic RNA are too low. Based on competition hybridization, 60–70% of the sequences in AAV-1, -2, and -3 DNA have been found to be homologous (Koczot and Rose, in preparation). Low estimates with synthetic RNA could have been due to incomplete copying of DNA template molecules (Rose *et al.*, 1968).

When heteroduplexes are prepared with any combination of AAV-1, -2, or -3 DNA, regions containing heterology melt out progressively as denaturing conditions are increased (Koczot and Rose, in preparation). This shows that partial homology is present within these regions and suggests that sequence differences among the serotypes have arisen by retention of random base changes (Davis and Hyman, 1971). In addition, interserotypic homologies physically map at similar locations, indicating that the same sequences are conserved in all three serotypes. The heteroduplex molecules shown in Fig. 3B demonstrate

TABLE 6

Molecular Weights of Parvovirus Polypeptides

Virus	Gel component,[a] mol. wt. $\times 10^{-3}$			Reference
	A	B	C	
RV	72 (9)	62 (60)	55 (10)	Salzman and White (1970)
H-1	92 (8)	72 (52)	56 (9)	Kongsvik and Toolan (1972)
Haden	85.5	76.8	66.8	Johnson and Hoggan (1973)
MVM	92 (9)	72 (54)	69 (8)	Tattersall, A. Shatkin, and D. Ward, personal communication
AAV (types 1, 2, and 3)	87 (3.5)	73 (2.7)	62 (52)	Rose *et al.* (1971)
AAV 3	91.6 (7)	79.3 (8)	65.9 (72)	Johnson *et al.* (1971)

[a] Figures in parentheses represent calculated number of molecules per virion. The A, B, and C components have also been named VP3, VP2, and VP1, respectively (Johnson *et al.*, 1971; Johnson and Hoggan, 1973).

the locations and extent of regions containing heterology which are revealed in a high concentration of formamide (denaturing conditions approximating $T_m - 6°C$); no melting was observed in homoduplex molecules spread under similar conditions. Because single-stranded circles remain essentially intact under the same spreading conditions, open heteroduplex ends may mean that the terminal self-complementary sequences are not completely conserved among different serotypes.

Nucleic acid-hybridization tests have failed to reveal homologies between AAV and adenovirus or herpesvirus helpers (Rose *et al.*, 1968; Rose and Koczot, 1972; Carter and Rose, 1972; Carter *et al.*, 1973). Because only adenoviruses and herpesviruses have AAV helper activity, it is also notable that Ad 2 and HSV-1 share no apparent homology (Rose and Koczot, unpublished results).

3.3. Protein

3.3.1. Number and Size of Components

Structural proteins of RV, H-1, Haden, MVM, and AAV-1, -2, and -3 have been analyzed in SDS-polyacrylamide gels (Salzman and White, 1970; Rose *et al.*, 1971; Johnson *et al.*, 1971; Kongsvik and Toolan, 1972; Johnson and Hoggan, 1973; Tattersall, A. Shatkin, and D. Ward, personal communication). The protein of each virus is resolved into three polypeptide components (A, B, and C), and the molecular weights of respective components are fairly comparable among all the viruses (Table 6). The C protein of AAV overlaps the electrophoretic position of the adenovirus fiber-penton component, but a lack of either antigenic or genetic relatedness between viruses suggests that this finding is coincidental (Rose *et al.*, 1968, 1971; Johnson *et al.*, 1972). The intermediate-sized polypeptide, B, is the major component of RV, H-1, and MVM, whereas the smallest polypeptide, C, predominates in Haden and AAV protein. These major components would appear to be the repeating units of capsid structure because the estimated number of their molecules per virion (52–72, see Table 6) approximates the minimum number, 60, required to construct an icosahedral capsid (Caspar and Klug, 1962). Further analysis of MVM protein, however, has revealed that the amounts of B and C polypeptides may vary inversely, with C increasing as the time allowed for nuclease digestion of crude lysates (virus purification step) is lengthened (P. Tattersall, A. Shatkin, and D. Ward, personal communication). Also, a quantitative conversion of B polypeptide to C polypeptide was seen when purified virus particles were treated with trypsin. Since the foregoing suggests that C polypeptides could be artifacts due to proteolysis during virus isolation, the question must be raised as to whether the C polypeptide is actually the primary unit of Haden or AAV capsid structure.

The possibility that one or more of the structural polypeptides might be derived from a larger precursor would explain why the coding capacity of parvovirus genomes falls as much as 35% short of that needed to account for the total molecular weight of observed protein components (Rose *et al.*, 1971). It has recently been found that peptide maps of the separated, ^{35}S-methionine-labeled MVM polypeptides show considerable sequence overlap among all the polypeptides (P. Tattersall and A. Shatkin, personal communication). In addition, sequence similarities between the B and C polypeptides of AAV have been suggested by the finding that antiserum prepared against SDS-dissociated B polypeptide cross-reacts with C polypeptide; it was not certain, however, that the purified B-protein immunogen was entirely free of C protein (Johnson *et al.*, 1972). Other data also imply that the structural polypeptides of AAV should share extensive common sequences. Only one messenger RNA species appears to be synthesized *in vivo*, and this RNA has a molecular weight of about 9×10^5 Carter and Rose, 1972; Carter *et al.*, 1972; Carter and Rose, 1974). A message of this size could not specify all three observed polypeptides if the sequence of each was unique. Its size, in fact, is sufficient to account for little else expect the largest polypeptide. This might mean that the AAV genome only codes for a single structural protein. Based on the molecular weights of A polypeptides, the nondefective parvoviruses should expend at least half their genetic content on the coding of structural protein. These deductions assume that estimates of polypeptide molecular weight are correct and that the A polypeptide is virus-specified and not constructed from a smaller basic unit. A preexisting cell protein is probably not incorporated into virions since MVM purified from cells labeled with ^{14}C-amino acids before infection and ^{3}H-amino acids during infection shows no enrichment for the prelabel in any of its polypeptides (P. Tattersall and A. Shatkin, personal communication).

3.3.2. Composition

Amino acid compositions of the total protein of AAV-2 and H-1 have been reported (Rose *et al.*, 1971; Kongsvik and Toolan, 1972).

The proteins of both viruses are rich in acidic amino acids, with acidic: basic amino acid ratios of 2.4 and 1.8, respectively. Based on the comparative labeling of AAV polypeptides with mixed ^{14}C-amino acids and ^{14}C-arginine alone, there is no AAV structural component relatively rich in arginine (Rose *et al.*, 1971) as is the case with internal components of adenovirus virions (Maizel *et al.*, 1968; Prage *et al.*, 1968).

3.3.3. Enzymatic Activity

Salzman (1971) has described a DNA polymerase activity associated with purified preparations of RV. Reactions require all four deoxynucleoside triphosphates, $MgCl_2$, and exogenous DNA. Native salmon sperm DNA was ten times more active than when denatured, and the reaction product was double-stranded on the basis of hydroxyapatite chromatography. An attempt to confirm this finding, and possibly demonstrate a similar activity with H-1 virus, failed (Rhode, 1973). Whether the observed activity was due to a cell contaminant or represents the presence of an enzyme involved in virus replication is still to be settled. It should be noted that Rhode (1973) propagated his RV in hamster embryo cells, whereas Salzman (1971) grew her virus in rat nephroma cells.

4. MULTIPLICATION OF NONDEFECTIVE PARVOVIRUSES

Nondefective parvoviruses replicate best in cultures of actively dividing cells (Johnson, 1967; Mayr *et al.*, 1968; Tennant *et al.*, 1969; Parker *et al.*, 1970*b*; Hallauer *et al.*, 1972; Tattersall, 1972*b*). This relationship between virus yield and cell physiological state is apparently due to viral dependence on one or more functions supplied by the cell during or immediately following the period of cellular DNA synthesis (Tennant *et al.*, 1969; Hampton, 1970; Tennant and Hand, 1970; Tattersall, 1972*b*; Siegl and Gautschi, 1973*a*; Rhode, 1973). Virus infection, in turn, inhibits cell mitosis (Johnson, 1967; Hampton, 1970; Tennant, 1971; Siegl and Gautschi, 1973*a*; Tattersall *et al.*, 1973) by a mechanism which may involve a noninfectious viral component (Siegl and Gautschi, 1973*a*). The parvovirus requirement for dividing cells seems to explain why selective injury to marrow, lymphatic, and intestinal mucosal cells is seen in FPV-infected cats (Johnson, 1969). In addition, developmental abnormalities induced by certain parvoviruses probably result from their affinity for proliferating generative cells in embryonic and newborn animals (Toolan *et al.*, 1960; Kilham, 1961*b*; Kilham and Margolis, 1964, 1965; Kilham *et al.*, 1967). A need for dividing cells also explains why tumor cells have been the source of numerous parvovirus isolates (Table 2).

4.1. Infectious Cycle

In principal studies, RV was grown in rat embryo (RE), hamster embryo (HE), or rat nephroma (RN) cells (Tennant *et al.*, 1969; Fields and Nicholson, 1972; Salzman *et al.*, 1972); H-1 in newborn human kidney (NB), Salk monkey heart, RE, or HE cells (Al-Lami *et al.*, 1969; Hampton, 1970; Rhode, 1973); MVM in RE, mouse embryo (ME), mouse L cells (Parker *et al.*, 1970*a,b*; Tattersall, 1972*a,b*; Tattersall *et al.*, 1973); and LuIII in HeLa cells (Siegl and Gautschi, 1973*a,b*). Growth cycles of FPV, PPV, HER, KBSH, and Haden were analyzed in kitten kidney, pig kidney, RE, KB, and fetal bovine cells, respectively (Johnson, 1967; Mayr *et al.*, 1968; Cole and Nathanson, 1969; Siegl *et al.*, 1972; Bates and Storz, 1973). All these viruses share two major properties: (1) Their latent periods are about the same and (2) their synthesis seems to depend on actively dividing cells. Owing to this latter feature, most studies have been performed with actively dividing cells, i.e., cultures of nonconfluent or tumor cells, or cells induced to synchronously divide by a serum pulse (Tennant *et al.*, 1969; Tattersall, 1972*b*; Rhode, 1973), or by reversing an inhibition of DNA synthesis (Hampton, 1970; Siegl and Gautschi, 1973*a*). In studies cited the infection multiplicities used (2–200 PFU or $TCID_{50}$/cell) were generally sufficient to provide reasonably efficient infections.

4.1.1. Adsorption, Latent Period, and Maturation

Relatively large fractions of added infectious virus were adsorbed in the few instances where measured. Approximately 75% of an 8-PFU/cell inoculum of Haden was adsorbed in 2 hours (Bates and Storz, 1973), and 90% of a 5-PFU/cell inoculum of RV was taken up in 1 hour (Salzman *et al.*, 1972). In H-1 infections of human lung cells, 60–70% of the virus was adsorbed with or without helper Ad 12 (Ledinko and Toolan, 1968). Latent periods of 10–16 hours have been observed with FPV, H-1, HER, RV, LuIII, and Haden (Johnson, 1967; Al-Lami *et al.*, 1969; Cole and Nathanson, 1969; Tennant and Hand,

1970; Fields and Nicholson, 1972; Salzman *et al.*, 1972; Siegl and Gautschi, 1973*b*; Bates and Storz, 1973). The ensuing synthesis of virus progeny is exponential over the next 8–12 hours, and virus maturation is usually complete by 24–30 hours after infection (Johnson, 1967; Tennant and Hand, 1970; Salzman *et al.*, 1972; Siegl and Gautschi, 1973*b*; Rhode, 1973). Virus is assembled in the cell nucleus (Mayor and Jordon, 1966; Mayr *et al.*, 1968; Al-Lami *et al.*, 1969; Siegl *et al.*, 1972) and remains mostly cell-associated until late after infection (Johnson, 1967; Salzman *et al.*, 1972; Siegl and Gautshi, 1973*b*; Bates and Storz, 1973). Final yields of infectious virus are usually 1–4 logs greater than cell infectivity during eclipse (Johnson, 1967; Al-Lami *et al.*, 1969; Tennant and Hand, 1970; Salzman *et al.*, 1972; Bates and Storz, 1973; Rhode, 1973). For MVM, at least, the infectious unit is a single particle (Tattersall, 1972*b*).

4.1.2. Efficiency of Infection

The number of cells competent for virus synthesis during an initial cycle can be estimated from the fraction of cells synthesizing viral protein as determined by FA staining. For RV in actively dividing RE cells percentages of 30% and 50% were achieved with multiplicities of 15 and 50 PFU/cell, respectively (Tennant *et al.*, 1969; Fields and Nicholson, 1972). After HER infection with 100 $TCID_{50}$/cell, 60% of RE cells contained viral antigen (Cole and Nathanson, 1969). At a multiplicity of 10 $TCID_{50}$/cell 95% of LuIII-infected HeLa cells synthesized viral antigen whether or not cells had been synchronized (Siegl and Gautschi, 1973*a*). When equating virus synthesis with the presence of viral protein, it should be noted that in abortive infections of human lung cells with H-1 about one-third of the cells contain viral antigen, whereas little, if any, virus multiplication occurs (Ledinko *et al.*, 1969). With RV, however, the number of virus-synthesizing cells based on FA staining was approximately proportional to the amount of progeny virus synthesized (Tennant and Hand, 1970).

4.2. Viral DNA Synthesis

After RV infection of RN cells, the synthesis of DNA incorporated into virions was first detected at 8–10 hours, and this synthesis continued until 23 hours, about the time virus maturation was completed (Salzman *et al.*, 1972). The onset of viral DNA synthesis thus preceeds the appearance of newly made virus by at least 4–6 hours, and progeny production seems to require concomitant viral DNA synthesis. Similarly, LuIII virion DNA was replicated 8–17 hours post-infection, preceding infectious virus synthesis by 2–4 hours and ending with the cessation of maturation (Siegl and Gautschi, 1973*b*). The fate of single-stranded parental RV DNA in RN cells has been followed by analyzing Hirt extracts of cells which were infected with particles containing radioactive DNA (Salzman and White, 1973). Between 30 and 60 minutes after infection about 30–40% of the total parental DNA in the supernatant extract was converted to a double-stranded, linear form on the basis of sucrose sedimentation, hydroxyapatite chromatography, and buoyant density in CsCl. The relative proportions of parental DNA in single- and double-stranded components were unchanged up to 20 hours after infection. These experiments suggest that the parental DNA is converted to a linear, duplex replicative form soon after infection. The reason for the 7 to 9-hour interval between the formation of these molecules and the observed onset of viral DNA synthesis is not clear. There is a possibility that such molecules might be transcribed before progeny DNA is synthesized since virus-specific antigen has been detected in the cytoplasm of HER-, KBSH-, and LuIII-infected cells as early as 4–6 hours after infection (Cole and Nathanson, 1969; Siegl *et al.*, 1972; Siegl and Gautschi, 1973*a*). In studies with RV-infected cells, however, cytoplasmic antigen was not found so early (Fields and Nicholson, 1972) and, in fact, RV cytoplasmic antigen may not be seen at all (Tennant *et al.*, 1969).

A double-stranded form of MVM DNA has also been found in modified Hirt extracts of infected mouse L cells (Tattersall, 1972a; Dobson and Helleiner, 1973; Tattersall *et al.*, 1973). Tattersall *et al.* observed that 25% of the total DNA synthesized in infected cells by 20 hours appears as supernatant DNA, as compared to only 0.8% with uninfected cells. Infected-cell supernatant DNA can be resolved into two approximately equal components on hydroxyapatite columns: single-stranded progeny molecules and double-stranded replicative intermediates. Progeny strands are first detected at 8 hours, and the time course of their synthesis is similar to that seen with RV (Salzman *et al.*, 1972) and LuIII (Siegl and Gautschi, 1973*b*). Double-stranded molecules are linear according to alkaline sucrose sedimentation and banding in ethidium bromide-CsCl, and they do not represent DNA–RNA hybrids on the basis of density in neutral CsCl. Further analysis of the double-stranded molecules indicates that 20–40% of this DNA renatures intramolecularly, suggesting either interstrand cross-linking or a hairpin structure. Chromatography of the double-stranded DNA on benzoylated diethylaminoethyl cellulose yields two components, one with and one without single-stranded regions. The former class of DNA is predominantly labeled by a short ^{3}H-thymidine pulse late in infection, indicating that it contains replicating molecules. Although the mechanism of DNA replication cannot be specified from these experiments, it is possible that self-complementary sequences might play a role in priming the synthesis of viral DNA.

4.3. Viral Protein Synthesis

Synthesis of virus protein has been followed both by specific FA staining and assay of cell-associated hemagglutinating viral antigen. The latter probably represents either assembled or unassembled capsid units since the purified B protein of RV strongly hemagglutinates (Salzman and White, 1970). As expected, the appearance and increase of cell-associated hemagglutinin parallels the rise of infectious titer (Mayr *et al.*, 1968; Siegl and Gautschi, 1973*b*; Bates and Storz, 1973; Rhode, 1973). Additionally, elevations of extracellular hemagglutinin coincide with release of virus from cells (Bates and Storz, 1973; Rhode, 1973).

FA staining has been carried out by both direct and indirect methods, utilizing hyperimmune mouse, rat, or rabbit antisera prepared with crude or partially purified virus as antigen (Cole and Nathanson, 1969; Tennant *et al.*, 1969; Parker *et al.*, 1970*a*; Hallauer *et al.*, 1971; Fields and Nicholson, 1972). Staining reactions with HER-, KBSH-, and LuIII-infected cells have revealed an early cytoplasmic fluorescence occurring between 4 and 6 hours, and a late nuclear fluorescence appearing at 8–10 hours after infection (Cole and Nathanson, 1969; Siegl *et al.*, 1972; Siegl and Gautschi, 1973*a*). After HER infection, cytoplasmic antigen appeared in up to 60% of cells, and 50% of cells eventually developed nuclear fluorescence. In LuIII infections the fractions of cells with cytoplasmic and nuclear antigen eventually reached 95% and 80%, respectively. Cytoplasmic antigen has been found in RV-infected cells, but it was first detected at 8 hours (Fields and Nicholson, 1972). Parker *et al.* (1970*a*) have also observed cytoplasmic antigen following MVM infection, but it appeared at 14 hours after infection, 2 hours after nuclear antigen could be seen. In HER- and RV- infected cells, and to a lesser extent in LuIII-infected cells, the number of cells with cytoplasmic antigen decreased as cells with nuclear antigen increased. Assuming that cytoplasmic fluorescence is due to virus structural protein, its decrease might represent the migration of capsid precursors to the nucleus where they are assembled into mature virions (Cole and Nathanson, 1969; Fields and Nicholson, 1972). Of interest in this regard is the finding that although HER or

RV do not multiply in mouse L cells, both induce cytoplasmic antigen in at least 70% of cells, whereas little or no nuclear antigen can be found (Cole and Nathanson, 1969; Fields and Nicholson, 1972). This could mean that the L-cell restriction involves an inability to transport viral protein to the nucleus. Since the appearance of nuclear antigen correlates with production of infectious virus, it probably corresponds to capsid protein or hemagglutinin. Because the synthesis of capsid protein is most probably linked to the synthesis of both cellular and viral DNA (see Sect. 4.4), it is not surprising that fluorescent nuclear antigen or hemagglutin do not appear when infected cells are treated with inhibitors of DNA synthesis (Parker *et al.*, 1970*a*; Siegl and Gautschi, 1973*b*; Rhode, 1973). Somewhat surprising, however, was the finding that DNA synthesis inhibitors did not block the appearance of early cytoplasmic antigen in LuIII-infected cells (Siegl and Gautschi, 1973*b*). In addition, production of early antigen was diminished by 75% by inhibiting RNA synthesis with 10 μg/ml of actinomycin D. These experiments led the authors to conclude that single-stranded (parental) viral DNA would have to be transcribed or act as a message itself to account for synthesis of this antigen. Another possibility is that early cytoplasmic antigen might represent a cell protein which is modified (or even induced) after virus infection. Further characterization of the early antigen will be necessary to determine whether or not it represents virus structural protein. The differential effects of α-amanitin and rifampicin on cytoplasmic- and nuclear-antigen formation were also examined (Siegl and Gautschi, 1973*b*). Neither inhibited cytoplasmic antigen, but both interfered with nuclear antigen. These results are difficult to interpret because of the prolonged presence of relatively high concentrations of inhibitor prior to nuclear antigen assay (15 hours) as opposed to the interval (5 hours) before cytoplasmic antigen was measured.

In contrast to nuclear antigen, accumulation of the early cytoplasmic antigen is not decreased when cells with reduced mitotic activity are infected with LuIII (Siegl and Gautschi, 1973*a*). This would indicate, at least, that production of the cytoplasmic antigen does not depend on the physiological state of the cell. It has been recently suggested that the early cytoplasmic fluorescence may in fact represent antigen from input virus particles (G. Siegl, personal communication).

4.4. Cellular Requirement

Studies with infected synchronized cells have strongly indicated that multiplication of nondefective parvoviruses requires one or more cellular functions generated during the S or G2 periods of cell growth (Tennant *et al.*, 1969; Hampton, 1970; Tennant and Hand, 1970; Tattersall, 1972*b*; Siegl and Gautschi, 1973*a*; Rhode, 1973). Because minimum latent-period and maximum single-cycle yields are achieved when cells are infected during the early S period, it has been concluded that the needed cell function(s) is not expressed before late S (Hampton, 1970; Siegl and Gautschi, 1973*a*; Rhode, 1973). Rhode (1973) has offered evidence that the formation of H-1 hemagglutinin depends on a DNA synthetic event which is initiated 1–2 hours before hemagglutinin can be detected, and that this hemagglutinin-dependent DNA synthesis does not commence until cells have been in S phase for at least 5 hours. Initiation of the hemagglutinin-dependent DNA synthesis probably coincides with initiation of viral progeny DNA synthesis, and, because this DNA synthesis appears to begin (late) in S phase, it seems likely that one or more components of the cell DNA-synthesizing apparatus are utilized for the replication of viral DNA. One could further speculate that the needed cell factor(s) may be involved in regulating the synthesis of a specific portion of the cell DNA which is made only at or near the end of S phase.

4.5. Cellular Impairment

FPV, RV, LuIII, and MVM are potent inhibitors of cell mitosis (Johnson, 1967; Hampton, 1970; Tennant, 1971; Siegl and Gautschi, 1973*a*; Tattersall *et al.*, 1973). After RV infection mitosis is even inhibited in cells which do not immediately synthesize virus (Tennant, 1971). The effect on mitosis in a given cell seems to depend on when infection occurs during the cell cycle. Using synchronized cells, Siegl and Gautschi (1973*a*) found that LuIII infection in early S appeared to completely prevent mitosis, whereas late S infection delayed its onset, and infection in G2 caused a substantial decrease in the number of cells entering mitosis, although the mitotic wave was not delayed. When the same experiment was repeated with UV-inactivated virus, results were similar except that addition of virus in early S only delayed the onset of mitosis. These findings suggest that capsid protein alone may be capable of retarding mitosis, but a complete blockade may require viral DNA replication.

In view of viral interference with mitosis and the probable virus requirement for a function(s) mediating cell DNA synthesis, the effect of infection on cell DNA synthesis seems especially important. Available data, however, are limited and conflicting. Following RV infection total DNA synthesis was increased in RN cells between 10–24 hours post-infection (Salzman *et al.*, 1972), but was decreased in RE and rat kidney cells during the same period (Tennant, 1971). Infection of rapidly dividing ME cells with MVM did not greatly alter their rate of DNA synthesis (Tattersall, 1972*b*). Concerning viral effects on total RNA and protein synthesis, RV infection produced little change (Tennant, 1971; Salzman *et al.*, 1972), but a selective decrease in 28 S RNA and a stimulation of 4 S RNA has been observed following H-1 infection of NB cells (Fong *et al.*, 1970).

5. MULTIPLICATION OF DEFECTIVE PARVOVIRUSES

AAV multiplication occurs only in cells concurrently infected with an adenovirus (Sect. 2.2.2). Herpesviruses, however, are capable of inducing an abortive AAV replication sequence and are therefore known as partial or incomplete AAV helpers (Atchison, 1970). Coinfections with AAV and HSV-1 lead to the synthesis of AAV DNA, RNA, FA-staining nuclear antigen, and possibly empty capsids, but production of infectious AAV cannot be detected (Atchison, 1970; Blacklow *et al.*, 1970, 1971*a*; Boucher *et al.*, 1971; Rose and Koczot, 1972; Carter and Rose, 1972; Henry *et al.*, 1972; Carter *et al.*, 1972; Dolin and Rabson, 1973). All other herpesviruses so far tested have been found capable of eliciting the formation of AAV nuclear antigen without an associated synthesis of AAV progeny. Included are bovine rhinotracheitis virus, EB virus, cytomegalovirus, varicella-zoster virus, and herpesvirus saimiri (Atchison, 1970; Blacklow *et al.*, 1970; Dolin and Rabson, 1973). Even though the percentage of cells containing FA antigen is often higher in coinfections with herpesviruses than with adenoviruses, AAV CF antigen can be detected only when adenoviruses are helpers (Atchison, 1970; Blacklow *et al.*, 1970; Dolin and Rabson, 1973). The morphology of the FA-staining antigen produced with herpesviruses is similar to that produced with adenoviruses, and antigen presumably represents capsid protein (Atchison, 1970; Blacklow *et al.*, 1971*a*).

Although the number and specific biochemical activities of the helper functions required for AAV multiplication are not definitively known, it now appears that adenoviruses at least provide separate helper functions for AAV DNA and RNA synthesis (Carter *et al.*, 1973). Herpesviruses may provide similar functions but might not support AAV multiplication because another needed function(s) is lacking, because a faulty AAV product is made, or because AAV synthesis is blocked as a consequence of herpesvirus replication (e.g., a herpesvirus protein might inhibit AAV assembly) in spite of the fact that all necessary helper functions can be provided (Blacklow *et al.*, 1971*a*; Boucher *et al.*, 1971; Rose and Koczot, 1972; Carter and Rose, 1972; Carter *et al.*, 1973).

5.1. Infectious Cycle

5.1.1. Adsorption, Latent Period, and Maturation

After KB cell infections with 10 $TCID_{50}$/cell of purified AAV-2, 30% of virus was adsorbed, and adsorption was complete within 1–2 hours (Rose and Koczot, 1972). The presence of adenovirus had no effect on adsorption or subsequent penetration of the AAV genome to the cell nucleus. These findings, plus the fact that AAV DNA infectiousness requires a helper adenovirus (Hoggan *et al.*, 1968), suggest that adenoviruses assist AAV replication only after uncoating of AAV DNA. Following a simultaneous infection of KB cells with AAV-2 and Ad 2, the AAV latent period lasts about 17 hours and is succeeded by an exponential synthesis of virus which usually ends 12–15 hours later (Fig. 4). Similar results were obtained when AGMK cells were si-

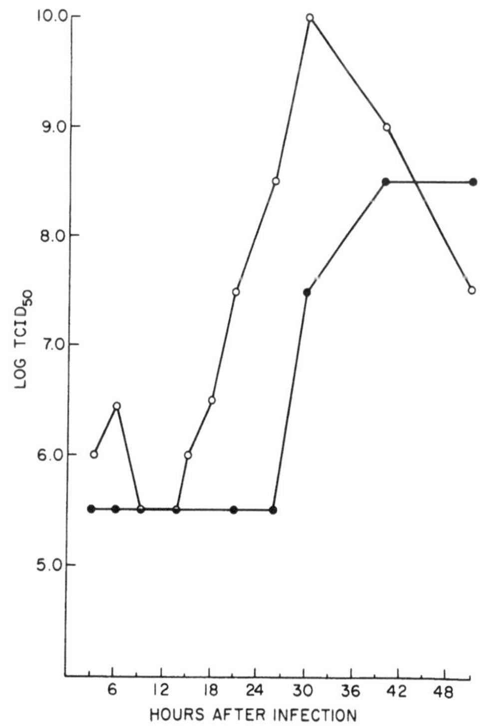

Fig. 4. Time course of intracellular and extracellular AAV infectivity. KB cells in suspension culture were simultaneously infected with 10 $TCID_{50}$/cell of AAV-2 and 50 $TCID_{50}$/cell of Ad 2. (O——O) infectivity of cells/ml of culture. (●——●) infectivity/ml of supernatant medium.

multaneously infected with AAV-4 and SV15 (Parks *et al.*, 1967*a*). Figure 4 also shows that virus is not released from cells until late in the maturation cycle. The AAV latent period can be shortened if cells are preinfected with adenovirus. Based on an increase in AAV particles and the appearance of AAV FA antigen, a minimum latent period of 4–6 hours is achieved with adenovirus preinfections of at least 10–12 hours (Parks *et al.*, 1967*a*; Ito *et al.*, 1967; Blacklow *et al.*, 1967*a*). All necessary adenovirus requirements are therefore present by about 10 hours after adenovirus infection, a relatively late time in the adenovirus cycle. With HSV-1 as helper, the minimum latent period between AAV-1 infection and the appearance of AAV FA antigen is 6 hours (Blacklow *et al.*, 1971*a*). In contrast to preinfections with adenovirus, preinfections with herpes simplex virus delay the appearance of AAV antigen, suggesting that helper virus replication may deplete the cell of some product needed for AAV antigen synthesis (Blacklow *et al.*, 1971*a*).

With helper adenoviruses AAV yields can reach 10^{10} $TCID_{50}$/ml, and greater than 10^{10} particles/ml in crude cell-culture harvests (Fig. 4; Parks *et al.*, 1967*a*; Rose and Koczot, 1972). AAV particles are often present in a 10- to 1000-fold excess over helper adenovirus particles (Hoggan *et al.*, 1966; Smith *et al.*, 1966; Parks *et al.*, 1967*a*). Particle counts indicate that on the average each AGMK cell may yield more than 2.5×10^5 AAV-4 virions (Parks *et al.*, 1967*a*). In addition, based on the average amount of AAV DNA synthesized per cell, it has been calculated that $2–4 \times 10^5$ AAV-2 genomes are replicated per KB cell, a number 10–100 times greater than that accounting for cellular infectivity (Rose and Koczot, 1972). The number of particles or genomes produced in a given cell probably exceeds these estimates because not all cells appears to be competent for virus synthesis (Ito *et al.*, 1967; Blacklow *et al.*, 1967*a*; Atchison, 1970).

5.1.2. Efficiency of Infection

On the basis of FA staining, up to 60% of cells can produce AAV protein after coinfection with HSV-1 or adenovirus (Ito *et al.*, 1967; Atchison, 1970). However, both cell sensitivity for developing antigen and the maximum percentage of cells capable of yielding antigen depend upon the cell type used (Ito *et al.*, 1967; Blacklow *et al.*, 1967*a*; Atchison, 1970). With human embryonic kidney cells, for example, only about 15% of cells are able to synthesize AAV antigen after coinfection with adenovirus (Blacklow *et al.*, 1967*a*). The reason why some cells are at least initially unable to support AAV replication is not known. AAV replication is clearly linked to specific events in the adenovirus cycle (Carter *et al.*, 1973), and the adenovirus cycle does not appear to depend on the physiological state of the cell (Hodge and Scharff, 1969). Dose–response studies with AAV-1 and either Ad 7 or HSV-1 have shown that a single infectious AAV particle and a single infectious helper virus particle are sufficient to initiate AAV antigen synthesis (Blacklow *et al.*, 1967*a*; 1971*a*).

5.2. Helper Requirement

Experiments which defined the minimum AAV latent period by detection of FA antigen suggested that helper virus was at least needed for the synthesis of AAV capsid protein, and that some critical event required for AAV growth occurred after the onset of helper virus DNA synthesis (Ito *et al.*, 1967; Blacklow *et al.*, 1967*a*, 1971). It was then found that neither AAV-2 DNA nor RNA synthesis could be detected by nucleic acid hybridization when cells were infected with AAV alone, but that AAV DNA and RNA production were readily detected after coinfections with either Ad 2 or HSV-1 (Rose and Koczot, 1972). Thus, both AAV DNA replication and transcription must depend upon one or more functions supplied by helper viruses. The net amount of AAV DNA synthesized with Ad 2 as helper is equivalent to 3% of total extracted DNA, whereas about half as much AAV DNA was synthesized in the presence of HSV-1 (Rose and Koczot, 1972). The observation that infectious AAV DNA could be extracted from cells coinfected with AAV-1 and HSV-1 suggests that all AAV DNA sequences are replicated when a herpesvirus is the helper (Boucher *et al.*, 1971).

5.2.1. Comparative Kinetics of Viral DNA and RNA Synthesis

Some insight has been gained concerning the number and possible roles of adenovirus helper functions by analyzing the time course of adenovirus and AAV DNA and RNA synthesis in KB cells after simultaneous or prior infection with adenovirus (Rose and Koczot, 1972; Carter *et al.*, 1973). After simultaneous infection, AAV and adenovirus DNA synthesis are first detected after 6.5–7.5 hours, but AAV RNA synthesis does not begin until about 2.5 hours later (Fig. 5). The onset of AAV RNA synthesis is associated with a marked increase in adenovirus RNA synthesis which actually commences by at least 2–3 hours after infection. This lag between AAV DNA and RNA synthesis is eliminated when cells are infected with adenovirus for 10 hours before infection with AAV (Fig. 6). Furthermore, both AAV DNA and RNA synthesis are detected within 3–4 hours after AAV infection, much earlier than after simultaneous infection. The onset of AAV RNA synthesis is closely associated with the onset of AAV DNA synthesis, and the synthesis of both proceed together (Fig. 6). Since the adenovirus requirement for AAV DNA and RNA synthesis is no longer limiting 10 hours after adenovirus infection, this 3- to 4-hour interval represents the minimum delay between AAV infection and the onset of AAV DNA replication and transcription. A similar minimum

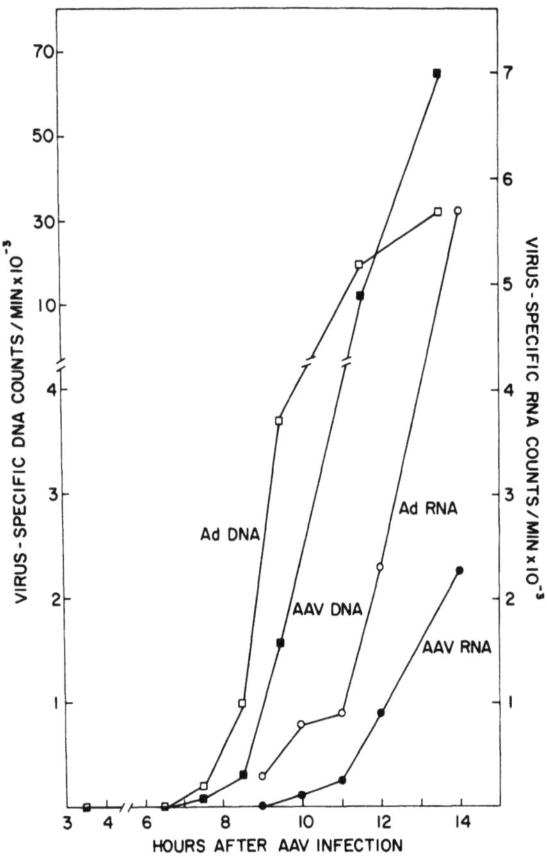

Fig. 5. Viral DNA and RNA synthesis in KB cells simultaneously infected with AAV-2 and Ad 2 at multiplicities noted in Fig. 4. Portions of the infected culture were labeled with ^{3}H-thymidine or ^{3}H-uridine for 1-hour periods, and the total cell DNA or RNA was isolated as described previously (Rose and Koczot, 1971, 1972). DNA labeling was carried out in the presence of 2×10^{-5} M 5-fluorodeoxyuridine. ^{3}H-labeled viral DNA and RNA were measured by hybridization with AAV-2 or Ad 2 DNA on nitrocellulose filters. Input of ^{3}H-DNA was 2×10^5 cpm; ^{3}H-RNA input, 2×10^6 cpm. (□——□) ^{3}H-labeled DNA bound to filters containing 2 μg Ad 2 DNA, (■——■) ^{3}H-labeled DNA bound to filters containing 2 μg AAV-2 DNA, (○——○) ^{3}H-labeled RNA bound to filters containing 2 μg Ad 2 DNA, (●——●) ^{3}H-labeled RNA bound to filters containing 2 μg AAV-2 DNA. Reproduced with permission from *Nature* (Carter *et al.*, 1973).

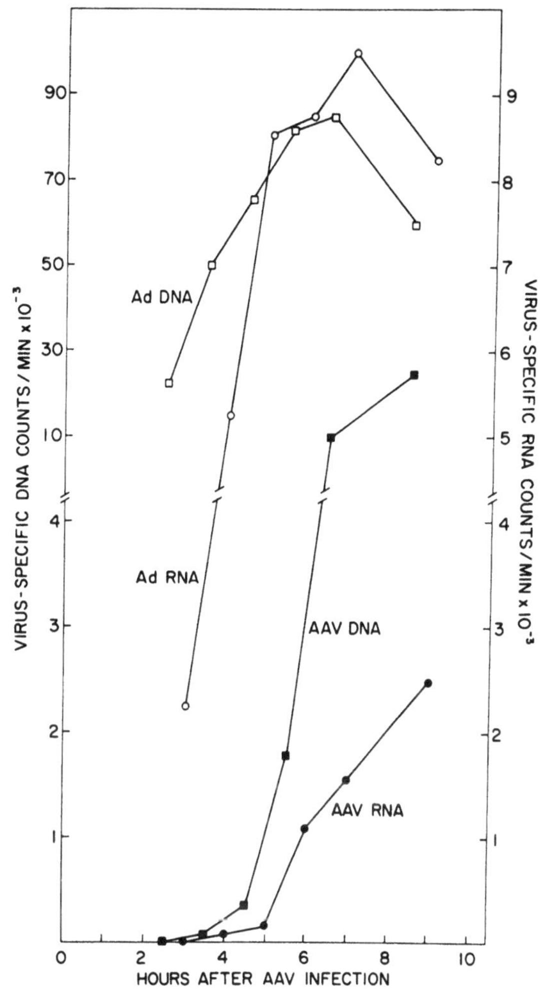

Fig. 6. Viral DNA and RNA synthesis in KB cells preinfected with Ad 2 for 10 hours before infection with AAV-2. Multiplicities are as in Fig. 4. Virus-specific DNA and RNA synthesized during 1-hour periods were measured as described in the legend to Fig. 5. (□——□) ^{3}H-labeled DNA bound to filters containing 2 μg Ad 2 DNA, (■——■) ^{3}H-labeled DNA bound to filters containing 2 μg AAV-2 DNA, (○——○) ^{3}H-labeled RNA bound to filters containing 2 μg Ad 2 DNA, (●——●) ^{3}H-labeled RNA bound to filters containing 2 μg AAV-2 DNA. Reproduced with permission from *Nature* (Carter *et al.*, 1973).

delay between AAV infection and the detection of AAV antigen synthesis (Ito *et al.*, 1967; Blacklow *et al.*, 1967*a*) suggests that AAV protein can be made soon after AAV transcription begins. In simultaneously infected cells, FUDR is found to block both AAV RNA synthesis and the late rise in adenovirus RNA synthesis. Furthermore, Ad 2 T antigen is not inhibited, but the synthesis of adenovirus structural proteins is prevented (Rowe *et al.*, 1965; Carter *et al.*, 1973). Thus, AAV RNA synthesis as well as transcription of messages coding for adenovirus structural proteins seem to depend upon viral DNA synthesis. Plausible interpretations of the kinetic data are: (1) since the lag between AAV DNA and RNA synthesis is abolished by adenovirus preinfection, AAV DNA and RNA synthesis require separate helper functions which appear sequentially after adenovirus infection, (2) the function(s) needed for initiating AAV DNA synthesis might also be required for initiating adenovirus DNA synthesis because the DNA of both viruses seem to be synthesized concomitantly after simultaneous infection, and (3) the temporal relationship between the initiation of AAV RNA synthesis and the late increase in adenovirus RNA synthesis suggests that AAV transcription may depend upon either the same factor(s) responsible for this late adenovirus RNA synthesis or upon some product of the late adenovirus transcription itself.

5.2.2. Mechanism of DNA Replication

Little can be specifically said about how AAV DNA is replicated since AAV replicative intermediates have not yet been characterized. There are, however, two findings which suggest that the mechanisms of AAV and adenovirus DNA synthesis may share common features: (1) the occurrence of inverted terminal repetitions in the DNA of both viruses (Garon *et al.*, 1972; Wolfson and Dressler, 1972; Koczot *et al.*, 1973) and (2) evidence that the initiation of AAV DNA synthesis may be linked to that of adenovirus DNA synthesis (Carter, *et al.*, 1973). The inverted terminal repetition might play a role in DNA replication since adenovirus DNA seems to replicate from an end of the template molecule (Sussenbach *et al.*, 1972; Sussenbach and Van Der Vliet, 1973) and since hybridization data indicate that the self-complementary sequences in AAV DNA are not transcribed *in vivo* (Carter *et al.*, 1972). Furthermore, double-stranded adenovirus DNA molecules as well as the putative duplex replicative form of AAV would be expected to possess ends which are identical, a circumstance that could provide a structural basis for initiating DNA replication from either end of the parent molecule (Koczot *et al.*, 1973). With AAV, at least, this kind of bidirectional mechanism for synthesizing plus or minus strands by strand displacement from a linear duplex molecule is an attractive possibility (Koczot *et al.*, 1973). In this regard it should be noted that although adenovirus DNA replication appears to involve displacement of a strand from a linear duplex template, strand displacement may be initiated only from one specific end of the template molecule (Sussenbach *et al.*, 1972; Sussenbach and Van Der Vliet, 1973; Van Der Eb, 1973).

5.3. Viral RNA

5.3.1. DNA-Strand Specificity and Sequence Representation

AAV-2 RNA synthesized *in vivo* with either Ad 2 or HSV-1 as helper (1) hybridizes only with the minus strand of AAV DNA (Rose and Koczot, 1971; Carter and Rose, 1972; Carter *et al.*, 1972), (2) contains similar nucleotide sequences on the basis of hybridization inhibition tests (Carter and Rose, 1972), and (3) has a mean molecular weight of about 9×10^5 as determined by sedimentation in DMSO-sucrose or electrophoresis in polyacrylamide gels in 98% formamide (Carter and Rose, 1972; Carter and Rose, 1974). Furthermore, only 70–80% of the AAV genome is transcribed in the presence of Ad 2 or HSV-1, as measured by hydroxyapatite analysis of hybrids formed between fragmented AAV minus strands and RNA from infected cells (Carter *et al.*, 1972). These data suggest that the restricted helper function of HSV-1 is not due to aberrant transcription of the AAV genome. Whether the "incomplete" transcription of AAV DNA relates to defectiveness or merely reflects a transcriptional pattern common to all parvoviruses remains to be determined. Present evidence strongly suggests that the self-complementary terminal sequences of AAV DNA are not transcribed, and that they comprise no more than half of the total nontranscribed sequences (Carter *et al.*, 1972; Koczot and Rose, in preparation; Sect. 3.2.5).

5.3.2. RNA Species

AAV-2 RNA in both the nucleus and polysomes of KB cells coinfected with Ad 2 contains the same nucleotide sequences and appears to consist of a single, main, 19–20 S species having an estimated molecular weight of about 9×10^5 (Carter and Rose, 1974). These RNA molecules approximate 70% of the size of the AAV DNA minus strand (1.35×10^6; Table 4) and might thus account for the entire region of the genome which is stably transcribed. A population of smaller, very heterogeneous, AAV RNA molecules is additionally found in the nucleus and nonpolysomal cytoplasmic fraction. Such molecules may result from degradation of the 20 S RNA. It is also possible that small RNA molecules could arise from the transcription of incomplete DNA template molecules. All these data suggest that (1) a single species of AAV RNA is transcribed *in vivo*, (2) it is not cleaved after synthesis, in contrast to adenovirus RNA which is cleaved following transcription and prior to transport to the polysomes (Parsons and Green, 1971; Parsons *et al.*, 1971; Carter and Rose, 1972), and (3) the AAV message may only specify coat protein(s) since its size is about that required to code for the largest structural polypeptide (see Sect. 3.3.1). The AAV polysomal message also appears to contain poly A, but the poly A sequence has not been characterized (J. Newbold, personal communication; B. Carter and J. Rose, unpublished results).

6. CONCLUSIONS

Parvoviruses are ubiquitous in nature, and numerous species have been found to infect a variety of animal hosts. For the most part, nondefective parvoviruses exhibit considerable host specificity. The host specificity of defective parvoviruses, at least *in vitro*, is largely determined by whether cells can support reproduction of the coinfecting adenovirus. It is expected that new members of the parvovirus group will continue to be discovered.

The biological significance of most of the parvoviruses is still obscure. Only the feline and related viruses are known to cause disease naturally, but the pathogenicity of rodent viruses under experimental conditions indicates that other disease potentials may exist. It is worth noting that two viruses presently under study, acute infectious gastroenteritis virus and hepatitis A virus, have major properties in common with parvoviruses (i.e., heat and ether stability, size, morphology, and buoyant density in CsCl; Kapikian *et al.*, 1973; Provost *et al.*, 1973). They are, therefore, possible candidates for pathogenic parvoviruses in man. Because of their latency and hardiness, parvoviruses must always be considered as potential contaminants of cells, other virus stocks, and certain vaccines.

Elucidation of the biochemistry of parvovirus replication should provide some insight into mechanisms involved in the initiation and regulation of cellular and (in the case of the AAV) helper-virus DNA synthesis. Hopefully, such information might also explain the basis for the antimitotic activity of nondefective parvoviruses, an activity with obviously important biological implications. The biochemical interactions which result in viral interference and, possibly, viral latency seem especially significant and deserve continued study. It may well be that parvoviruses will join the list of viruses found to integrate their DNA within the cellular genome.

It seems probable at present that the AAV synthesize only a single message species *in vivo*, and there is reason to think that this message might only specify coat protein. Whether or not a similar situation will be the case for the nondefective parvoviruses is yet to be determined.

Finally, although the data are far from complete, one is led to suspect that the replicative forms of parvovirus DNA are duplex, linear molecules. The detection, thus far, of an inverted terminal repetition in only AAV DNA and the DNA of helper adenoviruses suggests that this structure may relate to AAV dependence. If so, it is conceivable that the DNA of herpesviruses might also contain a similar type of terminal repetition. The possibility that these terminal sequences play a role in both AAV and adenovirus DNA synthesis is intriguing, but still a matter of speculation.

7. REFERENCES

Abinanti, F. R., and Warfield, M. S., 1961, Recovery of a hemadsorbing virus (HADEN) from the gastrointestinal tract of calves, *Virology* **14,** 288.

Al-Lami, F., Ledinko, N., and Toolan, H., 1969, Electron microscope study human NB and SMH cells infected with the parvovirus, H-1, *J. Gen. Virol.* **5,** 485.

Andrewes, C., 1970, Generic names of viruses of vertebrates, *Virology* **40,** 1070.

Archetti, I., Bereczky, E., and Bocciarelli, D., 1965, A small virus associated with the simian adenovirus SV11, *Virology* **29,** 642.

Atchison, R. W., 1970, The role of herpesviruses in adenovirus-associated virus replication *in vitro*, *Virology* **42,** 155.

Atchison, R. W., Casto, B. C., and McD. Hammon, W., 1965, Adenovirus-associated defective virus particles, *Science (Wash., D.C.)* **149,** 754.

Atchison, R. W., Casto, B. C., and McD. Hammon, W., 1966, Electron microscopy of adenovirus-associated virus (AAV) in cell cultures, *Virology* **29,** 353.

Barwise, A. H., and Walker, I. O., 1970, Studies on the DNA of a virus from *Galleria mellonella*, *FEBS (Fed. Eur. Biochem. Soc.) Lett.* **6,** 13.

Bates, R. C., and Storz, J., 1973, Host cell range and growth characteristics of bovine parvoviruses, *Infect. Immun.* **7,** 398.

Baum, S. G., Reich, P. R., Hybner, C. J., Rowe, W. P., and Weissman, S. M., 1966, Biophysical evidence for linkage of adenovirus and SV40 DNAs in adenovirus 7–SV40 hybrid particles, *Proc. Natl. Acad. Sci. USA* **56,** 1509.

Berns, K. I., and Adler, S., 1972, Separation of two types of adeno-associated virus particles containing complementary polynucleotide chains, *J. Virol.* **9,** 394.

Berns, K. I., and Rose, J. A., 1970, Evidence for a single-stranded adenovirus-associated virus genome: Isolation and separation of complementary single strands, *J. Virol.* **5,** 693.

Binn, L. N., Lazar, E. C., Eddy, G. A., and Kajima, M., 1968, Minute viruses of canines, *Abstr. 68th Annu. Meet. Am. Soc. Microbiol.* **98,** 161.

Binn, L. N., Lazar, E. C., Eddy, G. A., and Kajima, M., 1970, Recovery and characterization of a minute virus of canines, *Infect. Immun.* **1,** 503.

Blacklow, N. R., Hoggan, M. D., and Rowe, W. P., 1967*a*, Immunofluorescent studies of the potentiation of an adenovirus-associated virus by adenovirus 7, *J. Exp. Med.* **125,** 755.

Blacklow, N. R., Hoggan, M. D., and Rowe, W. P., 1967*b*, Isolation of adenovirus-associated viruses from man, *Proc. Natl. Acad. Sci. USA* **58,** 1410.

Blacklow, N. R., Hoggan, M. D., and Rowe, W. P., 1968*a*, Seriologic evidence for human infection with adenovirus-associated viruses, *J. Natl. Cancer Inst.* **40,** 319.

Blacklow, N. R., Hoggan, M. D., Kapikian, A. Z., Austin, J. B., and Rowe, W. P., 1968*b*, Epidemiology of adenovirus-associated virus infection in a nursery population, *Am. J. Epidemiol.* **88,** 368.

Blacklow, N. R., Hoggan, M. D., and McClanahan, M. S., 1970, Adenovirus-associated viruses: Enhancement by human herpesviruses, *Proc. Soc. Exp. Biol. Med.* **134,** 952.

Blacklow, N. R., Dolin, R., and Hoggan, M. D., 1971*a*, Studies of the enhancement of an adenovirus-associated virus by herpes simplex virus, *J. Gen. Virol.* **10,** 29.

Blacklow, N. R., Hoggan, M. D., Sereno, M. S., Brandt, C. D., Kim, H. W., Parrott, R. H., and Chanock, R. M., 1971*b*, A seroepidemiologic study of adenovirus-associated virus infection in infants and children, *Am. J. Epidemiol.* **94,** 359.

Boggs, J. D., Melnick, J. L., Conrad, M. E., and Felsher, B. F., 1970, Viral hepatitis clinical and tissue culture studies. *JAMA* **214,** 1041.

Boucher, D. W., Melnick, J. L., and Mayor, H. D., 1971, Nonencapsidated infectious DNA of adeno-satellite virus in cells coinfected with herpesvirus, *Science* (*Wash., D.C.*) **173,** 1243.

Boucher, D. W., Parks, W. P., and Melnick, J. L., 1969, Failure of a replicating adenovirus to enhance adeno-associated satellite virus replication, *Virology* **39,** 932.

Breese, S. S., Jr., Howatson, A. F., and Chany, Ch., 1964, Isolation of virus-like particles associated with Kilham rat virus infection of tissue cultures, *Virology* **24,** 598.

Carter, B. J., and Rose, J. A., 1972, Adenovirus-associated virus multiplication. VIII. Analysis of *in vivo* transcription induced by complete or partial helper viruses, *J. Virol.* **10,** 9.

Carter, B. J., and Rose, J. A., 1974, Transcription *in vivo* of a defective parvovirus: Sedimentation and electrophoretic analysis of RNA synthesized by adenovirus-associated virus and its helper adenovirus, *Virology,* in press.

Carter, B. J., Khoury, G., and Rose, J. A., 1972, Adenovirus-associated virus multiplication. IX. Extent of transcription of the viral genome *in vitro, J. Virol.* **10,** 1118.

Carter, B. J., Koczot, F. J., Garrison, J., Dolin, R., and Rose, J. A., 1973, Separate helper functions provided by adenovirus for adenovirus-associated virus multiplication, *Nat. New Biol.* **244,** 71.

Cartwright, S. F., and Huck, R. A., 1967, Viruses isolated in assocition with herd infertility, abortions and stillbirths in pigs, *Vet. Rec.* **81,** 196.

Cartwright, S. F., Lucas, M., and Huck, R. A., 1969, A small haemagglutinating porcine DNA virus. I. Isolation and properties, *J. Comp. Pathol.* **79,** 371.

Caspar, D. L. D., and Klug, A., 1962, Physical principles in the construction of regular viruses, *Cold Spring Harbor Symp. Quant. Biol.* **27,** 1–24.

Casto, B. C., and Goodheart, C. R., 1972, Inhibition of adenovirus transformation *in vitro* by AAV-1, *Proc. Soc. Exp. Biol. Med.* **140,** 72.

Casto, B. C., Atchison, R. W., and McD. Hammon, W., 1967*a*, Studies on the relationship between adeno-associated virus type 1 (AAV-1) and adenoviruses. I. Replication of AAV-1 in certain cell cultures and its effect on helper adenovirus, *Virology* **32,** 52.

Casto, B. C., Armstrong, J. A., Atchison, R. W., and Hammon W. McD., 1967*b*, Studies on the relationship between adeno-associated virus type 1 (AAV-1) and adenoviruses. II. Inhibition of adenovirus plaques by AAV; its nature and specificity, *Virology* **33,** 452.

Cheong, L., Fogh, J., and Barclay, R., 1965, Some properties of the H-1 virus and nucleic acid, *Fed. Proc.* **24,** 596.

Cole, G. A., and Nathanson, N., 1969, Immunofluorescent studies of the replication of rat virus (HER strain) in tissue culture, *Acta Virol.* **13,** 515.

Crawford, L. V., 1966, A minute virus of mice, *Virology*, **29,** 605.

Crawford, L. V., Follett, E. A. C., Burdon, M. G., and McGeoch, D. J., 1969, The DNA of a minute virus of mice, *J. Gen Virol.* **4,** 37.

Cross, S. S., and Parker, J. C., 1972, Some antigenic relationships of the murine parvoviruses: Minute virus of mice, rat virus, and H-1 virus, *Proc. Soc. Exp. Biol. Med.* **139,** 105.

Dalldorf, G., 1960, Viruses and human cancer, *Bull. N.Y. Acad. Med.* **36,** 795.

Davis, R. W., and Hyman, R. W., 1971, A study in evolution: The DNA base sequence homology between coliphages T7 and T3, *J. Mol. Biol.* **62,** 287.

Dobson, P. R., and Helleiner, C. W., 1973, A replicative form of the DNA of minute virus of mice, *Can. J. Microbiol.* **19,** 35.

Dolin, R., and Rabson, A. S., 1973, Herpesvirus saimiri: Enhancement of adenovirus-associated virus, *J. Natl. Cancer Inst.* **50,** 205.

Domoto, K., and Yanagawa, R., 1969, Properties of a small virus associated with infectious canine hepatitis virus, *Jap. J. Vet. Res.* **17,** 32.

Dutta, S. K., and Pomeroy, B. S., 1967, Electron microscopic studies of quail bronchitis virus, *Am. J. Vet. Res.* **28,** 296.

El Dadah, A. H., Smith, K. O., Squire, R. A., Santos, G. W., and Melby, E. C., 1967, Viral hemorrhagic encephalopathy of rats, *Science* (*Wash., D.C.*) **156,** 392.

El Mishad, A. M., Yates, V. J., McCormick, K. J., and Trentin, J. J., 1973, Avian adeno-associated virus (AAAV): Characterization and detection in avian adenovirus stocks, *Abstr. 73rd Annu. Meet. Am. Soc. Microbiol.* **217,** 230.

Feldman, L. A., Melnick, J. L., and Rapp, F., 1965, Influence of SV40 genome on the replicataion of an adenovirus–SV40 "hydrid" population, *J. Bacteriol.* **90,** 778.

Fields, H. A., and Nicholson, B. L., 1972, The replication of Kilham rat virus (RV) in various host systems: Immunofluorescent studies, *Can. J. Microbiol.* **18,** 103.

Fong, C. K. Y., Toolan, H. W., and Hopkins, M. S., 1970, Effect of H-1 on RNA synthesis in NB cells, *Proc. Soc. Exp. Biol. Med.* **135,** 585.

Freifelder, D., Kleinschmidt, A., and Sinsheimer, R., 1964, Electron microscopy of single-stranded DNA: Circularity of DNA of bacteriophage ϕX174, *Science* **146,** 254.

Garon, C. F., Berry, K. W., and Rose, J. A., 1972, A unique form of terminal redundancy in adenovirus DNA molecules, *Proc. Natl. Acad. Sci. USA* **69,** 2391.

Gerry, H. W., Kelly, T. J., Jr., and Berns, K. I., 1973, Arrangement of nucleotide sequences in adeno-associated virus DNA, *J. Mol. Biol.* **79,** 207.

Gorham, J. R., Hartsough, G. R., Sato, N., and Lust, S., 1966, Studies on cell culture adapted feline panleukopenia virus, *Vet. Med.* **61,** 35.

Green, M., Piña, M. Kines, R., Wensink, P., MacHattie, L. A., and Thomas, C. A., Jr., 1967, Adenovirus DNA. I. Molecular weight and conformation, *Proc. Natl. Acad. Sci. USA* **57,** 1302.

Greene, E. L., 1965, Physical and chemical properties of H-1 virus. I. *p*H and heat stability of the hemagglutinating property, *Proc. Soc. Exp. Biol. Med.* **118,** 973.

Greene, E. L., and Karasaki, S., 1965, Physical and chemical properties of H-1 virus. II. Partial purification, *Proc. Soc. Exp. Biol. Med.* **119,** 918.
Hallauer, C., Kronauer, G., and Siegl, G., 1971, Parvoviruses as contaminants of permanent human cell lines. I. Virus isolations from 1960–1970, *Arch. Ges. Virusforch.* **35,** 80.
Hallauer, C., Siegl, G., and Kronauer, G., 1972, Parvoviruses as contaminants of permanent human cell lines. III. Biological properties of the isolated viruses, *Arch. Ges. Virusforsch.* **38,** 366.
Hampton, E. G., 1970, H-1 virus growth in synchronized rat embryo cells, *Can. J. Microbiol.* **16,** 266.
Henry, C. J., Merkow, L. P., Pardo, M., and McCabe, C., 1972, Electron microscope study on the replication of AAV-1 in herpes-infected cells, *Virology* **49,** 618.
Hodge, L. D., and Scharff, M. D., 1969, Effect of adenovirus on host cell DNA synthesis in synchronized cells, *Virology* **37,** 554.
Hoggan, M. D., 1970, Adenovirus associated viruses, *in* "Progress in Medical Virology" (J. L. Melnick, ed.), Vol. 12, pp. 211–239, Karger, Basel.
Hoggan, M. D., 1971, Small DNA viruses, *in* "Comparative Virology" (K. Maramorosch and E. Kurstak, eds.), pp. 43–74, Academic Press, New York.
Hoggan, M. D., 1973, "Continuous Carriage of Adenovirus-Associated Virus Genome in Cell Culture in the Absence of Helper Adenovirus, Proceedings of the Forth Lepetit Colloquium," North Holland, Amsterdam.
Hoggan, M. D., Blacklow, N. R., and Rowe, W. P., 1966, Studies of small DNA viruses found in various adenovirus preparations: Physical, biological, and immunological characteristics, *Proc. Natl. Acad. Sci. USA* **55,** 1467.
Hoggan, M. D., Shatkin, A. J., Blacklow, N. R., Koczot, F. J., and Rose, J. A., 1968, Helper-dependent infectious deoxyribonucleic acid from adenovirus-associated virus, *J. Virol.* **2,** 850.
Horzinek, M., Mussgay, M., Maess, J., and Petzoldt, K., 1967, Nachweis dreier virussarten (Schweinepest-, adeno-, picodna-virus) in einem als cytopathogen bezeichneten schweinpest-virusstamn, *Arch. Ges. Virusforsch.* **21,** 98.
Huang, A., and Wagner, R., 1966, Defective T particles of vesicular stomatitis virus. II. Biologic role in homologous interference, *Virology* **30,** 173.
Ishibashi, M. and Ito, M., 1971, The potentiation of type 1 adeno-associated virus by temperature-sensitive conditional-lethal mutants of CELO virus at the restrictive temperature, *Virology* **45,** 317.
Ito, M., and Mayor, H. D., 1968, Hemagglutinin of type 4 adeno-associated satellite virus, *J. Immunol.* **100,** 61.
Ito, M., and Suzuki, E., 1970, Adeno-associated satellite virus growth supported by a temperature-sensitive mutant of human adenovirus, *J. Gen. Virol.* **9,** 243.
Ito, M., Melnick, J. L., and Mayor, H. D., 1967, An immunofluorescence assay for studying replication of adeno-satellite virus, *J. Gen. Virol.* **1,** 199.
Johnson, F. B., and Hoggan, M. D., 1973, Structural proteins of Haden virus, *Virology* **51,** 129.
Johnson, F. B., Ozer, H. L., and Hoggan, M. D., 1971, Structural proteins of adenovirus-associated virus type 3, *J. Virol.* **8,** 860.
Johnson, F. B., Blacklow, N. R., and Hoggan, M. D., 1972, Immunological reactivity of antisera prepared against the sodium dodecyl sulfate-treated structural polypeptides of adenovirus-associated virus, *J. Virol.* **9,** 1017.

Johnson, R. H., 1965*a*, Feline panleucopaenia. I. Identification of a virus associated with the syndrome, *Res. Vet. Sci.* **6,** 466.
Johnson, R. H., 1965*b*, Feline panleucopaenia. II. Some features of the cytopathic effects in feline kidney monolayers, *Res. Vet. Sci.* **6,** 472.
Johnson, R. H., 1967, Feline panleucopaenia virus. IV. Methods for obtaining reproducible *in vitro* results, *Res. Vet. Sci.* **8,** 256.
Johnson, R. H., 1969, A search for parvoviridae (picodnaviridae), *Vet. Rec.* **82,** 19.
Johnson, R. H., and Cruickshank, J. G., 1966, Problems in the classification of feline panleucopaenia virus, *Nature (Lond.)* **212,** 622.
Johnson, R. H., and Halliwell, R. E. W., 1968, Natural susceptibility to feline panleucopaenia of the coati-mundi, *Vet. Rec.* **82,** 582.
Johnson, R. H., Margolis, G., and Kilham, L., 1967, Identity of feline ataxia virus with feline panleucopenia virus, *Nature (Lond.)* **214,** 175.
Kapikian, A. Z., Gerin, J. L., Wyatt, R. G., Thornhill, T. S., and Channock, R. M., 1973, Density in cesium chloride of the 27-nm "8 FIIa" particle associated with acute infectious nonbacterial gastroenteritis: Determination by ultracentrifugation and immune electron microscopy, *Proc. Soc. Exp. Biol. Med.* **142,** 874.
Karasaki, S., 1966, Size and ultrastructure of the H-viruses as determined with the use of specific antibodies, *J. Ultrastruct. Res.* **16,** 109.
Kassanis, B., 1962, Properties and behavior of a virus depending for its multiplication on another, *J. Gen. Microbiol.* **27,** 477.
Kelly, T. J., Jr., and Rose, J. A., 1971, Simian virus 40 integration site in an adenovirus 7–simian virus 40 hybrid DNA molecule, *Proc. Natl. Acad. Sci. USA* **68,** 1037.
Kilham, K., 1961*a*, Rat virus (RV) infections in hamsters, *Proc. Soc. Exp. Biol. Med.* **106,** 825.
Kilham, L., 1961*b*, Mongolism associated with rat virus (RV) infection in hamsters, *Virology* **13,** 141.
Kilham, L., 1966, "Viruses of Laboratory and Wild Rats," National Cancer Institute Monograph No. 20, p. 117, National Cancer Institute, Bethesda, Md.
Kilham, L., and Maloney, J. B., 1964, Association of rat virus and Maloney leukemia virus in tissues of inoculated rats, *J. Natl. Cancer Inst.* **32,** 523.
Kilham, L., and Margolis, G., 1964, Cerebellar ataxia in hamsters inoculated with rat virus, *Science (Wash., D.C.)* **143,** 1047.
Kilham, L., and Margolis, G., 1965, Cerebellar disease in cats induced by inoculation of rat virus, *Science (Wash., D.C.)* **148,** 224.
Kilham, L., and Margolis, G., 1970, Pathogenicity of minute virus of mice (MVM) for rats, mice, and hamsters, *Proc. Soc. Exp. Biol. Med.* **133,** 1447.
Kilham, L., and Oliver, L. J., 1959, A latent virus of rats isolated in tissue culture, *Virology* **7,** 428.
Kilham, L., Margolis, G., and Colby, E., 1967, Congenital infections of cats and ferrets by feline panleukopenia virus manifested by cerebellar hypoplasia, *Lab. Invest.* **17,** 465.
Kirschstein, R. L., Smith, K. O., and Peters, E. A., 1968, Inhibition of adenovirus 12 oncogenicity by adeno-associated virus, *Proc. Soc. Exp. Biol. Med.* **128,** 670.
Koczot, F. J., Carter, B. J., Garon, C. F., and Rose, J. A., 1973, Self-complementarity of terminal sequences within plus or minus strands of adenovirus-associated virus DNA, *Proc. Natl. Acad. Sci. USA* **70,** 215.

Kongsvik, J. R., and Toolan, H. W., 1972, Capsid components of the parvovirus H-1, *Proc. Soc. Exp. Biol. Med.* **139,** 1202.

Kurstak, E., 1971, Small DNA densonucleosis virus (DNV), *in* "Comparative Virology" (K. Maramorosch and E. Kurstak, eds.), pp. 207–241, Academic Press, New York.

Ledinko, N., 1967, Plaque assay of the effects of cytosine arabinoside and 5-iodo-2′-deoxyuridine on the synthesis of H-1 virus particles, *Nature* (*Lond.*) **214,** 1346.

Ledinko, N. and Toolan, H., 1968, Human adenovirus type 12 as a "helper" for growth of H-1 virus, *J. Virol.* **2,** 155.

Ledinko, N., Hopkins, S., and Toolan, H., 1969, Relationship between potentiation of H-1 growth by human adenovirus 12 and inhibition of the "helper" adenovirus by H-1, *J. Gen. Virol.* **5,** 19.

Luchsinger, E., and Wellemans, G., 1971, A study on the presence of antibodies against adenovirus and associated virus X_7 in the sera of cattle from Eynatten and of calves with respiratory disease, *Arch. Ges. Virusforsch.* **35,** 203.

Luchsinger, E., Strobbe, R., Wellemans, G., Dekegel, D., and Sprecher-Goldberger, S., 1970, Hemagglutinating adeno-associated virus (AAV) in association with bovine adenovirus type 1, *Arch. Ges. Virusforsch.* **31,** 390.

Luchsinger, E., Strobbe, R., Dekegel, D., and Wellemans, G., 1971, Use of B-IV zonal rotor centrifugation as a simple tool for the separation of adeno-associated X_7 virus (AAV X_7) from helper adenoviruses, *Arch. Ges. Virusforsch.* **33,** 251.

Lum, G. S., 1970, Serological studies of rat viruses in relation to tumors, *Oncology* **24,** 335.

Lum, G. S., and Schreiner, A. W., 1963, Study of a virus isolated from a chloroleukemic Wistar rat, *Cancer Res.* **23,** 1742.

Lwoff, A., and Tournier, P., 1966, The classification of viruses, *in* "Annual Review of Microbiology" (C. Clifton, ed.), Vol. 20, pp. 45–74, Annual Reviews, Palo Alto Calif.

McGeoch, O. J., Crawford, L. V., and Follett, E. A. C., 1970, The DNAs of three parvoviruses, *J. Gen. Virol.* **6,** 33.

Maizel, J. V., Jr., White, D. O., and Sharff, M. D., 1968, The polypeptides of adenovirus. I. Evidence for multiple protein components in the virion and a comparison of types 2, 7A, and 12, *Virology* **36,** 115.

May, P., and May, E., 1970, The DNA of Kilham rat viruses, *J. Gen. Virol.* **6,** 437.

Mayor, H. D., and Jordon, L. E., 1966, Electron microscopic study of the rodent "picodnavirus" X14, *Exp. Mol. Pathol.* **5,** 580.

Mayor, H. D., and Melnick, J. L., 1966, Small deoxyribonucleic acid-containing viruses (picodnavirus group), *Nature* (*Lond.*) **210,** 331.

Mayor, H. D., and Ratner, J. D., 1972, Conditionally defective helper adenoviruses and satellite virus replication, *Nature* (*Lond.*) *New Biol.* **239,** 20.

Mayor, H. D., and Ratner, J., 1973, Analysis of adenovirus-associated satellite virus DNA, *Biochem. Biophys. Acta* **299,** 189.

Mayor, H. D., Jamison, R. M., Jordon, L. E., and Melnick, J. L., 1965, Structure and composition of a small particle prepared from a simian adenovirus, *J. Bacteriol.* **90,** 235.

Mayor, H. D., Ito, M., Jordon, L. E., and Melnick, J. L., 1967, Morphological studies on the replication of a defective satellite virus and its helper adenovirus, *J. Natl. Cancer Inst.* **38,** 805.

Mayor, H. D., Jordan, L., and Ito, M., 1969*a*, Deoxyribonucleic acid of adeno-associated satellite virus, *J. Virol.* **4,** 191.

Mayor, H. D., Torikai, K., Melnick, J. L., and Mandel, M., 1969*b*, Plus and minus single-stranded DNA separately encapsidated in adeno-associated satellite virions, *Science* (*Wash., D.C.*) **166,** 1280.
Mayor, H. D., Houlditch, G. S., and Mumford, D. M., 1973, Influence of adeno-associated satellite virus on adenovirus-induced tumors in hamsters, *Nat. New Biol.* **241,** 44.
Mayr, A., and Mahnel, H., 1966, Weitere untersuchungen über die Zuchtung von schweinepestvirus in zellkulturen mit cytopathogenem effekt, zentrablatt fur bakteriologie, parasitenkunde, *Zbl. Bakt. I. Abt. Orig.* **199,** 399.
Mayr, A., Bachmann, P. A., Siegl, G., Mahnel, H., and Sheffy, B. E., 1968, Characterization of a small porcine DNA virus, *Arch. Ges. Virusforsch.* **25,** 38.
Melnick, J. L., 1971, Classification and nomenclature of animal viruses, 1971, *in* "Process in Medical Virology" (J. L. Melnick, ed.), pp. 462–484, Karger, Basel.
Melnick, J. L., and Parks, W. P., 1966, Identification of multiple defective and noncytopathic viruses in tissue culture, *Rel. VI Congr. Int. Patol. Clin., Roma (October, 1966)*, 237–262.
Meynadier, G., Vago, C., Planteoin, G., and Atger, P., 1964, Virose d'un type inhabituel chez le lepidoptere *Galleria mellonella* L., *Rev. Zool. Agr. Appl.* **63,** 207.
Mirkovic, R. R., Adamova, V., Boucher, D. W., and Melnick, J. L., 1971, Identification of the Kirk "hepatitis" virus as a member of the parvovirus (picodnavirus) group, *Proc. Soc. Exp. Biol. Med.* **138,** 626.
Moore, A. E., 1962, Characteristics of certain viruses isolated from transplantable tumors, *Virology* **18,** 182.
Nathanson, N., Cole, G. A., Santos, G. W., Squire, R. A., and Smith, K. O., 1970, Viral hemorrhagic encephalopathy of rats. I. Isolation, identification, and properties of the HER strains of rat virus, *Am. J. Epidemiol.* **91,** 328.
Onuma, M., 1971, Distribution of antibodies in dogs against adeno-associated satellite virus associated with infectious canine hepatitis virus and serological typing of the satellite virus, *Jap. J. Vet. Res.* **19,** 40.
Parker, J. C., Cross, S. S., Collins, M. J., Jr., and Rowe, W. P., 1970*a*, Minute virus of mice. I. Procedures for quantitation and detection, *J. Natl. Cancer Inst.* **45,** 297.
Parker, J. C., Collins, M. J., Jr., Cross, S. S., and Rowe, W. P., 1970*b*, Minute virus of mice. II. Prevalence, epidemiology, and natural occurrence as a contaminant of transplanted tumors, *J. Natl. Cancer Inst.* **45,** 305.
Parks, W. P., Melnick, J. L., Rongey, R., and Mayor, H. D., 1967*a*, Physical assay and growth cycle studies of a defective adeno-satellite virus, *J. Virol.* **1,** 171.
Parks, W. P., Green, M., Piña, M., and Melnick, J. L., 1967*b*, Physicochemical characterization of adeno-associated satellite virus type 4 and its nucleic acid, *J. Virol.* **1,** 980.
Parks, W. P., Casazza, A. M., Alcott, J., and Melnick, J. L., 1968, Adeno-associated satellite virus interference with the replication of its helper adenovirus, *J. Exp. Med.* **127,** 91.
Parks, W. P., Boucher, D. W., Melnick, J. L., Taber, L. H., and Yow, M. D., 1970, Seroepidemiological and ecological studies of the adenovirus-associated satellite viruses, *Infect. Immun.* **2,** 716.
Parsons, J. T., and Green, M., 1971, Biochemical studies on adenovirus multiplication. XVIII. Resolution of early virus-specific RNA species in Ad 2-infected and -transformed cells, *Virology* **45,** 154.
Parsons, J. T., Gardener, J., and Green, M., 1971, Biochemical studies on adenovirus

multiplication. XIX. Resolution of late viral RNA species in the nucleus and cytoplasm, *Proc. Natl. Acad. Sci. USA* **68,** 557.

Payne, F. E., Shellabarger, C. J., and Schmidt, R. W., 1963, A virus from mammary tissue of rats treated with X-ray or methylcholanthrene (MC), *Proc. Am. Assoc. Cancer Res.* **4,** 51.

Payne, F. E., Beals, T. F., and Preston, R. E., 1964, Morphology of a small DNA virus, *Virology* **23,** 109.

Piña, M., and Green, M., 1965, Biochemical studies on adenovirus multiplication. IX. Chemical and base composition analysis of 28 human adenoviruses, *Proc. Natl. Acad. Sci. USA* **54,** 547.

Prage, L., Pettersson, U. and Philipson, L., 1968, Internal basic proteins in adenovirus, *Virology* **36,** 508.

Provost, P. J., Ittensohn, O. L., Villarejos, V. M. Arquedas, J. A., and Hilleman, M. R., 1973, Etiologic relationship of marmoset-propagated CR326 hepatitis: A virus to hepatitis in man, *Proc. Soc. Exp. Biol. Med.* **142,** 1257.

Rabson, A. S., O'Conor, G. T., Berezesky, I. K., Paul, F. J., 1964, Enhancement of adenovirus growth in African green monkey kidney cell cultures by SV40, *Proc. Soc. Exp. Biol. Med.* **116,** 187.

Rapoza, N. P., and Atchison, R. W., 1967, Association of AAV-1 with simian adenoviruses, *Nature* (*Lond.*) **215,** 1186.

Rapp, F., 1969, Defective DNA animal viruses, *in* "Annual Review of Microbiology" (C. E. Clifton, ed.), Vol. 23, pp. 293–316, Annual Reviews, Palo Alto, Calif.

Reich, P. R., Baum, S. G., Rose, J. A., Rowe, W. P., and Weissman, S. M., 1966, Nucleic acid homology studies of adenovirus type 7–SV40 interactions, *Proc. Natl. Acad. Sci. USA* **55,** 336.

Reichmann, M. E., 1964, The satellite tobacco necrosis virus: A single protein and its genetic code, *Proc. Natl. Acad. Sci. USA* **52,** 1009.

Rhode, S. L., III, 1973, Replication process of the parvovirus H-1. I. Kinetics in a parasynchronous cell system, *J. Virol.* **11,** 856.

Robey, R. E., Woodman, D. R., and Hetrick, F. M., 1968, Studies on the natural infection of rats with Kilham rat virus, *Am. J. Epidemiol.* **88,** 139.

Robinson, D. M., and Hetrick, F. M., 1969, Single-stranded DNA from the Kilham rat virus, *J. Gen Virol.* **4,** 269.

Rose, J. A., and Koczot, F. J., 1971, Adenovirus-associated virus multiplication. VI. Base composition of the deoxyribonucleic acid strand species and strand-specific *in vitro* transcription, *J. Virol.* **8,** 771.

Rose, J. A., and Koczot, F. J., 1972, Adenovirus-associated virus multiplication. VII. Helper requirement for viral deoxyribonucleic acid and ribonucleic acid synthesis, *J. Virol.* **10,** 1.

Rose, J. A., Hoggan, M. D., and Shatkin, A. J., 1966, Nucleic acid from an adeno-associated virus: Chemical and physical studies, *Proc. Natl. Acad. Sci. USA* **56,** 86.

Rose, J. A., Hoggan, M. D., Koczot, F. J., and Shatkin, A. J., 1968, Genetic relatedness studies with adenovirus-associated viruses, *J. Virol.* **2,** 999.

Rose, J. A., Berns, K. I., Hoggan, M. D., and Koczot, F. J., 1969, Evidence for a single-stranded adenovirus-associated virus genome: Formation of a DNA density hybrid on release of viral DNA, *Proc. Natl. Acad. Sci. USA* **64,** 863.

Rose, J. A., Maizel, J. V., Jr., Inman, J. K., and Shatkin, A. J., 1971, Structural proteins of adenovirus-associated viruses, *J. Virol.* **8,** 766.

Rowe, W. P., and Baum, S. G., 1964, Evidence for a possible genetic hybrid between adenovirus type 7 and SV40 viruses, *Proc. Natl. Acad. Sci. USA* **52,** 1340.
Rowe, W. P., Baum, S. G., Pugh, W. E., and Hoggan, M. D., 1965, Studies of adenovirus SV40 hydrid viruses. I. Assay system and further evidence for hydridization, *J. Exp. Med.* **122,** 943.
Salzman, L. A., 1971, DNA polymerase activity associated with purified Kilman rat virus, *Nat. New Biol.* **231,** 174.
Salzman, L. A., and Jori, L. A., 1970, Characterization of the Kilham rat virus, *J. Virol.* **5,** 114.
Salzman, L. A., and White, W. L., 1970, Structural proteins of Kilham rat virus. *Biochem. Biophys. Res. Commun.* **41,** 1551.
Salzman, L. A., and White, W. L., 1973, *In vivo* conversion of the single-stranded DNA of the Kilham rat virus to a double-stranded form, *J. Virol.* **11,** 299.
Salzman, L. A., White, W. L., and Kakefuda, T., 1971, Linear, single-stranded deoxyribonucleic acid isolated from Kilham rat virus, *J. Virol.* **7,** 830.
Salzman, L. A. White, W. L., and McKerlie, L., 1972, Growth characteristics of Kilham rat virus and its effect on cellular macromolecular synthesis, *J. Virol.* **10,** 573.
Siegl, G., 1972, Parvoviruses as contaminants of permanent human cell lines. V. The nucleic acid of KBSH-virus, *Arch. Ges. Virusforsch.* **37,** 267.
Siegl, G., and Gautschi, M., 1973*a*, The multiplication of parvovirus LuIII in a synchronized culture system. I. Optimum conditions for virus multiplication, *Arch. Ges. Virusforsch.* **40,** 105.
Siegl, G., and Gautschi, M., 1973*b*, The multiplication of parvovirus LuIII in a synchronized culture system. II. Biochemical characteristics of virus replication, *Arch. Ges. Virusforsh.* **40,** 119.
Siegl, G., Hallauer, C., Novak, A., and Kronauer, G., 1971, Parvoviruses as contaminants of permanent human cell lines. II. Physicochemical properties of the isolated viruses, *Arch. Ges. Virusforsch.* **35,** 91.
Siegl, G., Hallauer, C., and Novak, A., 1972, Parvoviruses as contaminants of permanent cell lines. IV. Multiplication of KBSH-Virus in KB cells, *Arch. Ges. Virusforsch.* **36,** 351.
Sinsheimer, R. L., 1959*a*, Purification and properties of bacteriophage ϕX174, *J. Mol. Biol.* **1,** 37.
Sinsheimer, R. L., 1959*b*, A single-stranded deoxyribonucleic acid from bacteriophage, ϕX174, *J. Mol. Biol.* **1,** 43.
Sinsheimer, R. L., Starman, B., Nagler, C., and Guthrie, S., 1962, The process of infection with bacteriophage ϕX174. I. Evidence for a "replicative form," *J. Mol. Biol.* **4,** 142.
Smith, K. O., and Gehle, W. D., 1967, Replication of an adeno-associated virus in canine and human cells with infectious canine hepatitis virus as a "helper," *J. Virol.* **1,** 648.
Smith, K. O., and Thiel, J. F., 1967, Adeno-associated virus studies employing a fluorescent focus assay technique, *Proc. Soc. Exp. Biol. Med.* **125,** 887.
Smith, K. O., Gehle, W. D., and Thiel, J. F., 1966, Properties of a small virus associated with adenovirus type 4, *J. Immunol.* **97,** 754.
Smith, K. O., Gehle, W. D., and Montes de Oca, H., 1968, Persistence of adeno-associated virus in human amnion cells without adenovirus potentiation, *Bacteriol Abstr. 68th Annu. Meet. Am. Soc. Microbiol.* **52,** 153.

Spahn, G. J., Mohanty, S. B., and Hetrick, F. M., 1966, Experimental infection of calves with hemadsorbing enteric (HADEN) virus, *Cornell Vet.* **56,** 377.

Storz, J., and Warren, G. S., 1970, Effect of antimetabolites and actinomycin D on the replication of HADEN, a bovine parvovirus, *Arch. Ges. Virusforsch.* **30,** 271.

Storz, J., Bates, R. C., Warren, G. S., and Howard, H., 1972, Distribution of antibodies against bovine parvovirus 1 in cattle and other animal species. *Am. J. Vet. Res.* **33,** 269.

Studier, F. W., 1965, Sedimentation studies of the size and shape of DNA, *J. Mol. Biol.* **11,** 373.

Sussenbach, J. S., and Van Der Vliet, P. C., 1973, Studies on the mechanism of replication of adenovirus DNA, *Virology* **54,** 299.

Sussenbach, J. S., Van Der Vliet, P. C., Ellens, D. J., and Jansz, H. S., 1972, Linear intermediates in the replication of adenovirus DNA, *Nat. New Biol.* **239,** 47.

Szybalski, W., 1968, Use of cesium sulfate for equilibrium density gradient centrifugation, *in* "Methods in Enzymology" (L. Grossman and K. Moldave, eds.), pp. 330–360, Academic Press, New York.

Tattersall, P., 1972*a*, Replication of the single-stranded DNA of a minute virus of mice (MVM), *Fed. Proc.* **31,** 3973.

Tattersall, P., 1972*b*, Replication of the parvovirus MVM. I. Dependence of virus multiplication and plaque formation on cell growth, *J. Virol.* **10,** 586.

Tattersall, P., Crawford, L. V., and Shatkin, A. J., 1973, Replication of the parvovirus MVM. II. Isolation and characterization of intermediates in the replication of the viral deoxyribonucleic acid, *J. Virol.,* **12,** 1446.

Tennant, R. W., 1971, Inhibition of mitosis and macromolecular synthesis in rat embryo cells by Kilham rat virus, *J. Virol.* **8,** 402.

Tennant, R. W., and Hand, R. E., Jr., 1970, Requirement of cellular synthesis for Kilham rat virus replication, *Virology* **42,** 1054.

Tennant, R. W., Layman, K. R., and Hand, R. E., Jr., 1969, Effect of cell physiological state on infection by rat virus, *J. Virol.* **4,** 872.

Thomas, C. A., Jr., and MacHattie, L. A., 1967, The anatomy of viral DNA molecules, *in* "Annual Review of Biochemistry" (P. D. Boyer, ed.), Vol. 36, Part II, pp. 485–518, Annual Reviews, Palo Alto, Calif.

Toolan, H. W., 1964, Studies on the H-Viruses, *Proc. Am. Assoc. Cancer Res.* **5,** 64.

Toolan, H., 1966, Susceptibility of the rhesus monkey (macaca mulatta) to H-1 virus, *Nature* (*Lond.*) **209,** 833.

Toolan, H., 1967, Lack of oncogenic effect of the H-viruses for hamsters, *Nature* (*Lond.*) **214,** 1036.

Toolan, H. W., 1968, The picodna viruses: H, RV, and AAV, *in* "International Review of Experimental Pathology," Vol. VI, pp. 135–180, Academic Press, New York.

Toolan, H. W., 1972, The parvoviruses, *in* "Progress in Experimental Tumor Research" (F. Homburger, ed.), Vol. 16, pp. 410–425, Karger, Basel.

Toolan, H., and Ledinko, N., 1965, Growth and cytopathogenicity of H-viruses in human and simian cell cultures, *Nature* (*Lond.*) **208,** 812.

Toolan, H., and Ledinko, N., 1968, Inhibition by H-1 virus of the incidence of tumors produced by adenovirus type 12 in hamsters, *Virology* **35,** 475.

Toolan, H. W., Dalldorf, G., Barclay, M., Chandra, S., and Moore, A. E., 1960, An unidentified, filtrable agent isolated from transplanted human tumors, *Proc. Natl. Acad. Sci. USA* **46,** 1256.

Toolan, H., Saunders, E. L., Southam, C. M., Moore, A. E., and Levin, A. G., 1965, H-1 virus viremia in the hamsters, *Proc. Soc. Exp. Biol. Med.* **119,** 711.

Torikai, K., and Mayor, H. D., 1969, Interference between two adeno-associated satellite viruses: A three-component system, *J. Virol.* **3,** 484.

Torikai, K., Ito, M., Jordan, L. E., and Mayor, H. D., 1970, Properties of light particles produced during growth of type 4 adeno-associated satellite virus, *J. Virol.* **6,** 363.

Torpier, G., D'Halluin, J., and Boulanger, P., 1971, Electron microscopic observations on KB cells infected with adeno-associated satellite virus, *J. Microscop.* **11,** 259.

Tromans, W. J., and Horne, R. W., 1961, The structure of bacteriophage ϕX174, *Virology* **15,** 1.

Usategui-Gomez, M., Toolan, H. W., Ledinko, N., Al-Lami, F., and Hopkins, M. S., 1969, Single-stranded DNA from the parvovirus, H-1, *Virology* **39,** 617.

Vasquez, C., and Brailovsky, C., 1965, Purification and fine structure of Kilham's rat virus, *Exp. Mol. Pathol.* **4,** 130.

Vasquez, C., and Tournier, P., 1964, New interpretation of the reovirus structure, *Virology* **24,** 128.

Van Der Eb, A. J., 1973, Intermediates in type 5 adenovirus DNA replication, *Virology* **51,** 11.

Vernon, S. K., Stasny, J. T., Neurath, A. R., and Rubin, B. A., 1971, Electron microscopy of DNA from adeno-associated virus type I, *J. Gen. Virol.* **10,** 267.

Wills, C. G., 1952, Notes on infectious enteritis of mink and its relationship to feline enteritis, *Can. J. Comp. Med.* **16,** 419.

Wolfson, J., and Dressler, D., 1972, Adenovirus-2 DNA contains an inverted terminal repetition, *Proc. Natl. Acad. Sci. USA* **69,** 3054.

CHAPTER 2

Reproduction of Papovaviruses

Norman P. Salzman
and
George Khoury

Laboratory of Biology of Viruses
National Institutes of Allergy and Infectious Diseases
National Institutes of Health
Bethesda, Maryland 20014

1. GENERAL PROPERTIES OF PAPOVAVIRUSES

The principal members of the papova group are polyoma virus (Stewart *et al.*, 1957), simian virus 40 (SV40), which is a vacuolating virus of monkeys (Sweet and Hilleman, 1960), and the papilloma viruses (Melnick, 1962). The name for this group of viruses is derived from the first two letters of the names of each of the viruses that were first included in the group, *pa*pilloma, *po*lyoma, *va*cuolating virus (Melnick, 1962). The viruses are 40–57 nm in diameter and, as determined by negative staining, the outer shell has symmetry of the T = 7 icosahedral surface lattice and is composed of 72 morphological subunits (Finch and Klug, 1965; Anderer *et al.*, 1967). The viruses contain no lipids and therefore are resistant to ether. Polyoma and SV40 do not share common antigens, nor is there evidence for the existence of any homology between their DNAs. The papovaviruses are capable of initiating a lytic cycle of replication or a latent infection. For the papilloma viruses it is difficult to obtain a suitable cell line in which the lytic cycle can be studied and for this reason studies that we will discuss concerning viral replication will deal exclusively with SV40 and polyoma.

In general, studies of the lytic cycle of virus replication of polyoma have been carried out with whole mouse embryo cultures or embryonic mouse kidney cultures or with 3T3 cells (a continuous mouse cell line), while hamster cultures have been used for studies of viral transformation. For SV40, primary African green monkey kidney (AGMK) cell cultures or various continuous monkey kidney cell lines, e.g., BSC-1, CV-1, or Vero, have been used to study the lytic cycle, and mouse, hamster, and human cell cultures have been used in studies dealing with transformation.

The papovaviruses contain DNA within the core of the virus particle which is present as a covalently closed duplex molecule. Each of the papilloma DNAs that have been examined have molecular weights close to 5×10^6 daltons, which is higher than the molecular weights of SV40 (3.6×10^6) (Tai *et al.*, 1972) and polyoma (3.0×10^6) (Weil and Vinograd, 1963). The DNAs of the papovaviruses are infectious (McCutchan and Pagano, 1968) and are able to transform cells *in vitro* (Crawford *et al.*, 1964; Bourgaux *et al.*, 1965; Aaronson and Todaro, 1969; Aaronson and Martin, 1970). The small size of the viral genomes, and their ability to transform cells, have made them of great interest as models for studying the mechanism of transformation of a normal to a malignant cell. In this chapter we will discuss those studies which deal with transformation only when they are related to, or help to clarify aspects of, the lytic cycle of replication. It seems likely that precise definition of the lytic cycle will be required in order to define the mechanism of cell transformation.

1.1. Initiation of the Replication Cycle—Adsorption, Penetration, and Uncoating of the Virus

We have already mentioned those virus–cell systems that are presently used for studies of viral replication. One reason that a cell is resistant to infection is that there is a block which prevents virus adsorption and/or penetration. The adsorption of polyoma virus occurs in two stages. The first stage of adsorption can be reversed by changing either the ionic conditions or the *p*H. The virus is then irreversibly adsorbed, and the subsequent virus replication cycle is no longer affected by the addition of antiserum. A similar two-stage mechanism of adsorption has been observed for many other animal viruses and bacteriophages. Adsorption of polyoma virus does not occur if the cells have been treated with neuraminidase which destroys the cell receptors that contain sialic acid (Crawford, 1962; Fried, 1970).

An interesting biological observation which relates to virus adsorption and/or penetration is the frequency of transformation by SV40 of human cell lines derived from individuals with certain metabolic diseases. Cell lines established from skin biopsies of patients with Fanconi's anemia or Down's syndrome can be transformed by SV40 with a much higher frequency than control human skin cultures (Todaro and Martin, 1967; Todaro *et al.*, 1966). However, these differences in transformation frequency are not observed when cells are transformed using SV40 DNA instead of virions (Aaronson and Martin, 1970). Thus, the frequency with which SV40 virions can transform cells is determined by cellular differences that affect the rate of adsorption, penetration, or uncoating of the virus. Virion capsid proteins and receptor sites on the cell surface are two of the determinants which define the process of virus adsorption. While a number of chemical studies have provided a partial molecular characterization of the capsid proteins, thus far there are no comparable studies on cell receptor sites.

The changes in cell surfaces that are associated with viral transformation have provided the stimulus for additional studies in this general area. In many cases, cells that are permissive for SV40 replication can also be transformed by SV40. These transformed cells become resistant to superinfection with SV40 virions, but remain susceptible to infection with SV40 DNA (Swetly *et al.*, 1969; Rapp and Trulock, 1970; Shiroki and Shimojo, 1971; Reznikoff *et al.*, 1972). There are, however, transfomed cell lines which become either partially or completely resistant to superinfection by SV40 DNA as well as by SV40 virions. Onc such example is a line of monkey kidney cells which are transformed by a defective virion fraction (a T fraction; Uchida *et al.*, 1968) of SV40. When these cells are reinfected with SV40 DNA, a yield of 10^{-2}–10^{-3} PFU (plaque forming units)/cell is obtained, as compared with a yield of 10 PFU/cell when SV40 DNA is used to infect independently isolated transformed cells (Shiroki and Shimojo, 1971). These results suggest that only one transformed cell in 10^3–10^4 may be permissive, even when a cycle of replication is initiated with viral DNA. Still other transformed lines have been obtained which do not produce any infectious virus after superinfection with either SV40 virions or SV40 DNA (Shiroki and Shimojo, 1971; Butel *et al.*, 1971). The mechanism by which certain transformed clones become resistant to superinfection with either virions or DNA is still unclear. In bacterial systems, there are restriction enzymes which degrade heterologous DNA and which can block a cycle of virus replication (Arber and

Linn, 1969; Boyer, 1971). It is possible that viral nucleic acid modification and mammalian restriction enzymes play some role in the determination of susceptibility or resistance of animal cells to infection.

Many of the properties that are used to select for transformed cells depend on changes in the cell surface properties that occur following exposure to oncogenic viruses. Most transformed cells show a loss of contact inhibition; they are able to grow in soft agar and to form colonies in depleted growth medium. These surface properties, as well as the enhanced agglutination of transformed cells with wheat germ agglutinin or concanavalin A, would suggest extensive changes in the cell surface. Therefore, it may not be surprising that transformed cells are resistant to superinfection. However, in another well-characterized transformed cell system only limited surface changes have been demonstrated at the molecular level. Chick embryo fibroblasts (CEF) have been infected with a Rous sarcoma virus which is temperature sensitive (*ts*) in its ability to transform cells. Among the many proteins that can be resolved when isolated plasma membranes of uninfected CEF cells are examined is a polypeptide of about 45,000 molecular weight. This polypeptide is present in normal amounts in cells infected with the *ts* virus and cultivated at the restrictive temperature. It is present in reduced amounts or absent in cells infected with the *ts* mutant virus and cultivated at the permissive temperature. There are also reduced amounts of the polypeptide in plasma membranes isolated from CEF cells transformed with the wild-type Schmidt Ruppin RSV-A (Wickus and Robbins, 1973). Clearly, there are wide biologic differences between the avian tumor viruses and the papovaviruses, and there is no basis for predicting that they will behave in a parallel way. However, it will be interesting to see if there is a class of SV40 transformants in which such minimal membrane changes occur and if such transformants are resistant to superinfection with SV40 virions. It is hoped that biochemical studies in these systems will provide more precise data on the role of the cell membrane in virus adsorption and penetration.

Studies to define the events that occur during adsorption, penetration, and uncoating have been carried out using purified radioactive virion preparations. When the process of virus uptake was studied using purified SV40 virus (Hummeler *et al.*, 1970), the most frequently observed mode of entry of virus into the cell was by pinocytosis of single particles. Virus particles were transported through the cytoplasm to the nucleus where they were observed as early as 1 hour after infection. Since virus particles seen at 1 and 2 hours post-infection (p.i.) were no longer present at 4 hours p.i., it is assumed that they were uncoated by this time. When studied with radiolabeled virus, 50% of input virus is adsorbed within 2 hours after infection (Ozer and Takemoto, 1969; Barbanti-Brodano *et al.*, 1970). Input viral DNA can be found in the nucleus many hours before any detectable events in the replication cycle are observed.

In these studies, which employ biochemical or cytological procedures to define the fate of the input virus, it is not possible to relate the experimental observations to that small fraction of the particles which are the infectious ones or which are involved in transformation. In general, there are at least 100 physical particles for every particle that causes an infection. In studies with viral DNA, 10,000–100,000 DNA molecules are needed for each infection; to produce one transformant (with a susceptible strain), approximately 10^7 physical particles are required. In view of this, biological conclusions which are derived from studies with purified virions must be considered almost speculative in nature.

A temperature-sensitive mutant of SV40, *ts* 101, which at the restrictive temperature cannot induce synthesis of either T or U antigen, can adsorb to and penetrate the cell. This may be useful in defining the very early events in a lytic cycle of virus replication (Robb and Martin, 1972).

1.2. Time Course of Synthesis of Viral Macromolecules

The replication cycle can be considered to occur in two phases. Viral macromolecules are either synthesized early, prior to the time of initiation of viral DNA replication, or they are synthesized late, subsequent to viral DNA synthesis. In the early phase, there is synthesis of early mRNA and T and U antigens. In the late phase of replication, both early and late mRNA and capsid proteins are synthesized. When DNA synthesis is blocked by a metabolic inhibitor such as cytosine arabinoside or 5-fluorodeoxyuridine, the early events occur but viral DNA and capsid proteins are not synthesized. Both SV40 and polyoma show a lag phase of 6–8 hr. after virus infection before the initial events in virus replication are detected. Factors that determine the length of this lag phase have not been defined. In a lytic cycle of replication, the synthesis of early mRNA can be detected as early as 6 hours post-infection (p.i.), and T-antigen synthesis, which is observed to start at 8–9 hours p.i., is next observed. Viral DNA synthesis can be detected at

15–18 hours p.i., and at this time one also observes the induction of enzyme synthesis and late viral mRNA; two or three hours later synthesis of viral capsid proteins is observed. There is considerable asynchrony within an infected population. When the rates of viral RNA, DNA, or protein synthesis are measured using uptake of radioactive precursors, the extent of incorporation reflects the fraction of the population which is carrying out synthesis as well as the true rate of synthesis. As a consequence of the asynchronous induction of the infectious cycle, synthesis of viral macromolecules and infectious virus formation (which is seen by 20 hours p.i.) continues for a prolonged period (during the next 40 hours).

While the kinetics of macromolecular synthesis depend on the virus stock and the cell line used in a particular study, the sequence of events noted above should be the same when studied during the lytic cycle. With regard to the time of induction of synthesis, this depends on the sensitivity of the detection procedure used; thus there can be considerable uncertainty when the induction of macromolecular synthesis occurs. For example, the detection of viral mRNA at an earlier time than the detection of T antigen is consistent with T antigen being a virus-coded function. Since procedures with different sensitivities are used for the detection of T antigen and early mRNA, however, there remains considerable uncertainty about the exact times of induction of these two processes.

2. DNA REPLICATION

There are two general areas that have been investigated during the replication of papovaviruses. One series of studies has attempted to describe the mechanism of chain growth, i.e., to provide a description of events that occur at the replication fork. The second area of interest concerns the mechanism of semiconservative replication of a covalently closed molecule. Both areas are important in almost all studies of DNA replication. Besides the papovavirus DNAs, a number of DNAs exist as covalently closed structures within cells, and these include the chromosomes of bacteria, several bacterial and animal viruses, mitochondrial DNA, and bacterial plasmids.

2.1. DNA Configurations

A ring form for DNA was first observed for ϕX174 (Fiers and Sinsheimer, 1962). While the circular DNA within this virion is single-stranded, during a cycle of virus replication the DNA is converted to a covalently closed duplex molecule (Burton and Sinsheimer, 1963). Soon after this initial observation, it was demonstrated that the DNA of polyoma virus was in the form of a covalently closed circular duplex molecule (Dulbecco and Vogt, 1962; Weil and Vinograd, 1963). Subsequently, SV40, rabbit papilloma, and human papilloma were all shown to contain covalently closed circular DNA (Crawford and Black, 1964; Crawford, 1964, 1965). These interesting and important studies with circular DNA have been reviewed (Vinograd and Lebowitz, 1966).

Two forms of viral DNA are obtained when DNA is extracted from virions. There is DNA I, a covalently closed duplex, and DNA II, a form which can be generated from DNA I by breaking a single phosphodiester bond in either DNA strand. The forms in which papova DNA molecules exist are seen in Fig. 1. DNA III has been used to

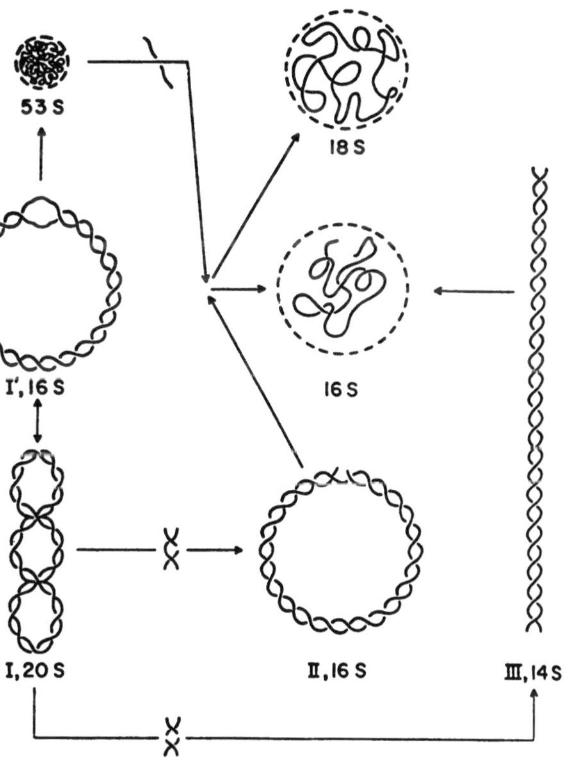

Fig. 1. Diagrammatic representation of the several forms of polyoma and SV40 DNA. The dashed circles around the denatured forms indicate the relative hydrodynamic diameters. The sedimentation coefficients were measured in neutral and alkaline NaCl solutions. The twist in I should be right-handed (Vinograd and Lebowitz, 1966).

refer either to double-stranded cellular DNA of about 2–3 × 10^6 daltons which is found in varying amounts in purified polyoma preparations and has been found only rarely in preparations of SV40, or it may refer to linear duplex molecules that are formed when DNA I or II are cleaved by restriction enzymes. DNA I is the predominant type of DNA that is found in virions. There are wide variations in the amounts of DNA II and DNA III that are present in virions. The significance of the DNA II which is obtained from virions is not clear. It can be considered to arise by nicking of DNA I during the extraction procedure. The amount of DNA II that is obtained from virions has been observed to change when conditions for the extraction of DNA have been varied (Vinograd *et al.*, 1965). However, there are no apparent steric reasons why DNA II could not be encapsidated. In support of this idea is the finding that DNA III, a cellular linear duplex, does occur in an encapsidated form. Recently DNA II has been shown to be an intermediate during SV40 DNA replication (Fareed *et al.*, 1973*b*). With methods now available for determining if the nick in DNA II occurs at a specific site in the molecule, it will be possible to determine whether the DNA II that is present in virion-extracted DNA is synthesized and encapsidated as such, or whether it is simply generated from DNA I during preparation of viral DNA.

DNA I forms of polyoma, SV40, and rabbit and human papilloma (Crawford and Black, 1964; Crawford, 1964, 1965) possess common structural features. Each DNA is a covalently closed duplex in which the DNA contains no free ends. Each of these molecules also contains superhelical turns in which the DNA double helix winds on itself. This latter property results from constraints within the DNA molecule which prevent a change in the pitch of the double helix since the molecule contains neither nicks nor free ends. If, at the time DNA I is formed, the pitch of the helix differs from the pitch which the molecule will assume *in vitro*, then this molecule will generate superhelical turns under *in vitro* conditions. A nicked circular duplex molecule or linear duplex DNA may have a different helical pitch *in vivo* and *in vitro*, but since there are no constraints in the molecule (one strand being free to wind around the other) the molecule will not generate superhelical turns under *in vivo* or *in vitro* conditions. Both, polyoma and SV40 DNAs contain a deficiency of about 19 turns in the double helix at the time of formation of DNA I, and they both contain about 19 negative superhelical turns. The presence of superhelical turns in polyoma and SV40 provides the molecule with a decreased intrinsic viscosity (Opschoor *et al.*, 1968) and, therefore, its rate of sedimentation (21 S) is more rapid than that of DNA II (16 S) during velocity gradient analysis in neutral conditions. All naturally occurring, covalently closed DNAs similarly contain negative superhelical turns, although the superhelix density (the number of superhelical turns per 10 base pairs) is somewhat variable among different DNAs. A number of reagents can change the average pitch of the duplex, and thus the superhelix density. The binding of intercalative dyes, changes in ionic conditions, or partial alkaline denaturation may all produce these changes (Helinski and Clewell, 1971). It can be seen (Fig. 1) that when DNA I is partially denatured by alkali it loses superhelical turns and assumes a configuration equivalent to DNA II. A similar effect is observed in the presence of critical concentrations of ethidium bromide. When the two strands in DNA I are completely denatured by alkali, a compact structure which sediments rapidly in alkali is obtained.

After denaturation at *p*H values of 12.1–12.6, a rapidly sedimenting DNA (53 S) is obtained; but if these DNA preparations are neutralized, the process of denaturation is reversible and 21 S DNA I is reformed. However, at *p*H values above 12.6, denaturation is no longer a reversible process (Westphal, 1970; Salzman *et al.*, 1973*a*). Similar behavior is observed for the replicative form of ϕX174 (Rush and Warner, 1970).

When DNA II is treated with alkali, it gives rise to a linear single strand of DNA (16 S) and a single-stranded circle (18 S).

2.2. Structure of Replicating Molecules

A number of procedures have facilitated studies of SV40 and polyoma DNA replication. The method in which cells are lysed by the addition of SDS, and high-molecular-weight DNA is separated from low-molecular-weight DNA by selective salt extraction is of particular value (Hirt, 1967). In general, cellular DNA is precipitated, and SV40 DNA and intermediates involved in SV40 DNA replication are found in the low-molecular-weight (LMW) supernatant fluid. The absence of the large amounts of cellular DNA greatly facilitates subsequent sedimentation and electron microscopic analyses. A second important technical advantage in studies with papovaviuses derives from the unique properties of covalently closed, duplex molecules which enable them to be separated from nicked molecules in either alkaline or neutral velocity gradients or by isopycnic centrifugation in the presence of ethidium bromide.

To determine if polyoma DNA replication occurred by a semiconservative mechanism, infected mouse kidney cell cultures were labeled with ^3H-thymidine for 20 minutes at 24 or 36 hours p.i. to generate a pool of ^3H-labeled LL (light, light) DNA I (d = 1.709). These cells were then incubated in the presence of ^{14}C-BUDR and FUDR for an additional 2 hours. The presence of hybrid HL (heavy, light; d = 1.753) and HH (d = 1.795) polyoma DNA I and DNA II containing ^3H-thymidine was observed. This HL and HH polyoma DNA was encapsidated. These data show that polyoma DNA replication proceeds by a semiconservative mechanism (Hirt, 1969).

Intermediates in polyoma DNA replication were seen after short (3.5–5 minute) pulses (Bourgaux *et al.*, 1969). These intermediates seemed to sediment like DNA I in neutral velocity gradients, but they banded in cesium chloride-ethidium bromide (CsCl-EtBr) like DNA II, and they were designated DNA II*. When DNA II* was sedimented in an alkaline gradient, the labeled, newly synthesized DNA sedimented at a rate equal to or slower than intact single polyoma DNA strands (16 S). These data are consistent with replicating polyoma DNA possessing a Cairn's-type structure (Cairns, 1963). They are not consistent with a rolling-circle model for polyoma DNA replication (Gilbert and Dressler, 1968) which has been reported to be the mechanism by which ϕX174 replicates. One prediction of the rolling-circle model is the presence of a covalent link beteen parental DNA and one strand of newly synthesized DNA. This would require that a fraction of the newly synthesized DNA sediment more rapidly than 16 S, the length of an intact single strand, and no newly synthesized DNA has been observed with these properties.

A more accurate description of replicating molecules was provided in studies of SV40 DNA replication in primary African green monkey kidney cells (Levine *et al.*, 1970). In this system, synthesis of SV40 DNA I is observed to start at 15 hours p.i., and a maximum rate of synthesis is observed at 30 hours p.i. DNA synthesis continues until 70 hours p.i. Since the cells have been infected at a rather high input multiplicity (25–100 PFU/cell), the extended time of synthesis is not a consequence of the failure to rapidly infect each cell. There is asynchrony within the population as to when macromolecular synthesis commences. This study showed clearly that isotope is first incorporated (after a 2.5-minute pulse) into replicating molecules which sediment in neutral sucrose gradients with a mean S value of 25–26 S. Newly synthesized DNA I is not detected until the labeling time has been extended to 10 minutes, and by pulse-chase experiments these 25 S replicating forms can be shown to be precursors of DNA I. The structures of replicating molecules appear in the electron microscope as Cairn's-type structures, with two branch points, three branches, and no free ends, findings similar to those previously observed for polyoma (Hirt, 1969). It was subsequently shown (Sebring *et al.*, 1971) that while these structures are replicating forms of SV40 DNA, only a small fraction of the replicative molecules are seen to have this configuration. Most of the replicating molecules (80–90%) have a structure in which there are two branch points, three branches, and no free ends. In addition, one branch contains superhelical turns (Fig. 2). From length measurements of replicative molecules it is clear that the two branches, L1 and L2, are of equal length and correspond to the replicated portion of the molecule. Superhelical turns are contained in the unreplicated part of the molecule. The reason that these structures contain superhelical turns is that the parental template strands are covalently closed. While the replicative intermediates contain covalently closed structures, during the replication cycle nicks must be introduced into the parental strands. This is discussed more fully below.

Since the superhelical turns are contained in the unreplicated part of the molecule, it follows that as replication proceeds the size of the unreplicated region will decrease and the total number of superhelical turns in the replicative intermediates will decrease. This feature of replicative intermediates provides a simple way of sorting molecules as a function of the extent of replication. Covalently closed molecules bind less ethidium bromide than the corresponding nicked DNA II structures, and consequently, the two can be easily separated by isopycnic banding in CsCl-EtBr (Bauer and Vinograd, 1968). The DNA II which binds more dye bands at a lower density. It is clear that young replicating molecules which contain almost the same number of superhelical turns as DNA I will band close to the position of DNA I. As the DNA molecules progress through a replicative cycle, the number of superhelical turns decreases. This results in a gradual shift in the position at which they band to a lower density, so that almost fully replicated molecules band with DNA II. Once they give rise to progeny DNA I molecules, they then band at the position of DNA I. The iospycnic banding of DNA I, DNA II, and replicating molecules is seen in Fig. 3. DNA I and II band sharply and are well resolved, while replicative intermediates band heterogeneously in the region from DNA I to DNA II. For most studies involving electron microscopy of replicating molecules, that region between DNA I and DNA II is most readily studied since this fraction will be richest in replicative inter-

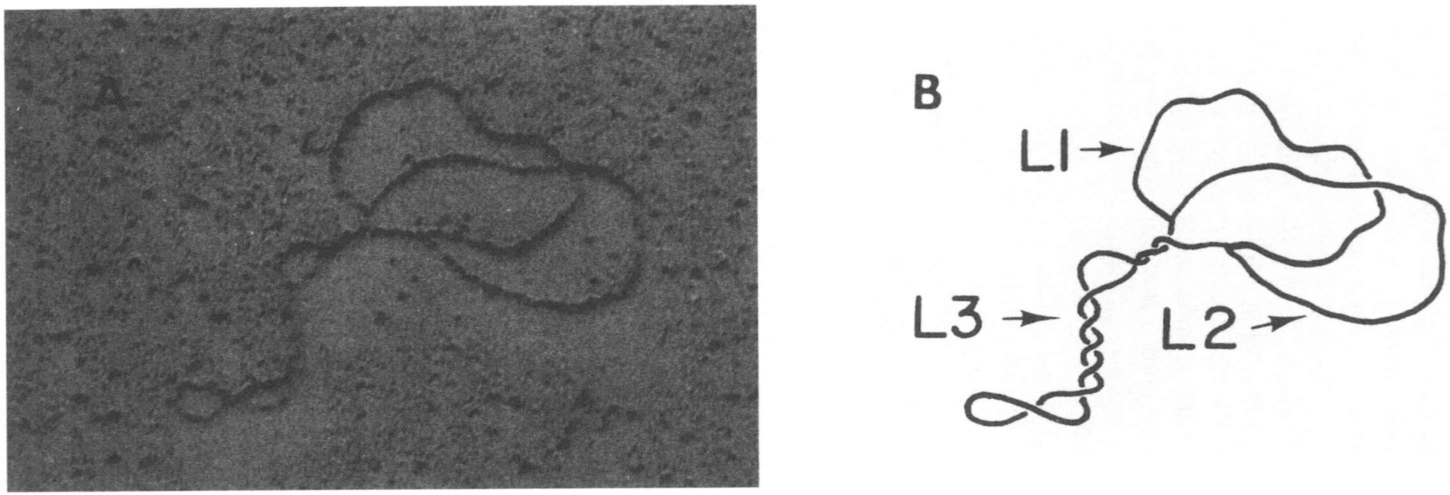

Fig. 2. A twisted SV40 replicating DNA molecule. (A) Electron micrograph obtained by modification of the technique described by Davis *et al.* (1971). The hypophase was distilled water and the DNA was contrasted by shadowing from one direction with 80% platinum–20% paladium. In some regions the individual strands which comprise the superhelical branch can be seen. Magnification is 1.5×10^5. (B) An interpretive drawing of the molecule. The branches of the molecule measured are indicated. The two branches that were not superhelical were designated L1 and L2. The superhelical branch was designated L3 (Sebring *et al.*, 1971).

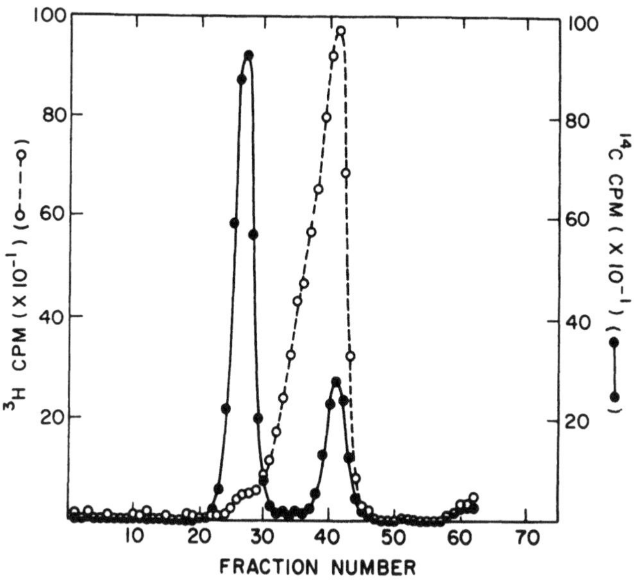

Fig. 3. CsCl-EtBr isopycnic banding of DNA contained in a Hirt supernatant fluid. At 30 hours after infection with SV40, an African green moneky kidney cell monolayer was pulsed for 2 minutes with medium containing 50 μCi of ^{3}H-thymidine/ml. The Hirt supernatant fluid was prepared and dialyzed. A sample of this fluid and purified ^{14}C-SV40 DNA marker were centrifuged to equilibrium in a CsCl-EtBr gradient (volume, 6 ml; CsCl density, 1.564; ethidium bromide, 200 μg/ml). Fractions were collected directly into scintillation vials and counted (Sebring *et al.*, 1971).

mediates. In contrast, regions close to DNA I will be obscured by the large amount of DNA I which is an end product of replication and therefore accumulates in great quantities, while in the region where DNA II bands, even trace contamination with cellular DNA and DNA II will make the examination of mature replicating molecules more difficult. As was observed for polyoma (Bourgaux *et al.*, 1969), all of the newly synthesized SV40 DNA in replicative intermediates is also released by alkaline treatment (Sebring *et al.*, 1971; Jaenisch *et al.*, 1971). The evidence which established that replicative intermediate molecules are effectively separated by CsCl-EtBr was obtained by alkaline sedimentation of replicative intermediates which had banded at different densities. It can be seen that the size of the newly synthesized DNA strands released by alkaline treatment is that predicted if molecules band at progressively lower densities as they go through the replication cycle (Fig. 4). In this experiment molecules were obtained that had, on the average, completed 23, 36, 57, and 85% of the replication cycle. Since there is a unique site for the initiation of replication (see below), the strands of DNA released by alkali cor-

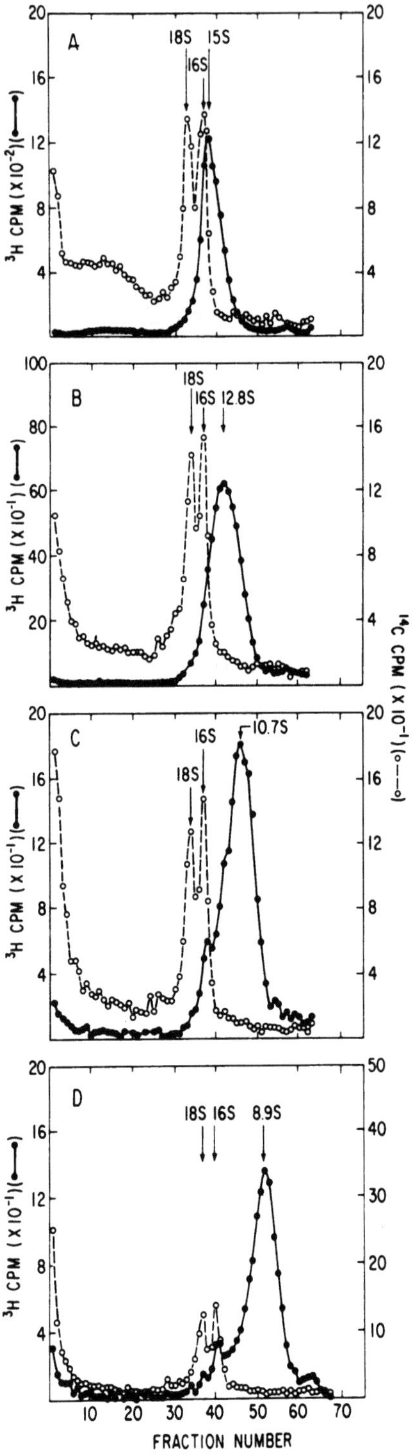

Fig. 4. Velocity sedimentation in alkaline sucrose of molecules taken from CsCl-EtBr densities of (A) 1.570, (B) 1.562, (C) 1.555, and (D) 1.548. Each fraction was extracted three times with CsCl-saturated isopropanol to remove the ethidium bromide, dialyzed against 0.01 M Tris, 0.01 M EDTA (*p*H 7.2) buffer, and concentrated. A sample was layered onto an 11.6-ml 10–30% alkaline sucrose gradient and sedimented at 10°C for 13 hours at 40,000 rpm in an SW 41 rotor. Fractions were collected directly into scintillation vials, neutralized by the addition of two drops of glacial acetic acid, and counted. Sedimentation is from right to left. The ^{14}C cpm reflect not only the small amount of ^{14}C present in the double-labeled sample, but also a purified ^{14}C-SV40 DNA that was added to each gradient as a marker. (Sebring *et al.*, 1971).

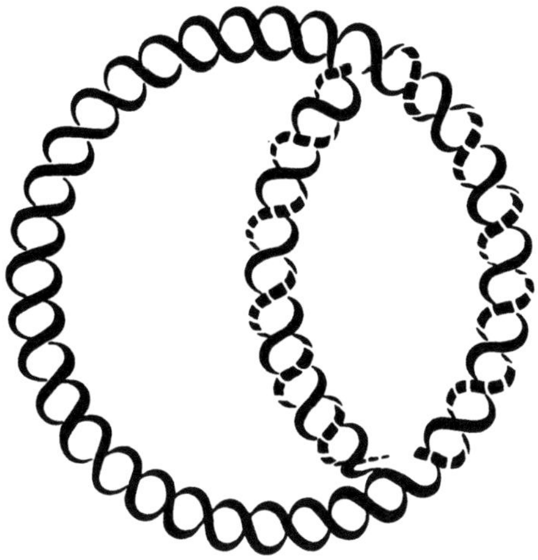

Fig. 5. Diagrammatic representation of replicating SV40 DNA. The salient features of the molecule are: (1) both parental DNA strands (solid lines) are covalently closed, and (2) the two newly synthesized DNA strands (broken lines) are not covalently linked to the parental DNA nor are they linked together (Sebring *et al.*, 1971).

respond to particular regions of the viral genome. The structure of replicating molecules is shown schematically in Fig. 5.

The features possessed by replicating molecules that are shared commonly by polyoma and SV40 are the following:

1. The parental strands of the replicative intermediates exist in a covalently closed form (Sebring *et al.*, 1971; Jaenisch *et al.*, 1971; Bourgaux and Bourgaux-Ramoisy, 1972*a*).
2. The dissociation of newly synthesized strands from parental strands establishes the absence of a covalent linkage between them, and also the absence of a covalent linkage between the two newly synthesized strands.
3. There are single-stranded regions at each of the two replicating forks. This was suggested by the binding of replicative intermediates to benzoylated naphthoylated DEAE-cellulose and their subsequent elution by caffeine (Levine *et al.*, 1970). These binding properties depend on the presence of single-stranded DNA regions (Kiger and Sinsheimer, 1969). More direct evidence has been obtained by demonstrating the susceptivity of replicative intermediates to cleavage by a single-strand-specific nuclease from *Neurospora crassa* which generates, as one product of digestion, ring-shaped structures with tails (Bourgaux and Bourgaux-Ramoisy, 1972*a*).

2.3. Cleavage of SV40 DNA by Bacterial Restriction Endonucleases

Studies of restriction and modification in bacterial systems (Arber and Linn, 1969; Boyer, 1971) have provided a number of enzymes which have proven extremely powerful reagents for studies with SV40 and polyoma. Those enzymes which have been purified, although not in each case to a homogeneous state, include enzymes from *Hemophilus influenzae* (R · *Hin*) (Smith and Wilcox, 1970), from *Hemophilus aegipticus* (R · *Hae*) (Middleton *et al.*, 1972), from *Escherichia coli* carrying an R factor (R · *Eco* R_I and R · *Eco* R_{II}), and from *Hemophilus parainfluenzae* (R · *Hpa*) (Gromkova and Goodgal, 1972). Subsequent studies with R · *Hin* (Danna *et al.*, 1973) and with R · *Hpa* (Sharp *et al.*, 1973) have revealed that in each case the original isolation procedures had yielded at least two restriction activities which could be further resolved. The enzymes recognize a particular sequence of nucleotides in the DNA molecule, and in this region they cleave both DNA strands. The first cleavage site to be sequenced, R · *Hin* (Kelly and Smith, 1970), possessed a twofold axis of symmetry. Its structure is shown in Fig. 6, as are the structures for the R · *Eco* R_I and R · *Eco* R_{II} cleavage sites, which have also been characterized (Hedgpeth *et al.*, 1972; Bigger *et al.*, 1973, Boyer *et al.*, 1973). They have the same symmetry as first observed for the R · *Hin* cleavage site; however, the R · *Eco* R_I and R · *Eco* R_{II} cleavage products possess single-stranded ends; these molecules can thus be recyclized at low temperatues, and then covalently closed, duplex rings can be generated by the sealing action of ligase. It was first reported that the R · *Hin* endonuclease would cleave SV40 into 11 fragments (Danna and Nathans, 1971) and that R · *Hpa* would cleave SV40 into 4 fragments (Sack and Nathans, 1973). Later it was demonstrated that the single cleaveage of SV40 by R · *Eco* R occurred at a specific site in the molecule (Morrow and Berg, 1972; Mulder and Delius, 1972; Fareed *et al.*, 1972). Cleavage of SV40 with several restriction enzymes has been used

THE DNA CLEAVAGE SITES OF THREE RESTRICTION ENDONUCLEASES

5′.... GpTpPy ↓ pPupApC .... 3′
3′.... CpApPup ↑ PypTpG .... 5′
Hemophilus influenza **endonuclease**

5′.... A/TpG ↓ pApApTpTpCpT/A .... 3′
3′.... T/ApCpTpTpApAp ↑ GpA/T .... 5′
E. coli R_I **endonuclease**

5′.... N ↓pCpCpApGpGpN′.... 3′
3′.... NpGpGpTpCpCp ↑ N .... 5′
E. coli R_{II} **endonuclease**

Fig. 6. The recognition sequences of three restriction endonucleases: *H. influenzae* (Kelly and Smith, 1970), *E. coli* R_I (Hedgpeth *et al.*, 1972), and *E. coli* R_{II} (Bigger *et al.*, 1973; Boyer *et al.*, 1973). Both the *Eco* · R_I and *Eco* · R_{II} endonucleases produce a staggered cleavage which results in cohesive termini, while the *Hin* endonucleases does not. The substrate sites for all three enzymes possess a twofold rotational axis of symmetry.

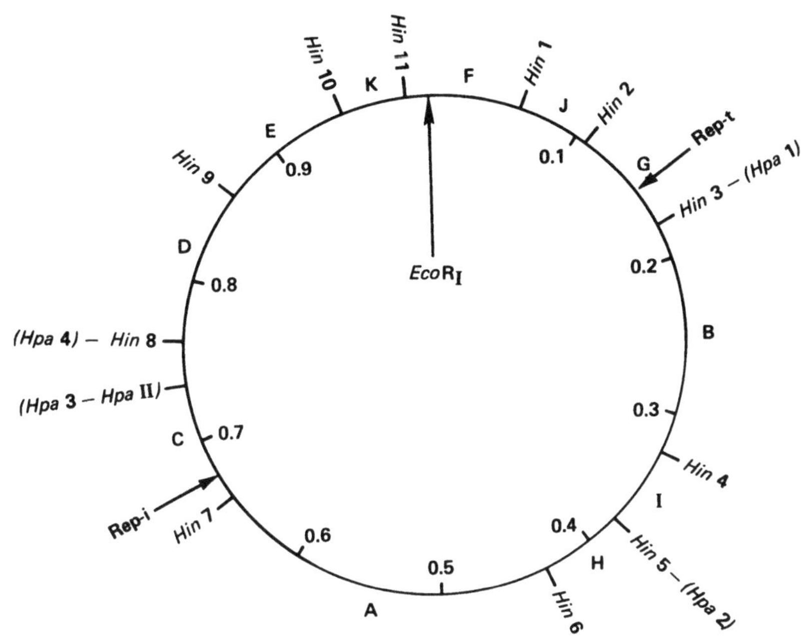

Fig. 7. A cleavage map of the SV40 genome (Danna and Nathans, 1972; Danna *et al.*, 1973). The R · *Eco* R$_I$ site (Morrow and Berg, 1972; Mulder and Delius, 1972) is used as a reference for the *Hin* and *Hpa* cleavage sites. Note that the *Hpa* 3 cleavage site appears to be recognized by a unique enzyme, *Hpa* II (Sharp *et al.*, 1973). The positions on the map for initiation (Rep-i) and termination (Rep-t) of DNA replication are indicated (Danna and Nathans, 1972; Fareed *et al.*, 1972).

recently (Danna *et al.*, 1973) in a study which has defined an impressively detailed cleavage map of the SV40 genome (Fig. 7). In some of the studies described below, which relate to DNA replication and transcription, the usefulness of these enzymes will be illustrated. Clearly, they represent extraordinarily useful reagents, and their use has made possible the rapid progress during the past three years in our understanding of the papovaviruses.

There are a number of other methods which have been used to gain information about the anatomy of the viral genome. Several studies have reported denaturation maps for both polyoma (Bourguignon, 1968; Follett and Crawford, 1968) and SV40 (Yoshiike *et al.*, 1972; Mulder and Delius, 1972). Those regions rich in adenine and thymine are the first to become denatured, and the denaturation map gives some idea of the clustering of nucleotides within the molecule. The combined action of restriction enzymes and determination of both base composition and of purine and pyrimidine tracts on various fragments will be an alternative method to obtain similar data in a more precise manner.

The T4 gene-32 protein binds to supercoiled SV40 DNA I but not to SV40 DNA II (Mulder and Delius, 1972; Morrow and Berg, 1972). The site(s) at which the gene-32 protein binds to SV40 DNA I is in a region corresponding to 0.44–0.48 SV40 map units from the *Eco* · R_I cleavage site (Morrow and Berg, 1973). This is the same region which is preferentially denatured at alkaline *p*H values (Mulder and Delius, 1972).

The finding that supercoiled molecules have unpaired bases is based on the binding of formaldehyde to the replicative form of φX174 and to PM2 (Dean and Lebowitz, 1971). It is estimated that 3–4% of the bases in φX174–RF can bind methylmercury and are, therefore, in an unpaired state (Beerman and Lebowitz, 1973). Single-strand regions in covalently closed molecules can be cleaved by the single-strand-specific nuclease from *N. crassa* which converts the supercoiled molecule to the relaxed form (Kato *et al.*, 1973) or by S_1 nuclease of *Aspergillus oryzae* which produces unit-length, linear, duplex, molecules from supercoiled SV40 DNA (Beard *et al.*, 1973). In this latter case, cleavage occurs predominantly at 0.45 and 0.55 SV40 map units, the same regions at which denaturation and gene-32 protein binding occurs. At present, the biological significance of this single-stranded region is not known.

2.4. Site of Initiation and Direction of DNA Replication

Two separate experimental approaches have been used to demonstrate that DNA synthesis is initiated at a specific site. The first depends on the cleavage of replicating SV40 molecules or newly synthesized SV40 DNA by the bacterial restriction endonuclease from *H. influenzae*. This enzyme preparation cleaves SV40 and yields 11 fragments, which can be resolved by polyacrylamide gels; they range in size from 6.5×10^5 to 7.4×10^4 daltons (Danna and Nathans, 1971). If there is a preferred site for DNA initiation, then in molecules labeled for a period slightly longer than the time required to complete one round of DNA replication, the fragment which corresponds to the DNA initiation site will be most highly labeled in the replicative intermediates. However, when the pulse time is shorter than that required for a round of replication and newly synthesized DNA I is examined, the region where termination occurs will be preferentially labeled. Based on the rates of labeling, it was concluded that replication of

SV40 DNA starts in *Hin* fragment C and terminates in *Hin* fragment G (see Fig. 7) (Danna and Nathans, 1972).

A preferred initiation site for SV40 DNA replication was also demonstrated in experiments based on the reassociation rates of the strands of replicating molecules. When the rate of duplex formation of strands of newly synthesized DNA, isolated from molecules which had undergone different extents of replication, was studied, the time required to effect 50% renaturation increased at a rate proportional to the increase in the length of the newly synthesized strands. These results are those predicted if there is a specific site for initiation of DNA synthesis (Thoren *et al.*, 1972).

In order to determine if DNA replication, which was initiated at a unique • site, proceeds unidirectionally or bidirectionally SV40 replicative intermediates were cleaved with R · *Eco* R_I (Fareed *et al.*, 1972). The structures that are predicted, depending on the mode of DNA replication, are shown in Fig. 8. After cleavage, linear structures

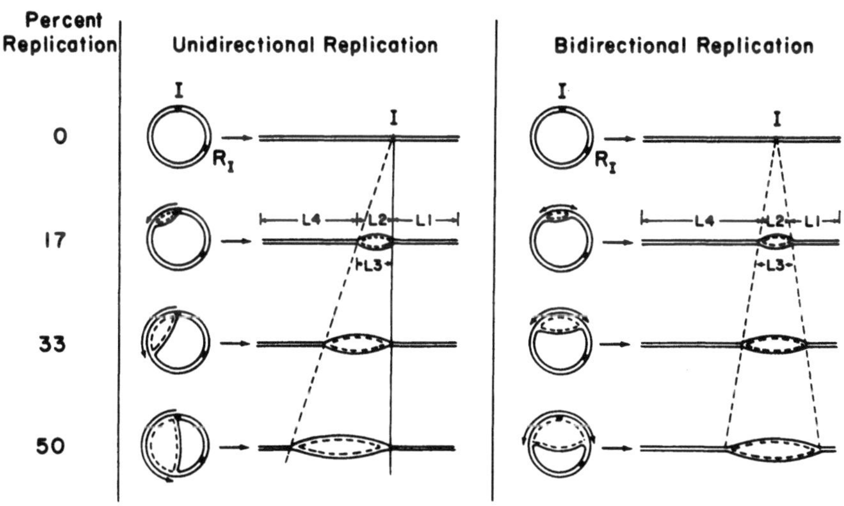

Fig. 8. Unreplicated SV40 molecules or molecules which have replicated to different extents are schematically represented, as are the molecules which are obtained after cleavage with the R · *Eco* R_I endonuclease. The solid lines represent the parental strands and the broken lines are the newly synthesized strands. I is the site at which DNA replication is initiated. Molecules are cleaved by the R · *Eco* R_I endonuclease at site R_I. The arrow indicates the direction of replication. It can be seen that during unidirectional replication the length of one branch remains constant while the second branch decreases in length. Since the R · *Eco* R_I cleavage site is 17% of the genome length from the termination site of DNA synthesis, it is the short arm L1, which would remain constant during unidirectional replication. During bidirectional replication, both L1 and L4 decrease in length as replication proceeds (Salzman *et al.*, 1973*a*).

containing one bubble will be generated. If replication is unidirectional, all molecules, regardless of their extent of replication, will have one branch L1, of constant length. With bidirectional replication, both L1 and L4 will decrease as replication proceeds. In the case of bidirectional replication, if each replication fork moves at the same rate, the lengths of both L1 and L4 will decrease at the same rate as replication proceeds.

The structures that were observed demonstrated that DNA synthesis proceeds bidirectionally, and that the two replication forks move at the same rate. This study also located the R · *Eco* R_I cleavage site at 0.33 (or 0.67) genome lengths from the initiation site for DNA replication. Recently, a parallel study has been carried out with polyoma DNA in which the R · *Eco* R_I endonuclease also introduces a single break. Bi-directional replication from a unique origin was also observed. (Crawford *et al.*, 1973). When the distribution of ^{3}H-thymidine within R · *Hin*-produced fragments of newly completed SV40 DNA I molecules was measured, a similar conclusion was reached, i.e., that DNA replication proceeds in a bidirectional manner (Danna and Nathans, 1972).

2.5. Mechanism of Chain Growth

When SV40-infected cells are pulsed for short time periods (15 seconds to 1 minute) and replicating molecules are isolated in neutral sucrose gradients, subsequent examination in alkaline sucrose gradients demonstrates two populations of newly synthesized DNA strands. One population consists of growing SV40 DNA chains ranging from 6 S to almost 16 S, while the second consists of a discrete peak sedimenting at 4 S. These 4 S fragments, which correspond to DNA chains of 50, 000 daltons, are present in molecules at all stages of replication; by pulse-chase experiments, it is seen that these short fragments are precursors of growing SV40 chains. Thus, a discontinuous mechanism of synthesis of DNA, as first proposed by Okazaki *et al.* (1968) for bacterial DNA, is also observed for this small circular genome (Fareed and Salzman, 1972). A similar discontinuous mechanism of synthesis for polyoma DNA is observed during *in vitro* incubation with nuclei from infected cells. In addition, the presence of RNA which is covalently linked at the 5′ end of the 4 S fragments has been observed during these *in vitro* studies (Magnusson *et al.*, 1973).

It is likely that attachment of the 4 S fragments to growing chains requires the action of a DNA polymerase which may be different from that which carries out the synthesis of the 4S fragment. The basis for this theory emerges from three separate studies (Salzman and Thoren, 1973; Magnusson, 1973; Laipis and Levine, 1973). For both SV40 and polyoma, under conditions where DNA synthesis is first inhibited by FUdR and then restarted by the addition of thymidine, or where DNA synthesis is partially blocked by hydroxyurea, there is an accumulation of 4 S fragments, both free and within replicating molecules. These 4 S fragments in replicative intermediates cannot be joined *in vitro* to growing chains by the action of ligase, but the combined action of T4 DNA polymerase and ligase does effect a joining to growing chains (Laipis and Levine, 1973). The synthesis of 4 S DNA fragments, together with the failure to continue chain elongation, suggests that two polymerase activities are involved in this process. Since the RNA which has been found joined to growing chains is excised prior to the incorporation of the 4S fragment into the growing chain, one can speculate that gaps between the 4 S fragment and the growing chain are located at the site where RNA previously existed. The presence of enzymes in eukaryotic cells which excise RNA from RNA–DNA hybrids has been described (Keller and Crouch, 1972), and it is possible that such an enzyme removes the RNA primer from the 4 S DNA fragments.

When 4 S fragments are isolated from replicating molecules and then allowed to reanneal, 70–90% are converted to a double-stranded form (Fareed *et al.*, 1973*a*). Similarly, when 4 S fragments are isolated from hydroxyurea-treated polyoma-infected cells, a lower but still sigificant percentage of the 4 S DNA is converted to a double-stranded form (Magnusson, 1973). These are the results that are expected only if both strands, at each of the replication forks, are made in a discontinuous manner. Based on our present understanding of SV40 the following series of events can be specified:

1. DNA synthesis is initiated at a specific site which is located 0.67 genome lengths from the R · *Eco* R_I cleavage site.
2. Replication then proceeds bidirectionally with an equal rate of chain growth at each of the replication forks.
3. Chain growth occurs by a discontinous mechanism in which 4 S DNA fragments are first synthesized. A fraction of the 4 S fragments are covalently linked to RNA. The RNA must be removed rapidly since the association of 4 S DNA with RNA can only be demonstrated during *in vitro* studies or *in vivo* when DNA synthesis has been modified by metabolic inhibitors.
4. Each of the chains at each of the replication forks is made in a discontinuous manner.

5. There is a gap between the 4 S fragment and the growing chain. It seems likely that an enzyme other than the one which synthesizes the 4 S DNA is involved in filling in the gap.
6. The newly synthesized fragment is joined to the growing chain by a covalent link which is probably formed by ligase, an enzyme present in animal cells (Sambrook and Shatkin, 1969).

While these facts seem quite likely, based on present experimental evidence, it may also be worthwhile to mention some general areas of DNA replication which are not at all understood. While there is a specific site for initiation of DNA synthesis, there is no evidence as to the precise mechanism of DNA initiation. Is nicking involved as the first step in DNA replication? If initiation involves an RNA polymerase which recognizes and transcribes a region that serves as an RNA primer to which DNA is linked, why is this region recognized preferentially compared with the other sites where RNA synthesis presumably occurs? When 4 S DNA fragments are linked to an RNA primer, any one of four deoxynucleotides can lie adjacent to the terminal 3′ ribonucleotide (Magnusson *et al.*, 1973); what are the signals, then, within the DNA which determine initiation of RNA synthesis and termination of 4 S DNA synthesis? It should also be noted that there has been only a partial characterization of the enzymes involved in DNA replication in mammalian cells and a large number of questions about the detailed mechanism of chain growth still remain unanswered. Temperature-sensitive mutants of SV40 have been isolated which are able to carry out chain elongation at the restrictive temperature but fail to initiate new rounds of DNA synthesis (Tegtmeyer, 1972). These mutants should be of value in defining the mechanism of initiation. Furthermore, the virus mutants which fail to initiate new rounds of DNA replication are also unable to transform cells at the restricted temperatures and this makes their characterization of considerable interest.

2.6. Termination of DNA Synthesis

Kinetic analysis of the time course of formation of replicative intermediates, as well as of DNA I and II, suggests that DNA II is the first product to arise by segregation of parental strands (Fareed *et al.*, 1973*b*). The properties of this DNA II, which can only be observed after short pulses with ^{3}H-thymidine, are distinct from a pool of DNA

II seen after extended periods (1–3 hours) of labeling. The latter arises by endonucleolytic cleavage of DNA I and the nick is randomly located. In contrast, precursor DNA II (pDNA II) contains all of the newly synthesized DNA in the discontinuous (16 S) strand. It can also be shown by cleavage with *Eco* · R_I endonuclease that the nick in the newly synthesized strand is located 0.5 genome length from the initiation site for DNA replication (Fareed *et al.*, 1973*b*). This is the expected site for termination since replication has been shown to be bidirectional and each of the two replication forks were observed to move at the same rate (Fareed *et al.*, 1972).

2.7. Mechanism for Effecting Semiconservative Replication of Covalently Closed Duplex DNA

For each covalently closed duplex DNA molecule, the topological winding number is fixed; it can only be changed by introducing a break into one of the two strands, allowing unwinding to occur, and then sealing the break. The topological winding number is related to the superhelix winding number by the relationship

$$\tau = \alpha - \beta$$

where the topological winding number, α, is the number of revolutions made by one strand about the duplex axis when the axis is constrained to lie in a plane; the duplex winding number, β, is the number of revolutions made by one strand about the duplex axis in the unconstrained molecule; and the superhelix winding number, τ, is the number of revolutions made by the duplex about the superhelix axis (Vinograd *et al.*, 1968). During DNA replication, unwinding of the parental strands must occur and there will be a decrease in α, the topological winding number. After displacement of the newly synthesized strands from these replicative intermediate structures, the parental molecule will have the potential to form a Watson-Crick base-paired structure in which the value β is the same as in DNA I. Since α decreases as a result of replication, it is clear that these structures will have an increased number of negative superhelical turns, which can be quantitated experimentally.

Parental DNA strands have been obtained from replicating molecules after dissociation of the newly synthesized strands by treatment at *p*H 12.2. When examined by alkaline velocity gradient analysis or isopycnic banding in CsCl-EtBr, they provide direct experimental confirmation that the topological winding number of the parental template strands does progressively decrease as replication proceeds (Salzman *et al.*, 1973*b*; Bourgaux and Bourgaux-Ramoisy, 1972*a*).

A DNA untwisting enzyme has been found in mammalian cells. This activity breaks one parental strand, allows the DNA strands to unwind, and then restores the covalently closed structure (Champoux and Dulbecco, 1972). A similar type of enzymatic activity has been obtained from bacterial cells and has been extensively purified (Wang, 1971). Such an enzyme is a good candidate to carry out the unwinding that must occur as DNA replication proceeds.

2.8. SV40 DNA Synthesis in Heterokaryons of SV40-Transformed Cells and Cells Permissive for SV40

Induction of extremely low levels of virus has been noted after treatment of SV40-transformed cells with chemical agents (Rothschild and Black, 1970). In general, however, a cell line which is transformed by SV40 cannot be induced to synthesize infectious virus after treatment with agents which cause induction of lysogenic bacteria. A finding of great interest and importance was that by cocultivation of an SV40-transformed mouse line with monkey kidney cells, which are permissive for SV40 replication, there was "activation" of the integrated genome and synthesis of infectious virus (Gerber, 1966). The amount of virus that is rescued is higher when the two cell lines, the susceptible and the transformed, are treated with inactivated Sendai virus in order to enhance the extent of cell fusion (Gerber, 1966; Koprowski *et al.*, 1967; Watkins and Dulbecco, 1967; Burns and Black, 1968; Dubbs *et al.*, 1967). It is difficult to use this system for biochemical studies since only a fraction of the cells can be induced to synthesize virus, even after fusion with inactivated Sendai virus. The time of appearance of infectious SV40 DNA is about the same after cell fusion as it is during a lytic cycle. Thus, viral DNA synthesis was detected 19 hours after fusion (Kit *et al.*, 1968). When cell fusion was carried out between a transformed mouse and a transformed human cell line, infectious virus was rescued. By using plaque mutants it was shown that only that virus which had been used to transform the mouse cell line was released and no SV40 virus was released from the transformed human line.

SV40-transformed hamster cells have been fused with susceptible CV-1 cells, and nuclei were isolated from the heterokaryons. The nuclei from the two cell species could be separated by sucrose gradient analysis. Virus was first found in the transformed nucleus (40 hours) and later (68–72 hours) was found associated with both nuclei (Wever *et al.*, 1970). Additional studies with this system may provide some interesting insight into factors which determine a permissive state for DNA synthesis and virus replication starting with an integrated viral genome.

2.9. SV40 DNA-Containing Cellular DNA Sequences

A field of current interest has been the characterization of covalently closed SV40 DNA molecules in which a fraction of the viral genome has been deleted and/or in which there is an insertion of cellular sequences. The first suggestion that cellular DNA is present in SV40 was the finding that SV40 DNA I was able to hybridize to an appreciable extent with DNA from BSC-1 cells (Aloni *et al.*, 1969). However, when the hybridization reaction between SV40 DNA I and cellular DNA was studied in various laboratories, there were considerable quantitative differences in the degree of hybridization. At that time the reasons for these differences were not understood. It was finally shown that incorporation of cellular DNA into covalently closed molecules occurred when virus was passaged at a high input multiplicity. Infection of BSC-1 cells with a plaque-purified virus at either low or high multiplicities, or with low multiplicities of a virus pool that was not plaque-purified, yielded virions which did not hybridize with cellular DNA. However, infection with high inputs of non-plaque-purified virus yielded viral DNA that hybridized to cellular DNA as well as to viral DNA. Similar results were obtained after multiple passages at high input multiplicities of plaque-purified SV40 (Lavi and Winocour, 1972). At the same time that these homology studies of high- and low-passage SV40 were reported, SV40 DNA was obtained from progeny virus that was purified after infection of BSC-1 cells with either high or low virus input multiplicities (Tai *et al.*, 1972). The DNA I was converted to DNA II which was denatured and then reannealed. Using formamide-spreading of the DNA to prepare the grids, both single- and double-stranded DNA could be distinguished. The low-input-multiplicity DNA was of uniform size (1.7 μm) while the high-input DNA showed considerable size heterogeneity, and a significant fraction of the molecules were smaller than 1.7 μm. Similarly, the heteroduplexes from low-input-multiplicity DNA gave evidence for only a very low level of DNA deletions (2.5%). In contrast, high-input-multiplicity DNA contained 13% deletions (seen as single-stranded loops) and 7–12% substitutions (which are seen as a region where two single strands of DNA replace the duplex structure). The conclusion from both of the above studies is that under conditions of high input multiplicity there is recombination between viral and cellular DNA, and one consequence is the incorporation of host DNA sequences into covalently closed SV40 DNA molecules. This may involve the integration of the viral DNA into the cell genome and its subsequent excision. The findings cited above are clearly related to earlier studies where it was first reported that at high input multiplicities a population of defective particles was formed and that these contained covalently closed DNA molecules smaller than SV40 DNA I (Yoshiike and Furano, 1969). A consequence of the insertion of cellular DNA and deletion of viral sequences is the production of particles which are noninfectious. It is only under conditions where cells are coinfected with a defective and nondefective virus particle that the defective virus particle can replicate, presumably as a result of complementation. While defective particles are unable to replicate, they can support a partial cycle of virus replication (Yoshiike, 1968*a,b*). The loss of infectivity which is a result of a loss of a random segment of the viral genome suggests that the entire viral genome is required for infectivity, and this agrees with UV and X-ray inactivation studies (Basilico and DiMayorca, 1965; Benjamin, 1965). In these studies, infectivity was lost with one-hit kinetics, i.e., a single break anywhere in the virus genome inactivates the virus.

These studies of alterations in the viral genome are important for a number of reasons. It is clear, for example, that stock virus pools must be prepared at low input multiplicities. Thus, the significance of early studies with purified viral DNA that involved nucleic acid homology, nearest-neighbor analysis, etc., in which the purity of the viral DNA was not known, is in question.

The answers to questions of whether the covalent insertion of cellular DNA into SV40 DNA I is biologically significant and whether this insertion occurs at a specific site in the host cell genome are unknown. In an attempt to answer the latter question, the progeny SV40 DNA molecules from a set of serial passages were examined by digestion with R · *Hin* endonuclease (Rozenblatt *et al.*, 1973). The cleavage fragments were then examined in hybridization experiments in an attempt to analyze the incorporated cellular DNA sequences. The results of this study showed that independent series of passages produced different defective progeny viral DNA. It will, of course, be necessary to examine other serial-passage populations in order to determine if there is a consistent pattern either in the incorporated host cell sequences or in the deleted viral sequences. The defective virus particles containing host DNA sequences are presumed to have been formed by integration and excision, and they are then replicated at a rate which assures their presence at significant levels in the final virus preparation. However, there exist no data at present to support the hypothesis that integration and excision seen at high input multiplicities are related to integration which occurs when cells are transformed with polyoma or SV40. Future studies with the high-passage viral DNA may provide information not only about the mechanisms of integration and excision but also about the types of molecules which replicate rapidly in defective populations. In one recent study, it was observed that after multiple passages at high input multiplicities, certain DNA sequences which include the initiation site for DNA replication are selectively preserved (Brockman *et al.*, 1973). In another group of studies (Fareed *et al.*, 1974; Khoury *et al.*, 1974), viral DNA was generated in which a single DNA molecule contained three regions corresponding to the site for the initiation of DNA replication. Detectable levels of substituted particles do not arise after infection with low input multiplicities. However, under these conditions high virus titers are finally achieved. This suggests that integration and excision may not be necessary events during the lytic cycle.

2.10. The Role of Proteins in DNA Replication

Several groups have studied the effect of an inhibition of protein synthesis on SV40 or polyoma DNA replication. These various studies differed in the time after infection when the inhibitor (cycloheximide) was added and in the length of the pulse time (1–4 hours) for labeling newly synthesized DNA. It is difficult to study the fate of intermediates in replication in cases where such long periods of labeling have been used since the time for one round of viral DNA replication is 10–25 minutes (Danna and Nathans, 1972; Fareed *et al.*, 1973*b*). It is clear, however, in each of these studies, that the presence of cycloheximide resulted in a decrease in the rate of viral DNA synthesis (Branton *et al.*, 1970; Branton and Sheinin, 1973; Kit and Nakajima, 1971; Kang *et al.*, 1971). These observations are consistent with a requirement for a protein to initiate DNA synthesis. What is not clear is whether a second protein is needed in the conversion of mature replicating molecules to pDNA II or in the conversion of pDNA II to

DNA I. The various studies do not agree on this point. Initiation of viral DNA synthesis does not occur when polyoma replication is carried out in an *in vitro* system (Winnacker *et al.*, 1972). However, chain elongation and synthesis of DNA I can occur in this system, evidence which may provide the means to determine the role of proteins in virus maturation.

The role of a virus-coded protein for initiation of DNA synthesis has been demonstrated using temperature-sensitive mutants which are blocked in viral DNA synthesis (Tegtmeyer, 1972). These mutants fail to initiate new rounds of replication but can, at the restrictive temperature, continue the process of chain elongation and formation of DNA molecules. Thus, a virus-coded protein is apparently required to initiate viral DNA synthesis. If there is a second protein that is required for chain termination, it is likely to be a cellular protein and may also be involved in cellular DNA synthesis.

When cells are exposed to puromycin (Bourgaux and Bourgaux-Ramoisy, 1972*b*), an inhibition of polyoma DNA replication is observed. There is a reduced rate of synthesis of covalently closed DNA molecules in the presence of the inhibitor, and, in addition, these molecules do not seem to contain superhelical turns as judged by the fact that they band at a slightly higher density than DNA I in CsCl-EtBr. This suggests that specific proteins affect the configuration of DNA, perhaps by complexing with them during replication.

3. TRANSCRIPTION OF SV40 AND POLYOMA DNA

Synthesis of virus-specific RNA during the lytic cyle of polyoma replication was first demonstrated using a filter hybridization technique (Benjamin, 1966). Polyoma virus RNA was synthesized in very small amounts early after a productive infection of mouse kidney cells, and in significantly greater concentrations (*ca.* 100-fold greater) after the onset of polyoma DNA synthesis. Subsequently it was shown that virus-specific RNA was produced in two phases during the lytic cycle of SV40 (Aloni *et al.*, 1968; Oda and Dulbecco, 1968*b*; Sauer and Kidwai, 1968; Carp *et al.*, 1969). Prior to viral DNA replication (within the first 12 hours after infection) "early" SV40 RNA is synthesized. Using competition-hybridization experiments it was shown that the early RNA continues to be produced late in the lytic cycle and represents approximately 30–40% of the total transcribed gene sequences. Furthermore, the fact that late, lytic SV40 RNA could completely compete against the early RNA indicated that all of the early sequences are transcribed late in the cycle. A similar division of the lytic cycle of polyoma virus into early and late phases of transcription has also been reported (Hudson *et al.*, 1970).

In order to determine the extent of transcription of SV40 (Martin and Axelrod, 1969*a*) or polyoma DNA (Martin and Axelrod, 1969*b*), a saturation hybridization technique was employed (see Fig. 9). A known amount of ^{14}C-labeled viral DNA was bound on filters, and increasing amounts of ^{32}P-labeled lytic RNA preparations of known specific activity were hybridized to the DNA until a plateau was reached. From the amount of RNA required to saturate the DNA on each filter it was determined that approximately 50%, or the equivalent of one full strand of the polyoma or SV40 DNA, was bound in a hybrid. Since the total transcription product of SV40 or polyoma late in infection is equivalent to one full strand of the genome and the early gene sequences represent about one-third of a strand, the true late gene sequences (those transcribed only after the onset of DNA replication) represent about two-thirds of the coding capacity of the genome.

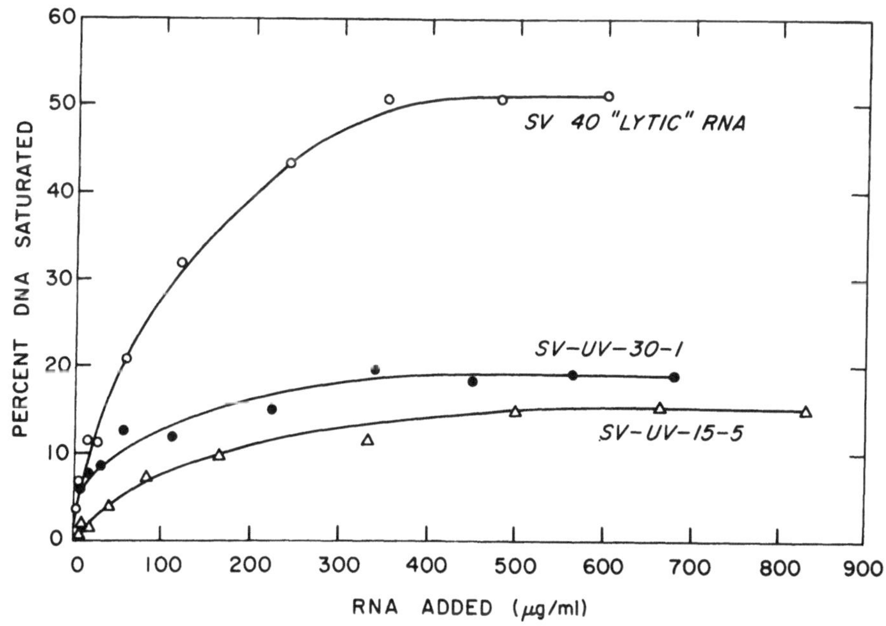

Fig. 9. A saturation hybridization experiment in which increasing amounts of ^{32}P-labeled lytic (or SV-UV-transformed cell) RNA was hybridized with a fixed amount of ^{14}C-SV40 DNA on filters. Since the specific activities of the RNAs and DNA were known, the percentage of DNA saturated could be determined. The plateau of 50% saturation with SV40 late lytic RNA indicates that the equivalent of one full DNA strand is transcribed during productive infection. Lower values were obtained for the two transformed cell lines examined (Martin and Axelrod, 1969*b*).

3.1. Strand Orientation of Transcription

Because of the similar base composition of the two strands of SV40 DNA, strand separation could not be effected by labeling with a heavy base analogue or by selective binding to polyribonucleotides. In 1970 it was shown that *in vitro* transcription of supercoiled SV40 DNA (DNA I) with *E. coli* DNA-dependent RNA polymerase results in an asymmetric RNA product which is complementary to only one of the two viral DNA strands (Westphal, 1970). This observation provided a means of separating the viral DNA strands on a preparative scale. In order to achieve this separation, hydroxyapatite chromatography (HA) was employed. This method allows for a rapid and large-scale separation of single-stranded DNA from DNA–RNA (or DNA–DNA) hybrid molecules (Britten and Kohne, 1968). Intact or sheared ^{32}P-labeled SV40 DNA was denatured and allowed to anneal with a great excess of *in vitro* SV40 cRNA (complementary RNA). Since the RNA binds almost exclusively to one DNA strand (the minus, or E, strand), passing the renaturation product over HA with subsequent purification effectively provides a strand separation of SV40 DNA. Once separated, the plus and minus strands of the viral DNA were then used in hybridization studies to determine the strand orientation and extent of transcription of early and late lytic SV40 RNA (Khoury and Martin, 1972; Khoury *et al.*, 1972; Sambrook *et al.*, 1972), as well as that of RNA synthesized in abortively infected (Khoury *et al.*, 1972) or transformed cells (Sambrook *et al.*, 1972; Khoury *et al.*, 1973*a*; Ozanne *et al.*, 1973). The results of these studies indicated that 30–40% of the minus, or E, strand (the strand which is transcribed *in vitro* with *E. coli* RNA polymerase) is transcribed early in the lytic cycle; in addition, 60–70% of the plus, or L, DNA strand is transcribed late in infections. The extent of transcription and strand orientation for early and late viral mRNA, as determined with the separated strands of SV40, agrees with results of the previous competition-hybridization experiments (Aloni *et al.*, 1968; Oda and Dulbecco, 1968*b*; Sauer and Kidwai, 1968; Carp *et al.*, 1969). These results are also in agreement with studies in which the early or late RNAs from SV40-infected BSC-1 cells were annealed with asymmetric SV40 cRNA (Lindstrom and Dulbecco, 1972); on the basis of ribonuclease resistance it was concluded that early lytic RNA was synthesized from approximately 40% of the same DNA strand as is the *in vitro* RNA, whereas late was synthesized from about 60% of the opposite strand.

In SV40-transformed cell lines there is little or no stable RNA transcribed from the plus strand. Transcription of the minus strand

(35–80%) appears to be more extensive than in the lytic cycle, and, presumably, the "anti-sense" sequences (those homologous to the late lytic RNA) do not code for functional proteins (see Fig. 10).

On the basis of RNA hybridization to the separated strands of polyoma DNA (R. Kamen, personal communication) and RNA–RNA annealing studies similar to those described above (Mueller *et al.*, 1973), it appears that there is also a "strand switch" in transcription during the lytic cycle of polyoma.

The fact that in the lytic cycle early and late virus-specific RNAs are transcribed from opposite strands of SV40 DNA is especially interesting, and it allows one to propose several novel hypotheses for the control of transcription in lytically infected cells (see below).

The lytic SV40 RNAs studied in the experiments described above (Khoury and Martin, 1972; Lindstrom and Dulbecco, 1972; Khoury *et al.*, 1972; Sambrook *et al.*, 1972) are predominantly the stable or abundant RNA species in the population. In order to determine whether the initial transcription product might be more extensive than the stable species detected above, RNA has been examined after a brief pulse-label of monkey cells with ^{3}H-uridine, 48 hours after infection

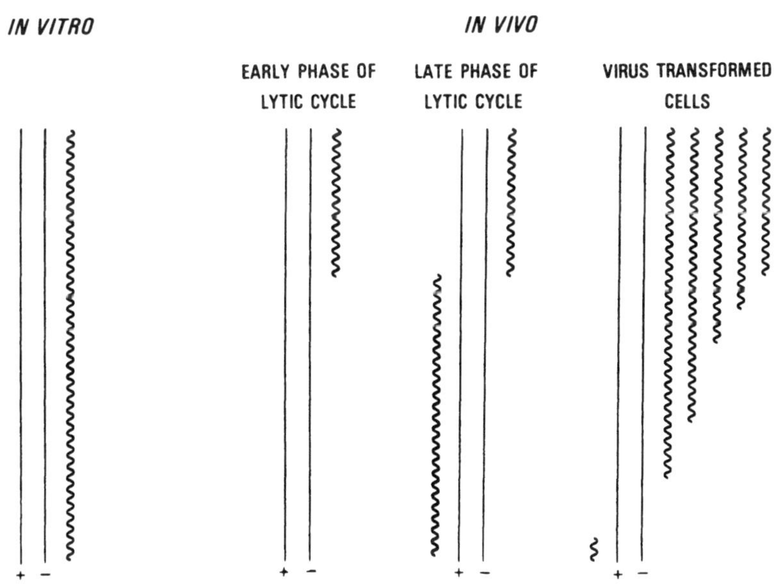

Fig. 10. The pattern of transcription of SV40 DNA *in vitro* and *in vivo*. SV40 DNA molecules (═) are represented as linear structures for convenience. The strand orientation and extent of transcription (∿∿) is based on the abundant, or stable, mRNA species as determined by hybridization experiments described in the text.

(Aloni, 1972, 1973). These studies suggested that extensive regions of the SV40 genome are transcribed symmetrically. If the brief pulse-label is followed by a 1-hour chase or if the 2-minute ^{3}H-uridine pulse is extended to 20 minutes, the labeled RNA exhibits significantly less symmetry. Thus, it appears that a substantial portion of the genome is symmetrically transcribed, and, subsequently, specific RNA sequences are rapidly degraded, resulting in the stable transcripts described above. A similar conclusion has been reached in analogous studies with pulse-labeled polyoma RNA (Aloni and Locker, 1973). The important implication of this work is that the control mechanism for selection of stable message occurs at a post-transcriptional level, i.e., the degradation of transcribed nonfunctional regions of RNA. Furthermore, this post-transcriptional control appears to be operative prior to or during transport of mRNA to the cytoplasm, since only the asymmetric SV40 sequences are detected in cytoplasmic RNA (Khoury, Martin, and Nathans, unpublished).

3.2. Control of Late Transcription

A consistent finding in the studies described above is the dependence of transcription of the late gene sequences of SV40 and polyoma on viral DNA synthesis. In the presence of inhibitors of DNA replication, transcription of the late genes does not occur (Carp *et al.*, 1969; Hudson *et al.*, 1970; Sauer, 1971). Current studies have shown that late transcription depends only on the initiation and not the continuation of viral DNA synthesis (Cowan *et al.*, 1973). Cells are infected at the permissive temperature with a temperature-sensitive mutant of SV40, defective in a function necessary for DNA replication. After the initiation of DNA synthesis, the cultures are shifted to the restrictive temperature, effectively terminating viral DNA replication. Subsequent pulses with ^{3}H-uridine, nevertheless, demonstrate the continued synthesis of late mRNA. However, the control mechanism which links the transcription of late SV40 and polyoma genes to the initiation of DNA replication is still unclear.

Several possibilities for regulation of late transcription are suggested by existing data. These include:

1. The physical state of the viral genome may determine the strand orientation or region of transcription. For example, transcription of late SV40 DNA sequences may occur on replicative intermediates (Girard *et al.*, 1974), thus requiring initiation of viral DNA replication. This hypothesis is based

on kinetic data which suggest that transcription of the late viral genes occurs on templates which build up in infected cells in direct relationship to viral DNA replication.

2. Alternatively, virion-associated histones which are associated with infecting SV40 or polyoma DNA may prevent late transcription (Huang *et al.*, 1972*b*). As these core proteins are removed during DNA replication, late transcription may occur. It is also possible that newly synthesized viral DNA, prior to the addition of core protein, may also act as a late transcriptional template.
3. The specificity of the RNA polymerase may be the controlling factor. Since there appears to be no virion-associated RNA polymerase, the early virus-specific RNA is almost certainly synthesized by a host enzyme. This enzyme may be specific for the minus (early) SV40 or polyoma DNA strand. The early viral gene products may provide or induce a new polymerase capable of late transcription. A model system in which the switch from early to late transcription is effected by a "late" RNA polymerase coded for by the early viral genes is the lytic cycle of bacteriophage T7 in *E. coli* (Chamberlain *et al.*, 1970; Summers and Siegel, 1970).
4. Evidence has been cited for extensive symmetrical transcription of the SV40 genome late in the lytic cycle with subsequent degradation of the RNA sequences which are not translated (Aloni, 1972, 1973). It is conceivable that there is extensive transcription early in the lytic cycle and that the late RNA sequences are efficiently degraded prior to the onset of DNA replication.

While the above hypotheses are highly speculative, each, at least in part, can be tested with techniques presently available. Since the absence of both DNA synthesis and late RNA are hallmarks of SV40- and polyoma-transformed cells, it is hoped that a better understanding of controls which are operative in the lytic cycle will provide an insight into the mechanism of cell transformation.

3.3. Size of the Papovavirus-Specific RNA

Several recent studies have been concerned with the size and potential processing of SV40- and poloyma-specific RNA, synthesized in productively infected cells. Perhaps the most surprising result of these investigations was the detection of intranuclear SV40-specific (Tonegawa *et al.*, 1970; Martin, 1970; Jaenisch, 1972; Rozenblatt and Winocour, 1972) or polyoma-specific (Acheson *et al.*, 1971) RNA molecules which were considerably larger than a unit length of the genome. It seems unlikely that these molecules are simply artifacts due to aggregation since the sizing technique employed in several of these studies was centrifugation in DMSO, a compound which removes secondary structure and aggregation. These large, intranuclear, virus-specific molecules could arise from multiple rounds of transcription of viral DNA. In certain investigations, however, it was shown that nonviral sequences are transcribed in tandem with the SV40 RNA (Jaenisch, 1972; Rozenblatt and Winocour, 1972), suggesting the possibility of transcription from integrated viral genomes.

High-molecular-weight RNA (HMW RNA) obtained late in the lytic cycle has been hybridized to SV40 DNA on filters (Jaenisch, 1972). The presence of RNA tails which remained susceptible to RNase after the hybridization to SV40 DNA filters is consistent with the idea that these tails contain nonviral sequences, presumably host-specific, which are covalently linked to the virus-specific RNA. In a more extensive analysis, it was shown that HMW RNA from a lytic infection which binds specifically to SV40 DNA filters could be eluted and hybridized to host cell DNA filters (Rozenblatt and Winocour, 1972). It is unlikely that the host RNA sequences were transcribed from host DNA incorporated into the virions since the SV40 inoculum was produced by low multiplicity passage, a procedure which selects against host incorporation. Since present evidence points to integration of some viral DNA molecules into host DNA in the lytic cycles of SV40 (Hirai and Defendi, 1972) and polyoma (Babiuk and Hudson, 1972; Ralph and Colter, 1972), the most likely explanation of covalently linked host and viral RNA sequences is that they result from linked transcription of integrated viral DNA and adjacent cellular DNA.

On the other hand, the presence of host-specific sequences in HMW virus-specific RNA in polyoma-infected mouse kidney cells was not observed (Acheson *et al.*, 1971). In this study it was concluded that the HMW virus-specific RNA may represent viral sequences transcribed in tandem or viral RNA linked to unique host RNA sequences which would not be detected by the usual hybridization techniques. Whether these hybrid or nonhybrid HMW RNA molecules have any function in the lytic cycle (as well as whether integration serves any necessary function in a productive infection) is still in question. In other systems, however, most or all of the high-molecular-weight nuclear RNA never gets to the cytoplasm, which suggests processing or at least selection of the virus-specific RNA which is eventually translated.

3.4. Cytoplasmic Viral RNA

RNA species which are found in the cytoplasm (and presumably on the polysomes) of SV40-infected cells (Weinberg *et al.*, 1972*a*) have been investigated using a formamide hybridization procedure to isolate intact virus-specific RNA. Similar studies have been carried out for cytoplasmic and polysomal polyoma RNA (E. Buetti and R. Weil, personal communication). Prior to viral DNA replication one can detect a 19 S virus-specific moiety; a similar species is found in mouse cells abortively infected with SV40 (May *et al.*, 1973). Subsequent to DNA replication, there are both a 19 S and a 16 S peak of SV40- or polyoma-specific RNA. While evidence suggests that the 19 S and 16 S RNAs differ in base compositions as well as in T1 ribonuclease fingerprints (Warnaar and deMol, 1973), there exist as yet no data which exclude the possibility that part of the 16 S peak results from cleavage of a 19 S precursor or that a distinct 19 S species is synthesized late. In fact, there is preliminary evidence for a distinct 19 S late RNA molecule (G. Khoury and B. Carter, unpublished results; Weinberg *et al.*, 1974).

In several studies, it has been demonstrated that a fraction of the SV40 RNA molecules both in the nucleus and in the cytoplasm contain terminal regions of poly A about 150–200 nucleotides in length (Weinberg *et al.*, 1972*b*; Aloni, 1973). A number of the observations mentioned above suggested some form of post-transcriptional processing of the SV40 RNA between synthesis in the nucleus and transport to the cytoplasm. If this processing occurs after the addition of the poly A sequences, it probably occurs from the 5′ end of the molecules, since it has been shown that the poly A is located at the 3′ terminus of mRNA molecules.

3.5. Concentration of Virus-Specific RNA

Extremely small amounts of SV40- and polyoma-specific RNA are synthesized early in the lytic cycle; present estimates suggest that only 0.001–0.01% of the total cell RNA is virus-specific prior to SV40 DNA replication. After the onset of viral DNA replication, this fraction increases to 0.1–1% of the total cell RNA (Benjamin, 1966; Khoury and Martin, 1972; Sambrook *et al.*, 1972), in agreement with data which suggested that there is a 40- to 50-fold increase in the amount of virus-specific RNA late after productive infection (Aloni *et al.*, 1968; Carp *et al.*, 1969). The increase in SV40-specific RNA which occurs after the initiation of viral DNA synthesis includes a commensurate stimulation in the production of early SV40 sequences; this early SV40 RNA appears to account for 20–40% of the total late virus-specific RNA (Martin and Khoury, 1973) yet only 1 to 5% of the cytoplasmic late virus-specific RNA. Since infection by SV40 (unlike adenovirus) stimulates the production of host cell RNA (Oda and Dulbecco, 1968*a*), the virus-specific sequences, even late in the lytic cycle, represent a small fraction of the total cellular RNA. For this reason, studies requiring the isolation and purification of virus-specific RNA have proven difficult. Recently, several methods have been developed for the selection of messenger RNA.

3.6. Selection of Viral mRNA

3.6.1. Hybridization and Elution

As noted above, SV40 mRNA can be purified by hybridization of infected cellular RNA to SV40 DNA on filters, with subsequent elution of the bound RNA (Weinberg *et al.*, 1972*a*). In order to prevent fragmentation of the RNA, annealing is carried out at 37°C in the presence of 50% formamide. Two disadvantages of this technique are a relatively low efficiency of recovery and the introduction of some breaks in the RNA in spite of the precautions taken. The procedure, however, does allow the isolation of a relatively pure virus-specific RNA. This method is currently being used in several laboratories to obtain virus mRNA for translation studies.

3.6.2. Selection of RNA Sequences on the Basis of Poly(A) Tracts

A number of studies have now shown that almost all mammalian cell and animal virus messenger RNAs contains poly(A)-rich regions, about 150–250 nucleotides in length. These tracts are covalently linked to the 3′ ends of the RNA molecules after the message has been synthesized. The poly(A) region contained in SV40 mRNA is not coded for by the viral genome (Weinberg *et al.*, 1972*b*). Several techniques have been developed for the isolation of mRNA based on specific binding of the poly(A)-rich regions. These methods include preferential selection of mRNA binding to oligo dT-cellulose columns (Gilham, 1964), to poly(U) which has been fixed on nitrocellulose filters (Sheldon *et al.*, 1972), or simply to nitrocellulose filters in the presence of high salt (Lee *et al.*, 1971). The separation of mRNA from total cell RNA by any of these methods is relatively simple; the techniques are efficient and provide large quantities of message. Such procedures, however, do not provide for a separation of virus-specific from host-specific messages. Furthermore, transcriptional products which either do not contain poly(A) tracts or have lost them due to breakage or degradation are excluded from these selection procedures.

3.6.3. Immunoprecipitation

The technique of immunoprecipitation relies on the fact that nascent polypeptides, being synthesized on polysomes, can react with specific antibodies [see Palacios and Schinke (1973)]. After precipitation of the antigen–antibody complexes, one can isolate and purify the particular mRNA which codes for the precipitated antigen. This procedure has recently been used to obtain the mRNA coding for a mouse immunoglobulin L chain (Schechter, 1973). The mRNA isolated by this procedure retained its biological activity and was translated in a cell-free system forming recognizable L-chain precursors. While this technique requires the use of relatively pure antibodies, it promises to provide specific mRNA species which should prove invaluable both for sequencing experiments and for *in vitro* translational studies.

3.7. Mapping of Transcriptional Sites on the SV40 Genome

Although previous studies had determined the proportions and strand orientation of SV40 transcription early and late in the lytic cycle, none of the experiments localized these gene sequences on the genome. In order to locate the topographical positions of the early and late gene sequences, the 11 specific fragments of SV40 DNA generated by the restriction endonuclease isolated from *H. influenzae* (R · *Hin*) have been used (Khoury *et al.*, *1973b*). The order of these fragments, which vary in size from 4% to 22.5% of the length of SV40, has been determined (Danna and Nathans, 1971; Danna *et al.*, 1973), as has their relationship to other restriction endonuclease sites (Fig. 7). By reacting unlabeled RNA from SV40-infected monkey cells with the separated strands of the 11 ^{32}P-labeled R · *Hin* fragments, the early and late gene sequences were localized. Stable species of RNA were complementary to the minus (E) strands of the contiguous fragments A, H, I, and B. The late SV40 RNA is transcribed predominantly from the plus (L) strands of fragments A, C, D, E, K, F, J, G, and B which also form a continuous set on the physical map (Fig. 11). The findings of these experiments were in agreement with previous results which suggested that the early gene sequences are localized on one-third of the minus strand, while the late gene sequences occupy the other two-thirds of the plus strand.

In this same study, it was found that some of the late, lytic SV40 RNA reacted partially with the plus strands of fragments H and I (Khoury *et al.*, 1973*b*), which are located on the physical map in the middle of the early-gene region. This result may reflect an incomplete post-transcriptional degradation of RNA complementary to the plus strands of these fragments.

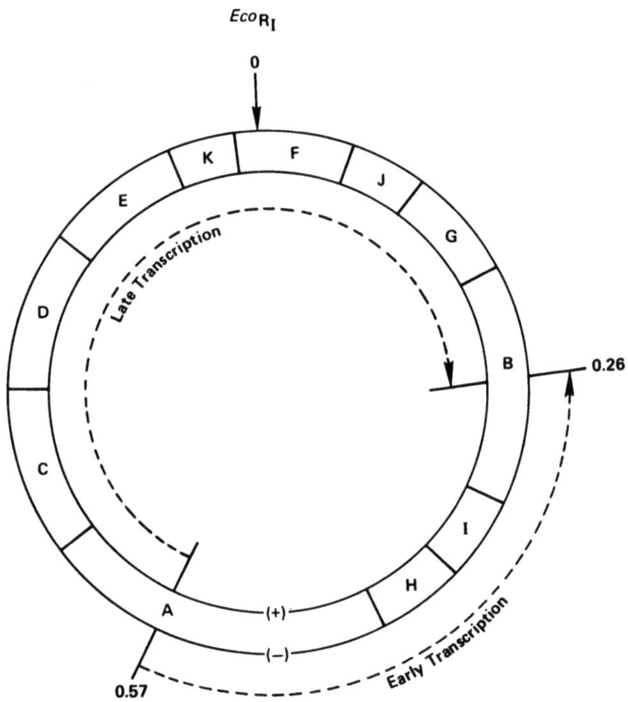

Fig. 11. Diagrammatic representation of SV40 transcription in productively infected cells. Arrows indicate the template strand and direction of transcription early and late in the lytic cycle (Khoury *et al.*, 1973*b*).

In fact, recent results of hybridization experiments with late lytic *cytoplasmic* RNA indicate that the early SV40 template is intact and includes most if not all of *Hin* fragments A and B as well as H and I, nearly half of the SV40 genome. It now appears that previous experiments, in which total cellular RNA was employed, underestimated the size of the early SV40 gene region. Initially, the genome appears to be symmetrically transcribed and subsequently portions of each RNA strand are degraded. Since the plus (or late) strand is transcribed at a much higher frequency during the late stages of the lytic cycle, the "anti-early" sequences present in the nuclear component of total cellular RNA lead to an overestimate of the late template. (Khoury, Martin, and Nathans, in preparation).

3.8. The Direction of SV40 DNA Transcription

It appears to be a general rule that transcription of DNA proceeds in a 5′ to 3′ direction with respect to the synthesis of the messenger RNA molecule, or 3′ to 5′ along the DNA template strand. Since the strands of DNA in a duplex molecule have an antiparallel orientation, and since early and late SV40 messages are transcribed from opposite DNA strands (Lindstrom and Dulbecco, 1972; Khoury *et al.*, 1972; Sambrook *et al.*, 1972), then it seems apparent that the transcription of the early and late SV40 sequences occurs in opposite directions. There are several prerequisites for determining the direction of SV40 transcription:

1. It is necessary to obtain a unique population of linear molecules.
2. One must be able to distinguish the 3′ and 5′ ends of each DNA strand.
3. The strand which codes for the early SV40 RNA must be distinguished from the strand which codes for the late message.

A diagram of the experimental approach used to determine the direction of SV40 transcription is shown in Fig. 12. ^{32}P-SV40 DNA I was first cleaved with the R · *Eco* R_I endonuclease to obtain unique, full-length linear molecules. Since the R_I enzyme cleaves within fragment F (Danna *et al.*, 1973), these molecules have the map order F_1-J-G-B-I-H-A-C-D-E-K-F_2 (see Fig. 7; letters refer to the fragments produced by the R · *Hin* endonuclease as mentioned above). In the next step, the linear molecules were digested with *E. coli* exonuclease III, which removes the 3′ halves of each strand, leaving 5′ "half-

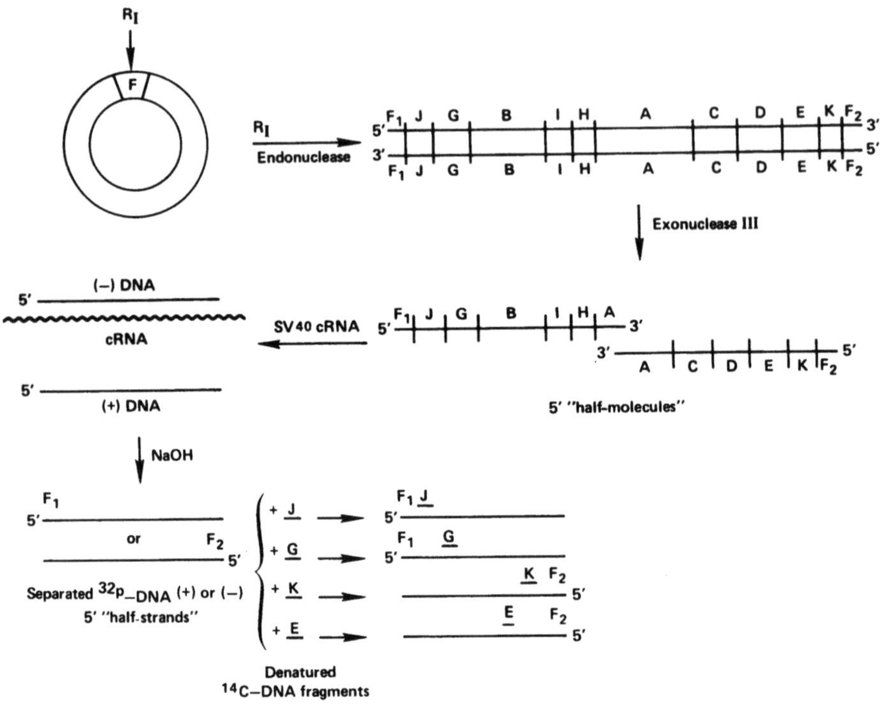

Fig. 12. Scheme for defining the direction of transciption in SV40-infected cells. This method is based on a determination of the 5′→3′ orientation of SV40 DNA strands. See the text for a description of each step. F_1 and F_2 are the parts of fragment F resulting from cleavage by the R_I restriction enzyme, cRNA is SV40 complementary RNA (Khoury *et al.*, 1973*b*).

molecules." The individual minus and plus 5′ "half-strands" were next isolated by annealing with SV40 cRNA. It was then determined which 5′ "half-strand" (plus or minus) contained fragment-J and fragment-G sequences and which contained fragment-K and fragment-E sequences by annealing each "half-strand" with denatured ^{14}C-labeled fragment J, G, K, or E. (The sequences corresponding to these four fragments are nearest to sequence F, which is present at each end of the R_I linear molecules, Fig. 12.) The results indicated that fragments E and K hybridized preferentially with the *plus* 5′ "half-strand," while fragments G and J hybridized preferentially with the *minus* 5′ "half-strand" of SV40 DNA. Since the minus strand is the template for "early" SV40 RNA, one can conclude that the orientation of the "early" DNA template strand is 5′-F-J-G-B-I-H-A-····-3′, i.e., 3′ to 5′ counterclockwise on the cleavage map shown in Fig. 7, and the orientation of the plus strand ("late" template strand) is 5′-F-K-E-D-

C-A- · · · · -3′, i.e., 3′ to 5′ clockwise on the cleavage map. Therefore, transcription of early genes proceeds from A to B *counterclockwise* on the minus DNA strand and transcription of late genes proceeds from A to B in a *clockwise* direction on the plus DNA strand (see Fig. 11).

3.9. *In Vitro* Studies of Transcription

While much is known about the quantitative aspects of transcription of the SV40 and polyoma genomes, the factors which control transcription are still poorly understood. This is in large part related to the difficulty of studying biochemical events which occur against the complex background of cell-specific functions. Although it has been shown that transcription of SV40 with the *E. coli* RNA polymerase differs from transcription of the virus *in vivo*, it is possible that the study of viral transcription in cell-free systems will permit investigators to control many of the factors which at present make the study of *in vivo* systems so complex.

The number and location of promoter sites for the *in vitro* transcription of SV40 DNA has been the subject of several recent investigations. It is clear that the results of such studies depend on the form of the DNA template, the enzyme, and the conditions used for *in vitro* transcription.

Both the strand orientation and the efficiency of transcription with the *E. coli* RNA polymerase depend on the form of SV40 DNA employed in the reaction mixture. When supercoiled SV40 DNA (DNA I) is used as a template, the RNA product is highly asymmetric, heterogeneous in size, and representative of the entire minus (early) DNA strand (Westphal, 1970, 1971; Fried and Sokol, 1972). When the cyclic-coil or nicked form of SV40 DNA (DNA II) is transcribed, a considerable fraction of the RNA is symmetric, as demonstrated by its ability to self-anneal (Westphal, 1970).

In order to evaluate the relative *in vitro* transcriptional rates of SV40 DNA components I and II, *E. coli* RNA polymerase was allowed to associate with each template and then incubated with a substrate solution containing rifampicin (the drug was added to prevent reinitiation). The kinetics of synthesis of RNA, as determined by the incorporation of ^{3}H-ribonucleotides under conditions of enzyme excess, indicated that the efficiency of transcription of the supercoiled SV40 DNA was significantly greater than that of the component-II DNA template (Westphal, 1971).

Estimates of the rate of *in vitro* RNA synthesis of SV40 component I range from 37 nucleotides/sec (Westphal, 1971) to 5 nucleotides/sec (Fried and Sokol, 1972), perhaps dependent on the reaction conditions. Most studies agree, however, that the RNA product is asymmetric and contains some molecules greater in length than unit SV40 (Westphal, 1970; Fried and Sokol, 1972; Delius *et al.*, 1973). This latter finding suggests that the transcriptional complex can, at times, pass its initiation site, which in turn may indicate the absence of, or lack of recognition of a termination site. There is, however, little evidence at present to suggest multiple rounds of transcription *in vivo*.

The number of polymerase-binding sites on a viral DNA molecule is inversely proportional to the salt concentration (Crawford *et al.*, 1965; Pettijohn and Kamiya, 1967). *E. coli* DNA-dependent RNA polymerase binds to several sites on the SV40 molecule (Herzberg and Winocour, 1970; Delius *et al.*, 1973). Based on an electron microscopic study, it was concluded that these sites are not clustered (Delius *et al.*, 1973). However, when similar experiments were performed using linear SV40 DNA produced with the R · *Eco* R_I endonuclease, there appeared to be one preferred promoter, in addition to several weaker initiation sites (Westphal *et al.*, 1973). This preferred promoter was localized at a position 0.16 unit from the R · *Eco* R_I cleavage site, and transcription proceeded toward the short arm of the linear molecule. When transcription of SV40 DNA by *E. coli* RNA polymerase was investigated at reduced temperatures with RNA sequencing studies (Zain *et al.*, 1973), a strong *in vitro* promoter site was found 0.16 unit clockwise from the SV40 R · *Eco* R_I site. Since transcription on the minus DNA strand (the strand which is transcribed by the *E. coli* polymerase) occurs in a counterclockwise direction (see Fig. 11), *in vitro* transcription almost certainly progresses from the strong promoter toward the R · *Eco* R_I site. These results are in good agreement with those obtained by electron microscopy as described above (Westphal *et al.*, 1973). The strong promoter for the *E. coli* RNA polymerase does not coincide with that predicted for early RNA synthesized *in vivo* (Khoury *et al.*, 1973*b*). As has been pointed out, however, if lytic messenger RNA represents the product of post-transcriptional degradation, the transcriptional promoters and terminators cannot be definitely localized to the ends of the final product.

Much less is known about the transcription of polyoma DNA with *E. coli* RNA polymerase. It appears, however, that a considerable fraction of the RNA product is symmetricaL unless the template has been prepared from a high-multiplicity viral stock (Lindstrom, personal communication). By using polyoma-specific RNA prepared on such a template with preincubation of the RNA to remove symmetrical regions, it is possible to obtain an RNA which will permit strand separation and transcriptional mapping.

3.10. The Applications of *In Vitro* Virus-Specific RNA

In addition to providing insight into the mechanism of transcription, the *in vitro* product obtained from transcription of SV40 DNA with the *E. coli* DNA-dependent RNA polymerase has proven to be a valuable reagent for a number of other studies.

1. The radiolabeled *in vitro* RNA has been used as a molecular probe in studies designed to determine the number of copies of the viral genome in transformed cells (Westphal and Dulbecco, 1968; Levine *et al.*, 1970) and to establish the integration of the viral DNA within the host DNA of transformed cell lines (Sambrook *et al.*, 1968). It has also been used to determine the transcriptional pattern of the lytic cycles of SV40 (Lindstrom and Dulbecco, 1972), as well as the pattern of transcription from the chromatin of SV40-transformed cells (Astrin, 1973).
2. Since the RNA obtained from transcription of DNA I with the *E. coli* RNA polymerase is essentially asymmetric, it has been a valuable tool for separating the strands of SV40 DNA (Westphal, 1970; Khoury and Martin, 1972; Sambrook *et al.*, 1972). These separated DNA strands have subsequently been used as probes to determine the transcriptional pattern in lytically infected cells (Khoury *et al.*, 1972; Sambrook *et al.*, 1972), transformed cells (Sambrook *et al.*, 1972; Khoury *et al.*, 1973*a*; Ozanne *et al.*, 1973), or from the chromatin of transformed cells (Shih *et al.*, 1973). The separated SV40 DNA strands have also been used to map specific viral deletion mutants (Khoury *et al.*, 1974).
3. The RNA transcribed from unique segments of SV40 DNA have been employed in nucleotide-sequencing studies (Dhar *et al.*, 1973). A sequence of more than 180 nucleotides located within the region of *Hin* fragment G (see Fig. 11) has already been determined.

3.11. Transcription of SV40 DNA by Mammalian Polymerases

The *in vitro* analysis of transcription of viral DNA is directed toward an understanding of the mechanism and control factors which operate *in vivo*. Since *in vivo* transcription of SV40 and polyoma DNA most likely requires a host cell RNA polymerase, it would seem preferable to use mammalian cell polymerases in the *in vitro* studies. Until recently, however, most of our information has been obtained in experiments with the *E. coli* RNA polymerase, primarily because of the difficulty associated with isolating and purifying mammalian enzymes.

It has now been shown that there are two principle DNA-dependent RNA polymerases found in mammalian cells which can be separated by ion-exchange chromatography (Roeder and Rutter, 1969). RNA polymerase I is found in nucleoli and has been associated with the synthesis of ribosomal RNA (Reeder and Roeder, 1972). RNA polymerase II is found in the cell nucleoplasm and is distinguished from polymerase I by its sensitivity to α-amanitin, a toxin extracted from the mushroom *Amanita phalloides* Stirpe and Fuime, 1967; Roeder and Rutter, 1970).

Using isolated nuclei from SV40-infected cells (since cellular membranes are not uniformly permeable to α-amanitin), transcription of the SV40 sequences were shown to be sensitive to α-amanitin. This evidence suggests that RNA polymerase II is responsible for transcription of viral sequences *in vivo* Jackson and Sugden, 1972).

While purified DNA-dependent RNA polymerase II from HeLa and KB cells actively transcribes SV40 DNA, the enzymes appear to prefer a single-stranded to a double-stranded DNA template (Sugden and Keller, 1973). The significance of that finding and its relationship to the specificity and activity of mammalian cell RNA polymerase II *in vivo* remain to be determined.

4. THE PROTEINS OF SV40 AND POLYOMA

A number of virus-specific antigens which are synthesized early after infection have been superficially characterized; these are probably proteins and may be products of viral and/or cellular genes. The most fully characterized of the papovavirus proteins are the structural proteins which are synthesized late in the lytic cycle. In addition, during the lytic cycle there is a general stimulation of synthetic enzymes which are probably coded for by the host cell.

4.1. Early Antigens

The early antigens of SV40 and polyoma are synthesized prior to viral DNA replication and also appear in cells transformed by these viruses. These antigens have been detected primarily by immunological methods, and past attempts to purify the early antigens have met with only partial success (Kit *et al.*, 1967; Lazarus *et al.*, 1967; Potter *et al.*, 1969; Del Villano and Defendi, 1973). Nevertheless, several lines of experimentation lead to the conclusion that the early papovavirus antigens are virus-specific and, perhaps, virus-coded:

1. When cells from different species are infected (Rapp *et al.*, 1964*a*; Hoggan *et al.*, 1965) or transformed by SV40 or polyoma (Black and Rowe, 1963; Black *et al.*, 1963; Sabin *et al.*, 1964; Rapp *et al.*, 1964*c*; Habel, 1965; Habel *et al.*, 1965), antigens are synthesized which appear to be immunologically identical.
2. UV irradiation of SV40 (Carp and Gilden, 1965) and polyoma (Benjamin, 1965; Basilico and DiMayorca, 1965; Latarjet *et al.*, 1967) sequentially inhibits the ability of the virus to produce infectious progeny, late antigens, and early antigens. Temperature-sensitive (*ts*) mutants of polyoma virus which are defective at an early stage (prior to DNA replication) either fail to make T antigen, or make reduced amounts of it (Oxman *et al.*, 1972). Similarly, early *ts* mutants of SV40 (Tegtmeyer, 1972) also appear to be incapable of making the normal T antigen, although a somewhat altered product with similar antigenicity is detected (Tegtmeyer, personal communication; M. Osborn and K. Weber, personal communication).
3. In the SV40 lytic cycle, the synthesis of early SV40 RNA (Oxman and Levin, 1971) and T antigen (Oxman *et al.*, 1967) are inhibited when cultures are treated with interferon. Interferon is considered to be a specific inhibitor of viral protein synthesis, but it does not block cellular protein synthesis. Neither viral RNA nor T antigen is sensitive to inhibitors of DNA synthesis (Gilden *et al.*, 1965; Melnick and Rapp, 1965; Butel and Rapp, 1965), which indicates that they are independent of viral DNA replication. Both, however, are inhibited in the presence of actinomycin D, an agent which blocks viral transcription.

While these experiments appear to make a strong case for the virus-specific nature of early papovavirus antigens, they do not prove that the antigens are virus-coded. Such proof will probably require the *in vitro* synthesis of defined viral proteins. Ideally, such experiments would be performed in a coupled system with transcription of the viral

DNA and translation of the newly synthesized message. Since transcription of the SV40 and polyoma genomes by *E. coli* RNA polymerase results in a product which differs considerably from stable lytic RNA (see Sect. 3), future studies may require better characterization of transcriptional intermediates and mammalian RNA polymerases. Until then, preliminary experiments will probably rely on the translation of purified *in vivo* viral mRNA. These experiments are now in progress in a number of laboratories and should provide important answers concerning the nature of viral-coded products.

4.1.1. T Antigen

The best characterized of the early antigens are the SV40 and polyoma T ("tumor") antigens. They were first detected by a complement-fixation (CF) method (Black *et al.*, 1963; Takemoto and Habel, 1965) and are localized in the cell nucleus using immunofluorescence (IF) techniques (Pope and Rowe, 1964; Rapp *et al.*, 1964*a*; Gilden *et al.*, 1965). Antibodies for the detection of T antigen are present in sera from animals bearing SV40 or polyoma tumors. The T antigen can be detected as early as 10–18 hours after infection (Rapp *et al.*, 1964*a*; Hoggan *et al.*, 1965) and persists throughout the lytic cycle. The time of appearance of SV40 T antigen and its rate of accumulation have been shown to depend in part on the line of cells infected and the temperature of incubation (Khoury, 1970; Kitahara and Melnick, 1965).

T antigen is heat labile and is sensitive to trypsin, but not to DNase (Gilden *et al.*, 1965). SV40 T antigen does not cross-react with the polyoma T antigen; however, it does share immunological properties with antigens induced by some of the recently discovered papovaviruses isolated from human sources (Takemoto and Mullarkey, 1973).

The synthesis of T antigen is not affected by inhibitors of DNA replication (Gilden *et al.*, 1965; Melnick and Rapp, 1965; Butel and Rapp, 1965), but its synthesis is inhibited by actinomycin D and interferon (Oxman *et al.*, 1967) and by cycloheximide (Gilden and Carp, 1966), thus suggesting that *de novo* transcription and subsequent translation of this message are required for T-antigen production. The presence of T antigen has served as one of the important criteria for determining whether cells are infected or transformed by SV40 or polyoma. Yet, it is still not known whether this antigen has a function in the lytic cycle. Some preliminary data suggest that an altered T antigen may be made at the restrictive temperature, by an early SV40 *ts* mutant (M. Osborn and K. Weber, personal communication). Since this mutant is blocked in SV40 DNA synthesis (Tegtmeyer, 1972), one interpretation would be that T antigen is in some way required for viral DNA synthesis.

One of the major problems in determining the function of T antigen has been an inability to adequately purify this antigen. A number of investigators however have recently made considerable progress toward that goal (P. Tegtmeyer; D. Livingston; M. Osborn and K. Weber; and R. Carrol, personal communications) and it appears the T antigen may be a DNA-binding protein. A considerable amount of data concerning the structure and function of T antigen should be forthcoming in the next few years, and we may hope that this will result in a fuller understanding of the mechanisms responsible for lytic infection and transformation.

4.1.2. U Antigen

The U antigen has been detected by CF and IF methods in SV40-infected and -transformed cells (Lewis and Rowe, 1971). Like T antigen, the U antigen reacts with serum from SV40-tumor-bearing hamsters. Unlike T antigen, however, U antigen is found at the nuclear membrane and is heat stable (Lewis and Rowe, 1971). While most antisera for T antigen contain anti-U antibodies, there are some batches of anti-T sera which do not. Furthermore, there is a hybrid deletion mutant of adenovirus and SV40 ($Ad2^{+}ND_1$; see Sect. 5.4) which induces U antigen but not T-antigen, thus suggesting that the two antigens differ, at least in part. Nothing is known about the structure or function of this antigen.

4.1.3. Tumor-Specific Transplantation Antigen (TSTA)

The polyoma (Habel, 1961; Sjogren *et al.*, 1961) and SV40 (Habel and Eddy, 1963; Khera *et al.*, 1963; Koch and Sabin, 1963; Defendi, 1963) TSTAs have been studied primarily in transformed cells and have been demonstrated by *in vivo* immunological methods such as transplantation rejection. In these tests, animals immunized with cells containing the TSTA become resistant to a subsequent challenge with transplantable SV40- or polyoma-transformed cells or tumor cells. TSTA appears to be localized at the cell surface. Newborn hamsters inoculated with membranes from SV40 tumor cells develop a tolerance to immunization as adults against SV40 tumor transplantation (Tevethia and Rapp, 1966). Furthermore, adult animals can be protected against transplantable tumors by immunization with membranes from SV40 tumor cells (Coggin *et al.*, 1969). In more recent studies, TSTA has been shown to be synthesized during the lytic cycle. The ability to block its production with inhibitors of DNA-dependent-RNA synthesis and inhibitors of protein synthesis suggests that TSTA appears as a virus-induced protein during productive infection (Girardi and Defendi, 1970). Whether it is virus-coded is not known. The antigen has not been purified and its physical and chemical structures are unknown.

4.2. Induction of Host Cell Proteins

The synthesis of a number of enzymes, as well as host cell DNA and RNA synthesis, is stimulated after infection of permissive cells by SV40 and polyoma. Most of the enzymes have a function in DNA-synthetic pathways. The relationship between the early viral antigens and many of the induced enzymes is not clear, but the latter often resemble host enzymes in their properties. It is obvious that a genome of the size of SV40 or polyoma could not code for all of these proteins. Therefore, most, if not all, are presumed to be host enzymes, stimulated by viral infection (Basilico *et al.*, 1969). A more complete discussion of the induction of cellular nucleic acids or enzymes can be found in a number of compreehensive reviews (Kit *et al.*, 1966; Kit, 1967, 1968).

4.3. Virion-Associated Endonucleases

An endonuclease capable of introducing a single-strand endonucleolytic cleavage in the host genome has been found associated with purified polyoma (Cuzin *et al.*, 1971) and SV40 virions (Kaplan *et al.*, 1972; Kidwell *et al.*, 1972). Since a productive infection can be established with viral DNA alone (Crawford *et al.*, 1964), these enzymes, in a virus-associated form, are not essential to the lytic cycle. If the endonuclease could be shown to be virus-specific or site-specific, however, the implications with respect to an integration function either in lytic or transforming cell interactions would be extremely important.

There is considerable uncertainty at present, whether the SV40-associated enzyme is virus-specific; it is not unlikely that association of the endonuclease with the virion can occur as a result of the extraction procedure.

The endonuclease associated with SV40 virions is still present when virus preparations from both early and late temperature-sensitive mutants are grown at permissive temperatures and then incubated at the restrictive temperature for 1–2 hours. This suggests that the *ts* functions are not required for the presence or activity of the enzyme (Kidwell *et al.*, 1972). Furthermore, the data at present do not suggest specificity with respect to the site of cleavage in the SV40 DNA, and prolonged incubation of SV40 DNA in the presence of the endonuclease leads to extensive digestion.

On the other hand, the endonuclease from polyoma virus is not associated with the early *ts* mutant, *Ts*A (Cuzin *et al.*, 1970). Furthermore, this enzyme appears to introduce a specific break into the polyoma genome (F. Cuzin, A. Parodi, D. Blangy, O. Croissant, and P. Rouget, unpublished results). Determination of the source and activity of these virion-associated endonucleases will, of course, require further investigation.

4.4. Structural Proteins of SV40 and Polyoma

Late in the lytic cycle of SV40 and polyoma, subsequent to viral DNA replication, a new intranuclear antigen (V antigen) can be detected by immunofluorescent staining (Mayor *et al.*, 1962). This antigen reacts specifically with hyperimmune antisera against intact virions and almost certainly represents the viral coat protein(s). Since it is likely that the capsid protein is synthesized in the cytoplasm, its presence in the nucleus is probably the result of a rapid transport of synthesized coat protein to the nucleus for viral assembly. It is possible, however, that V antiserum is specific for assembled coat protein(s) which may be present only in cell nuclei. Whether the anti-V antisera are directed against one or more proteins is not known.

Since the capsid proteins of polyoma and SV40 can be obtained in large quantities and are relatively stable, they have been well characterized (Fine *et al.*, 1968; Girard *et al.*, 1970; Barban and Goor, 1971; Estes *et al.*, 1971; Hirt and Gesteland, 1971; Roblin *et al.*, 1971; Huang *et al.*, 1972*a*; Frearson and Crawford, 1972; Friedmann and David, 1972) (see Table 1). In the case of SV40, the major capsid polypeptide (VP1) has been shown by SDS-polyacrylamide gel electrophoresis to have a molecular weight between 43,000 and 45,000 and to constitute more than 75% of the total protein. Two other viral structural proteins (VP2, VP3) with molecular weight of approximately 30,000–38,000 and 20,000–23,000 daltons, respectively, have also been described which are present in smaller amounts than the major polypeptide; VP2 appears to be a capsid protein, while the location of VP3 within the virion is uncertain.

TABLE 1

The Proteins of SV40 and Polyoma

Protein	Mol. wt. $\times 10^{-3}$	Predicted origin
VP1	43–48[a]	Viral capsid
VP2	30–38	Viral capsid
VP3	20–23	?
VP4	14–16	Host cell histone (F3)[b]
VP5	12–14	Host cell histone (F2b)
VP6	12–13	Host cell histone (F2a2)
VP7	10–12	Host cell histone (F2a1)

[a] In addition, some studies of polyoma proteins have demonstrated the presence of a larger polypeptide (molecular weight = 86,000) which is thought to be a dimer of VP1 [see, for example, Friedmann (1974)].

[b] The correspondence between the small viral polypeptides and particular host cell histones is based on studies by Lake *et al.*, (1973), B. Hirt (personal communication), and D. Pett and M. K. Estes (personal communication).

The major capsid proteins of polyoma are similar in size to those of SV40 (VP1, 45,000–50,000; VP2, 30,000–35,000; and VP3, 20,000–25,000). In addition, there is a large polypeptide (P1) of approximately 80,000 daltons (Friedmann, 1974), which is thought to be a dimer of VP1. Peptide mapping of the tryptic digests of polyoma structural proteins suggests that some may share common polypeptide sequences and probably have arisen by cleavage of a common precursor (Friedmann, 1974). Alternatively, the proteins may have been translated from overlapping segments of a common mRNA molecule. There appears to be no similarity in the peptide maps of the respective tryptic digests of polyoma and SV40 proteins, suggesting that the coat proteins of the two viruses are significantly different.

SDS polyacrylamide gels have also been used to compare the proteins synthesized in SV40-infected monkey cells with those present in mock-infected cells. Among the new proteins detected in virus-infected cells were two corresponding in mobility to the major and minor capsid proteins (Fischer and Sauer, 1972; Ozer, 1972; Anderson and Gesteland, 1972; Walter *et al.*, 1972). A tryptic digest of the larger protein resulted in a peptide map quite similar to that obtained from the major SV40 capsid protein (Anderson and Gesteland, 1972). The capsid proteins are virus-specific, and they are probably coded for by the late region of the viral genomes.

While viral DNA is itself infections (Crawford *et al.*, 1964), the capsid proteins appear to provide a valuable if not essential function in the reproduction of these viruses. Infection of cells with intact virions is clearly much more efficient than infection with DNA. This may be related to an increased efficiency of absorption and/or penetration in the presence of the coat protein. It is also possible that the capsid protein protects the viral DNA from cellular nucleases.

The amino acid composition of the capsid proteins has been determined for both SV40 (Schlumberger *et al.*, 1968; Greenaway and LeVine, 1973) and polyoma virus (Murakami *et al.*, 1968). Using an isoelectric focusing technique, differences are seen in the capsid proteins of a large-plaque SV40 strain as compared to small- or minute-plaque virus (Barbon, 1963). Various plaque mutants also differ in their host cell restriction, temperature sensitivity, oncogenicity, and antigenicity (Takemoto *et al.*, 1966; Takemoto and Martin, 1970). Which of these properties, if any, is related to the differences in structural polypeptides remains to be determined. Differences in the polypeptides of polyoma plaque mutants have also been reported (Murakami *et al.*, 1968; Thorne *et al.*, 1968). While two large-plaque strains of polyoma have similar amino acid compositions, both appear to differ from a small-plaque polyoma strain (Murakami *et al.*, 1968).

Recent studies have shown that all of the SV40 structural proteins are phosphoproteins (Tan and Sokol, 1972) and that the phosphate groups are linked only to serine residues (Tan and Sokol, 1973). The significance of this finding with relationship to the structure or function of the virion polypeptides is still unknown.

4.5. Internal Proteins

A group of low-molecular-weight proteins (10,000–16,000 daltons; Table 1) have been found associated with the nucleic acid cores of both SV40 (Girard *et al.*, 1970; Barban and Goor, 1971; Estes *et al.*, 1971; Hirt and Gesteland, 1971; Huang *et al.*, 1972*a*; Frearson and Crawford, 1972) and polyoma (Roblin *et al.*, 1971; Frearson and Crawford, 1972; Hirt, unpublished results) virions. These polypeptides remain associated with the viral nucleic acid after mild alkaline treatment of intact virions (Estes *et al.*, 1971), and they are not found in empty viral capsids (Frearson and Crawford, 1972). Using SDS-polyacrylamide gel electrophoresis, three low-molecular-weight basic proteins are observed. But with Tris-acetate-SDS-polyacrylamide gel electrophoresis it appears that at least four proteins are associated with SV40 cores (Lake *et al.*, 1973; Fig. 13). A considerable amount of data has been accumulated which strongly suggest that these basic polypeptides are derived from the host cell histones:

1. When permissive cells are labeled with ^{3}H-lysine prior to polyoma infection, a greater amount of label appears in the core proteins of progeny virions than in the viral capsid proteins (Frearson and Crawford, 1972).
2. The mobility of the SV40 basic polypeptides on Tris-acetate-SDS-polyacrylamide gels corresponds directly to that of the evolutionarily conserved histones F3, F2b, F2a2, and F2a1 (Lake *et al.*, 1973; see Fig. 13). Similar results for SV40 proteins VP4–6 have been found in an independent study using similar methods (Pett and Estes, in preparation). Furthermore, the peptide maps of the core proteins of both polyoma and SV40 correspond to those of the basic proteins of the cell lines which support their permissive cycles, mouse and monkey kidney cultures (Hirt, unpublished results).
3. Histones are rich in two basic amino acids, arginine and lysine, but are deficient in tryptophan. When polyoma-infected mouse cells were labeled with ^{14}C-arginine and ^{3}H-tryptophan, both isotopes were found in the capsid protein of progeny virions, but only the ^{14}C-arginine was incorporated into a core polypeptides (Roblin *et al.*, 1971).

Recently, it has been suggested that although the small internal proteins of polyoma and SV40 may be host-derived, it is unlikely that they are histones, on the basis of their higher tryptophan content (Greenaway and LeVine, 1973). This report is in disagreement with the findings previously mentioned. Considering all of the data presently available, it seems most likely that the core polypeptides of SV40 and polyoma are derived from the histones of the host cell.

While the relative proportions of the core proteins are the same in both whole virions and in nuclear cores, there is clearly a predominance of VP4 (histone F3) in SV40 virions as compared to the concentration in host Vero cells (Lake *et al.*, 1973). This finding suggests a nonrandom incorporation of core proteins into virions and may be related to the histone's function. What role the core proteins perform in the virion, however, is not known. On the basis of *in vitro* transcription studies with SV40 (Huang *et al.*, 1972*b*), it has been suggested that the histonelike proteins may have a regulatory function in

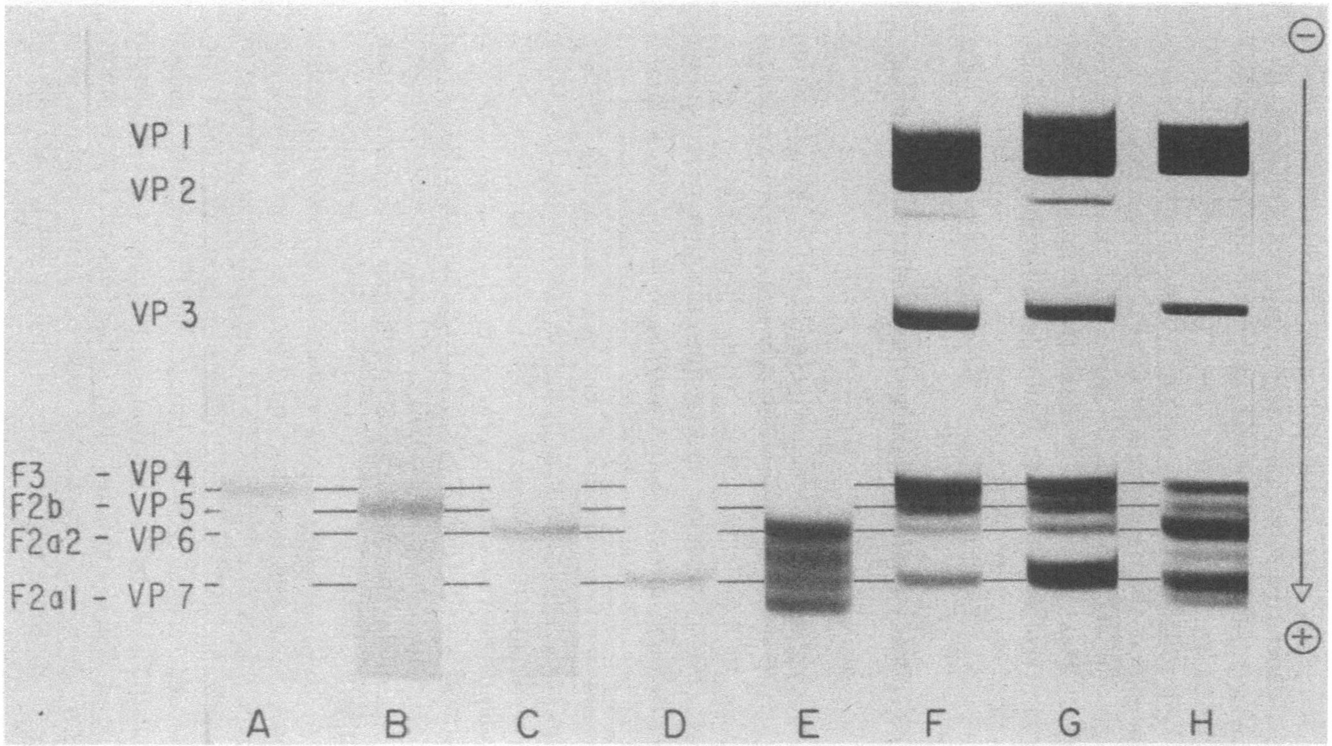

Fig. 13. A comparison of various calf thymus and Vero cell histone fractions with SV40 structural proteins by 13% Tris-acetate-SDS-polyacrylamide gel eletrophoresis. (A) calf thymus F3; (B) Vero cell F2b, (C) calf thymus F2a2, (D) calf thymus F2a1, (E) Vero cell histone fraction F2a contaminated with cleavage products of Vero F3, (F) SV40 complete, (G) SV40 plus calf thymus F2a1, (H) SV40 plus Vero cell fraction F2a as in (E) (Lake *et al.*, 1973).

the transcription of the genome. The kinetics and extent of transcription of an SV40 DNA template were compared with that of a deoxynucleoprotein template; The latter contained SV40 core proteins. Using the DNA-dependent RNA polymerase from either *E. coli* or a mammalian source, the rate of transcription of the SV40 deoxynucleoprotein complex (DNP I) was shown to be less than that of SV40 DNA I. Furthermore, competition-hybridization experiments showed that the extent of transcription of DNP I was less than that of SV40 DNA I. One interpretation of these data is that the core polypeptides exert a specific control over the transcription of the SV40 genome. It is conceivable that these histones also limit the gene sequences transcribed early after lytic infection.

4.6. The SV40 Helper Function for Adenovirus Replication in Monkey Cells*

While monkey kidney cells are semipermissive for the replication of adenoviruses, the ability of these viruses to replicate in such cells is greatly enhanced by coinfection with SV40 (Rabson *et al.*, 1964; O'Conor *et al.*, 1965). Whether or not this function relies on the presence of virus-coded protein is not known. It is known, however, that synthesis of new protein(s) is required since pretreatment of cells with cycloheximide prevents the enhancement of this ability to replicate (Friedman *et al.*, 1970). Coinfection with the nondefective (ND) adenovirus 2–SV40 hybrid virus [Ad $2^{+}ND_{1}$] can provide the helper function for adenovirus replication (Lewis *et al.*, in preparation). This hybrid virus contains only enough SV40 DNA (approximately 18% of the SV40 genome) to code for the SV40 U antigen (Lewis and Rowe, 1971); thus, one might speculate that if the helper function is virus-coded it might be provided by (or depend on) the SV40 U antigen.

A number of temperature-sensitive mutants of SV40 (including mutants in both early and late functions) were able to aid the replication of adenoviruses in monkey cells at the restrictive temperature (Jerkofsky and Rapp, 1973). This evidence suggests that the helper function is induced very early in the lytic cycle. In the same study it was shown that the SV40 helper function was ineffective in a particular continuous line of BSC-1 monkey kidney cells. This cell line is unique in that it fails to shown an induction of host cell DNA synthesis after infection with SV40. Since the stimulation of host cell DNA synthesis

* See also Chapter 3 in this volume.

is an event which occurs very early after SV40 infection of most monkey kidney cell lines, and does not occur in these BSC-1 cells, it is possible that the helper function for adenovirus replication is directly linked to the factor responsible for the stimulation of host cell DNA synthesis. Such a conclusion is supported by the demonstration that a clone of BSC-1 cells in which SV40 infection does stimulate host cell DNA synthesis would permit complementation of adenovirus replication by SV40 (Jerkofsky and Rapp, 1973).

The nature of the helper function is not understood, but it has been shown that in the absence of coinfection with SV40, adenovirus 2 induces the synthesis of both early and late adenovirus RNA (Fox and Baum, 1972). Furthermore, this RNA is polyadenylated and transported to the cytoplasm. While some investigators have detected the presence of adenovirus proteins after abortive infection of monkey cells by adenoviruses alone (Friedman *et al.*, 1970; Henry *et al.*, 1971; Baum *et al.*, 1972), they are clearly present in decreased quantities unless cells are coinfected with SV40. The block in adenovirus replication and the site of the SV40-related helper function may therefore reside at the level of the interaction of adenovirus mRNA with polyribosomes (Hashimoto *et al.*, 1973), in the translation of the message, or in the processing of the polypeptides. In any case, the block seems to occur at a post-transcriptional level. This conclusion is further supported by the finding of equivalent amounts of polysome-associated adenovirus-specific RNA in enhanced and unenhanced infections. Furthermore both populations of mRNA are translated *in vitro* with equal efficiency (L. Eron, H. Westphal, and G. Khoury, in preparation).

5. OTHER PROPERTIES OF PAPOVAVIRUSES

5.1. Induction of Cellular Processes in Infected Cells

As previously described, the lytic infection of either polyoma or SV40 is accompanied by the stimulation of a number of cellular processes; there is the induction of synthesis of cellular DNA (Dulbecco *et al.*, 1965; Hatanaka and Dulbecco, 1966; Ritzi and Levine, 1969) and mitochondrial DNA (Levine, 1971), certain enzymes (Kit, 1967), cellular RNA (Oda and Dulbecco, 1968*a*), and of histones (Shimono and Kaplan, 1969; Hancock and Weil, 1969; Winocour and Robbins, 1970; Rovera *et al.*, 1972). These phenomena are related to the exposure of cells to virus, and in general, higher levels of induction are seen after infection at higher input multiplicities (Basilico *et al.*, 1966). While these effects are the consequence of exposure to the virus, it is difficult to assess their importance in relation to virus replication. There is some implication, when you discuss the induction of cellular processes which results from virus infection, that viral gene products are acting in some immediate way as regulators of these processes. For example, it has been proposed that the induction of cellular DNA synthesis by SV40 is an *in vitro* parallel to the process of malignant cell transformation that occurs *in vivo*. However, the very broad stimulation of cellular processes results from the exposure of cells to a virus which codes for only six to ten proteins.

The induction of cellular DNA synthesis is seen following infection with either SV40 or polyoma virus in almost every cell line that has been examined. These cell cultures in which induction of cellular DNA is observed have been held in a contact-inhibited state for several days prior to infection. The cells are presumed to be arrested in the G1 phase, a phase which is normally characterized as a brief period (of approximately 4–12 hours duration) following mitosis and prior to the initiation of DNA synthesis. There has not yet been an adequate characterization of the changes that occur in a cell population which has been kept in a nondividing state for several days prior to infection. It is most likely that these cells may differ markedly from G1 cells even though both share certain common properties (e.g., a diploid DNA content and the absence of DNA synthesis). In a growing cell population, virus infection does not change the orderly progression of cells through the cycle, i.e., cells which are in different parts of the cell cycle when they are infected are not brought into synchrony in a single phase of the cycle (Ben-Porat and Kaplan, 1967). When growing cells are infected, it has been shown that some biosynthetic event occurs during late G1 or early S phase which is required for virus replication (Pages *et al.*, 1973; Thorne, 1973).

Given these two facts, (1) that events which are required for virus replication can occur only during G1 or S phase, and (2) that virus infection *per se* is not sufficient to change the phase of the cell cycle, it would seem likely that a necessary consequence of infection of a resting cell population is that it affects the cell like a mitogenic agent. This is likely to be a precondition for virus replication, since after resting cells are stimulated to divide they will move through the cell cycle and allow biosynthetic events to occurs which are required for virus replication and which are coupled to particular parts of the cell cycle. The general induction of cellular processes would then be a reflection of cell growth and division. There are significant differences between resting cell cultures in which cellular processes are induced as a consequence of infection and uninfected growth cells. Some of these differences may be a consequence of the extended time that cultured cells were held in a contact-inhibited state prior to infection.

5.2. Pseudovirions

Fragmented cellular DNA can be encapsidated within a virus protein coat giving rise to pseudovirions, noninfectious, viruslike particles which contain linear duplex fragments of cellular DNA. It is only cellular DNA that has replicated during the infectious cycle which breaks down to material of about the same molecular weight as viral DNA (Ben-Porat and Kaplan, 1967). In contrast with polyoma, where pseudovirions are commonly found after infection of primary or continuous monkey kidney cell cultures (Michel *et al.*, 1967), SV40 pseudovirions are observed only rarely and not in any consistent or predictable manner (Trilling and Axelrod, 1970, 1972). Three different types of monkey kidney cultures have been compared to determine the conditions in which pseudovirion formation occurs. When these three monkey kidney cell lines were compared following SV40 infection, there was a 13–23% breakdown of prelabeled cellular DNA of primary African green monkey kidney cell cultures within 96 hours post-infection, for CV-1 cells there was a 1–2% breakdown, and there was no breakdown of prelabeled BSC-1 cellular DNA (Ritzi and Levine, 1967). These results parallel other studies in the same laboratory which showed that pseudovirions were obtained after infection of primary AGMK cells but not in virus stocks prepared in CV-1 or BSC-1 cells (Levine and Teresky, 1970). While this would suggest that cleavage of cellular DNA is a precondition for pseudovirion formation, it has not been established that cleavage alone is sufficient for pseudovirion production. Treatment of polyoma-infected mouse embryo cells with phleomycin also enhances the incorporation of host cell DNA into virions (Iwata and Consigli, 1971). The three requirements for pseudovirion formation are: (1) conditions which permit cellular DNA replication, (2) breakdown of this cellular DNA, and (3) encapsidation of the DNA within a virus protein coat. It has also been suggested that the relative pool sizes of polyoma DNA and degraded cellular DNA at the time of virus assembly may determine the relative proportion of polyoma virions and pseudovirions (Yelton and Aposhian, 1972).

5.3. Nucleoprotein Complexes

The association of proteins with viral nucleic acid in nucleoprotein complexes may occur for a number of reasons. These complexes may be intermediates in the process of uncoating of the virus particle; they may represent the association of enzymes with DNA during replication of viral DNA or during transcription; or they may be intermediates in the *de novo* formation of virions which accumulate during virus replication. The components which make up the virion have been described above in some detail. These include the four histones which are now thought to be cellular in origin, the three viral proteins, and the DNA. Little is known, however, about the internal structure of the virion and the functional implications of this structure.

When polyoma- or SV40-infected cells were treated with Triton X-100, nucleoprotein complexes were obtained which sediment at 55 S and 44 S, respectively (Green *et al.*, 1971; White and Eason, 1971). The complexes contain DNA I bound to proteins which have not yet been characterized. More relevant to the process of virus replication are nucleoprotein complexes which contain 25 S DNA and which are more difficult to extract from cells than nucleoprotein complexes containing DNA I (White and Eason, 1971; Goldstein *et al.*, 1973; Hall *et al.*, 1973). After a pulse and chase, the replicating-polyoma nucleoprotein complexes are converted to complexes containing DNA I; this indicates that they are intermediates in replication. DNA molecules at all stages of replication are present in nucleoprotein complexes. Recent experiments (Seebeck and Weil, 1974) suggest that the proteins associated with replicating polyoma DNA in nucleoprotein complexes are cellular histones. This is in agreement with previous studies which have shown that the polyoma and SV40 core proteins are derived from the histones of the host cell (see Sect. 4.5). The complexes containing replicating DNA consist of an equal mass of protein and DNA. Based on the bouyant densities of complexes, it seems likely that protein is bound to both replicated and unreplicated portions of the DNA within the complex. A study of SV40 nucleoprotein complexes has yielded results that are similar in almost all major respects to those obtained for polyoma. The association of nucleoprotein complexes with capsid protein gives rise to mature virions, and it has been shown that capsid proteins become incorporated into virions within a short period (15 minute) after their synthesis (Ozer and Tegtemeyer, 1972). The mechanism of virus maturation however, remains to be determined.

5.4. Adenovirus–SV40 Hybrid Viruses

The adenovirus–SV40 hybrid viruses will be discussed at length in Chapter 3 or this volume. Nevertheless, they have played an important role in studies designed to elucidate the structure and function of SV40 itself, and for this reason, a preliminary discussion seems warranted.

5.4.1. PARA ($E46^+$)

PARA [particle aiding (and aided by) the replication of adenovirus], or $E46^+$, refers to a defective adenovirus 7–SV40 hybrid virus (Huebner *et al.*, 1964; Rowe and Baum, 1964; Rapp *et al.*, 1964*b*) which was isolated from a strain of adenovirus 7 grown for vaccine purposes (Hartley *et al.*, 1956) and which became contaminated with SV40. About 75% of the SV40 molecule appears to have been inserted into the adenovirus genome, approximately 5% from one adenovirus end (Kelly and Rose, 1971). These molecules also contain a deletion of 16% of the adenovirus genome. They are defective in that they require coinfection with a normal adenovirus genome to initiate a lytic cyle in human cells (Rowe and Baum, 1965); the normal genome presumably codes for adenovirus coat proteins into which the hybrid viral DNA is then encapsidated. The SV40 DNA present in this hybrid virus appears to contain all of the early SV40 genes. It is sufficient to induce the SV40 T and U antigens (Huebner *et al.*, 1964; Rowe and Baum, 1964; Rapp *et al.*, 1964*b*) and to provide the SV40 helper function (see above); therefore, PARA is able to replicate efficiently in monkey cells in the presence of a competent adenovirus genome. Although $E46^+$ appears to contain 75% of the SV40 genome, the predominant, if not the only, SV40 RNA sequences synthesized after $E46^+$ infection of monkey cells are the true early RNA sequences (Lebowitz and Khoury, 1974); this has been shown in hybridization studies between RNA from primary AGMK (African green monkey kidney) cells infected with $E46^+$ and the R · *Hin* fragments of SV40 DNA. This finding is in agreement with the data mentioned above, indicating that $E46^+$ induces the early SV40 antigens and suggesting that the SV40 helper function for replication of adenovirus in monkey cells is specified by the early SV40 genes. Variants of $E46^+$ (PARA) have been isolated in which the SV40 T antigen is localized in the cytoplasm (Butel *et al.*, 1969). On the basis of these studies, it is suggested that the variants are perhaps further defective in a function necessary for transport of T antigen to the nucleus.

5.4.2. Adenovirus 2–SV40 Nondefective Hybrid Viruses

A hybrid virus, $Ad2^{+}ND_1$ (Lewis *et al.*, 1969), was obtained from an isolate B55 of a stock of virus referred to as $Ad2^{++}$ which contained populations of both adenovirus 2 and SV40 (see Lewis *et al.*, 1973). Like the hybrid virus $E46^{+}$, $Ad2^{+}ND_1$ contains a segment of SV40 (18%) covalently inserted (Levin *et al.*, 1971) into the adenovirus genome at the site of an adenovirus deletion. Unlike $E46^{+}$, $Ad2^{+}ND_1$ is nondefective, i.e., it will replicate in human embryonic kidney cells without a helper adenovirus (Lewis *et al.*, 1969). This indicates that the adenovirus portion of the genome is capable of coding for its own capsid proteins. In addition, $Ad2^{+}ND_1$ can replicate in monkey kidney cells; thus, the integrated segment of SV40 seems sufficient to provide the helper function for replication of adenovirus.

From plaques of the B55 hybrid population, four additional nondefective hybrid viruses (Lewis *et al.*, 1973), $Ad2^{+}ND_2$–$Ad2^{+}ND_5$ have been isolated. By electron microscopic heteroduplex mapping techniques (Kelly and Lewis, 1973), the five hybrid viruses were shown to contain a segment of SV40 inserted at the same site in the adenovirus 2 molecule (14% from one adenovirus end). Furthermore, segments of SV40 contained in these hybrids had a common end point at 0.11 unit from the SV40 R · *Eco* R_I cleavage site (Morrow *et al.*, 1973) and contained a portion (7–48%) of the SV40 genome (Table 2; see Fig. 14).

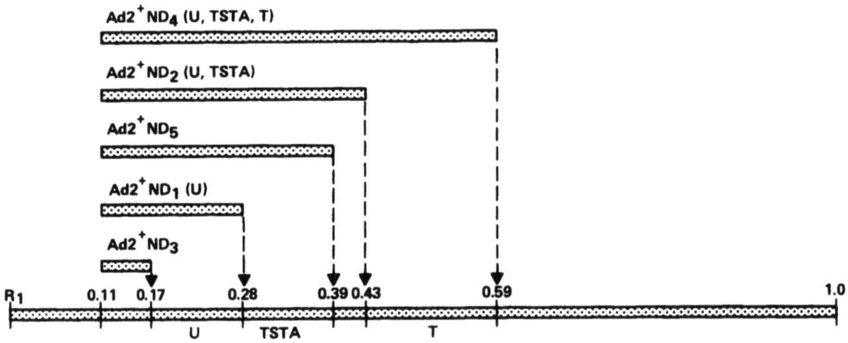

Fig. 14. Map positions of the SV40 DNA segments incorporated in the five Ad 2 SV40 nondefective hybrid viruses. The SV40-specific antigens induced after infection by certain of these viruses (T, U, or TSTA) are indicated. Note that the SV40 molecule is represented as linear for convenience, with the *Eco* · R_I restriction enzyme cleavage site at 0 map units (Lewis and Rowe, 1973).

TABLE 2

Properties of the Nondefective Ad 2–SV40 Hybrid Viruses

Virus	Host[a] range	SV40 Antigens[b]				Size of integrated SV40 DNA Segment,[c] SV40 units	Size of adenovirus DNA deletion,[d] Ad units	Extent of SV40 transcription,[e] SV40 units
		T	U	TSTA	V			
$Ad2^+ND_1$	AGMK/HEK	–	+	–	–	0.18	0.054	0.17
$Ad2^+ND_2$	AGMK/HEK	–	+	+	–	0.32	0.061	0.32
$Ad2^+ND_3$	HEK	–	–	–	–	0.06	0.053	0.07
$Ad2^+ND_4$	AGMK/HEK	+	+	+	–	0.43	0.045	0.45
$Ad2^+ND_5$	HEK	–	–	–	–	0.28	0.071	0.26

[a] Lewis *et al.* (1973). AGMK = African green monkey kidney cells; HEK = human embryonic kidney cells.

[b] Lewis *et al.* (1973).

[c] Estimates of integrated SV40 DNA segment (fraction of the SV40 genome), based on electron microscopic heteroduplex mapping experiments (Kelly and Lewis, 1973).

[d] Estimates of the deleted adenovirus DNA segment of each hybrid (fraction of the adenovirus genome), based on electron microscopic heteroduplex mapping experiments (Kelly and Lewis, 1973).

[e] Determined in RNA–DNA hybridization experiments with the separated strands of SV40 DNA (Khoury *et al.*, 1973*c*).

SV40-specific RNA is synthesized during the lytic cycle of each of the nondefective hybrid viruses (Levine *et al.*, 1973); Khoury *et al.*, 1973*c*) and as was the case for the SV40 DNA segments of these viruses, the specific RNAs have been shown by competition-hybridization experiments to be subsets of each other (Levine *et al.*, 1973; Table 2).

In subsequent studies with the separated strands of SV40 DNA, it was shown that transcription in all of the nondefective hybrid viruses is limited to the minus (or early) DNA strand. Under saturating conditions it was further demonstrated that the SV40 segment in each of the hybrids is transcribed in its entirety (Khoury *et al.*, 1973*c*). It had previously been shown that $Ad2^{+}ND_1$ (as well as the other nondefective hybrids) contain both early and late SV40 gene sequences (Patch *et al.*, 1972, 1974). Since the early and late SV40 messages are transcribed from opposite strands of the SV40 DNA, the extent and strand orientation of transcription in the nondefective hybrid viral lytic infections indicates that there is no synthesis of stable late SV40 mRNA, and that stable "anti-late" SV40 RNA sequences are synthesized (Khoury *et al.*, 1973*c*). More recent data confirm the presence of these "anti-late" transcripts and suggests that they are present in lower concentrations than the true early SV40-specific RNA sequences.

Since the direction of RNA synthesis on the minus (early) strand of SV40 DNA has been determined (Khoury *et al.*, 1973*b*, see Fig. 11), the direction of transcription of the SV40 sequences from the nondefective adenovirus–SV40 hybrids is known. It is clear that this transcription proceeds from the uncommon toward the common end of the SV40 segments (see Figs. 11 and 14). If transcription were to begin in the SV40 portion of the genomes, this would require a different promoter in each of the SV40 segments of the five nondefective hybrid viruses. Thus, it was suggested that transcription is initiated at a promoter in the adenovirus portion of the hybrid, proximal to the early SV40 region (Khoury *et al.*, 1973*c*). This prediction would also suggest that hybrid-adenovirus message is synthesized with adenovirus sequences on the 5′ end of the molecule. Adenovirus–SV40 hybrid mRNA has been detected by filter hybridization techniques (Oxman *et al.*, 1974). Although the orientation of the adenovirus and SV40 sequences has not yet been determined, it has been concluded that the hybrid promoter is in adenovirus DNA, since transcription from the hybrid viral DNA is resistant to interferon. In contrast, SV40 transcription (and/or translation) is quite sensitive to interferon (Oxman *et al.*, 1967).

The prototype of these hybrid viruses, $Ad2^+ND_1$ which contains 18% of the SV40 genome (see Table 2), does not induce the SV40 T antigen but does code for the SV40 U antigen. In addition to the U antigen, $Ad2^+ND_2$ (containing 32% of the SV40 genome) codes for the SV40 transplantation antigen (TSTA), while $Ad2^+ND_4$, which contains the largest SV40 segment (48%), codes for U, TSTA, and T antigen.

An analysis of the antigen-inducing ability of the hybrids and the location of the SV40 segments of each on the parental SV40 genome allows one to construct a map of the early SV40 antigens (see Fig. 14). The U antigen would be located at the common end of the hybrids, and between 0.11 and 0.28 unit. Similarly, TSTA and T antigens would be associated with 0.28–0.39 and 0.39–0.59 unit, respectively. It should be pointed out, however, that although the SV40 segments of the nondefective hybrid viruses have been carefully mapped, the early "gene" regions associated with these segments have been located only by inference. Furthermore, the three regions of the SV40 genome described above are only large enough to code for proteins of 15,000–25,000 daltons; the early SV40 antigens, on the other hand, may be considerably larger (see above). In spite of these reservations, the map positions of the early SV40 antigens as determined from the adeno–SV40 nondefective hybrid viruses and the early transcriptional region of the genome localized by hybridization experiments (Khoury *et al.*, 1973*b*; Fig. 11) are in good agreement.

Two of the nondefective hybrid viruses, $Ad2^+ND_3$ and $Ad2^+ND_5$, induce no known SV40 antigens (Lewis *et al.*, 1973; Table 2). Yet infection by each of these viruses results in the synthesis of SV40-specific RNA, which represents a complete transcript of the incorporated SV40 DNA segment (Khoury *et al.*, 1973*c*). Perhaps the SV40 gene sequences in $Ad2^+ND_3$ are too small (7% of SV40) to induce any SV40 antigens. Furthermore, most of the SV40-specific RNA transcribed from this hybrid is "anti-late" message (Patch *et al.*, 1974). On the other hand, $Ad2^+ND_5$ contains almost as much SV40 DNA as $Ad2^+ND_2$ (Kelly and Lewis, 1973), a hybrid which induces two of the three early SV40 antigens (Lewis *et al.*, 1973). There appears to be no inhibition of the transcription of $Ad2^+ND_5$; mRNA is synthesized (Levine *et al.*, 1973; Khoury *et al.*, 1973*c*); transported to the cytoplasm, and is found on polyribosomes (C. S. Crumpacker, Levin, and Lewis, unpublished results). Whether there is a block in the initiation of translation or a frame shift resulting in the translation of a nonsense protein is unclear. In addition, it remains to be determined if the fact that $Ad2^{+}ND_5$ contains the largest adenoviral DNA deletion among the nondefective hybrids relates to the lack of induction of early SV40 antigens.

Note. See Addendum on page 475.

6. REFERENCES

Aaronson, S. A., and Martin, M. A., 1970, Transformation of human cells with different forms of SV40 DNA, *Virology* **42,** 848.

Aaronson, S. A., and Todaro, G. J., 1969, Human diploid cell transformation by DNA extracted from the tumor virus SV40, *Science* (*Wash. D.C.*) **166,** 390.

Acheson, N. H., Buetti, E., Scherrer, K., and Weil, R., 1971, Transcription of the polyoma virus genome: Synthesis and cleavage of giant late polyoma-specific RNA, *Proc. Natl. Acad. Sci. USA* **68,** 2231.

Aloni, Y., 1972, Extensive symmetrical transcription of simian virus 40 DNA in virus-yielding cells, *Proc. Natl. Acad. Sci. USA* **69,** 2024.

Aloni, Y., 1973, Poly A and symmetrical transcription of SV40 DNA, *Nat. New Biol.* **243,** 2.

Aloni, Y., and Locker, H., 1973, Symmetrical *in vivo* transcription of polyoma DNA and the separation of self-complementary viral and cell DNA, *Virology* **54,** 495.

Aloni, Y., Winocour, E. and Sachs, L., 1968, Characterization of the simian virus 40-specific RNA in virus-yielding and -transformed cells, *J. Mol. Biol.* **31,** 415.

Aloni, Y., Winocour, E., Sachs, L., and Torten, J., 1969, Hybridization between SV40 DNA and cellular DNA's, *J. Mol. Biol.* **44,** 333.

Anderer, F. A., Schlumberger, H. D., Koch, M. A., Frank, H., and Eggers, H. J., 1967, Structure of simian virus 40. II. Symmetry and components of the virus particle, *Virology* **32,** 511.

Anderson, C. W., and Gesteland, R. F., 1972, Pattern of protein synthesis in monkey cells infected by simian virus 40, *J. Virol.* **9,** 758.

Arber, W. and Linn, S., 1969, DNA modification and restriction, *in* "Annual Review of Biochemistry," pp. 467–500, Annual Reviews, Palo Alto, California.

Astrin, S. M., 1973, *In vitro* transcription of simian virus 40 sequences in SV3T3 chromatin, *Proc. Natl. Acad. Sci. USA* **70,** 2304.

Babiuk, L. A., and Hudson, J. B., 1972, Integration of polyoma virus DNA into mammalian genomes, *Biochem. Biophys. Res. Commun.* **47,** 111.

Barban, S., 1973, Electrophoretic differences in the capsid proteins of simian virus 40 plaque mutants, *J. Virol.* **11,** 971.

Barban, S., and Goor, R. S., 1971, Structural proteins of simian virus 40, *J. Virol.* **7,** 198.

Barbanti-Brodano, G., Swetly, P., and Koprowski, H., 1970, Early events in the infection of permissive cells with simian virus 40: Adsorption, penetration, and uncoating, *J. Virol.* **6,** 78.

Basilico, C., and DiMayorca, G., 1965, Radiation target size of the lytic and the transforming ability of polyoma virus, *Proc. Natl. Acad. Sci. USA* **54,** 125.

Basilico, C., Marin, G., and DiMayorca, G., 1966, Requirement for the integrity of the viral genome for the induction of host DNA synthesis by polyoma virus, *Proc. Natl. Acad. Sci. USA* **56,** 208.

Basilico, C., Matsuya, Y., and Green, M. H., 1969, Origin of the thymidine kinase induced by polyoma virus in productively infected cells, *J. Virol.* **3,** 140.

Bauer, W., and Vinograd, J., 1968, The interaction of closed circular DNA with intercalative dyes. I. The superhelix density of SV40 DNA in the presence and absence of dye, *J. Mol. Biol.* **33,** 141.

Baum, S. G., Horwitz, M. S., and Maizel, J. V., Jr., 1972, Studies of the mechanism of enhancement of human adenovirus infection in monkey cells by simian virus 40, *J. Virol.* **10,** 211.

Beard, P., Morrow, J. F., and Berg, P., 1973, Cleavage of circular superhelical SV40 DNA to a linear duplex by S_1 nuclease, *J. Virol.* **12,** 1303.

Beerman, T. A., and Lebowtiz, J., 1973, Further analysis of the altered secondary structure of superhelical DNA. Sensitivity of methylmercuric hydroxide a chemical probe for unpaired bases, *J. Mol. Biol.* **79,** 451.

Benjamin, T. L., 1965, Relative target sizes for the inactivation of the transforming and reproductive abilities of polyoma virus, *Proc. Natl. Acad. Sci. USA* **54,** 121.

Benjamin, T. L., 1966, Virus-specific RNA in cells productively infected or transformed by polyoma virus, *J. Mol. Biol.* **16,** 359.

Ben-Porat, T., and Kaplan, A. S., 1967, Correlation between replication and degradation of cellular DNA in polyoma virus-infected cells, *Virology* **32,** 457.

Ben-Porat, T., Coto, C., and Kaplan, A. S., 1966, Unstable DNA synthesized by polyoma virus-infected cells, *Virology* **30,** 74.

Bigger, C., Murray, K., and Murray, N. E., 1973, Recognition sequence of a restriction enzyme, *Nat. New Biol.* **244,** 7.

Black, P. H., and Rowe, W. P., 1973, SV40-induced proliferation of tissue culture cells of rabbit, mouse and procine origin, *Proc. Soc. Exp. Biol. Med.* **114,** 721.

Black, P. H., Rowe, W. P., Turner, H. C., and Huebner, R. J., 1963, A specific complement-fixing antigen present in SV40 tumor and transformed cells, *Proc. Natl. Acad. Sci. USA* **50,** 1148.

Bourgaux, P., and Bourgaux-Ramoisy, D., 1972*a*, Unwinding of replicating polyoma virus DNA, *J. Mol. Biol.* **70,** 399.

Bourgaux, P., and Bourgaux-Ramoisy, D., 1972*b*, Is a specific protein responsible for the supercoiling of polyoma DNA? *Nature* (*Lond.*) **235,** 105.

Bourgaux, P., Bourgaux-Ramoisy, D., and Stoker, M., 1965, Further studies on transformation by DNA from polyoma virus, *Virology* **25,** 364.

Bourgaux, P., Bourgaux-Ramoisy, D., and Dulbecco, R., 1969, The replication of ring-shaped DNA of polyoma virus. I. identification of the replicative intermediate, *Proc. Natl. Acad. Sci. USA* **64,** 701.

Bourguignon, M. F., 1968, A denaturation map of polyoma virus DNA, *Biochim. Biophys. Acta* **166,** 242.

Boyer, H. W., 1971, DNA restriction and modification mechanisms in bacteria, *in* "Annual Review of Microbiology," p. 153–176, Annual Reviews, Palo Alto, Calif.

Boyer, H. W., Chow, L. T., Dugaiczyk, A., Hedgpeth, J., and Goodman, H. M., 1973, DNA substrate site for the Eco_{RII} restriction endonuclease and modification methylase, *Nat. New Biol.* **244,** 172.

Branton, P. E., and Sheinin, R., 1973, Studies on the replication of polyoma DNA: Physicochemical properties of viral DNA synthesized when protein synthesis is inhibited, *Can. J. Biochem.*, **51,** 305.

Branton, P. E., Cheevers, W. P., and Sheinin, R., 1970, The effect of cycloheximide on DNA synthesis in cells productively-infected with polyoma virus, *Virology* **42,** 979.

Britten, R. J., and Kohne, D. E., 1968, Repeated sequences in DNA, *Science* (*Wash., D.C.*) **161,** 529.

Brockman, W. W., Lee, T. N. H., and Nathans, D., 1973, The evolution of new species of viral DNA during serial passage of simian virus 40 at high multiplicity, *Virology* **54,** 384.

Burns, W. H., and Black, P. H., 1968, Analysis of simian virus 40-induced transformation of hamster kidney tissue *in vitro*. V. Variability of virus recovery from cell clones inducible with mitomycin C and cell fusion, *J. Virol.* **2,** 606.

Burton, A., and Sinsheimer, R. L., 1963, Process of infection with ϕX174: Effect of exonucleases on the replicative form, *Science* (*Wash., D.C.*) **142,** 962.

Butel, J. S., and Rapp, F., 1965, The effect of arabinofuranosylcytosine on the growth cycle of simian virus 40, *Virology* **27,** 490.

Butel, J. S., Guentzel, M. J., and Rapp, F., 1969, Variants of defective simian papovavirus 40 (para) characterized by cytoplasmic localization of simian papovavirus 40 tumor antigen, *J. Virol.* **4,** 632.

Butel, J. S., Richardson, L. S., and Melnick, J. L., 1971, Variation in properties of SV40-transformed simian cell lines detected by superinfection with SV40 and human adenoviruses, *Virology* **46,** 844.

Cairns, J., 1963, The chromosome of *Escherichia coli*, *Cold Spring Harbor Symp. Quant. Biol.* **28,** 43.

Carp, R. I., and Gilden, R. V., 1965, The inactivation of simian virus 40 infectivity and antigen-inducing capacity by ultraviolet light, *Virology* **27,**639.

Carp, R. I., Sauer, G., and Sokol, F., 1969, The effect of actinomycin D on the transcription and replication of simian virus 40 deoxyribonucleic acid, *Virology* **37,** 214.

Chamberlin, M., McGrath, J., and Waskell, L., 1970, New RNA polymerase from *Escherichia coli* infected with bacteriophage T7, *Nature* (*Lond.*) **228,** 227.

Champoux, J. J., and Dulbecco, R., 1972, An activity from mammalian cells that untwists superhelical DNA—A possible swivel for DNA replication. *Proc. Natl. Acad. Sci. USA* **69,** 143.

Coggin, J. H., Elrod, L. H., Ambrose, K. R., and Anderson, N. G., 1969, Induction of tumor-specific transplantation immunity in hamster with cell fractions from adenovirus and SV40 tumor cells, *Proc. Soc. Exp. Biol. Med.* **132,** 328.

Cowan, K., Tegtmeyer, P., and Anthony, D. D., 1973, Relationship of replication and transcription of simian virus 40 DNA, *Proc. Natl. Acad. Sci. USA* **70,** 1927.

Crawford, L. V., 1962, The adsorption of polyoma virus, *Virology* **18,** 177.

Crawford, L. V., 1964, A study of shope papilloma virus DNA, *J. Mol. Biol.* **8,** 489.

Crawford, L. V., 1965, A study of human papilloma virus DNA, *J. Mol. Biol.* **13,** 362.

Crawford, L. V., and Black, P. H., 1964, The nucleic acid of simian virus 40, *Virology* **24,** 388.

Crawford, L. Dulbecco, R., Fried, M., Montagneir, L., and Stoker, M., 1964, Cell transformation by different forms of polyoma virus DNA, *Proc. Natl. Acad. Sci. USA* **52,** 148.

Crawford, L. V., Crawford, E. M., Richardson, J. P., and Slayter, H. S., 1965, The binding of RNA polyomerase to polyoma and papilloma DNA, *J. Mol. Biol.* **14,** 593.

Crawford, L. V., Syrett, C., and Wilde, A., 1973, The replication of polyoma virus, *J. Gen. Virol.* **21,** 515.

Cuzin, F., Vogt, M., Dieckmann, M., and Berg, P., 1970, Induction of virus-multiplication in 3T3 cells transformed by a thermosensitive mutant of polyoma virus. II. Formation of oligomeric polyoma DNA molecules, *J. Mol. Biol.* **47,** 317.

Cuzin, F., Blangy, D., and Rouget, P., 1971, Activite endonucleastique de preparation purifie du virus polyome, *C. R. Hebd. Seances Acad. Sci. Ser. D Sci. Nat.* **273,** 2650.

Danna, K. and Nathans, D., 1971, Specific cleavage of simian virus 40 DNA by restriction endonuclease of *Hemophilus influenza*, *Proc. Natl. Acad. Sci. USA* **68,** 2913.

Danna, K. J. and Nathans, D., 1972, Bidirectional replication of simian virus 40 DNA, *Proc. Natl. Acad. Sci. USA* **69,** 3097.

Danna, K. J., Sack, G. H., Jr., and Nathans, D., 1973, Studies of simian virus 40 DNA. VII. A cleavage map of the SV40 genome, *J. Mol. Biol.* **78,** 363.

Davis, R., Simon, M., and Davidson, N., 1971, Electron microscope heteroduplex methods for mapping regions of base sequence homology in nucleic acids, *in* "Methods in Enzymology" (L. Grossman and K. Moldave, eds.), pp. 413–428, Academic Press, New York.

Dean, W. W., and Lebowitz, J., 1971, Partial alteration of secondary structure in native superhelical DNA, *Nat. New Biol.* **231,** 5.

Defendi, V., 1963, Effect of SV40 virus immunization on growth of transplantable SV40 and polyoma virus tumors in hamsters, *Proc. Soc. Biol. Med.* **113,** 12.

Delius, H., Westphal, H., and Axelrod, N., 1973, Length measurements of RNA synthesized *in vitro* by *E. coli* polymerase, *J. Mol. Biol.* **74,** 677.

Del Villano, B. C., and Defendi, V., 1973, Characterization of the SV40 T antigen, *Virology* **51,** 34.

Dhar, R., Zain, S., Weissman, S. M., Pan, J., and Subramanian, K., 1973, Nucleotide sequences of RNA transcribed in infected cells and by *E. Coli* polymerase from a segment of simian virus 40 DNA, *Proc. Natl. Acad. Sci. USA* **71,** 371.

Dubbs, D. R., Kit, S., deTorres, R. A., and Anken, M., 1967, Virogenic properties of bromodeoxyuridine-sensitive and bromodeoxyuridine-resistant simian virus 40-transformed mouse kidney cells, *J. Virol.* **1,** 968.

Dulbecco, R., and Vogt, M., 1963, Evidence for a ring structure of polyoma virus DNA, *Proc. Natl. Acad. Sci. USA* **50,** 236.

Dulbecco, R., Hartwell, L. H., and Vogt, M., 1965, Induction of cellular DNA synthesis by polyoma virus, *Proc. Natl. Acad. Sci. USA* **53,** 403.

Estes, M. K., Huang, E.-S., and Pagano, J. S., 1971, Structural polypeptides of simian virus 40, *J. Virol.* **7,** 635.

Fareed, G. C., and Salzman, N. P., 1972, Intermediate in SV40 DNA chain growth, *Nat. New Biol.* **238,** 274.

Fareed, G. C., Garon, C. F., and Salzman, N. P., 1972, Origin and direction of simian virus 40 deoxyribonucleic acid replication, *J. Virol.* **10,** 484.

Fareed, G. C., Khoury, G., and Salzman, N. P., 1973*a*, Self-annealing of 4 S strands from replicating simian virus 40 DNA, *J. Mol. Biol.* **77,** 457.

Fareed, G. C., McKerlie, M. L., and Salzman, N. P., 1973*b*, Characterization of simian virus 40 DNA component II during viral DNA replication, *J. Mol. Biol.* **74,** 95.

Fareed, G. C., Byrne, J. C., and Martin, M. A., 1974, Triplication of a unique genetic segment in an SV40-like virus of human origin and evolution of new viral genomes, *J. Mol. Biol.* **86,** in press.

Fiers, W., and Sinsheimer, R. L., 1962, The structure of the DNA of bacteriophage ϕX174, *J. Mol. Biol.* **5,** 408.

Finch, J. T., and Klug, A., 1965, The structures of viruses of the papilloma-polyoma type. III. Structure of rabbit papilloma virus, *J. Mol. Biol.* **13,** 1.

Fine, R., Mass, M., and Murakami, W. T., 1968, Protein composition of polyoma virus, *J. Mol. Biol.* **36,** 167.

Fischer, H., and Sauer, G., 1972, Identification of virus-induced proteins in cells productively infected with simian virus 40, *J. Virol.* **9,** 1.

Follett, E. A. C., and Crawford, L. V., 1968, Electron microscope study of the denaturation of polyoma virus DNA, *J. Mol. Biol.* **34,** 565.

Fox, R. I., and Baum, S. G., 1972, Synthesis of viral ribonucleic acid during restricted adenovirus infection, *J. Virol.* **10,** 220.

Frearson, D. M., and Crawford, L. V., 1972, Polyoma virus basic proteins, *J. Gen. Virol.* **14,** 141.

Fried, A. H., and Sokol, F., 1972, Synthesis *in vitro* by bacterial RNA polymerase of simian virus 40-specific RNA: Multiple transcription of the DNA template into a continuous polyribonucleotide, *J. Gen. Virol.* **17,** 69.

Fried, M., 1970, Characterization of a temperature-sensitive mutant of polyoma virus, *Virology* **40,** 605.

Friedman, M. P., Lyons, M. J., and Ginsberg, H. S., 1970, Biochemical consequences of type 2 adenovirus and simian virus 40 double infections of African green monkey kidney cells, *J. Virol.* **5,** 586.

Friedmann, T., 1974, Novel genetic economy of polyoma virus: Capsid proteins are cleavage products of same viral gene, *Proc. Natl. Acad. Sci. USA* **71,** 257.

Friedmann, T., and David, D., 1972, Structural roles of polyoma virus proteins, *J. Virol.* **10,** 776.

Gerber, P., 1966, Studies on the transfer of subviral infectivity from SV40-induced hamster tumor cells to indicator cells, *Virology* **28,** 501.

Gilbert, W. and Dressler, D., 1968, DNA replication: The rolling circle model, *Cold Spring Harbor Symp. Quant. Biol.* **33,** 473.

Gilden, R. V., and Carp, R. I., 1966, Effects of cycloheximide and puromycin on synthesis of simian virus 40 T antigen in green monkey kidney cells, *J. Bacteriol.* **91,** 1295.

Gilden, R. V., Carp, R. I., Taguchi, F., and Defendi, V., 1965, The nature and localization of the SV40-induced complement-fixing antigen, *Proc. Natl. Acad. Sci. USA* **53,** 684.

Gilham, P. T., 1964, The synthesis of polynucleotide-celluloses and their use in the fractionation of polynucleotides, *J. Am. Chem. Soc.* **86,** 4982.

Girard, M., Marty, L., and Suarez, F., 1970, Capsid proteins of simain virus 40, *Biochem. Biophys. Res. Commun.* **40,** 97.

Girard, M., Marty, L., and Manteuil, S., 1974, Viral DNA–RNA hybrids in simian virus 40 infected cells: The simian virus 40 transcriptional intermediates, *Proc. Natl. Acad. Sci. USA* **71,** 1267.

Girardi, A. J., and Defendi, V., 1970, Induction of SV40 transplantation antigen (TrAg) during the lytic cycle, *Virology* **42,** 688.

Goldstein, D. A., Hall, M. R., and Meinke, W., 1973, Properties of nucleoprotein complexes containing replicating polyoma DNA, *J. Virol.* **12,** 887.

Green, M. H., Miller, H. I., and Hendler, S., 1971, Isolation of a polyoma nucleoprotein complex from infected mouse cell cultures, *Proc. Natl. Acad. Sci. USA*, **68,** 1032.

Greenaway, P. J., and LeVine, D., 1973, Amino acid compositions of simian virus 40 structural proteins, *Biochem. Biophys. Res. Commun.* **52,** 1221.
Gromkova, R., and Goodgal, S. H., 1972, Action of *Haemophilus* endodeoxyribonuclease on biologically active deoxyribonucleic acid, *J. Bacteriol.* **109,** 987.
Habel, K., 1961, Resistance of polyoma virus immune animals to transplanted polyoma tumors, *Proc. Soc. Exp. Biol. Med.* **106,** 722.
Habel, K., 1965, Specific complement-fixing antigens in polyoma tumors and transformed cells, *Virology* **25,** 55.
Habel, K., and Eddy, B. E., 1963, Specificity of resistance to tumor challenge of polyoma and SV40 virus-immune hamsters, *Proc. Soc. Exp. Biol. Med.* **113,** 1.
Habel, K., Jensen, F. C., Pagano, J., and Koprowski, H., 1965, Specific complement-fixing tumor antigen in SV40-transformed human cells, *Proc. Soc. Exp. Biol. Med.* **118,** 4.
Hall, M. R., Meinke, W., and Goldstein, D. A., 1973, Nucleoprotein complexes containing replicating simian virus 40 DNA: Comparison with polyoma nucleoprotein complexes, *J. Virol.* **12,** 901.
Hancock, R., and Weil, R., 1969, Biochemical evidence for induction by polyoma virus of replication of the chromosomes of mouse kidney cells, *Proc. Natl. Acad. Sci. USA* **63,** 1144.
Hartley, J. W., Huebner, R. J., and Rowe, W. P., 1956, Serial propagation of adenoviruses (APC) in monkey kidney tissue cultures, *Proc. Soc. Exp. Biol. Med.* **92,** 667.
Hashimoto, K., Nakajima, K., Oda, K., and Shimojo, H., 1973, Complementation of translational defect for growth of human adenovirus type 2 in simian cells by an SV40-induced factor, *J. Mol. Biol.* **81,** 207.
Hatanaka, M., and Dulbecco, R., 1966, Induction of DNA synthesis by SV40, *Proc. Natl. Acad. Sci. USA* **56,** 736.
Hedgpeth, J., Goodman, H. M., and Boyer, H. W., 1972, DNA nucleotide sequence restricted by the R1 endonuclease, *Proc. Natl. Acad. Sci. USA* **69,** 3448.
Helinski, D. R., and Clewell, D. B., 1971, Circular DNA, *in* "Annual Review of Biochemistry," pp. 899–942, Annual Reviews, Palo Alto, Calif.
Henry, C. J., Slifkin, M., and Merkow, L., 1971, Mechanism of host cell restriction in African green monkey kidney cells abortively infected with human adenovirus type 2, *Nat. New Biol.* **233,** 39.
Herzberg, M. and Winocour, E., 1970, Simian virus 40 deoxyribonculeic acid transcription *in vitro:* Binding and transcription patterns with a mammalian ribonculeic acid polymerase, *J. Virol.* **6,** 667.
Hirai, K., and Defendi, V., 1972, Integration of SV40 DNA into the DNA of permissive monkey cells, *J. Virol.* **9,** 705.
Hirt, B., 1967, Selective extraction of polyoma DNA from infected mouse cell cultures, *J. Mol. Biol.* **26,** 265.
Hirt, B., 1969, Replicating molecules of polyoma DNA, *J. Mol. Biol.* **40,** 141.
Hirt, B., and Gesteland, R. F., 1971, Characterization of SV40 and polyoma virus, *Lepetit. Colloq. Biol. Med.* **2,** 98.
Hoggan, M. D., Rowe, W. P., Black, P. H., and Heubner, R. J., 1965, Production of "tumor-specific" antigens by oncogenic viruses during acute cytolytic infections, *Proc. Natl. Acad. Sci. USA* **52,** 12.
Huang, E.-S., Estes, M. and Pagano, J., 1972*a*, Structure and function of the polypeptides in simian virus 40. I. Existence of subviral deoxynucleoprotein complexes, *J. Virol.* **9,** 923.

Huang, E.-S., Nonoyana, M., and Pagano, J. S., 1972*b*, Structure and function of the polypeptides in simian virus 40. II. Transcription of subviral deoxynucleoprotein complexes *in vitro*, *J. Virol.* **9**, 930.

Hudson, J., Goldstein, D., and Weil, R., 1970, A study on the transcription of the polyoma viral genome, *Proc. Natl. Acad. Sci. USA* **65**, 226.

Huebner, R. J., Chanock, R. M., Rubin, B. A., and Casey, M. J., 1964, Induction by adenovirus type 7 of tumors in hamsters having the antigenic characteristics of SV40 virus, *Proc. Natl. Acad. Sci. USA* **52**, 1333.

Hummeler, K., Tomassini, N., and Sokol, F., 1970, Morphological aspects of the uptake of simian virus 40 by permissive cells, *J. Virol.* **6**, 87.

Iwata, A., and Consigli, R. A., 1971, Effect of phleomycin on polyoma virus synthesis in mouse embryo cells, *J. Virol.* **7**, 29.

Jackson, A. H., and Sugden, B. 1972, Inhibition by α-amanitin of simian virus 40-specific ribonucleic acid synthesis in nuclei of infected monkey cells, *J. Virol.* **10**, 1086.

Jaenisch, R., 1972, Evidence for SV40-specific RNA-containing virus and host-specific sequences, *Nat. New Biol.* **235**, 46.

Jaenisch, R., Mayer, A., and Levine, A. J., 1971, Replicating SV40 molecules containing closed circular template DNA strands, *Nat.* (*London*) **233**, 72.

Jerkofsky, M., and Rapp, F., 1973, Host cell DNA synthesis as a possible factor in the enhancement of replication of human adenoviruses in simian cells by SV40, *Virology* **51**, 466.

Kang, H. S., Eshbach, T. B., White, D. A., and Levine, A. J., 1971, Deoxyribonucleic acid replication in simian virus 40-infected cells. IV. Two different requirements for protein synthesis during simian virus 40 deoxyribonucleic acid replication, *J. Virol.* **7**, 112.

Kaplan, J. C., Wilbert, S. M., and Black, P. H., 1972, Endonuclease activity associated with purified simian virus 40 virions, *J. Virol.* **9**, 800.

Kato, A. C., Bartok, K., Fraser, M. J., and Denhardt, D. T., 1973, Sensitivity of superhelical DNA to a single-strand specific endonuclease, *Biochim. Biophys. Acta* **308**, 68.

Keller, W., and Crouch, R., 1972, Degradation of DNA RNA hybrids by ribonuclease H and DNA polymerases of cellular and viral origin, *Proc. Natl. Acad. Sci. USA* **69**, 3360.

Kelly, T. J., Jr., and Lewis, A. M., 1973, Use of nondefective adenovirus–simian virus 40 hybrids for mapping the simian virus 40 genome, *J. Virol.* **12**, 643.

Kelly, T. J., Jr., and Rose, J. A., 1971, Simian virus 40 integration site in an adenovirus 7–SV40 hybrid DNA molecule, *Proc. Natl. Acad. Sci. USA* **68**, 1037.

Kelly, T. J., Jr. and Smith, H. O., 1970, A restriction enzyme from *Hemophilus influenza*. II. Base sequence of the recognition site, *J. Mol. Biol.* **51**, 393.

Khera, K. S., Ashkenazi, A., Rapp, F. and Melnick, J. L., 1963, Immunity in hamsters to cells transformed *in vitro* and *in vivo* by SV40. Tests for antigenic relationships among the papovaviruses, *J. Immunol.* **91**, 604.

Khoury, G., 1970, An investigation of the properties of SV40-transformed human cells, Harvard Medical School Thesis.

Khoury, G., and Martin, M. A., 1972, Comparison of SV40 DNA transcription *in vivo* and *in vitro*, *Nat. New Biol.* **238**, 4.

Khoury, G., Byrne, J. C., and Martin, M. A., 1972, Pattern of simian virus 40 DNA transcription after acute infection of permissive and non-permissive cells, *Proc. Natl. Acad. SCi. USA* **69,** 1925.

Khoury, G., Byrne, J. C., Takemoto, K. K., and Martin, M. A., 1973*a*, Patterns of simian virus 40 deoxyribonucleic acid transcription. II. In transformed cells, *J. Virol.* **11,** 54.

Khoury, G., Martin, M. A., Lee, T. N. H., Danna, K. J., and Nathans, D., 1973*b*, A map of simian virus 40 transcription sites expressed in productively infected cells, *J. Mol. Biol.* **78,** 377.

Khoury, G., Lewis, A. M., Oxman, M. N., and Levine, A. S., 1973*c*, Strand orientation of SV40 transcription in cells infected by the nondefective adenovirus 2–SV40 hybrid viruses, *Nat. New Biol.*, **246,** 207.

Khoury, G., Fareed, G. C., Berry, K., Martin, M. A., Lee, T. N. H., and Nathans, D., 1974, Characterization of a rearrangement in viral DNA: Mapping of the circular SV40-like DNA containing a triplication of a specific one-third of the viral genome, *J. Mol. Biol.*, **86,** in press.

Kidwell, W. R., Saral, R., Martin, R. G., and Ozer, H. L., 1972, Characterization of an endonuclease associated with simian virus 40 virions, *J. Virol.* **10,** 410.

Kiger, J. A., Jr., and Sinsheimer, R. L., 1969, Vegetative lambda DNA IV. Fractionation of replicating lambda DNA on benzoylated-naphthoylated DEAE cellulose, *J. Mol. Biol.* **40,** 467.

Kit, S., 1967, Enzyme inductions in cell cultures during productive and abortive infections by papovavirus SV40, *in* "The Molecular Biology of Viruses" (J. S. Colter and W. Paranchyeh, eds.), pp. 495–525, Academic Press, New York.

Kit, S., 1968, Viral-induced enzymes and viral carcinogenesis, *Adv. Cancer Res.* **11,** 73.

Kit, S., and Nakajima, K., 1971, Analysis of the molecular forms of simian virus 40 deoxyribonucleic acid synthesized in cycloheximide-treated cell cultures, *J. Virol.* **7,** 87.

Kit, S., Dubbs, D. R., and Frearson, P. M., 1966, Enzymes of nucleic acid metabolism in cells infected with polyoma virus, *Cancer Res.* **26,** 638.

Kit, S., Melnick, J. L., Anken, M., Dubbs, D. R., deTorres, R. A., and Kitahara, T., 1967, Non-identity of some simian-virus 40-induced enzymes with tumor antigen, *J. Virol.* **1,** 684.

Kit, S., Kurimura, T., Salvi, M. L., and Dubbs, D. R., 1968, Activation of infectious SV40 DNA synthesis in transformed cells, *Proc. Natl. Acad. Sci USA* **60,** 1239.

Kitahara, T., and Melnick, J. L., 1965, Thermal separation of the synthesis of papovavirus SV40 tumor and virus antigen, *Proc. Soc. Exp. Biol. Med.* **120,** 709.

Koch, M. A., and Sabin, A. B., 1963, Specificity of virus-induced resistance to transplantation of polyoma and SV40 tumors in adult hamsters, *Proc. Soc. Exp. Biol. Med.* **113,** 4.

Koprowski, H., Jensen, F. C., and Steplewski, Z., 1967, Activation of production of infectious tumor virus SV40 in heterokaryon cultures, *Proc. Natl. Acad. Sci. USA* **58,** 127.

Laipis, P., and Levine, A. J., 1973, Deoxyribonucleic acid replication in SV40-infected cells. IX. The inhibition of a gap filling step during discontinuous synthesis of SV40 DNA, *Virology* **56,** 580.

Lake, R. S., Barban, S., and Salzman, N. P., 1973, Resolutions and identification of the core deoxynucleoproteins of the simian virus 40, *Biochem. Biophys. Res. Commun.* **54,** 640.

Latarjet, R., Cramer, R., and Montagnier, L., 1967, Inactivation by UV-, X-, and α-radiations of the infecting and transforming capacities of polyoma virus, *Virology* **33,** 104.

Lavi, S., and Winocour, E., 1972, Acquisition of sequences homologous to host deoxyribonucleic acid by closed circular simian virus 40 deoxyribonucleic acid, *J. Virol.* **9,** 309.

Lazarus, H. M., Sporn, M. B., Smith, J. M., and Henderson, W. R., 1967, Purification of T antigen from nuclei of simian virus 40-induced hamster tumors, *J. Virol.* **5,** 1093.

Lebowitz, P., and Khoury, G., 1974, The SV40 DNA segment of the adenovirus 7-SV40 hybrid, $E46^+$, and its transcription during permissive infection of monkey kidney cells, *J. Virol.*, in press.

Lee, Y., Mendecki, J., and Brawerman, G., 1971, A polynucleotide segment rich in adenylic acid in the rapidly-labeled polyribosomal RNA component of mouse sarcoma 180 ascites cells, *Proc. Natl. Acad. Sci. USA* **68,** 1331.

Levin, M. J., Crumpacker, C. S., Lewis, A. M., Oxman, M. N., Henry, P. H., and Rowe, W. P., 1971, Studies of nondefective adenovirus 2–simian virus 40 hybrid viruses. II. Relationship of adenovirus 2 deoxyribonucleic acid and simian virus 40 deoxyribonculeic acid in the $Ad2^+ND_1$ genome, *J. Virol.* **7,** 343.

Levine, A. J., 1971, Induction of mitochondrial DNA synthesis in monkey cells infected by SV40 and (or) treated with calf serum. *Proc. Natl. Acad. Sci. USA* **68,** 717.

Levine, A. J., and Teresky, A. K., 1970, Deoxyribonucleic acid replication in simian virus 40-infected cells. II. Detection and characterization of simian virus 40 pseudovirions, *J. Virol.* **5,** 451.

Levine, A. J., Kang, H. S., and Billheimer, F., 1970, DNA replication in SV40-infected cells. I. Analysis of replicating SV40 DNA, *J. Mol. Biol.* **50,** 549.

Levine, A. S., Levin, M. J. Oxman, M. N., and Lewis, A. M., 1973, Studies of nondefective adenovirus 2–simian virus 40 hybrid viruses. VII. Characterization of the simian virus 40 RNA species induced by five nondefective hybrid viruses, *J. Virol.* **11,** 672.

Lewis, A. M., and Rowe, W. P., 1971, Studies on nondefective adenovirus–simian virus 40 hybrid viruses, *J. Virol.* **7,** 189.

Lewis, A. M., Levin, M. J., Wiese, W. H., Crumpacker, C. S., and Henry, P. H., 1969, A nondefective (competent) adenovirus–SV40 hybrid isolated from the Ad2–SV40 hybrid population, *Proc. Natl. Acad. Sci. USA* **63,** 1128.

Lewis, A. M., Levine, A. S., Crumpacker, C. S., Levin, M. J., Samaha, R. J., and Henry, P. H., 1973, Studies of nondefective adenovirus 2–simian virus 40 hybrid viruses. V. Isolation of five hybrids which differ in their simian virus 40-specific biological properties, *J. Virol.* **11,** 655.

Linstrom, D. M., and Dulbecco, R., 1972, Strand orientation of simian virus 40 transcription in productively infected cells, *Proc. Natl. Acad. Sci. USA* **69,** 1517.

McCutchan, J. H., and Pagano, J. S., 1968, Enhancement of the infectivity of simian virus 40 deoxyronucleic acid with diethylamino-ethyldextran, *J. Natl. Cancer Inst.* **41,** 351.

Magnusson, G., 1973, Hydroxyurea-induced accumulation of short fragments during polyoma DNA replication. I. Characterization of fragments, *J. Virol.* **12,** 600.

Magnusson, G., Pigiet, V., Winnacker, E. L., Abrams, R., and Reichard, P., 1973, RNA-linked short DNA fragments during polyoma replication, *Proc. Natl. Acad. Sci. USA* **70**, 412.

Manteuil, S., Pages, J., Stehelin, D., and Girard, M., 1973, Replication of simian virus 40 deoxyriboncleic acid: Analysis of the one-step growth cycle, *J. Virol.* **11**, 98.

Martin, M. A., 1970, Characteristics of SV40 DNA transcription during lytic infection, abortive infection, and in transformed mouse cells, *Cold Spring Harbor Symp. Quant. Biol.* **35**, 833.

Martin, M. A., and Axelrod, D., 1969*a*, SV40 gene activity during lytic infection and in a series of SV40 transformed mouse cells, *Proc. Natl. Acad. Sci. USA* **64**, 1203.

Martin, M. A., and Axelrod, D., 1969*b*, Polyoma virus gene activity during lytic infection and in transformed animal cells, *Science* (*Wash. D.C.*) **164**, 68.

Martin, M. A., and Khoury, G., 1973, Transcription of SV40 DNA in lytically infected and transformed cells, *in* "Virus Research" (C. F. Fox and W. S. Robinson, eds.), pp. 33–50, Academic Press, New York.

May, E., May, P., and Weil, R., 1973, "Early" virus-specific RNA may contain information necessary for chromosome replication and mitosis induced by simian virus 40, *Proc. Natl. Acad. Sci. USA* **70**, 1658.

Mayor, H. D., Stinebaugh, S. E., Jamison, R. M., Jordan, L. E., and Melnick, J. L., 1962, Immunofluorescent, cytochemical, and microcytological studies on the growth of the simian vacuolating virus (SV 40) in tissue culture, *Exp. Mol. Pathol.* **1**, 397.

Melnick, J. L., 1962, Papova virus group, *Science* (*Wash. D.C.*) **135**, 1128.

Melnick, J. L., and Rapp, F., 1965, The use of antiviral compounds in analyzing the sequential steps in the replication of SV40 papovavirus, *Ann. N. Y. Acad. Sci.* **130**, 291.

Melnick, J. L., 1962, Papovavirus group, *Science*, **135**, 1128.

Michel, M. R., Hirt, B., and Weil, R., 1967, Mouse cellular DNA enclosed in polyoma viral capsids (pseudovirions), *Proc. Natl. Acad. Sci. USA* **58**, 1381.

Middleton, J. H., Edgell, M. H., and Hutchison C. A., III, 1972, Specific fragmentation of ϕX174 deoxyribonucleic acid produced by a restriction enzyme from *Haemophilus aegyticus*, endonuclease 2, *J. Virol.* **10**, 42.

Morrow, J. F., and Berg, P., 1972, Cleavage of simian virus 40 DNA at a unique site by a bacterial restriction enzyme, *Proc. Natl. Acad. Sci. USA* **69**, 3365.

Morrow, J. F., and Berg, P., 1973, The location of the T4 gene 32 protein binding site on SV40 DNA, *J. Virol.* **12**, 1631.

Morrow, J. F., Berg, P., Kelly, T. J., Jr., and Lewis, A. M., 1973, Mapping of simian virus 40 early functions on the viral chromosome, *J. Virol.* **12**, 653.

Mueller, N., Zemla, J., and Brandner, G., 1973, Strand switch during *in vivo* polyoma transcription, *FEBS* (*Fed. Eur. Biochem. Soc.*) **31**, 222.

Mulder, C., and Delius, H., 1972, Specificity of the break produced by restricting endonuclease R_1 in simian virus 40 DNA, as revealed by partial denaturation mapping, *Proc. Natl. Acad. Sci. USA* **69**, 3215.

Murakami, W. T., Fine, R., Harrington, M. R., and Ben Sassan, Z., 1968, Properties and amino acid composition of polyoma virus purified by zonal ultracentrifugation, *J. Mol. Biol.* **36**, 153.

O'Conor, G. T., Rabson, A. S., Malmgren, R. A., Berezesky, I. K., and Paul, F. J., 1965, Morphologic observations of green monkey kidney cells after single and double infection with adenovirus 12 and simian virus 40, *J. Natl. Cancer Inst.* **34**, 679.

Oda, K., and Dulbecco, R., 1968*a*, Induction of cellular mRNA synthesis in BSC-1 cells infected by SV40, *Virology* **35,** 439.

Oda, K., and Dulbecco, R., 1968*b*, Regulation of transcription of the SV40 DNA in productively infected and in transformed cells, *Proc. Natl. Acad. Sci. USA* **60,** 525.

Okazaki, R., Okazaki, T., Sakabe, K., Sugimoto, K., and Sugino, A., 1968, Mechanism of DNA chain growth. I. Possible discontinuity and unusual secondary structure of newly synthesized chains, *Proc. Natl. Acad. Sci. USA* **59,** 598.

Opschoor, A., Pouwels, P. H., Knijnenburg, C. M., and Aten, J. B. T., 1968, Viscosity and sedimentation of circular native deoxyribonucleic acid, *J. Mol. Biol.* **37,** 13.

Oxman, M. N., and Levin, M. J., 1971, Interferon and transcription of early virus-specific RNA in cells infected with simian virus 40, *Proc. Natl. Acad. Sci. USA* **68,** 299.

Oxman, M. N., Rowe, W. P., and Black, P. H., 1967, Differential effects of interferon on SV-40 and adenovirus T antigen formation in cells infected with SV40 virus, adenovirus, and adenovirus–SV40 hybrid virues, *Proc. Natl. Acad. Sci. USA* **57,** 941.

Oxman, M. N., Takemoto, K. K., and Eckhart, W., 1972, Polyoma T antigen synthesis by temperature-sensitive mutants of polyoma virus, *Virology* **49,** 675.

Oxman, M. N., Levin, M. J., and Lewis, A. M., 1974, Control of SV40 gene expression in adenovirus–SV40 hybrid viruses: The synthesis of hybrid adenovirus 2–SV40 RNA molecules in cells infected with a nondefective adenovirus 2–SV40 hybrid virus, *J. Virol.*, in press.

Ozanne, B., Sharp, P. A., and Sambrook, J., 1973, Transcription of simian virus 40. II. Hybridization of RNA extracted from different lines of transformed cells to the separated strands of simian virus 40 DNA, *J. Virol.* **12,** 90.

Ozer, H. L., 1972, Synthesis and assembly of simian virus 40. I. Differential synthesis of intact virions and empty shells, *J. Virol.* **9,** 41.

Ozer, H., and Takemoto, K. K., 1969, Site of host restriction of simian virus 40 mutants in an established African green monkey kidney cell line, *J. Virol.* **4,** 408.

Ozer, H. L., and Tegtmeyer, P., 1972, Synthesis and assembly of simian virus 40. II. Synthesis of the major capsid protein and its incorporation into viral particles, *J. Virol.* **9,** 52.

Pages, J., Manteuil, S., Stehelin, D., Fiszman, M., Marx, M., and Girard, M., 1973, Relationship between replication of simian virus 40 DNA and specific events of the host cell cycle, *J. Virol.* **12,** 99.

Palacios, R., and Schinke, R. T., 1973, Identification and isolation of ovalbumin-synthesizing polysomes, *J. Biol. Chem.* **248,** 1424.

Patch, C. T., Lewis, A. M., and Levine, A. S., 1972, Evident for a transcription-control region of simian virus 40 in the adenovirus 2–simian virus 40 hybrid, $Ad2^{+}ND_{1}$, *Proc. Natl. Acad. Sci. USA* **69,** 3375.

Pettijohn, D., and Kamiya, 1967, Interaction of RNA polymerase with polyoma DNA, *J. Mol. Biol.* **29,** 275.

Pope, J. H., and Rowe, W. P., 1964, Detection of a specific antigen in SV40 transformed cells by immunofluorescence, *J. Exp. Med.* **120,** 121.

Potter, C. W., McLaughlin, B. C., and Oxford, J. S., 1969, Simian virus 40-induced T and tumor antigens, *J. Virol.* **4,** 574.

Rabson, A. S., O'Conor, G. T., Berezesky, I. K., and Paul, F. J., 1964, Enhancement of adenovirus growth in African green monkey kidney cell cultures by SV40, *Proc. Soc. Exp. Biol. Med.* **116,** 187.

Ralph, R. K., and Colter, J. S., 1972, Evidence for the integration of polyoma virus DNA in a lytic system, *Virology* **48,** 49.

Rapp, F., and Trulock, S. C., 1970, Susceptibility to superinfection of simian cells transformed by SV40, *Virology* **40,** 961.

Rapp, F., Kitahara, T., Butel, J. S., and Melnick, J. L., 1964*a*, Synthesis of SV40 tumor antigen during replication of simian papovavirus (SV40), *Proc. Natl. Acad. Sci. USA* **52,** 1138.

Rapp, F., Melnick, J. L., Butel, J. S., and Kitahara, T., 1964*b*, The incorporation of SV40 genetic material into adenovirus 7 as measured by intranuclear synthesis of SV40 tumor antigen, *Proc. Natl. Acad. Sci. USA* **52,** 1348.

Rapp, F., Butel, J. S., and Melnick, J. L., 1964*c*, Virus-induced intranuclear antigen in cells transformed by papovavirus SV40, *Proc. Soc. Exp. Biol. Med.* **116,** 1131.

Reeder, R. H., and Roeder, R. G., 1972, Ribosomal RNA synthesis in isolated nuclei, *J. Mol. Biol.* **67,** 433.

Reznikoff, C., Tegtmeyer, P., Dohan, C., Jr., and Enders, J. F., 1972, Isolation of AGMK cells partially resistant to SV40: Identification of the resistant step, *Proc. Soc. Exp. Biol. Med.* **141,** 740.

Ritzi, E., and Levine, A. J., 1969, Deoxyribonucleic acid replication in simian virus 40-infected cells. III. Comparison of simian virus 40 lytic infection in three different monkey kidney cell lines, *J. Virol.* **5,** 686.

Robb, J. A., and Martin, R. G., 1972, Genetic analysis of simian virus 40. III. Characterization of a temperature-sensitive mutant blocked at an early stage of productive infection in monkey cells, *J. Virol.* **9,** 956.

Roblin, R., Harle, E., and Dulbecco, R., 1971, Polyoma virus proteins, I. Multiple virion components *Virology* **45,** 555.

Roeder, R. G., and Rutter, W. J., 1969, Multiple forms of DNA-dependent RNA polymerase in eukaryotic organisms, *Nature* (*Lond.*) **224,** 234.

Roeder, R. G., and Rutter, W. J., 1970, Specific nucleolar and nucleoplasmic RNA polymerases, *Proc. Natl. Acad. Sci. USA* **65,** 675.

Rothschild, H., and Black, P. H., 1970, Analysis of SV40-induced transformation of hamster kidney tissue *in vitro.* VII. Induction of SV40 virus from transformed hamster cell clones by various agents, *Virology* **42,** 251.

Rovera, G., Baserga, R., and Defendi, V., 1972, Early increase in nuclear acidic protein synthesis after SV40 infection, *Nat. New Biol.* **237,** 240.

Rowe, W. P., and Baum, S. G., 1964, Evidence for a possible genetic hybrid between adenovirus type 7 and SV40 viruses, *Proc. Natl. Acad. Sci. USA.* **52,** 1340.

Rowe, W. P., and Baum, S. G., 1965, Studies of adenovirus–SV40 hybrid viruses. II. Defectiveness of the hybrid particles, *J. Exp. Med.* **122,** 955.

Rozenblatt, S., and Winocour, E., 1972, Covalently linked cell and SV40-specific sequences in an RNA from productively infected cells, *Virology* **50,** 558.

Rozenblatt, S., Lavi, S., Singer, M. F., and Winocour, E., 1973, Acquisition of sequences homologous to host DNA by closed circular simian virus 40 DNA. III. Host sequences, *J. Virol.* **12,** 501.

Rush, M. G., and Warner, R. C., 1970, Alkali denaturation of covalently closed circular duplex deoxyribonucleic acid, *J. Biol. Chem.* **245,** 2704.

Sabin, A. B., Shein, H. M., Koch, M. A. and Enders, J. F., 1964, Specific complement-fixing tumor antigens in human cells morphologically transformed by SV40 virus, *Proc. Natl. Acad. Sci. USA* **52,** 1316.

Sack, G. H., and Nathans, D., 1963, Studies of SV40 DNA VI. Cleavage of SV40 DNA by restriction endonuclease from *Hemophilus parainfluenza, Virology* **51,** 517.

Salzman, N. P., and Thoren, M. M., 1973, Inhibition in the joining of DNA intermediates to growing simian virus 40 chains, *J. Virol.* **11,** 721.

Salzman, N. P., Fareed, G. C., Sebring, E. D., and Thoren, M. M., 1973*a*, The mechanism of replication of SV40 DNA, in *Virus Research* (C. F. Fox and W. S. Robinson, eds.), pp. 71–87, Academic Press, New York.

Salzman, N. P., Sebring, E. D., and Radonovich, M., 1973*b*, Unwinding of parental strands simian virus 40 DNA replication, *J. Virol.* **12,** 669.

Sambrook, J., and Shatkin, A. J., 1969, Polynucleotide ligase activity in cells infected with simian virus 40, polyoma virus, or vaccinia virus, *J. Virol.* **4,** 719.

Sambrook, J. F., Westphal, H., Srinivasan, P. R., and Dulbecco, R., 1968, The integrated state of viral DNA in SV40-transformed cells, *Proc. Natl. Acad. Sci. USA* **60,** 1288.

Sambrook, J., Sharp, P. A., and Keller, W., 1972, Transcription of simian virus 40 I. Separation of the strands of SV40 DNA and hybridization of the separated strands to RNA extracted from lytically infected and transformed cells, *J. Mol. Biol.* **70,** 57.

Sauer, G., 1971, Apparent differences in transcriptional control in cells productively infected and transformed by SV40, *Nat. New Biol.* **231,** 135.

Sauer, G., and Kidwai, J. R., 1968, The transcription of the SV40 genome in productively infected and transformed cells, *Proc. Natl. Acad. Sci. USA* **61,** 1256.

Schechter, I., 1973, Biologically and chemically pure mRNA coding for a mouse immunoglobin L-chain prepared with the aid of antibodies and immobilized oligothymidine. *Proc. Natl. Acad. Sci. USA* **70,** 2256.

Schlumberger, H. D., Anderer, F. A., and Koch, M. A., 1968, Structure of the simian virus 40. IV. The polypeptide chains of the virus particle, *Virology* **36,** 42.

Sebring, E. D., Kelly, T. J., Jr., Thoren, M. M., and Salzman, N. P., 1971, Structure of replicating simian virus 40 deoxyribonucleic acid molecules, *J. Virol.* **8,** 478.

Seebeck, T., and Weil, R., 1974, Polyoma viral DNA replicated as a nucleoprotein complex in close association with the host cell chromatin, *J. Virol.*, **13,** 567.

Sharp, P. A., Sugden, B., and Sambrook, J., 1973, Detection of two restriction endonuclease activities in *Haemophilus parainfluenzae* using analytical agaroser-ethidium bromide electrophoresis, *Biochemistry* **12,** 3055.

Sheldon, R., Jurale, C., and Kates, J., 1972, Detection of polyadenylic acid squences in viral and eukaryotic RNA, *Proc. Natl. Acad. Sci. USA* **69,** 417.

Shih, T. Y., Khoury, G., and Martin, M. A., 1973, *In vitro* transcription of the viral specific sequences present in the chromatin of SV40-transformed cells, *Proc. Natl. Acad. Sci. USA* **70,** 3506.

Shimono, H., and Kaplan, A. S., 1969, Correlation between the synthesis of DNA and histones in polyoma virus-infected mouse embryo cells, *Virology* **37,** 690.

Shiroki, K., and Shimojo, H., 1971, Transformation of green monkey kidney cells by SV40 genome: The establishment of transformed cell lines and the replication of human adenoviruses and SV40 in transformed cells, *Virology* **45,** 163.

Sjogren, H. O., Hellstrom, I., and Klein, G., 1961, Transplantation of polyoma virus-induced tumors in mice, *Cancer Res.* **21,** 329.

Smith, H. O., and Wilcox, K., 1970, A restriction enzyme from *Hemophilus influenza* I. Purification and general properties, *J. Mol. Biol.* **51,** 379.

Stewart, S. E., Eddy, B. E., Gochenour, A. M., Borgese, N. G., and Grubbs, G. E., 1957, The inductions of neoplasms with a substance released from mouse tumors by tissue culture, *Virology* **3,** 380.

Stirpe, F., and Fiume, L., 1967, Studies on the pathogenesis of liver necrosis by α-

amanitin. Effect of α-amanitin on ribonucleic acid synthesis and on ribonucleic acid polymerase in mouse liver nuclei, *Biochem. J.* **105,** 779.

Sugden, B., and Keller, W., 1973, Mammalian deoxyribonucleic acid-dependent ribonucleic acid polymerases I. Purification and properties of an α-amanitin-sensitive ribonucleic acid polymerase and stimulatory factors from HeLa and KB cells, *J. Biol. Chem.* **248,** 3777.

Summers, W. C. and Siegal, R. B., 1970, Transcription of late phage RNA by T7 RNA polymerase, *Nature* (*Lond.*) **228,** 1160.

Sweet, B. H., and Hilleman, M. R., 1960, The vacuolating virus, SV40, *Proc. Soc. Exp. Biol. Med.* **105,** 420.

Swetly, P., Barbanti-Brodano, G., Knowles, B., and Koprowski, H., 1969, Response of simian virus 40-transformed cell lines to superinfection with simian virus 40 and its deoxyribonucleic acid, *J. Virol.* **4,** 348.

Tai, H. T., Smith, C. A., Sharp, P. A., and Vinograd, J., 1972, Sequence heterogeneity in closed simian virus 40 deoxyribonucleic acid, *J. Virol.* **9,** 317.

Takemoto, K. K., and Habel, K., 1965, Hamster ascitic fluids containing complement-fixing antibody against virus-induced tumor antigens, *Proc. Soc. Exp. Biol. Med.* **120,** 124.

Takemoto, K. K., and Martin, M. A., 1970, SV40 thermosensitive mutant: Synthesis of viral DNA and virus-induced proteins at nonpermissive temperatures, *Virology* **42,** 938.

Takemoto, K. K., and Mullarkey, M. F., 1973, Human papovavirus, BK strain: Biological studies including antigenic relationship to simian virus 40, *J. Virol.* **12,** 625.

Takemoto, K. K., Kirschstein, R. L., and Habel, K., 1966, Mutants of simian virus 40 differing in plaque size, oncogenicity, and heat sensitivity, *J. Bacteriol.* **92,** 990.

Tan, K. B., and Sokol, F., 1972, Structural proteins of simian virus 40: Phosphoproteins, *J. Virol.* **10,** 985.

Tan, K. B., and Sokol, F., 1973, Phosphorylation of simian virus 40 proteins in a cell-free system, *J. Virol.* **12,** 676.

Tegtmeyer, P., 1972, Simian virus 40 deoxyribonucleic acid synthesis: The viral replicon, *J. Virol.* **10,** 591.

Tevethia, S. S., and Rapp, F., 1965, Demonstration of new surface antigens in cells transformed by papovavirus SV40 by cytotoxic tests, *Proc. Soc. Exp. Biol. Med.* **120,** 455.

Thoren, M. M., Sebring, E. D., and Salzman, N. P., 1972, Specific initiation site for simian virus 40 deoxyribonucleic acid replication, *J. Virol.* **10,** 462.

Thorne, H. V., 1973, Cyclic varation in susceptibility of Balb-C 3T3 cells to polyoma virus, *J. Gen. Virol.* **18,** 163.

Thorne, H. V., Evans, J., and Warden, D., 1968, Detection of biologically defective molecules in component I of polyoma virus DNA, *Nature* (*Lond.*) **219,** 728.

Todaro, G. J., and Martin, G. M., 1967, Increased susceptibility of Down's syndrome fibroblasts to transformation by SV40, *Proc. Soc. Exp. Biol. Med.* **124,** 1232.

Todaro, G. J., Green, H., and Swift, M. R., 1966, Susceptibility of human diploid fibroblast strains to transformation by SV40 virus, *Science* (*Wash. D.C.*) **153,** 1252.

Tonegawa, S., Walter, G., Bernardini, A. and Dulbecco, R., 1970, Transcription of the SV40 genome in transformed cells and during lytic infection, *Cold Spring Harb. Symp. Quant. Biol.* **35,** 833.

Trilling, D. M., and Axelrod, D., 1970, Encapsidation of free host DNA by simian virus 40: A simian virus 40 pseudovirus, *Science* (*Wash. D.C.*) **168,** 268.

Trilling, D., and Axelrod, D., 1972, Analysis of the three components of simian virus 40: Pseudo-, mature, and defective viruses, *Virology* **47,** 360.

Uchida, S., Yoshiike, K., Watanabe, S., and Furano, A., 1968, Antigen-forming defective viruses of simian virus 40, *Virology* **34,** 1.

Vinograd, J., and Lebowitz, J., 1966, Physical and topological properties of circular DNA, *J. Gen. Physiol.* **49,** 103.

Vinograd, J., Lebowitz, J., Radloff, R., Watson, R., and Laipis, P., 1965, The twisted circular form of polyoma viral DNA, *Proc. Natl. Acad. Sci. USA* **53,** 1104.

Vinograd, J., Lebowitz, J., and Watson, R., 1968, Early and late helix-coil transitions in closed circular DNA. The number of superhelical turns in polyoma DNA, *J. Mol. Biol.* **33,** 173.

Walter, G., Roblin, R., and Dulbecco, R., 1972, Protein synthesis in simian virus 40-infected monkey cells, *Proc. Natl. Acad. Sci. USA* **69,** 921.

Wang, J. C., 1971, Interaction between DNA and an *Escherichia coli* protein ω, *J. Mol. Biol.* **55,** 523.

Warnaar, S. O., and deMol, A. W., 1973, Characterization of two simian virus 40-specific RNA molecules from infected BSC-1 cells, *J. Virol.* **12,** 124.

Watkins, J. F., and Dulbecco, R., 1967, Production of SV40 virus in heterokaryons of transformed and susceptible cells, *Proc. Natl. Acad. Sci. USA* **58,** 1396.

Weil, R., and Vinograd, J., 1963, The cyclic helix and cyclic coil forms of polyoma viral DNA, *Proc. Natl. Acad. Sci. USA* **50,** 730.

Weinberg, R. A., Warnaar, S. O., and Winocour, E., 1972*a*, Isolation and characterization of simian virus 40 ribonucleic acid, *J. Virol.* **10,** 193.

Weinberg, R. A., Ben-Ishai, Z., and Newbold, J. E., 1972*b*, Poly A associated with SV40 messenger RNA, *Nat. N. Biol.* **238,** 111.

Weinberg, R. A., Ben-Ishai, Z., and Newbold, J. E., 1974, SV40-transcription in productively infected and transformed cells, *J. Virol.* **13,** 1263.

Westphal, H., 1970, SV40 DNA strand selection by *Escherichia coli* RNA polymerase, *J. Mol. Biol.* **50,** 407.

Westphal, H., 1971, Transcription of superhelical and relaxed circular SV40 DNA by *E. coli* RNA polymerase in the presence of rifampicin, *Lepetit. Colloq. Biol. Med.* **2,** 77.

Westphal, H., and Dulbecco, R., 1968, Viral DNA in polyoma-and SV40-transformed lines, *Proc. Natl. Acad. Sci. USA* **59,** 1158.

Westphal, H., Delius, H., and Mulder, C., 1973, Visualization of SV40 *in vitro* transcription complexes, *Lepetit, Colloq. Biol. Med.* **4,** 183.

Wever, G. H., Kit, S., and Dubbs, D. R., 1970, Initial site of synthesis of virus during rescue of simian virus 40 from heterokaryons of simian virus 40-transformed and susceptible cells, *J. Virol.* **5,** 578.

White, M., and Eason, R., 1971, Nucleoprotein complexes in simian virus 40-infected cells, *J. Virol.* **8,** 363.

Wickus, G. G., and Robbins, P. W., 1973, Plasma membrane proteins of normal and Rous sarcoma virus-transformed chick-embryo fibroblasts, *Nat. New Biol.* **245,** 65.

Winnacker, E. L., Magnusson, G., and Reichard, P., 1972, Replication of polyoma DNA in isolated nuclei. I. Characterization of the system from mouse fibroblast 3T6 cells, *J. Mol. Biol.* **72,** 523.

Winocour, E., and Robbins, E., 1970, Histone synthesis in polyoma and SV40-infected cell DNA, *Virology* **40,** 307.

Yelton, D. B., and Aposhian, H. V., 1972, Polyoma pseudovirions. I. Sequence of events in primary mouse embryo cells leading to pseudovirus production, *J. Virol.* **10,** 340.

Yoshiike, K., 1968*a*, Studies on DNA from low-density particles of SV40. I. Heterogeneous defective virions produced by successive undiluted passages, *Virology* **34,** 391.

Yoshiike, K., 1968*b*, Studies on DNA from low-density particles of SV40. II. Noninfectious virions associated with a large-plaque variant, *Virology* **34,** 402.

Yoshiike, K., and Furano, A., 1969, Heterogeneous DNA of simian virus 40, *Fed. Proc.* **28,** 1899.

Yoshiike, K., Furano, A., and Suzuki, K., 1972, Denaturation maps of complete and defective simian virus 40 DNA molecules, *J. Mol. Biol.* **70,** 415.

Zain, B. S., Dhar, R., Weissman, S. M., Lebowitz, P., and Lewis, A. M., 1973, Preferred site for initiation of RNA transcription by *Escherichia coli* RNA polymerase within the simian virus 40 DNA segment of the nondefective adenovirus-simian virus 40 hybrid viruses $Ad2^{+}ND_{1}$ and $Ad2^{+}ND_{3}$, *J. Virol.* **11,** 682.

CHAPTER 3

Reproduction of Adenoviruses

Lennart Philipson and Uno Lindberg
Department of Microbiology
The Wallenberg Laboratory
Uppsala University
Uppsala, Sweden

1. INTRODUCTION

The name adenovirus was coined in 1956 (Enders *et al.*, 1956) to designate a group of viruses isolated from the respiratory tracts of man and other animals. The first strains isolated were described in 1953 by Rowe *et al.* (1953) and in 1954 by Hilleman and Werner (1954). Today more than 80 different adenovirus serotypes have been isolated from a variety of animal species (Wadell, 1970), and all except the avian adenoviruses share one antigenic determinant. The human adenoviruses have been divided into subgroups on the basis of their ability to agglutinate rhesus monkey and rat erythrocytes (Rosen, 1960) and on the basis of their oncogenicity (Huebner *et al.*, 1965). Each subgroup contains several serotypes characterized by type-specific antigens present in their capsids, as evidenced by hemagglutination-inhibition or neutralization tests.

In humans, adenoviruses mainly cause milder respiratory disease, but cases of conjunctivitis, myocarditis, enteritis, and lymph node involvement have also been reported (Sohier *et al.*, 1965), and the epidemic character of the respiratory illness caused by some serotypes among military recruits has emphasized the need for the production of multivalent vaccines (Hilleman, 1966). One aspect of adenoviruses that has received much attention during the last ten years is that many of them are oncogenic. This was discovered in 1962 by Trentin *et al*

(1962), who found that human adenovirus type 12 (Ad 12) induces tumors in newborn hamsters. The same observation was reported by Huebner *et al.* (1962), who found that Ad 18 also had this property. Subsequently, a series of reports confirmed the oncogenicity in hamsters of human Ad 12 and also described the oncogenicity of several other human and nonhuman adenoviruses in several species of rodents (Huebner *et al.*, 1963, 1965; Rabson *et al.*, 1964*a*; Yabe *et al.*, 1964; Pereira *et al.*, 1965; Girardi *et al.*, 1964; Hull *et al.*, 1965; Darbyshire, 1966; Sarma *et al.*, 1965). The tumors formed usually have the characteristics of undifferentiated sarcomas, although malignant lymphomas were occasionally observed (Larsson *et al.*, 1965).

Transformation of *in vitro*-cultured cells by an adenovirus was first demonstrated by McBride and Wiener (1964), who showed that the highly oncogenic human Ad 12 could transform newborn hamster kidney cells. Subsequently, the transformation of rat embryo fibroblasts by the same adenovirus type was reported by Freeman *et al.* (1967*a*). Since then, several adenovirus types have been shown to cause *in vitro* transformation of rodent cells (Freeman *et al.*, 1967*b,c*; Van der Noorda, 1968*a,b*; McAllister *et al.*, 1969*a,b*; Riggs and Lennette, 1967; Casto, 1969), and the tumorigenic properties of the *in vitro*-transformed cells have been well established. The biochemical studies of the transformation of animal cells by oncogenic viruses *in vitro* have progressed during recent years, and much of our basic knowledge concerning growth regulation and carcinogenesis has been derived from studies on such *in vitro* systems (Pontén, 1971).

In addition to providing an insight into the changes involved in transformation of normally growing cells to tumor cells, the adenoviruses have become increasingly more important in providing convenient model systems for studies of DNA replication and gene expression in eukaryotic cells. In the transformed cells viral genes apparently have become integrated into the genome of the host cell (Green, 1970), and results of studies on the expression of these viral genes, the products of which can be identified, thus directly reflect the mechanisms used by the host cell in expressing its own genes. This was realized several years ago, and studies on SV40-transformed cells (Lindberg and Darnell, 1970; Tonegawa *et al.*, 1970) gave the first evidence suggesting that mRNAs in eukaryotes are synthesized as high-molecular-weight precursors. Recent developments in the biochemical analysis of the biogenesis of adenoviruses indicate that productively infected cells might also serve as a convenient model system for studies on the synthesis of macromolecules in the eukaryotic cell. Several observations strongly suggest that also under these conditions the adenovirus mimics the host cell in its gene expression. After the infection, the virus is rapidly established in the nucleus of the cell, where its DNA is transcribed and replicated (Green *et al.*, 1970). The viral DNA is transcribed into large RNA molecules (Green *et al.*, 1970; Wall *et al.*, 1972; McGuire *et al.*, 1972), which by post-transcriptional processes appear to be cleaved into the smaller mRNA molecules found on cytoplasmic polyribosomes (Green *et al.*, 1970; Philipson *et al.*, 1971; Lindberg *et al.*, 1972). Adenovirus mRNA is polyadenylated at the 3′ terminus in the same way as cell mRNA, and studies on this system have provided clear evidence that the polyadenylation of mRNA is a post-transcriptional event (Philipson *et al.*, 1971). Finally, it has been found recently that the messenger ribonucleoprotein particles (mRNP) containing viral mRNA isolated from polysomes during virus infection contain five major labeled polypeptides, four of which are identical in size to those found on normal cell mRNP (Lindberg and Sundquist, 1974).

Today, a large number of temperature-sensitive (*ts*) adenovirus mutants have been isolated (Williams *et al.*, 1971; Lundholm and Doerfler, 1971; Ensinger and Ginsberg, 1972; Shiroki *et al.*, 1972; Ishibashi, 1970, 1971). More than 50 *ts* mutants distributed between some 14 complementation groups have been described for Ad 5, and they are currently undergoing biochemical and genetic analysis (Williams and Ustacelebi, 1971; Russell *et al.*, 1972*a*; Wilkie *et al.*, 1973). The DNA of several adenovirus types has been cleaved by a restriction endonuclease into unique sets of fragments which can be separated (Pettersson *et al.*, 1973; Mulder *et al.*, 1974). The DNA strands of at least two adenovirus types have been separated (Landgraf-Leurs and Green, 1971; Patch *et al.*, 1972), and in the case of Ad 2 the separation of the two strands of each of the separated fragments can be achieved (Tibbetts and Pettersson, 1970). All together, this information means that the full armamentarium for a detailed mapping of the organization of the genes in this virus will soon be available.

This rapidly expanding field has been reviewed frequently during recent years, with emphasis on different aspects of adenovirus research. Relevant reviews of a general character are those of Green (1966, 1970), Schlesinger (1969), Ginsberg (1969), and Pettersson (1973*a*). The characteristics of the structural proteins (Philipson and Pettersson, 1973) and the biological properties of the capsid components (Norrby, 1968) have also been reviewed. Reviews on the oncogenic and transforming capacity of adenoviruses have also appeared (Green,

1970; Homburger *et al.*, 1973; Pettersson, 1973*a*). The present chapter will emphasize recent contributions which help to elucidate adenovirus reproduction and will focus on the possibility of using an animal DNA virus as a model for molecular events in replication, transcription, and translation in eukaryotic cells.

2. ARCHITECTURE AND COMPOSITION OF THE VIRION

The adenoviruses are nonenveloped viruses 65–80 nm in diameter. The DNA-containing core is surrounded by a protein capsid composed of 252 capsomers arranged into an icosahedron (20 triangular facets, 12 vertices), as originally described for Ad 5 in the classical electron microscopic study of Horne *et al.* (1959). Figure 1 shows a schematic drawing of an adenovirus with the major components of the capsid indicated.

Of the capsomers, 240 have 6 neighbors and are called hexons (Ginsberg *et al.*, 1966), whereas the 12 vertex capsomers have 5 neighbors and are called pentons (Ginsberg *et al.*, 1966). Each penton unit consists of a base anchored in the capsid and an outward projection (Valentine and Pereira, 1965; Norrby 1966). This latter structure is called the fiber. Two kinds of hexons may be defined: (1) those located in juxtaposition to the pentons, peripentonal hexons, which dif-

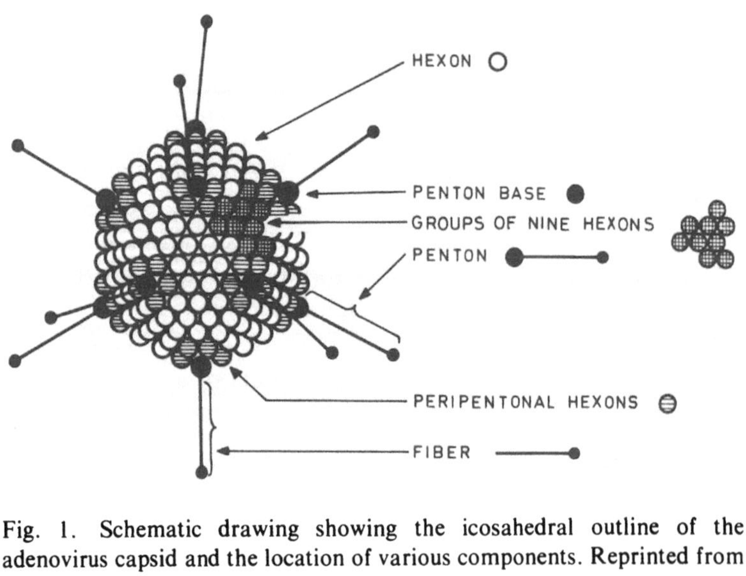

Fig. 1. Schematic drawing showing the icosahedral outline of the adenovirus capsid and the location of various components. Reprinted from Philipson and Pettersson (1973).

fer topologically from the remaining hexons in that they face the penton base as one of their 6 neighbors (Prage *et al.*, 1970), (2) and those, 180, which form the major part of the triangular facets and the edges of the icosahedron. Inside the capsid there is a core structure containing the DNA and additional protein, as originally revealed by thin-section electron microscopy (Epstein, 1959; Epstein *et al.*, 1960; Bernhard *et al.*, 1961).

Unlike the situation for most other viruses, the capsid proteins of adenoviruses are soluble under nondenaturing conditions. This has greatly faciliated the purification and characterization of the virus components. Several additional polypeptides have been discovered during the course of these investigations, and through the development of methods for sequential disintegration of the virion (Maizel *et al.*, 1968*b*; Laver *et al.*, 1969; Prage *et al.*, 1970) a rather detailed picture of the topography of the proteins in this virus has been obtained.

The pentons alone or together with peripentonal hexons may be selectively removed from virions after dialysis against water (Laver *et al.*, 1969) or Tris-maleate buffer *p*H 6.0–6.5 (Prage *et al.*, 1970). These pentonless virions are porous but stable structures, where the DNA is accessible to degradation with deoxyribonuclease. The hexons from the triangular facets can be isolated after dissociation of the virus particles with SDS, urea, or pyridine (Smith *et al.*, 1965; Maizel *et al.*, 1968*b*; Prage *et al.*, 1970) as groups of nine hexons in a defined structure with local threefold symmetry (Prage *et al.*, 1970; Crowther and Franklin, 1972). The core has been isolated from virions after treatment with heat (Russel *et al.*, 1967*a*, 1971), acetone (Laver *et al.*, 1967, 1968), 5M urea (Maizel *et al.*, 1968*b*), formamide (Stasny *et al.*, 1968), 10% pyridine, or repeated freezing and thawing (Prage *et al.*, 1968, 1970).

The complex polypeptide composition of adenoviruses has been studied by SDS-polyacrylamide gel electrophoresis (Maizel *et al.*, 1968,*a,b*; Pereira and Skehel, 1971; Everitt *et al.*, 1973; Anderson *et al.*, 1973). As shown in Fig. 2, Ad 2, which has been studied extensively, may contain as many as 15 polypeptides (Everitt *et al.*, 1973; Anderson *et al.*, 1973), but as has been pointed out the same "polypeptides" may be generated by proteolytic cleavage of the most abundant species (Pereira and Skehel, 1971). So far, 8 of the polypeptides have been shown to be antigenically distinct and to reside in different structures after sequential degradation of the virion (Everitt *et al.*, 1973, and unpublished). Five of the polypeptides are integral parts of the capsomers or the core structure. Polypeptide II is the constituent polypeptide of the hexon, III of the penton base, IV of the

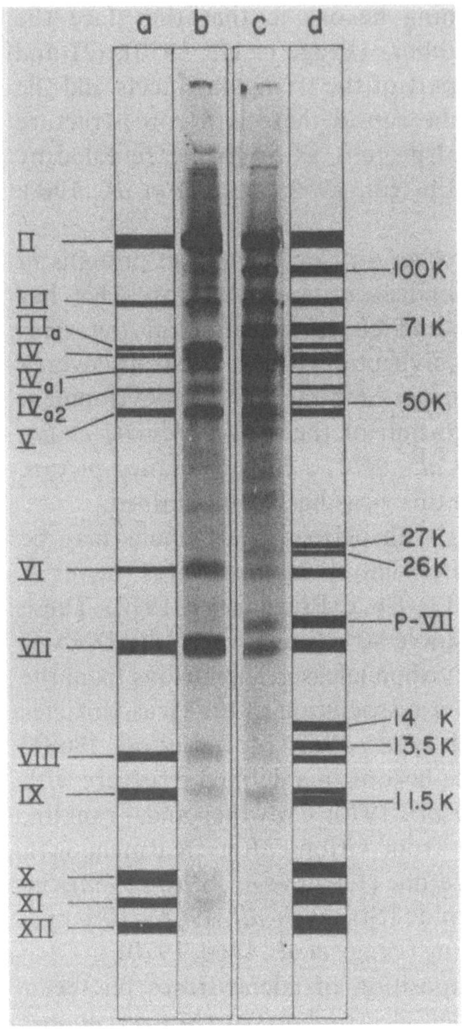

Fig. 2. SDS-polyacrylamide gel autoradiograms of ^{35}S-methionine-labeled purified virus and extract of infected whole cells. The gel contained 15% acrylamide and 0.08% bisacrylamide. The polypeptide pattern of purified adenovirus is shown in frame b and the polypeptide pattern of the cell extract is shown in frame c. The extract was obtained from cells labeled for 1 hour with ^{35}S-methionine 18 hours post-infection and then chased for 12 hours in the presence of 30 μg/ml of nonradioactive methionine. Frames a and d are drawings representing an idealized gel in which (a) all the virion components and (d) all the 22 virus-induced components are illustrated. The nomenclature used to identify the bands for the virus is reported in Everitt *et al.* (1973) and Anderson *et al.* (1973). The bands in the cell extract are identified by their molecular weight in daltons $\times\ 10^{-3}$ except for PVII which has been shown to be a precursor of polypeptide VII in the virus (Anderson *et al.*, 1973). Reprinted by permission from Anderson *et al.* (1973).

fiber, and V and VII are the two core proteins. The remaining polypeptides have been tentatively located in the virion structure as schematically indicated in Fig. 3, but we still lack rigorous evidence to establish that all polypeptides are unique.

The following polypeptides appear to be associated with the hexons. Polypeptide VI can be demonstrated in all fractions from disintegrated virus which contain hexons, and it cosediments with hexons in sucrose gradients at low salt concentration irrespective of whether the hexons are obtained by freezing and thawing or by pyridine treatment (Everitt *et al.*, 1973). There appear to be about 2

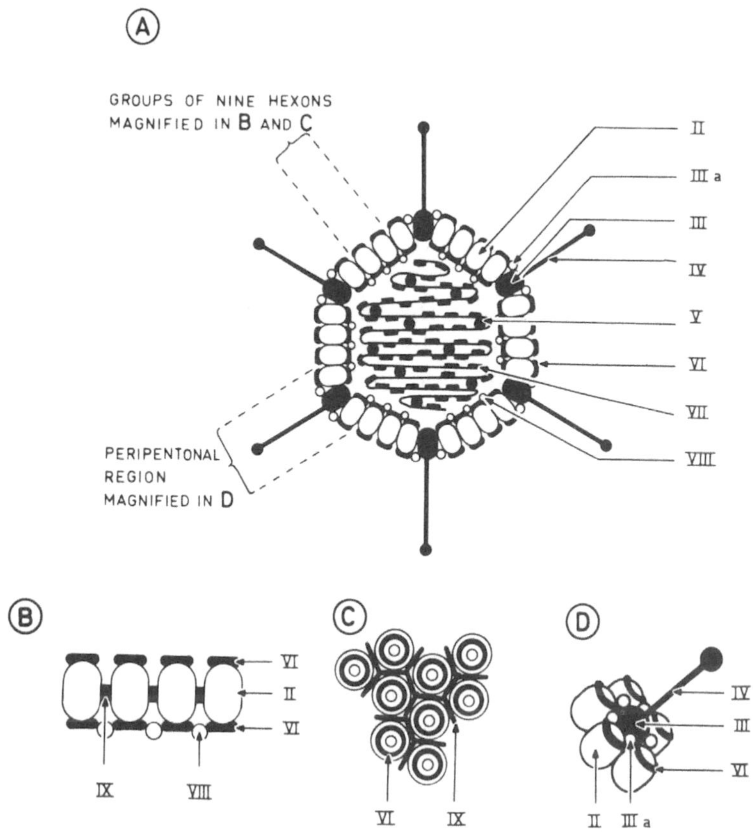

Fig. 3. A hypothetical model of the location of isolated proteins in the virus particle of Ad 2. The molar ratio between the internal basic proteins VII and V is 5, which is maintained in the schematic representation. Protein VII may cover about 50% of all phosphates of the DNA (Prage and Pettersson, 1971). The molar ratio of protein VI to hexons is about 2. Protein VI may, therefore, be located both at the inner and outer surface of the hexons. Peripentonal hexons also possess protein VI. The acidic, low-molecular-weight protein IX might be the cementing substance between the hexons since it is the only polypeptide, except for protein VI, associated with groups of nine hexons after pyridine treatment. Protein VIII, associated with the facet hexons after sequential disintegration and showing basic properties, may reside at the inner surface of the triangular facets to neutralize all or part of the residual DNA-phosphate groups. Polypeptide IIIa may be associated with material in the peripentonal region of the virion. The localization of protein X is unclear at present. (A) A schematic view of a vertical section of Ad 2. The different polypeptides are indicated by their Roman numerals according to Fig. 2. Magnified views of groups of nine hexons and the peripentonal region are given in B, C, and D. (B) A vertical section through a group of nine hexons showing the tentative location of proteins VI, VIII, and IX. (C) A horizontal view from the outside of the group of nine hexons showing the tentative location of proteins VI and IX. (D) A magnification of the peripentonal region showing the proteins II, III, IIIa, IV, and VI. Reprinted from Everitt *et al.* (1973).

molecules of polypeptide VI per molecule of hexon. Polypeptide VIII is recovered in association with hexons after freezing and thawing, but it is separated from the groups of nine hexons after degradation by pyridine. Thus it is possible that this polypeptide is associated with the hexon in the capsid, but that pyridine dissociates the hexon/polypeptide VIII complex (Maizel *et al.*, 1968*b*; Everitt *et al.*, 1973). Polypeptide IX appears to be present in preparations of groups of nine hexons (Maizel *et al.*, 1968*b*; Everitt *et al.*, 1973), but was not recovered in association with the hexons when the capsid was disintegrated into single capsomers by freezing and thawing. It is believed that this polypeptide is involved in keeping the hexons of the facets together. Polypeptide IIIa, appears to be preferentially released from the vertex region after pyridine treatment and together with the peripentonal hexons after dialysis against Tris-maleate buffer, *p*H 6.3; thus, it may be located in the peripentonal region (Everitt *et al.*, 1973). The origin of the smallest polypeptide X is not yet established and recently it was claimed that this band, which migrates with the front on 13% gels, might be resolved into 3 components—X, XI, and XII (Anderson *et al.*, 1973). In addition to these polypeptides, which each constitute as much as 0.1% or more of the total mass of protein in the virion based on radioactivity, five minor components can be observed in purified virions (Everitt *et al.*, 1973; Anderson *et al.*, 1973; see also Fig. 2). These, however, are present only in few copies per virus particle and may not represent polypeptides unique to the virion (Everitt *et al.*, 1973). Recent evidence suggests that some of these additional components might correspond to precursor polypeptides to polypeptides VI and VII of the virion, which normally are cleaved during assembly (Anderson *et al.*, 1973), as discussed in the section on assembly of adenovirus (Sect. 6.9).

3. BIOCHEMISTRY OF THE STRUCTURAL PROTEINS OF ADENOVIRUS

In the late phase of the infection, the synthesis of host cell proteins is turned off (Ginsberg *et al.*, 1967) and the protein-synthesizing machinery is mainly engaged in making large amounts of viral structural proteins. The viral polypeptides are made in excess of the amounts used for virus assembly; in fact, only 1–5% of the fiber and the penton base and 20–30% of the hexon synthesized are ever used for the assembly of mature virions (White *et al.*, 1969; Everitt *et al.*, 1971). The polypeptides produced in excess are present as multimeric proteins rather than polypeptide subunits (Horwitz *et al.*, 1969), the hexon, penton, and fiber proteins have been purified from this excess pool, and their immunological, chemical, and physical properties have been studied in detail. Some of the properties of the purified proteins will be summarized here.

The hexon is a hollow, cylinderlike object with a height of 12.5 nm and a diameter of 8–8.5 nm, as originally shown by electron microscopy (Wilcox *et al.*, 1963; Pettersson *et al.*, 1967) and recently confirmed by low-angle X-ray diffraction studies (Tejg-Jensen *et al.*, 1972). The protein from highly purified preparations has a molecular weight of 310,000–360,000 daltons (Franklin *et al.*, 1971*a*) and the constituent subunit polypeptide (II) moves on SDS-gels as having a molecular weight of around 120,000 (Maizel 1968*a*; Horwitz *et al.*, 1970). Estimation of the molecular weight of the monomer by X-ray crystallography, however, suggests a value around 95,000 (Cornick *et al.*, 1971, 1973), which has been corroborated by biochemical techniques involving determination of the unique cysteines in the hexon (Jörnvall *et al.*, 1974*b*). Thus, it is possible that the SDS-gel technique has overestimated the molecular weight, as was shown also to be the case for the polypeptide of tobacco rattle virus (Ghabrial and Lister, 1973). The subunit of the hexon contains 7 cysteines in a total of approximately 850 amino acid residues, and about 20% of the primary sequence has now been determined (Jörnvall, unpublished). The N terminus of the subunit is blocked, but by utilizing *in vivo* labeling with ^{14}C-acetate, it was possible to isolate a tetrapeptide (Acetyl-Ala-Thr-Pro-Ser) containing the acetylated N terminal (Jörnvall *et al.*, 1974*a*). The hexon from human Ad 2 and Ad 5 has been crystallized (Pereira *et al.*, 1968; Franklin *et al.*, 1971*a*; Cornick *et al.*, 1971). The crystals described were of a cubic type falling in the $P2_13$ space group (Franklin *et al.*, 1971*a*; Cornick *et al.*, 1971). This space group is characterized by 12 asymmetrical units per cell, and, since the unit cell contains four hexons, there must be three crystallographic asymmetrical units per hexon. This infers that there are only three structural units along the cylinder axis of the hexon, a situation which has also been confirmed by Patterson projection studies and by electron microscopy (Franklin *et al.*, 1971*b*; Crowther and Franklin, 1972). A dyad perpendicular to the threefold symmetry axis has been suggested (Franklin *et al.*, 1971*b*), but the structural implication of a twofold axis is uncertain, since there are only 3 polypeptides per hexon.

Immunologically, the hexon is a complex structure with one type-specific determinant (ϵ) (Köhler, 1965; Norrby, 1969*a*; Pettersson,

1971). This type specificity can be correlated with minor differences in the amino acid composition of hexons from different serotypes (Pettersson, 1971). In addition, the hexon carries a group-specific determinant, α, which shows marked cross-reactivity among all members of the adenovirus group, irrespective of the species of origin with one exception, the avian adenoviruses, which lack this determinant (Pereira *et al.*, 1963).

The fiber is an antennalike structure consisting of a rod terminated by a knob at the outer end; the latter has a diameter of 4 nm. The total length of the fiber varies among different subgroups. Viruses belonging to Rosen's subgroup I (Rosen, 1960) have the shortest length, 10 nm, and those of subgroup III have the longest length, 25–30 nm (Norrby, 1968, 1969*b*). The length seems to be correlated with the antigenic complexity of the fiber. The molecular weight of the fiber from human adenovirus type 2 appears to be 200,000 and SDS-polyacrylamide gel electrophoresis reveals only one polypeptide band corresponding to a molecular weight of 60,000–65,000 (polypeptide IV) (Sundquist *et al.*, 1973*a*). This would indicate that there are three subunits per fiber molecule for this virus, but it remains to be demonstrated that they are all identical. The fiber of Ad 5 has also been crystallized (Mautner and Pereira, 1971), but no structural studies have been reported. The type-specific determinants reside in the knob part of the fiber, and the long fibers of Rosen's subgroup II and III have additional antigenic determinants in the rod part, which are subgroup-specific (Norrby, 1968, 1969*b*; Pettersson *et al.*, 1968). The fibers appear to be attachment organs of the virion since purified fiber can block the attachment of intact virions to receptors on KB and HeLa cells (Philipson *et al.*, 1968). It has also been demonstrated that the fiber binds to erythrocyte receptors, but in the monomeric form it does not cause hemagglutination. On the other hand, dimers of fibers, dimers of pentons, aggregates of 12 pentons, so-called dodecons, as well as intact virions cause hemagglutination [for a review see Norrby (1969*b*)].

The penton base and the fiber are linked to each other probably by noncovalent bonds since they can be dissociated by 2.5 M guanidine hydrochloride or 8% pyridine (Norrby and Skaaret, 1967; Pettersson and Höglund, 1969). The molecular weight of the intact penton is about 500,000 (Pettersson and Höglund, 1969; Wadell, 1970), and the penton base then should have a molecular weight of about 300,000. It appears also to be composed of one type of subunit with a molecular weight of about 70,000 (polypeptide III) (Maizel *et al.*, 1968*a*). The penton base contains at least two different antigenic determinants, one group- and the other subgroup-specific (Wadell and Norrby, 1969; Pettersson and Höglund, 1969). Purified penton or isolated penton base can cause detachment of monolayer cells from their solid support (Pereira, 1958; Everett and Ginsberg, 1958; Rowe *et al.*, 1958; Pettersson and Höglund, 1969). The cells appear rounded and swollen, but there is no demonstrable shut off of macromolecular synthesis. This cytopathic effect is reversible, and its biological significance is not understood.

The major core protein (polypeptide VII) has a molecular weight of 17,000 and is rich in arginine (21%) and alanine (18%) (Laver, 1970; Prage and Pettersson, 1971; Russell *et al.*, 1971). Although it is similar to the arginine-rich histones, it differs from these in that it contains tryptophan and is precipitated by virus-specific antiserum (Prage and Pettersson, 1971). There is no evidence that this core protein is derived from the host (Maizel *et al.*, 1968*a*; Everitt *et al.*, 1973). It is synthesized late in the infection and apparently in a small excess of what is utilized for assembly of mature virions (White *et al.*, 1969; Everitt *et al.*, 1971). Since this virion polypeptide is derived from a slightly larger precursor (Anderson *et al.*, 1973), it is possible that the precursor is contained in the excess pool.

The second core protein (polypeptide V) has a molecular weight of 45,000 as estimated from SDS-polyacrylamide gels and is only moderately rich in arginine (Laver, 1970). The number of molecules of the two core proteins per viral DNA molecule has been estimated to be around 1000 for polypeptide VII and 180 for polypeptide V (Maizel *et al.*, 1968*b*; Everitt *et al.*, 1973). No details of the interactions of these proteins with the DNA have been revealed so far.

Polypeptide VI (molecular weight 24,000) presumably belongs to the hexon-associated proteins. It does not appear to be synthesized in excess in the cell, as concluded from SDS-polyacrylamide gel electrophoresis (Everitt, unpublished). However, recent evidence suggests that this polypeptide is derived from a considerably larger precursor (Anderson *et al.*, 1973) which could have other antigenic determinants and might have escaped detection in the immunological analysis. A summary of the different polypeptides of the virion and some of their properties is found in Table 1.

So far no DNA or RNA polymerase activity have been found associated with the purified preparations of virions, but an endonuclease activity which cleaves DNA into fragments of a limit size of about 1–5 $\times 10^6$ daltons has been isolated both from cells infected with

TABLE 1

Characteristics of Virion Proteins of Adenoviruses[a]

	Hexon	Penton base	Fiber	Core protein I	Core protein II	Hexon-associated I	Hexon-associated II	Hexon-associated III
Number per virion	240	12	12	~1000	~200	~450	NK	NK
Mol. wt.	310,000–360,000[b]	400,000–515,000[c]	200,000[d] (Ad 2)	17,500[e]	45,000[f]	50,000[g]	15,000[g]	NK
Polypeptide sizes[h]	90,000–120,000; presumably 3/hexon	70,000, presumably 5/penton base	60,000–65,000; presumably 3/fiber	18,500	48,500	24,000	13,000	12,000
Polypeptide number in Fig. 2	II	III	IV	VII	V	VI	VIII	IX
Morphology	8- × 12-nm[i] cylinder	8-nm sphere[c]	10- to 25-nm × 2-nm rod	NK	NK	NK	NK	NK
Antigenic specificity	Group and type	Group, subgroup, and intersubgroup	Type, subgroup, and intersubgroup	Group and type	NK	Group?	NK	NK
Associated biological activity	NF	Cell detaching, endonuclease	Inhibition of macromolecular synthesis, hemagglutination, attachment to cells	NF	NF	NF	NF	NF

[a] Four additional polypeptides (IIIa, X, XI, XII) have been identified on SDS-polyacrylamide gels (Everitt *et al.*, 1973; Andersson *et al.*, 1973); NK, not known; NF, none found.
[b] Franklin *et al.* (1971*a*).
[c] Pettersson and Höglund (1969), Wadell (1970).
[d] Sundquist *et al.* (1973*a*).
[e] Laver (1970), Prage and Pettersson (1971).
[f] Laver (1970).
[g] Everitt (unpublished).
[h] Estimated from SDS-polyacrylamide gel electrophoresis.
[i] Pettersson *et al.* (1967), Tejg-Jensen *et al.* (1972).

adenoviruses and from purified preparations of Ad 2 and Ad 12, and is absent in uninfected cells (Burlingham and Doerfler, 1972). Preparations of pentons purified from infected KB cells exhibit a similar activity (Burlingham *et al.*, 1971), but it appears that the majority of the endonuclease activity can be separated from the penton (Doerfler, personal communication). This endonuclease preferentially cleaves DNA in G+C-rich regions, but the nature of the cleavage site and the physiological function of this enzyme has not yet been determined.

4. THE ADENOVIRUS GENOME

The adenovirus chromosome contains a linear duplex DNA molecule (molecular weight 20–25 $\times$ 10^6; Green *et al.*, 1967*b*; Van der Eb *et al.*, 1969) which has no single-strand breaks.

The denaturation pattern of the DNA is unique for Ad 2, Ad 5, and Ad 12 (Doerfler and Kleinschmidt, 1970; Ellens *et al.*, 1974; Doerfler *et al.*, 1972), and the maps of the former two viruses are almost identical. In addition, terminal fragments of the DNA do not contain redundancies (Murray and Green, 1973), demonstrating, together with the unique maps, that the DNA is not circularly permuted.

Digestion of the double-stranded adenovirus DNA with exonuclease III does not generate sticky ends (Green *et al.*, 1967*b*), indicating that adenovirus DNA is not terminally redundant in the ordinary sense. Instead, the adenovirus DNA has another and novel property in that denaturation and renaturation of intact DNA molecules generates single-stranded circles, as recently described (Garon *et al.*, 1972; Wolfson and Dressler, 1972). Since the majority of the single strands of unit length could be recovered as rings, both strands can form circles. No cyclic structures could be formed when double-stranded linear molecules are digested with exonuclease III, showing that hydrogen bonding between the two ends of single strands is responsible for maintenance of the circles. The only likely interpretation of these findings is that the adenovirus DNA molecule contains an inverted terminal repetition, which probably forms a circle closed by a panhandlelike structure. A terminal inverted repetition has only been observed in double-stranded adenovirus DNA and in the single-stranded DNA of the adenovirus-associated viruses (AAV) belonging to the parvovirus group (Koczot *et al.*, 1973). The function of this inverted terminal repetition is still unclear.

Usually single-stranded DNA passes unadsorbed through hydroxyapatite (HA) in 0.2 M phosphate at 65°C. However, denatured adenovirus DNA similarly passed over hydroxyapatite is totally retained (Tibbetts *et al.*, 1973). An analysis of the dependence of the HA retention on the extent of fragmentation of the adenovirus DNA showed that the fraction retained decreased slowly until the length of the fragment approached 10% of the intact DNA. With more extensive degradation, the fraction retained dropped precipitously. The formation of single-stranded circles should contribute to the retention of intact single-stranded adenovirus DNA by hydroxyapatite, but the described behavior of the fragmented DNA on hydroxyapatite points at an additional feature of adenovirus DNA which is not yet fully understood. Possibly several short, proximally positioned, complementary, or partially complementary, sequences could be dispersed throughout the single strands of the adenovirus DNA.

The human adenoviruses have been divided into three subgroups, A, B, and C, based primarily on oncogenicity of the viruses (Huebner *et al.*, 1965), but later on the G+C content of the DNA (Piña and Green, 1965), and it has been shown by reciprocal DNA-DNA and DNA-RNA hybridization that the members of one group are more closely related to each other than to members of the other groups [for a review see (Green, 1970)]. A fourth subgroup, D, has been suggested on similar grounds (McAllister *et al.*, 1969*b*). Interestingly enough, the highly oncogenic adenoviruses all fall in subgroup A, which has the lowest G+C content (48–49%), while the so-called nononcogenic adenoviruses are found in subgroup C, which has the highest G+C content (57–59%). The weakly oncogenic adenoviruses finally belong to subgroup B, with a G+C content of 49–52%. An exception to this rule is found in the simian adenoviruses; the highly oncogenic SA7, for instance, has a G+C content of around 60% (Piña and Green, 1968; Goodheart, 1971). As determined by DNA–DNA hybridization, the genomes of selected members within each subgroup have up to 70% of the nucleotide sequences in common, whereas DNAs from serotypes of different subgroups differ markedly, showing only about 10% homology, as determined by filter hybridization techniques [reviewed by Green (1970), see also Table 3].

A recent report (Garon *et al.*, 1973) describes electron microscopic analysis of heteroduplexes formed between the DNAs from different adenoviruses. This analysis has confirmed the existence of extensive homologies between the nucleotide sequences of DNAs from viruses within the three subgroups (A, B, and C). In the heteroduplexes formed between DNAs from different serotypes of a subgroup, again around 70% of the sequences were found to be homologous, while the sequence homology between subgroups appears to amount only to about 10%. The greatest degree of interserotypic homology is found between DNAs of the subgroup B viruses (Ad 3, Ad 7, and Ad 21). The heteroduplexes formed between DNAs from viruses of subgroup B and also between DNAs of viruses of subgroup C (Ad 1, Ad 2, and Ad 5) all exhibit relatively simple patterns of short regions, which are not base-paired. These regions appear limited to two areas of the molecules, with minor variations located at fractional map positions ranging from 0.08 to 0.22 and from 0.35 to 0.50, respectively. This refers to analyses performed on heteroduplex DNA spread at 85% formamide, the maximum concentration of formamide which can be used in order not to melt native, homoduplex DNA molecules. Analysis at lower formamide concentrations shows that these regions consist of both heterologous sequences and sequences that are partially homologous. The latter are apparently sufficiently different to be melted out at 85% formamide. By varying the degree of denaturation, clear-cut differences can be recognized between the different interserotypic heteroduplexes. Heteroduplexes formed with DNA from members of the highly oncogenic subgroup A (Ad 12, Ad 18, and Ad 31), although homologous to a large extent, do contain more pronounced heterologies than the heteroduplexes of subgroups B and C. In some cases quite complex patterns of heterologous regions are observed for subgroup A viruses. However, the striking finding is that the pronounced heterologies always seem to appear at the same two positions indicated above. This would seem to imply, as pointed out by Garon *et al.* (1973) that there are two specific regions in adenovirus DNA which are especially prone to genetic "drift," and that it is possible that the sequences in these two regions correspond to the same genes in all serotypes.

The strands of adenovirus DNA have been separated by copolymer binding using poly (I, G) and poly (U, G) (Kubinski and Rose, 1967; Landgraf-Leurs and Green, 1971; Patch *et al.*, 1972), but the separation has proved difficult in practice. Not only do the conditions for the copolymer binding, but also the quality of the copolymer used, seem to be crucial. It is the experience of several investigators that most of the commercially obtained batches of polymers are inefficient in separating the adenovirus strands, but that occasional batches can give reproducible and near-complete separation (Tibbetts *et al.*, 1974). The reason for this variability is not known.

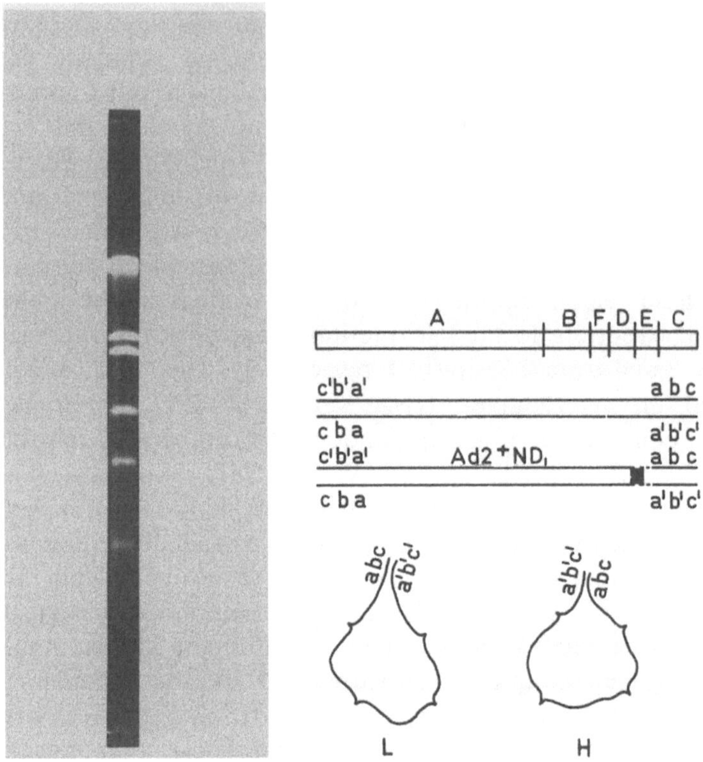

Fig. 4. Schematic drawing of the characteristic structures of Ad 2 DNA. Ad 2 DNA can be digested by a restriction enzyme endoR*Eco*RI from *E. coli* carrying a resistant transfer factor RTF-1 which specifies the enzyme. Six cleavage fragments, A–F, can be separated on agarose polyacrylamide gels, as shown in the insert to the left of the figure. The map for these fragments relative to the SV40 insertion in the nondefective hybrid virus $Ad2^{+}ND_1$ has been determined by heteroduplex formation (Sharp *et al.*, 1974). The fragment map is shown as a schematic drawing at the top of the figure, with fragment A to the left and fragment C to the right. Inverted terminal redundancies, giving rise to a DNA with similar structures at both ends, are schematically represented in the second structure from the top of the figure. The genome of the nondefective adeno-SV40 hybrid virus, $Ad2^{+}ND_1$, which has been used to map the fragments is also included, showing the position of the SV40 insertion which corresponds to an Ad 2 DNA deletion, which is about 50 times larger than the SV40 insertion but located at the same position. The circles formed of the single strands of adenovirus DNA because of the inverted terminal redundancy are illustrated for both the L and the H strands at the bottom of the figure. The evidence from hydroxyapatite chromatography also suggests intramolecular complementarities in the single strands. There appear to be 4–8 such regions per genome. These complementary sequences have also been illustrated in the schematic drawing of the single strands.

Separated adenovirus DNA strands have been used for the characterization of the transcription in productively infected and transformed cells, as will be discussed in Sects. 6.2 and 6.3 on transcription (Tibbetts *et al.*, 1974).

After controlled shearing of Ad 2 DNA, double-stranded half-molecules can be separated by CsCl gradient centrifugation (Kimes and Green, 1970; Doerfler and Kleinschmidt, 1970) and after cleavage by endonuclease endoR*Eco*RI (a restriction enzyme from *Escherichia coli* carrying a drug-resistant transfer factor RTF-1), six unique, double-stranded fragments can be separated by gel electrophoresis (Pettersson *et al.*, 1973). The fragments can be recovered in quantities sufficient for further biochemical studies and have already proved useful for the mapping of specific functions on the viral chromosome (Tibbetts and Pettersson 1974). Ad 5 and Ad 3 are cleaved in 3 fragments; Ad 12 in 6 fragments; and Ad 18 in 7 fragments (Mulder *et al.*, 1974).

Figure 4 schematically shows the structures of Ad 2 DNA and the cleavage fragments generated by the restriction endonuclease. The map of the fragments relative to the SV40 moiety of one adeno-SV40 hybrid virus, Ad2^{+}ND$_1$ (Lewis *et al.*, 1969), is also indicated.

After numerous failures to demonstrate infectivity of adenovirus DNA, it was thought to be noninfectious. However, 5 years ago the DNA of simian adenovirus 7 (SA7) was proven infectious and tumorigenic (Burnett and Harrington, 1968*a,b*), and infectivity, although with low efficiency, was also obtained with human Ad 1 DNA (Nicolson and McAllister, 1972). Recently, Graham and Van der Eb (1973*a,b*) introduced a novel technique for assay of the infectivity and transforming activity of adenovirus DNAs. The technique utilizes the adsorption of the DNA to calcium phosphate precipitates, which facilitates the uptake by the cell of infectious DNA. The efficiency of the technique was demonstrated using adenovirus type 5 DNA, and about 1–10 plaque forming units and one transforming focus was observed per microgram of DNA.

5. CLASSIFICATION OF ADENOVIRUSES

This section may be partially repetitious of previous data presented in this chapter, but it illustrates the slow and gradual accumulation of data required for a classification based on the characteristics of the genome of a virus.

More than 80 different adenovirus types have been isolated from a variety of animal species (Table 2). All types contain one group-

TABLE 2

Adenoviruses among Different Species

Natural host	Serological types	References
Human	28 accepted, 31 recognized	Béládi (1972)
Simian	23	Hillis and Goodman (1969)
Bovine	10	Mohanty (1971)
Ovine	1	McFerran *et al.* (1969)
Canine	2	Marusyk *et al.* (1970)
Murine	2	Hartley and Rowe (1960), Reeves *et al.* (1967)
Porcine	4	Bibrack (1969)
Avian	8	McFerran *et al.* (1972)
Opossum	1	Wilner (1969)

specific antigen determinant, which resides in the hexon of the outer capsid (Pereira *et al.*, 1963), except for avian adenoviruses, which lack this determinant. As stated, the general architecture of all adenoviruses is similar, including the avian group, although one type of avian adenovirus (celo virus) has a different arrangement of its fiber units at the vertices of the icosahedron (Laver *et al.*, 1971). The celo virus contains two fiber units of different length (42.5 nm and 8.5 nm) associated with each penton base.

The subgroup classification of human adenoviruses should primarily be based on similarities in the sequence of their DNA, as suggested by Lwoff *et al.*, (1962) for animal viruses in general. Two different subgroup classifications of human adenoviruses have been suggested, one based on the hemagglutination pattern (Rosen, 1960) and the other based on the oncogenicity in newborn hamsters (Huebner *et al.*, 1965). The first classification separates the human adenoviruses mainly on the basis of the agglutination patterns of monkey and rat erythrocytes. The viruses are subdivided into four groups: group I, which causes complete agglutination of monkey erythrocytes; group II, which causes complete agglutination of rat erythrocytes; group III, which causes incomplete agglutination of rat erythrocytes; and group IV for which no hemagglutination is detected. Later, it was reported (Schmidt *et al.*, 1965) that the group IV viruses, Ad 12, Ad 18, and Ad 31, could agglutinate rat erythrocytes with an incomplete pattern, in a way similar to that of group III viruses. Adenovirus types 20, 25, and 28, which were originally placed in hemagglutination group I (Rosen,

1960), probably belong in group II, both with respect to the pattern of hemagglutination and also with regard to the length of the fibers (Wigand, 1969). Thus, the original classification, based on hemagglutination pattern alone, has created some confusion about the subclassification of the oncogenic Ad 12, Ad 18, and Ad 31 serotypes and also of other serotypes. The classification based on oncogenicity (Huebner *et al.*, 1965), however, appears to correlate with the detailed structure of the adenovirus DNA. First, the overall G+C content of the viral DNA differs among groups A, B, and C, so that the highly oncogenic viruses, which have 48–49% G+C, fall in group A, the weakly oncogenic, with 50–52% G+C, fall in group B, and the nononcogenic, which contain 57–59% G+C, fall in group C [reviewed by Green (1970)]. Remaining human adenovirus serotypes have the same G+C content as group C viruses but have been placed in a separate group D for other reasons (McAllister *et al.*, 1969*b*). The subgroup classification of the human adenovirus in groups A–D correlates also with hybridization results, where the homology between the different subgroups have been determined by RNA–DNA or DNA–DNA hybridization (Lacy and Green, 1964, 1965, 1967). More recently this classification has been strengthened by heteroduplex mapping of the DNA of representative viruses within each subgroup (Garon *et al.*, 1973). It is, therefore, suggested that the subgroup classifications A–D of the adenoviruses, which have now been shown to be based on similarities in the sequences of the genome, should be accepted. It is evident from Table 3 that these classifications correlate well with classification based on hemagglutination, provided that the serotypes Ad 12, Ad 18, and Ad 31 are considered as a separate hemagglutination group, as originally suggested by Rosen (1960). Serotype 4 still has an anomalous position in this classification diagram with regard to its biological properties since this virus has fiber units which are intermediate in length to those of group B and C viruses. In several other respects, the Ad 4 serotype shares many features with serotype 16 of subgroup B [for a review see Norrby (1968)]. Intermediate adenovirus strains have been detected within subgroups B and D, which carry both hexon and fiber units that are distinct from only one of the parent viruses (Norrby, 1969*c*). It should finally be emphasized that not all types were studied with regard to both DNA homology and biological parameters, and the proposed classification in Table 3 is, therefore, only tentative. The group D viruses, especially, have not been carefully analyzed; members of this group resemble group C viruses in many respects except for hemagglutination pattern and biological properties.

TABLE 3

Classification of Human Adenoviruses

Subgroup	Serotypes	G + C in DNA,[a] %	DNA–DNA homology between types,[b] %	Hemaggluti-nation subgroup[c]	Length of fiber,[d] nm	Free dodecons[d]	Oncogenicity[e]
A	12, 18, 31	48–49	80–85	IV	28–31	–	High
B	3, 7, 11, 14, 16, 21	50–52	70–95	I	9–11	+	Weak
C	1, 2, 5, 6	57–59	85–95	III	23–31	–	None, but transform
	4				17–18	+	
D	8, 9, 10, 13, 15, 17, 19, 20, 22–24, 25, 26, 27, 28	57–61	NK[f]	II	12–13	+	None, but transform

[a] Data from Piña and Green (1965) and Green (1970).
[b] Data from Lacy and Green (1964, 1965, 1967) and also confirmed by heteroduplex mapping (Garon *et al.*, 1973).
[c] Data from Rosen (1960).
[d] Data from Noorby (1969*b*) and Liem *et al.* (1971).
[e] Highly oncogenic types cause tumors within 2 months, weak types cause tumors in some animals in 4–18 months.
[f] Not known.

6. THE PRODUCTIVE INFECTION

The biochemistry of the lytic cycle of adenoviruses has mainly been studied in suspension cultures of KB and HeLa cells. The time course of the infection with Ad 2 is shown in Fig. 5. When the cells are infected at high multiplicities (50–100 PFU/cell), a synchronous response is observed where two functionally different phases can be distinguished: an early phase, which precedes viral DNA replication, and a late phase, which begins with the onset of viral DNA synthesis around 6 hours after infection. The early phase is characterized by the expression of about half of the viral genome (Green *et al.*, 1970; Tibbetts *et al.*, 1974), resulting in the synthesis of the so-called T and P antigens (Gilead and Ginsberg, 1968*a,b*; Russell and Knight, 1967) whose functions have not yet been clarified. During the late phase at least 90% of the genome appears to be expressed (Fujinaga and Green, 1970; Tibbetts *et al.*, 1974). Between 8 and 12 hours after infection there is a dramatic change in the transcription pattern. The amount of

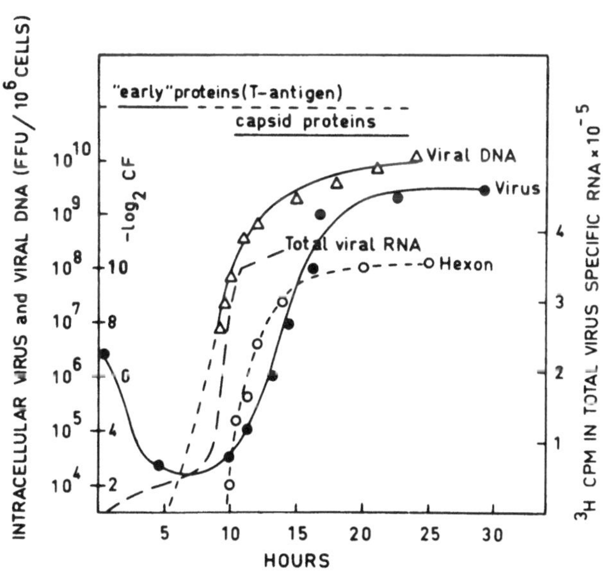

Fig. 5. Growth curve of Ad 2 in suspension culture of KB cells. (●——●) Intracellular virus measured as fluorescent focus forming units (FFU) per 10^6 cells, (O——O) hexon antigen monitored by complement fixation, (Δ——Δ) synthesis of viral DNA estimated from the results reported by Green *et al.* (1970), (- - - -) total virus-specific RNA formed during the infectious cycle expressed as radioactivity hybridizing to Ad 2 DNA. Time intervals for maximum synthesis of early and capsid proteins are indicated at the top of the figure. Modified from Philipson and Pettersson (1973).

viral RNA produced increases 5- to 10-fold and the polysomal mRNA pattern changes completely into the viral species. At this time the synthesis of host cell proteins is almost completely replaced by synthesis of viral products—mostly viral structural proteins (Bello and Ginsberg, 1967; White *et al.*, 1969; Russell and Skehel, 1972). The infectious cycle is completed within 20–25 hours.

Ad 2 and Ad 5, both members of subgroup C, have been used most commonly as model viruses, but similar time courses have been obtained with several other types (Green, 1970; Sundquist *et al.*, 1973*b*). The growth cycle of the highly oncogenic group A viruses is somewhat longer than for Ad 2 (Mak and Green, 1968), and when primary cells are used with Ad 2 the growth cycle is prolonged for at least 12–15 hours (Ledinko, 1970).

6.1. Early Interaction between Virus and Cell

The adenovirus eclipse has been studied mainly with Ad 2 and Ad 5. Two different techniques have been employed: (1) electron microscopy (Dales, 1962; Morgan *et al.*, 1969; Chardonnet and Dales, 1970*a,b*, 1972) and (2) analysis with the use of radioactively labeled virus (Lawrence and Ginsberg, 1967; Philipson, 1967; Sussenbach, 1967; Philipson *et al.*, 1968; Lonberg-Holm and Philipson, 1969). The ratio of particles to infectious units for these virus types is in the order of 10–30, and for other serotypes the ratio can be as high as 2000 (Green *et al.*, 1967*a*; Lonberg-Holm and Philipson, 1969). It should be pointed out, therefore, that the methods used for studying the early interaction reveal only the fate of the majority of the particles and that this might not reflect the true infectious pathway. Nevertheless, it appears that the virions are adsorbed to specific receptors of the cell surface. There are around 10^4 such receptors per cell. Uncoating of the virus occurs early after infection and seems to involve three steps. The *first* step gives rise to particles the DNA of which is partially accessible to DNase digestion, and chemical evidence suggests that these particles are devoid of penton capsomers (Sussenbach, 1967; Everitt, unpublished). This event occurs within a few minutes after attachment of the virus to the cell surface (Sussenbach, 1967; Lonberg-Holm and Philipson, 1969; Morgan *et al.*, 1969). Whether the uptake of the virus by the cells is brought about mainly by pinocytosis (Dales, 1962; Chardonnet and Dales, 1970*a,b*) or by direct penetration of the plasma membrane (Lonberg-Holm and Philipson, 1969; Morgan *et al.*, 1969), and exactly where this first step of uncoating takes place, are still unclear. It has been suggested that the product of this step is associated firmly with the pores in the nuclear membrane (Chardonnet and Dales, 1972). The *second* step seems to result in removal of the hexon capsomers from the DNA-protein core. As expected, this intermediate is even more sensitive to DNase. As visualized by electron microscopy, the cores enter into the nucleus, apparently leaving the capsid structure behind (Morgan *et al.*, 1969; Chardonnet and Dales, 1972). It has been suggested that microtubules carry the virions from the plasma membrane to the nuclear pores and that the transfer of the cores into the nucleus is an ATP-dependent process (Chardonnet and Dales, 1972). The *third* and final step in the uncoating of the adenovirus DNA occurs in the nucleus of the cell during the second hour after infection. The final product appears to be DNA, still intact, but almost free of virion proteins (Lonberg-Holm and Philipson, 1969). The time course of the uncoating process is dependent on the multiplicity of infection, but at high multiplicities the overall time required is 1–2 hours (Lonberg-Holm and Philipson, 1969). The mechanism and cell structures involved in the different steps are not known, but it is worth pointing out that the three intermediates seen during the *in vivo* uncoating appear to resemble the intermediates obtained in the *in vitro* sequential disintegration as discussed above. The whole process also appears to proceed normally in the presence of inhibitors of protein and nucleic acid synthesis; thus if enzymes are involved they seem to be pre-existing, brought by the infecting virus or present in the cell (Lawrence and Ginsberg, 1967; Philipson, 1967).

6.2. Early Transcription

Virus-specific RNA is usually assayed by hybridization of labeled RNA to filter-bound viral DNA (Rose *et al.*, 1965; Bello and Ginsberg, 1969; Thomas and Green, 1969). With this technique it has been demonstrated that transcription of the viral genome, established in the nucleus of the cell, starts immediately, and that the pattern of early transcription is maintained until the onset of viral DNA synthesis (Parsons and Green, 1971; Lucas and Ginsberg, 1971; Wall *et al.*, 1972; Philipson *et al.*, 1973). At this time the viral RNA constitutes only a minor fraction of the total RNA synthesized in the cell, but later in the early phase (3–5 hours post-infection) around 15% of the polysomal mRNA appears to be virus-coded (Lindberg *et al.*, 1972).

The nascent viral transcription products appear as heterogeneous as the heterogeneous nuclear RNA (HnRNA) of the uninfected cell, and the largest viral RNA molecules appear to be as large as 10^7 daltons (Wall *et al.*, 1972). The mRNAs on the polysomes, on the other hand, appear generally smaller, and a more-distinct pattern of RNA peaks are observed on gel electrophoresis (Parsons and Green, 1971; Lindberg *et al.*, 1972). These observations suggest that the viral mRNA is generated from high-molecular-weight precursors in the nucleus.

The viral transcripts are polyadenylated in the nucleus and since the poly(A) by itself does not hybridize significantly to virus DNA, it is concluded that this process occurs after transcription has taken place (Philipson *et al.*, 1971). The poly(A) segment recovered from viral RNA, selected by hybridization to virus DNA, has the same size as the poly(A) in mRNA from uninfected cells, i.e., 180–200 nucleotides long (Philipson *et al.*, 1971). Polysomal RNA from adenovirus-infected cells has been fractionated on the basis of poly(A) content by affinity chromatography on poly(U)-Sepharose and analyzed for virus-specific RNA (Lindberg *et al.*, 1972). The results of these experiments indicate that most if not all of the early viral mRNA species contain poly(A). As in the uninfected cell, the adenosine analogue cordycepin (3′-deoxyadenosine) inhibits both the polyadenylation reaction and the appearance of viral mRNA on polysomes (Philipson *et al.*, 1971). Thus, the conclusion that polyadenylation may be necessary for proper processing and/or nucleocytoplasmatic transport of mRNA (Darnell *et al.*, 1971*b*) appears to be valid also for the adenovirus system.

Three major size classes of early viral mRNA have been found (Parsons and Green, 1971; Lindberg *et al.*, 1972): one class of sediments around 15 S (apparently two components with approximate molecular weights of 0.35×10^6 and 0.43×10^6), a second class of sediments around 20 S (one major component of approximate molecular weight 0.78×10^6), and a third heterogeneous population of sediments in the range of 20–40 S (molecular weight $1.0–3.5 \times 10^6$). Figure 6 shows the polyacrylamide gel pattern of early mRNA.

Earlier reports claimed that not more than 10–20% of the genome in the case of Ad 2 is transcribed early after infection (Thomas and Green, 1969; Fujinaga and Green, 1970) and that some of the early sequences are not transcribed in the late phase (Lucas and Ginsberg, 1971). Recently, however, these questions have been reinvestigated (Tibbetts *et al.*, 1974) using separated DNA strands and hybridization in liquid with radioactively labeled DNA according to the method of Sambrook *et al.* (1972). The results of these experiments suggests that a much larger portion of the adenovirus genome, 40–50% (15–20% of

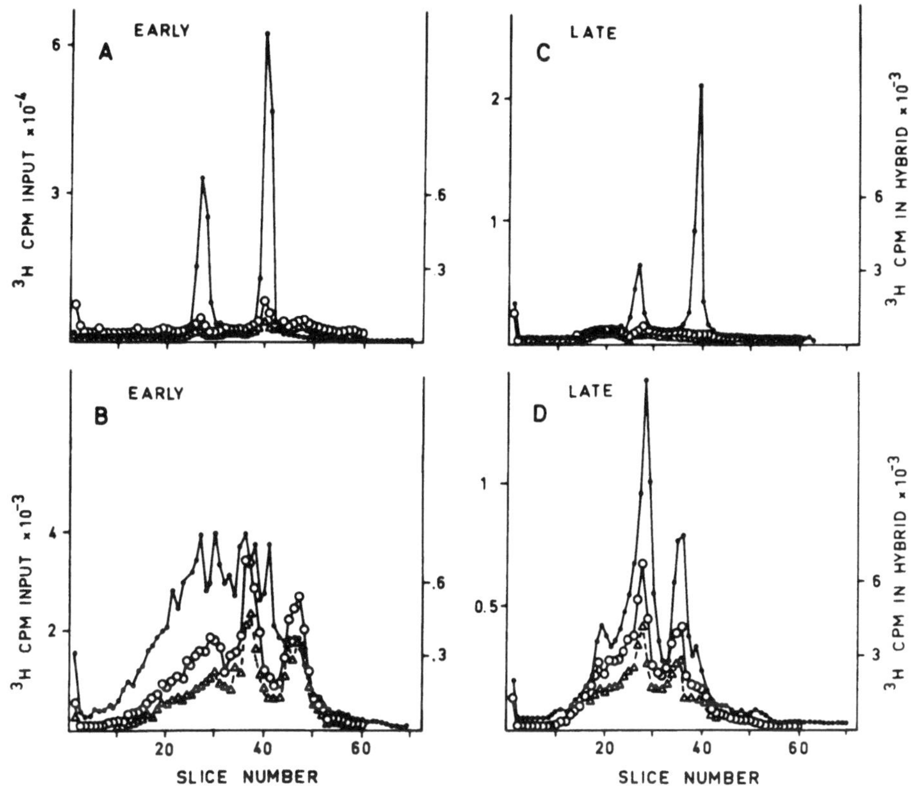

Fig. 6. Size distribution of early and late adenovirus mRNA from the two fractions of poly(U)-Sepharose fractionation. The RNA from the fraction not retained by the column (PUS I) and the fraction retained by the column (PUS II) were precipitated with ethanol at −20°C, redissolved in 0.1 × SSC (standard saline citrole) containing 0.5% sarcosyl, and analyzed by gel electrophoresis (Lindberg and Persson, 1972). Gels were analyzed for total radioactivity and virus-specific RNA by hybridization. Panels A and C, early and late PUS I RNA, respectively; panels B and D; early and late PUS II RNA, respectively. (●——●); Total ^{3}H-cpm (O——O) ^{3}H-cpm in viral mRNA before RNase digestion of hybrids; and (Δ- - -Δ) H^{3}-cpm in viral mRNA after RNase digestion of hybrids. Reprinted from Lindberg *et al.* (1972).

the H strand and 25–30% of the L strand) is represented as mRNA in the cytoplasm in the early phase, and that most if not all early viral RNA sequences persist but are not necessarily transcribed late in the infection.

The transcription of the adenovirus genome in the early phase is not sensitive to inhibitors of protein synthesis (Parsons and Green, 1971; Wall *et al.*, 1972). Since no RNA polymerase seems to be carried by the infecting virions, it is suggested that the early viral RNA is synthesized by a cellular enzyme. In accordance with this, several investigators (Ledinko, 1971; Price and Penman, 1972*a*; Wallace and Kates, 1972; Chardonnet *et al.*, 1972) have found that viral RNA is transcribed by an RNA polymerase which in some respects resembles one of the RNA polymerases of the host cell [polymerase II according to Roeder and Rutter (1970), which corresponds to polymerase B in the nomenclature used by Chambon *et al.* (1970)]. It is inhibited by α-amanitine and greatly stimulated by increasing ionic strength; in the purified state manganese ions are not stimulatory. The same enzyme seems to be responsible for the major part of the adenovirus transcription both early and late after the infection. In the uninfected cell this enzyme appears to be responsible for the synthesis of the major fraction of heterogeneous nuclear RNA, the assumed precursor to mRNA (pre-mRNA) (Zylber and Penman, 1971).

6.3. Late Transcription

The switch from early to late transcription encompasses both a qualitative and a quantitative change with regard to polysomal virus-specific mRNA. Large amounts of viral mRNA, about 5–10 times more than in the early phase, accumulate in the cytoplasm and appear to replace most of the host cell mRNA on the polysomes (Green *et al.*, 1970; Philipson *et al.*, 1973). The result is an almost total switch to synthesis of viral proteins, the major part of which constitutes the polypeptides of the structural proteins of the virus (Bello and Ginsberg, 1967; White *et al.*, 1969; Russell and Skehel, 1972). The transition to the late phase is prevented by inhibitors of both protein (Green *et al.*, 1970) and DNA synthesis (Flanagan and Ginsberg, 1962).

There is general agreement that also in the late phase of the infection a fraction of the heterogeneous nuclear viral RNA can be recovered as giant RNA molecules having sedimentation coefficients of 70–80 S (Green *et al.*, 1970; McGuire *et al.*, 1972; Wall *et al.*, 1972). The virus-specific RNA amounts to 10–20% of the newly synthesized RNA in the nucleus (Price and Penman, 1972*a*; McGuire *et al.*, 1972). The largest viral RNA molecules could by themselves represent the entire adenovirus genome. As in the early phase, much larger viral RNA molecules are found in the nucleus than are ever recovered from the cytoplasm, implying that also late viral mRNA is derived by cleavage from larger precursors in the nucleus. If this were so, the smaller nuclear viral RNA molecules should be thought of as the immediate products of the cleavage of the large precursors. Recently, evidence in favor of this interpretation was obtained by the use of the drug toyocamycin (an adenosine analogue) (McGuire *et al.*, 1972). When this compound is applied to adenovirus-infected cells at moderate concentrations (1.75 μg/ml), giant viral RNA is still made, but the accumulation of the smaller-molecular-weight classes of virus-specific RNA is almost completely prevented. This is directly analogous to the effect of the drug on the synthesis of ribosomal RNA (Tavitian *et al.*, 1968). In that case the 45 S precursor to ribosomal RNA is made, but it fails to undergo processing; this means that there is no cleavage of the precursor into mature 28 S and 18 S ribosomal RNA.

Analysis of nuclear viral RNA purified by hybridization to viral DNA has demonstrated that also in the late phase viral RNA molecules are polyadenylated (Philipson *et al.*, 1971), and, as pointed out before, this is a post-transcriptional step in the generation of mature viral mRNA.

In cells labeled 14–16 hours post-infection, more than 90% of the recovered, labeled, polysomal mRNA is virus-specific, and analysis of the polysomal RNA fractionated on poly(U)-Sepharose has demonstrated that most if not all the viral RNA contains poly(A) (Lindberg *et al.*, 1972). Electrophoretic analysis of poly(U)-selected polysomal mRNA has revealed two major viral mRNA classes, one at the 22 S position and the other at the 26 S position (approximate molecular weight 0.95×10^6 and 1.6×10^6, respectively), and, in addition, several minor mRNA peaks of higher and lower molecular weight (Lindberg *et al.*, 1972, Bhaduri *et al.*, 1972). Figure 6 shows the polyacrylamide gel pattern of viral mRNA from poly(U)-Sepharose-selected polysomal RNA.

With separated strands of adenovirus type 2 DNA, Green *et al.* (1970), using hybridization to filter-bound DNA, and Tibbetts *et al.* (1974), using hybridization in liquid according to the method of Sambrook *et al.* (1972), could show that the RNA sequences present in the cytoplasm are complementary to 65–70% of the sequences of the L strand and to 25–30% of the sequences in the H strand. Thus, apparently most of the virus genome is expressed in the late phase. Furthermore, most of the sequences expressed early are also expressed late, implying that the switch from early to late phase results from the "turning on" of additional genes, mainly on the L strand (Fujinaga and Green, 1970; Tibbetts *et al.*, 1974). That the picture will turn out to be more complicated than anticipated from these results is suggested by recent results of Tibbetts and Pettersson (1974), who by the use of the isolated fragments of adenovirus type 2 DNA in hybridization experiments with late RNA came to the conclusion that there must be at least 3 points in the DNA where the origin of stable transcripts shifts from one strand to the other. In addition, earlier findings made by several investigators (Wall *et al.*, 1972; McGuire *et al.*, 1972; Lucas and Ginsberg, 1972*a*) indicate that there are RNA sequences in the nucleus which cannot be "competed out" by cytoplasmic RNA in hybridization experiments using filter-bound DNA. It seems as if at least one-fourth of the nuclear sequences never reach the cytoplasm. Since most of the genome appears to be represented in the cytoplasm, this would mean that the additional nuclear sequences may constitute products from symmetrically transcribed regions in the genome and that the nonsense part of the pre-mRNA is removed during the post-transcriptional maturation of mRNA. It has been found that about 0.5–1.5% of the nuclear viral RNA can be recovered as double-stranded RNA after exposure of the RNA to annealing conditions (Lucas and Ginsberg, 1972*b*; Pettersson and Philipson, unpublished), but whether this is a reflection of molecular complementarity or a result of symmetric transcription has not yet been established.

In addition to the adenovirus-specific mRNA species, a small-molecular-weight virus-specific RNA is synthesized in large amounts late in the lytic cycle. This 5.5 S RNA, called virus-associated RNA (VA RNA), contains 156 nucleotides and their sequence has been determined (Ohe and Weissman, 1970; Ohe *et al.*, 1969). It is synthesized in larger numbers than any of the mRNAs, and it is apparently synthesized by an enzyme which is different from that which seems to be involved in the mRNA synthesis (Price and Penman, 1972*b*). VA RNA is transcribed from the adenovirus DNA and has been reported to be present in one copy in the DNA (Ohe, 1972). No functional role has as yet been ascribed to this RNA.

Obviously, many important questions concerning transcription of the adenovirus genome and its control remain unsolved. We do not know the exact nature of the large transcripts seen in the nucleus early after infection. Does their presence indicate that viral DNA is integrated into the cell genome early after infection as suggested from experiments on abortive infection with adenovirus type 12 in hamster cells (Doerfler, 1970) and that a fraction of the virus genome is transcribed into large RNA molecules consisting partly of viral and partly of cellular sequences? Is the latter cleaved off when mature mRNA is generated, or is the whole adenovirus genome transcribed both early and late after infection into giant pre-mRNA molecules, and is the control of the expression of the early and late genes a matter of post-transcriptional selection of the appropriate messenger sequences?

6.4. A Comparison between Cellular and Viral mRNA Production

In comparing recent results concerning the manufacturing of mRNA in uninfected mammalian cells in general with the corresponding findings in the adenovirus system, clear similarities as well as differences stand out. There is strong evidence suggesting that mRNA in animal cells is generated by cleavage of high-molecular-weight precursors in the nucleus and that the precursors are part of the metabolically active pool of RNA molecules called heterogeneous nuclear RNA (HnRNA). For a more detailed discussion of the experimental data concerning these conclusions, the papers by Darnell *et al.* (1972) and Jelinek *et al.* (1973) should be consulted. The first evidence for the presence in the nucleus of high-molecular-weight precursors to mRNA came from experiments on cells transformed by SV40 (Lindberg and Darnell, 1970; Tonegawa *et al.*, 1970), where the virus genome apparently is integrated, covalently linked, to host cell DNA [reviewed by Green (1970)]. These cells produce high-molecular-weight HnRNA molecules containing virus-specific sequences, while at the same time much smaller virus-specific mRNAs are found in protein-synthesizing polysomes.

During recent years, further strong evidence indicating the processing of HnRNA into mRNA has been collected. Polysomal mRNA as well as HnRNA have been shown to contain poly(A) at the 3′ terminus (Edmonds and Caramela, 1969; Kates 1970; Darnell *et al* 1971*a,b*; Edmonds *et al.*, 1971; Lee *et al.*, 1971; Philipson *et al.*, 1971; Mendecki *et al.*, 1972; Molloy *et al.*, 1972). A substantial fraction of HnRNA molecules of all size classes are polyadenylated and a significant fraction of poly(A) synthesized in the nucleus eventually appears to be transported to the cytoplasm as part of mRNA molecules (Jelinek *et al.*, 1973). Altogether, this evidence strongly supports the idea of processing of HnRNA into mRNA. Furthermore, the addition of poly(A) to nuclear RNA appears to be a prerequisite for the appearance of mRNA on polysomes (Darnell *et al.*, 1971*b*). In all these respects the adenovirus system resembles the host cell.

In the animal cell 80–90% of the HnRNA seems to be turning over in the nucleus, leaving a maximum of 10–20% to be transported as

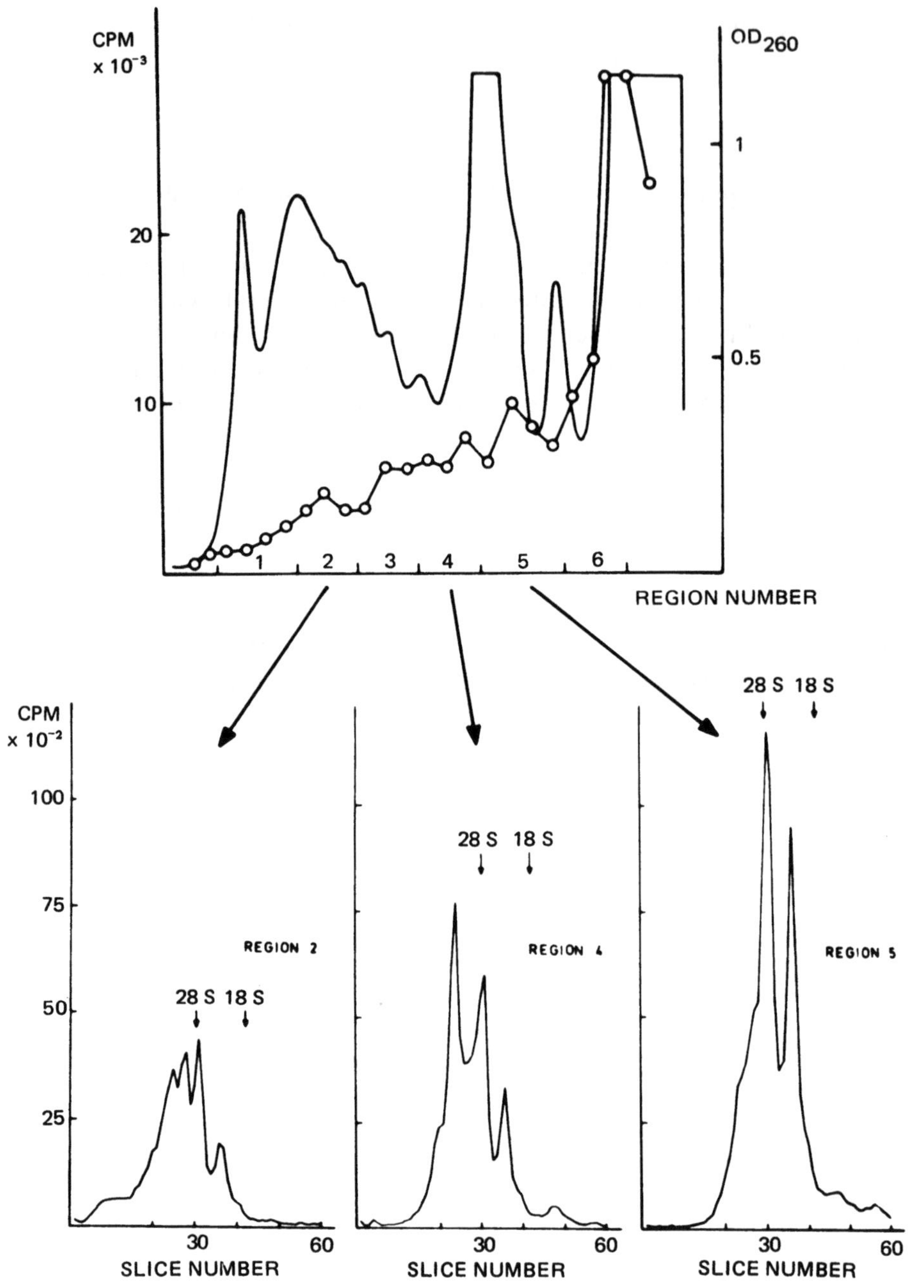

Fig. 7. Localization of late viral mRNA in different regions of a sucrose gradient of cytoplasmic extracts from adenovirus-infected cells. Cells were labeled with 1 mCi of ^{3}H-uridine from 14–16 hours after adenovirus infection and the cytoplasm was subjected to centrifugation in a 7–47% sucrose gradient in isotonic buffer as described

mRNA (Soiero *et al.*, 1968). In the case of the adenovirus nuclear RNA, however, much more seems to be preserved as mRNA. Late in the infection, the major part of the L-strand transcript (up to 70%) (Tibbetts *et al.*, 1974) after post-transcriptional modification is recovered as mRNAs of smaller sizes in the cytoplasm (Green *et al.*, 1970; Lindberg *et al.*, 1972). This infers that, in this case at least, the high-molecular-weight mRNA precursor is cleaved into a number of specific fragments, the majority of which becomes polyadenylated and transported as mRNA to the cytoplasm. The data on the turnover of cellular HnRNA suggest that, in general this, is not the case for the eukaryotic cell, but it has not been excluded that some HnRNA molecules are processed and utilized to this large extent.

The polysomal mRNA molecules of eukaryotic cells appear to have proteins associated with them (Spirin and Nemer, 1965; Perry and Kelley, 1968; Henshaw, 1968; Kumar and Lindberg, 1972; Lebleu *et al.*, 1971; Blobel, 1973; Morel *et al.*, 1971, 1973; Lindberg and Sundquist, 1974), and there is evidence suggesting that the HnRNA in the nucleus also exists in the form of ribonucleoprotein complexes (Samarina *et al.*, 1968; Perry and Kelley, 1968). Whether the nuclear and polysomal RNA molecules have some proteins in common is not clear. Polysomal mRNP particles in uninfected cells appear to contain at least 2 polypeptides (of molecular weights around 78,000 and 50, 000) (Morel *et al.*, 1973; Blobel, 1973), but at least one additional polypeptide has been found upon infection (Lebleu *et al.*, 1971). When polysomal mRNP particles were selected on oligo dT cellulose from uninfected cells, four major polypeptides were detected in labeling experiments with radioactive amino acids (Lindberg and Sundquist, 1974). The polysomal mRNP particles from adenovirus-infected cells in the late phase seem to contain the same four polypeptides, and, in addition, they contain a fifth polypeptide which appears to be virus-specified or -induced (Lindberg and Sundquist, 1974). The functional role of these polypeptides as well as their absolute specificity for the mRNA remain to be established.

---

earlier (Lindberg and Persson, 1972). The top panel shows the optical density and the radioactivity patterns [(O——O) ^{3}H-cpm] for the sucrose gradient. From each of the six indicated regions RNA was dissociated from protein with sarcosyl and EDTA and the poly(A)-containing RNA isolated by poly(U)-Sepharose chromatography. After precipitation with ethanol, the redissociated RNA was analyzed by polyacrylamide electrophoresis as shown in the lower frames. The poly(A)-containing RNA from regions 1–5 were found to hybridize to adenovirus DNA quantitatively. Reprinted from Philipson *et al.* (1973).

In the animal cell, 30–40% of the total poly(A)-containing RNA in the cytoplasm is not present in polyribosomes, but it is present instead in ribonucleoprotein particles which sediment slower than polysomes (Jelinek *et al.*, 1973; Lindberg and Persson, 1972). It has been suggested that this nonpolysomal poly(A)-containing RNA is in fact mRNA on its way to being engaged in translation (Schochetman and Perry, 1972), but it has been difficult to obtain direct proof for this assumption. In the case of the adenovirus-infected cells in the late phase, 40–50% of the cytoplasmic, virus-specific, poly(A)-containing RNA is found in this nonpolysomal pool (Philipson *et al.*, 1973); this is in accord with the findings of Raskas and Okubo (1971) that 50–70% of the virus-specific RNA in the cytoplasm is not associated with polysomes. In addition, this pool has been shown to contain the same viral RNA components as the polysomes. Figure 7 shows the adenovirus-specific RNA on polyacrylamide gels in different regions of a sucrose gradient of a cytoplasmic extract from late infection. Furthermore, kinetic experiments show that the labeled viral RNA molecules appear more rapidly in the nonpolysomal pools than in the polysome fraction (Philipson *et al.*, 1973). These data suggest that the nonpolysomal pool of RNP particles contains mRNA, which might be en route to translation.

6.5. Adenovirus DNA Replication

During one-step growth conditions, viral DNA synthesis begins in the nucleus 6–8 hours after infection. With Ad 2 and Ad 5, the maximum rate of viral DNA synthesis is reached about 13 hours after infection, and at this time host cell DNA synthesis has drastically been turned off; 90% or more of newly synthesized DNA is viral (Ginsberg *et al.*, 1967; Piña and Green, 1969). No information is available about the enzymes involved in synthesis of adenovirus DNA, and no virus-specific DNA polymerase has been identified. There is a clear-cut difference between synthesis of host cell and viral DNA with regard to the requirement for protein synthesis. Horwitz *et al.* (1973) demonstrated that after viral DNA synthesis has begun it is no longer sensitive to cycloheximide, whereas continued host cell DNA synthesis is highly susceptible.

Most studies on adenovirus DNA synthesis have been carried out with the group C viruses, mainly because the difference in G+C content between the DNA of these viruses and the host cell DNA facilitates the separation of the two kinds of DNA by equilibrium centrifugation in CsCl (Piña and Green, 1969) or by chromatography on columns of methylated albumin-kieselguhr (Ginsberg *et al.*, 1967).

6.5.1. Characteristics of Adenovirus DNA During *In Vivo* Replication

When DNA pulse-labeled *in vivo* or in isolated nuclei from infected cells is subjected to centrifugation in neutral sucrose gradients, two classes of molecules containing labeled adenovirus DNA are observed (Bellett and Younghusband, 1972; Van der Vliet and Sussenbach, 1972; Van der Eb, 1973; Pettersson, 1973). One class sediments with the same rate as the native double-stranded DNA (32 S for human adenoviruses and 35 S for celo virus DNA). The second class sediments faster (40–80 S) than native virus DNA. Data obtained from analyses by CsCl gradient centrifugation and BND-cellulose chromatography, as well as through electron microscopic observations, show that the major class consists mainly of unit-length duplex DNA molecules, while the fast-sedimenting molecules have characteristics of both double- and single-stranded DNA (Bellett and Younghusband, 1972; Van der Vliet and Sussenbach, 1972; Van der Eb, 1973; Pettersson, 1973). These latter molecules are believed to be replicative intermediates in the synthesis of adenovirus DNA. Electron microscopic investigations of the 40–80 S structures extracted from isolated nuclei have demonstrated the presence of double-stranded DNA with branches which are mainly single-stranded. In CsCl the presumed replicative intermediates band at a density 9–10 mg/ml higher than native adenovirus DNA. The higher density of the replicating DNA probably does not depend on the presence of RNA in association with the DNA since the replicative intermediates are observed also after treatment with RNase (Bellet and Younghusband, 1972; Van der Eb, 1973; Pettersson 1973). No covalent circles have been found during virus replication. In studies of celo virus DNA replication, Bellet and Younghusband (1972) observed radioactivity in DNA fragments after 1- to 3-minute pulses with ^{3}H-thymidine. These fragments sedimented at 12 S or less in alkaline CsCl gradients, and appeared to be intermediates in replication but larger than typical "Okasaki fragments" (Okasaki *et al.*, 1968). Evidence for such fragments has not been obtained with Ad 2 (Horwitz, 1971). With Ad 2, Ad 5, and celo viruses evidence has been secured for the presence of extensive single-stranded regions of DNA in the replicative intermediates (Pearson and Hanawalt, 1971; Sussenbach *et al.*, 1972; Bellet and Younghusband, 1972; Van der Eb, 1973; Pettersson, 1973).

By following the fate of parental DNA or by continous labeling *in vivo* Burlingham and Doerfler (1971) demonstrated three different classes of intracellular viral DNA from Ad 2 and Ad 12 in neutral sucrose gradients; one class larger than, one class the same size as, and one class smaller than viral DNA. The fast-sedimenting component may represent viral DNA integrated in cellular DNA or a transcription complex (Doerfler, personal communication). A complex of viral RNA and DNA was isolated by CsCl-propidium iodide gradients (Doerfler *et al.*, 1973) in which the RNA was of similar size to mRNA and susceptible to RNase. The relationship of this complex to the rapidly sedimenting paternal DNA is, however, unclear. No conclusive evidence that the fast-sedimenting DNA is integrated in cellular DNA has been obtained. The viral DNA molecules present in the slow-sedimenting fraction appear to be smaller than unit-length virus DNA. As inferred from the studies of Burlingham and Doerfler (1971) on adenovirus type 2- and 12-infected cells, these smaller molecules of viral DNA could possibly be products of degradation caused by the virion-associated endonuclease; Sussenbach (1971), however, has provided evidence suggesting that the small molecules in Ad 12 may replicate independently, and they could not be demonstrated in Ad 5-infected cells (Sussenbach and Van der Vliet, 1972).

6.5.2. Replication of Adenovirus DNA in Isolated Nuclei

Isolated nuclei have one advantage in comparison to intact cells in that they incorporate nucleoside triphosphates into nucleic acids (Friedman and Mueller, 1968; Winnacker *et al.*, 1971). In addition, pulse-chase experiments are facilitated in these systems and they also allow the use of the inhibitors of macromolecular synthesis, which do not penetrate the plasma membrane. In such nuclear preparations, already initiated DNA molecules appear to be elongated normally, but initiation of new DNA synthesis does not seem to occur.

An extensive characterization of the DNA structures made in isolated nuclei from cells infected with Ad 5 has been carried out by Sussenbach and co-workers (Sussenbach and Van der Vliet, 1972,

1973; Van der Vliet and Sussenbach, 1972; Sussenbach *et al.*, 1972, 1973; Ellens *et al.*, 1974). The replication of adenovirus DNA in the isolated nuclei is semiconservative (Van der Vliet and Sussenbach, 1972) as was also shown to be the case *in vivo* for celo virus DNA (Bellet and Younghusband, 1972).

Three classes of replicating molecules have been observed by electron microscopy (Sussenbach *et al.*, 1972). The *first* class consists of Y-shaped molecules with a double-stranded arm and a single-stranded arm of nearly equal length. Y-shaped molecules with some double-stranded regions on the otherwise single-stranded arm were also present. The *second* class consisted of nonbranched linear molecules, partly double-stranded and partly single-stranded, with varying ratios of double-stranded to single-stranded DNA. The *third* class also contains nonbranched linear molecules, either purely double-stranded or purely single-stranded. On the basis of these studies, Sussenbach *et al.* (1972) suggested the following model for the replication of adenovirus DNA: Replication starts from the end of one of the parental strands, displacing the other parental strand and giving rise to the branched structures. Eventually this displacement and concurrent replication is assumed to yield a single-stranded and a double-stranded linear molecule. The released single-stranded molecules then are supposed to be converted to a duplex molecule by synthesis of the complementary strand in a discontinuous fashion leading to the formation of DNA segments that are later joined. It is possible that the synthesis of the strand complementary to the displaced one could start before the displacement is completed. This model is compatible with the results obtained with celo and Ad 5 virus DNA replication *in vivo* (Bellett and Younghusband, 1972; Van der Eb, 1973).

Recently Sussenbach *et al.* (1973) have presented evidence that the displaced strand preferentially is the one which in alkaline CsCl gradients bands at a higher density. In the Ad 2 system this strand corresponds to the L strand obtained after separation by copolymer binding (Tibbetts *et al.*, 1974). The implication is that one of the ends of adenovirus DNA is unique in that it furnishes the starting point for replication. Further evidence corroborating this notion has been obtained by electron microscopic investigations of partially denatured replicating virus DNA (Sussenbach *et al.*, 1973). In this report, the end where replication starts, i.e., the end containing the 5′ terminus of the displaced strand, is identified as the A+T-rich end of the DNA. This is the molecular right end defined in the denaturation map of the Ad 5 DNA (Ellens *et al.*, 1974).

Although these data appear quite compelling, it is appropriate to stress that the DNA synthesis occurring in the isolated nuclei reflects only elongation of already initiated DNA molecules and that no reinitiation takes place. Furthermore, only 10% of the parental DNA which is present in the nucleus appears to be active in elongation. This means that there is some important factor(s) missing in this system and the structures observed could differ from those present *in vivo*. So far no information is available with regard to the initiation process. If it involves a temporary RNA primer, for which there is ample evidence in several other systems (Sugino *et al.*, 1972; Wickner *et al.*, 1972; Magnusson *et al.*, 1973), and if this primer later is digested and replaced by DNA through "repair synthesis," the problem arises as to how to complete the 5′ ends of the daughter strands after digestion of the primer. As pointed out (Bellet and Younghusband, 1972), it is difficult to conceive how this is accomplished since a linear duplex DNA like adenovirus DNA has no primer for the repair synthesis at the 5′ ends of the daughter strands. There is no terminal redundancy in adenovirus DNA, but instead there is an inverted terminal repetition as discussed above in Sect. 4. Thus, concatemers like those existing in replicating DNA of the *E. coli* bacteriophage T7 cannot form and the completion of the newly synthesized strands cannot take place, as suggested for phage T7 (Watson, 1972). Bellet and Younghusband (1972), however, have suggested a model for the completion of the 5′ ends of nascent molecules involving a concatemer with staggered nicks in both strands at genome intervals.

6.6. Translation

The adenoviruses are assembled in the nucleus, but all the proteins appear to be synthesized in the cytoplasm. Adenovirus-specific mRNA and newly synthesized viral polypeptides are predominantly found in association with nonmembrane-bound polyribosomes (Thomas and Green, 1966; Velicer and Ginsberg, 1968; Horwitz *et al.*, 1969). Shortly after synthesis, the polypeptides are assembled into viral structural proteins in the cytoplasm and are then rapidly transported to the nucleus for assembly of virions (Velicer and Ginsberg, 1968). Since after infection with some virus mutants most of the viral proteins accumulate in the cytoplasm and only little accumulates in the nucleus, the protein transport may be governed by virus-coded products (Ishibashi, 1970; Russell *et al.*, 1972*a*).

6.6.1. Early Products

The majority of the proteins synthesized early after infection are host proteins, which makes it difficult to detect the early viral products. However, with immunological techniques, complement fixation, and immunofluorescence with sera from hamsters bearing adenovirus-induced tumors, synthesis of viral antigens has been detected early after infection. The so-called T antigen is the first early viral product to appear (Rouse and Schlesinger, 1967; Russell *et al.*, 1967*b*; Pope and Rowe, 1964; Shimojo *et al.*, 1967). In contrast to the papovavirus system, where T antigens have been detected only in the nucleus, the T antigens of adenoviruses are found both in the nucleus and in the cytoplasm. These antigens are synthesized also when the infection proceeds in the presence of 5-fluorodeoxyuridine (Gilead and Ginsberg, 1965) or cytosine arabinoside (Feldman and Rapp, 1966), which prohibit viral DNA synthesis, and are therefore to be considered as true early products. It is possible that the T antigens of adenoviruses involve several polypeptides (Tockstein *et al.*, 1968), but they have not been purified to a degree which has made a detailed characterization possible. Another antigen, called the P antigen, has also been detected early after infection (Russell and Knight, 1967). This antigen appears relatively late in the early phase of the infection and may constitute a mixture of T antigen and one of the viral core proteins (Russell and Skehel, 1972).

Recently Russell and Skehel (1972) were able to resolve the virus-induced proteins by SDS-polyacrylamide gel electrophoresis of extracts from cells labeled with ^{35}S-methionine of high specific activity. Five polypeptide bands could be detected which appeared specific for the infected cells, and at least one of these polypeptides may be an early product; the pattern was resolved better when host cell protein synthesis had been suppressed by a concomitant poliovirus infection in the presence of guanidine. A polypeptide with a molecular weight of 71,000 appears to be synthesized early in infection (Bablanian and Russell, 1974); however, this early polypeptide has not yet been correlated with T antigen and it has not been unequivocally demonstrated that the T antigens are actually coded for by the viral genome.

Early after infection in growth-arrested human embryo kidney cells or monkey kidney cells, adenovirus infection causes an increase in the activity of at least two of the enzymes (thymidine kinase and DNA polymerase) involved in DNA synthesis. This increase appears to be the result of an increase in synthesis of these proteins and it is concurrent to the increase in cell DNA synthesis observed in these cells after virus infection. However, a similar effect cannot be detected after adenovirus infection of cells grown in suspension cultures (Takahashi *et al.*, 1966; Ledinko, 1967; Kit *et al.*, 1967; Bresnik and Rapp, 1968; Ogino and Takahashi, 1969).

No virus-specific enzyme has been identified early in infection, although there is genetic evidence for the presence of genes coding for catalytic functions (Williams, personal communication). There are several adenovirus mutants which in complementation tests give rise to the same yields regardless of large variations in the ratio between the mutiplicities of infection of the two virus mutants.

6.6.2. Late Products

The shift from early to late viral gene expression occurs about 8–10 hours after a highly synchronized infection in suspension cultures of KB cells. At the same time the first capsid proteins can be detected, and 2–3 hours later the first progeny virus appears (Wilcox and Ginsberg, 1963; Polasa and Green, 1965, Russell *et al.*, 1967*b*; Everitt *et al.*, 1971, Russell and Skehel, 1972). The synthesis of the late proteins requires concomitant viral DNA synthesis, and no late proteins can be observed in the presence of inhibitors of DNA synthesis, e.g., cytosine arabinoside, hydroxyurea, and 5-fluorodeoxyuridine (Green, 1962*a*; Flanagan and Ginsberg, 1962; Wilcox and Ginsberg, 1963). The polypeptides comprising the major capsid units, hexon, penton, and fiber, are synthesized in large excess, while the synthesis of some of the minor polypeptides and the core proteins appears to be less extensive (White *et al.*, 1969, Everitt *et al.*, 1971). Only 5–10% of the viral capsid proteins synthesized during the infectious cycle become incorporated into mature virions (Green, 1962*b*; White *et al.*, 1969; Horwitz *et al.*, 1969). After synthesis of the polypeptides, the viral proteins are rapidly assembled into structural units which within a few minutes after completion are transported to the nucleus (Velicer and Ginsberg, 1970; Horwitz *et al.*, 1969).

Earlier results gave no indication that the structural proteins were produced by cleavage of larger precursors (Horwitz *et al.*, 1969; White *et al.*, 1969). Recent results, however, suggest that minor cleavage of precursor polypeptides occurs shortly after synthesis or during assembly (Anderson *et al.*, 1973; B. Edvardsson, unpublished). The polypeptide VII, which is the major arginine-rich core protein (molecular weight 18,000) (Laver, 1970; Prage and Pettersson, 1971), is generated from a precursor of molecular weight 21,000 (PVII in Fig. 2) so that one of five methionine-containing tryptic peptides is missing in the final product (Anderson *et al.*, 1973). The hexon-derived polypeptide VI (molecular weight 23,000) may be generated in a complex manner from a precursor of molecular weight 27,000 (marked 27K in Fig. 2).

In addition to the polypeptides, which are either utilized to form structural units of the virion directly or serve as precursors for virion proteins, at least five but possibly as many as seven virus-induced polypeptides can be observed in cytoplasmic extracts from infected cells labeled with ^{35}S-methionine (Russell and Skehel, 1972; Anderson *et al.*, 1973). No data concerning the biological function of these proteins are available. However, preliminary results suggest that the virus-induced protein of molecular weight 100,000 might be involved in mRNP structures in infected cells (Lindberg and Sundquist, 1974) and the polypeptide of molecular weight 71,000 may possibly be a DNA-binding protein (Van der Vliet and Levine, 1973).

Late in infection viral proteins accumulate in the nucleus, where assembly of the virions is taking place. As a reflection of this, large crystalline structures, so-called paracrystals, appear in the nucleus of the infected cell (Kjellén *et al.*, 1955; Morgan *et al.*, 1957). The paracrystalline structures may be of different kinds, and noncrystalline, irregular, intranuclear inclusions may have been inadvertently identified as crystalline structures. Crystalline areas of partially assembled virus (Boulanger *et al.*, 1973) and large crystalline areas of structural proteins, probably hexons or pentons (Boulanger *et al.*, 1970), have been observed. Finally, crystalline structures have been discerned which accumulate after the peak of viral synthesis in the productive cycle and may contain the major core protein (Marusyk *et al.*, 1972). Wills *et al.* (1973) recently showed that the paracrystals in Ad 5 did not appear at the nonpermissive temperature in cells infected with a *ts* mutant defective in fiber synthesis.

In addition to complete particles and structural proteins, several other aggregated virus structures may be found in extracts of infected cells. Penton structures, which have assembled into starlike arrangements with 12 pentons, so-called dodecons, may be recovered from cells infected with adenovirus from subgroups B and D (Norrby, 1966, 1969*b*; Liem *et al.*, 1971). Empty capsids and incomplete particles, the latter containing reduced amounts of viral DNA, are synthesized preferentially in cells infected with group A and B viruses (Mak, 1971;

Prage *et al.*, 1972; Wadell *et al.*, 1973). The empty capsids lack the two core proteins, polypeptides V and VII, but they have other polypeptides not present in the intact virion (Maizel *et al.*, 1968*b*; Prage *et al.*, 1972; Sundquist *et al.*, 1973*b*). They contain the precursor to polypeptide VI of the virion and some additional unidentified polypeptides (Sundquist *et al.*, 1973*b*), which will be discussed further in Sect. 6.9.

6.6.3. Modification of Viral Products

There is evidence suggesting that the fiber polypeptide is both phosphorylated (Russell *et al.*, 1972*b*) and glucosylated (Ishibashi and Maizel, 1972), and the modification of the polypeptide probably occurs subsequent to translation. No sugars other than N-acetyl-glucosamine have been found associated with the fiber polypeptide. Two or three of the nonstructural polypeptides induced by the virus also appear to be phosphorylated and the major phosphorylated component late in the infectious cycle is the polypeptide 71K (see Fig. 2) (Russell *et al.*, 1972; B. Edvardsson, unpublished).

6.6.4. Functional Role of Late Viral Proteins

The functional roles of some of the virion components have been reviewed in Sect. 2. At present only little is known about the functional role of the nonstructural virus-induced polypeptides. Recently, however, evidence was obtained that the virus-induced polypeptides 71K and 45K (Fig. 2) might have some function in relation to either DNA replication or transcription since they can be purified by DNA-cellulose chromatography (Van der Vliet and Levine, 1973). This hypothesis is supported also by the finding that these proteins are not synthesized at nonpermissive temperatures in cells infected with a DNA negative mutant, *ts* 225, of Ad 5 (Ensinger and Ginsberg, 1972), which has been renumbered as *ts* 125 (Ginsberg *et al.*, 1973).

The endonuclease recovered from the virus particles in association with the penton (Burlingham *et al.*, 1971) has not yet been identified among the virus-induced proteins, and it remains to be established whether this endonuclease is a cellular, virus-modified, or a virus-coded component.

6.7. Host Cell Macromolecular Synthesis

Both primary cells in monolayer cultures and established cell lines in suspension cultures fail to divide after infection with adenoviruses. This effect can probably be ascribed to the generalized interference with the synthesis of host cell macromolecules seen in the late phase of the infection. In growth-arrested monolayer cultures, adenovirus infection, like SV40 and polyoma virus infection, results in early stimulation of cell DNA synthesis (Ledinko, 1967, 1970). No such effect is seen after infection of cells grown in suspension cultures. Except for this transient enhancement of synthesis of DNA and proteins involved in DNA replication, it appears that synthesis of ribosomal RNA is also increased during the early phase of the infection, at least in primary cells (Ledinko, 1972), and possibly also in suspension cultures (Raskas *et al.*, 1970). In the late phase, on the other hand, the synthesis of host cell macromolecules is drastically inhibited.

At 6–8 hours after infection with adenovirus the synthesis of cellular DNA begins to decline, and by 10–13 hours post-infection around 90% of the newly synthesized DNA is viral (Ginsberg *et al.*, 1967; Piña and Green, 1969). Hodge and Scharff (1969) have investigated the effect of adenovirus infection on host cell DNA synthesis in synchronized HeLa cells. When viral DNA synthesis occurs during the G1 phase of the growth cycle, cellular DNA replication is not induced, and its subsequent initiation is prevented. When viral DNA synthesis is timed to begin during the S phase the cellular DNA synthesis goes on but the round of cellular DNA replication apparently fails to go to completion. The implication is that under the latter conditions already initiated cell DNA synthesis can continue but that the initiation of new synthesis of cellular DNA is inhibited.

Host cell protein synthesis is shut off at about the same time as the cell RNA synthesis (Ginsberg *et al.*, 1967). In pulse-labeling experiments with radioactive amino acids at or later than 12 hours post-infection the label enters virus-coded or -induced proteins almost exclusively (Russell and Skehel, 1972; Andersson *et al.*, 1973). A possible mechanism for the shutoff of host protein synthesis will be discussed later.

Concurrent with the reduction in host protein synthesis the accumulation in the cytoplasm of newly synthesized ribosomal RNA is suppressed. At around 12 hours after infection only 10–20% of the normal amount of ribosomal RNA is transported to the cytoplasm (Raskas *et al.*, 1970; Philipson *et al.*, 1973). Since a significant amount of the 45 S precursor to ribosomal RNA is still synthesized (Raskas *et al.*, 1970; Ledinko, 1972), the interference with ribosome synthesis appears confined to the cleavage of the ribosomal precursor RNA. It is not clear whether the interference with the processing of ribosomal RNA is an indirect result of the inhibition of host cell protein synthesis or whether there is some specific inhibitory mechanism. It is known, however, that conditions which suppress host cell protein synthesis in uninfected cells also lead to drastic reduction of the maturation of ribosomal RNA (Maden *et al.*, 1969; Willems *et al.*, 1969).

The rate of transcription of host cell HnRNA does not appear to be affected during the lytic cycle. The major part of the newly synthesized HnRNA is cellular and only 10–20% of nuclear RNA is of viral origin (Price and Penman, 1972*a*; McGuire *et al.*, 1972). Nevertheless, almost all of the mRNA transported to the cytoplasm appears to be viral (Lindberg *et al.*, 1972). This would imply that there is either a block in the processing of HnRNA to cell mRNA or a preferential inhibition of the synthesis of the mRNA "portion" of the host cell HnRNA.

Some host cell mRNA appears to be functioning even late in the infectious cycle since the protein moiety of the mRNP appears to be synthesized at the same rate late in infection as in uninfected cells (Lindberg and Sundquist, 1974). Synthesis of cellular tRNA occurs unabatedly throughout the infectious cycle (Ginsberg *et al.*, 1967; Mak and Green, 1968) and no virus-induced or virus-coded tRNA has been recognized (Raska *et al.*, 1970; Kline *et al.*, 1972).

It has been suggested that the structural proteins of the virus could be involved in the inhibition of host macromolecular synthesis, and some results have been obtained which could be interpreted to mean that the fiber protein somehow causes the inhibition of host cell DNA synthesis (Levine and Ginsberg, 1967, 1968). It seems unlikely, though, that newly synthesized fiber should have this effect since inhibition of host DNA synthesis in the productive cycle occurs prior to detectable synthesis of the fiber protein (Russell *et al.*, 1967*b*). Futhermore *ts* mutants negative for the fiber protein and/or also for viral DNA synthesis still inhibit host cell DNA synthesis at nonpermissive temperatures (Wilkie *et al.*, 1973).

6.8. Control Mechanisms

The mechanims controlling the switch from early to late phase during productive infection are largely unexplored. A variety of effects seen in the interplay between the virus and the host cell and in the expression of the viral genome during this transition period indicate that several different control mechanisms must be operating. The key event, which makes the transition to the late phase possible, is the triggering of viral DNA replication. If this is prevented there is no switch from early to late phase expression of the viral genome. Concomitant with increased viral DNA synthesis, host cell DNA synthesis is shut off. Appearance of increasing amounts of late-phase transcription products coincide in time with a decreased manufacture of mature ribosomes and host mRNA. In all these steps specific control mechanisms are implied. In addition to this there is evidence for regulation also at the translational level.

6.8.1. Transcription

It was pointed out above that the adenovirus genome appears to be transcribed in large units, giving rise to the giant viral RNA molecules found in the nucleus. The assumption is that these mRNA precursors are cleaved at specific points along the polynucleotide chains to give rise eventually to the mature mRNAs found functioning on polyribosomes in the cytoplasm. Virtually nothing is known about the mechanisms governing the concerted expression of the early and then the late adenovirus genes.

An intriguing finding is that more than 90% of the mRNA transported to the polysomes late in the infection is adenovirus-specific (Lindberg *et al.*, 1972), whereas in the nucleus only 10–20% of the newly synthesized nonribosomal RNA is virus-specific (Price and Penman, 1972*a*; McGuire *et al.*, 1972). The implication is that there is a virus-controlled interference with post-transcriptional processing of host cell HnRNA into mature host mRNA, and that this leads to a preferential selection of viral mRNA for transport to the cytoplasm. Alternatively, a transcriptional control may be preventing synthesis of the mRNA portion of the host cell mRNA.

6.8.2. Translation

With regard to translation one can discern at least two control mechanisms: one governing the switch in the late phase to a preferential translation of viral mRNA and the other modulating the synthesis of the viral polypeptides.

In the late phase of the infection protein synthesis goes on at an undiminished rate, and the polysomes, instead of making mainly host cell proteins, now become engaged almost exclusively in the synthesis of viral proteins. This suggests that viral mRNA has replaced most of the host mRNA on the polysomes, in accordance with the finding that more than 90% of the mRNA transported to the cytoplasm in the late phase is virus-specific. However, in uninfected cells, the mRNA appears to turn over slowly, on the average once per cell generation (Greenberg, 1972; Singer and Penman, 1972), whereas the replacement of host mRNA by viral mRNA in the late phase of the infection takes only a couple of hours. This seems to suggest that the switch from early to late translation could not be accomplished merely by a preferential export of viral mRNA from the nucleus, but that there is an additional mechanism ensuring the preferential utilization of the viral mRNA once it has come to the cytoplasm. Whether this means that there are virus-specific initiation factors for the translation of the viral mRNAs or that there is an enhanced degradation of host mRNA is not clear. The fact that vaccinia virus transcription is unaffected but vaccinia virus translation is deficient in coinfection with vaccinia and adenovirus suggests a preferential initiation of adenovirus mRNA (Giorno and Kates, 1971).

It is worth noting that some host proteins appear to be made also in the late phase of the infection. This is illustrated by the continued appearance of amino acid label in the polypeptides constituting the protein part of the mRNP particles. Four of the five major polypeptides of the adenovirus mRNP appear to be host proteins (Lindberg and Sundquist, 1974). The delineation of the control mechanism discussed here, however, must await an *in vitro* protein-synthesizing system where initiation factors from virus-infected and uninfected cells can be compared. Although two reports have appeared (Caffier and Green, 1971; Wilhelm and Ginsberg, 1972) suggesting continued synthesis of viral proteins *in vitro* in extracts from infected cells, it is not clear whether these crude systems from infected cells initiate a new round of translation. Isolated viral mRNA has not yet been translated successfully in a purified *in vitro* protein-synthesizing system, although several investigators are presently engaged in this problem.

It is evident from the work of several investigators, including Perlman *et al.* (1972), that the hexon polypeptide is preferentially synthesized at 37°C. This may reflect the presence of different amounts of mRNA for different polypeptides. However, the relative amounts of the different viral polypeptides synthesized in the presence of low concentrations of cycloheximide is significantly altered, as compared to a control without the drug. Since cycloheximide slows elongation without affecting initiation of translation, this would mean that under normal conditions there are differences in the rate of initiation of the different polypeptides. The most pronounced effect is seen in the synthesis of the hexon polypeptide, which normally is synthesized at a rate 2–4 times faster than the fiber polypeptide (Everitt *et al.*, 1971).

Some of the polysomal mRNA molecules are much larger than one would expect judging from the size of the polypeptides for which they code. The largest polypeptide, the hexon polypeptide, has a molecular weight of 100,000–120,000 (Jörnvall *et al.*, 1974*b*; Horwitz *et al.*, 1970) and the largest mRNAs have molecular weights of approximately 3×10^6 (Lindberg *et al.*, 1972). The predominant peak in the late-mRNA pattern has a molecular weight of about 1.6×10^6 (26 S). In addition, this mRNA is larger than would be required to synthesize any of the viral polypeptides. This could mean that at least some of the viral mRNA are polycistronic, but there is no direct evidence for this assumption yet. In fact, there is no evidence for polycistronic mRNAs in any eukaryotic system. In the case of polio and encephalomyocarditis (EMC) viruses, the mRNA appears to be identical to the virus RNA (molecular weight 2.6×10^6). This mRNA is translated into one gigantic protein precursor molecule which through an intricate cleavage mechanism is processed to generate the mature viral proteins (Jacobson and Baltimore, 1970). The primary steps in this processing are carried out by cellular enzymes, and later more specific cleavages by virus-induced enzymes give rise to the final polypeptides (Korant, 1973). No evidence has been obtained in the adenovirus system for the presence of larger precursor proteins of the type seen in the polio and EMC virus systems. However, minor proteolytic trimming occurs with the two precursors for viral polypeptides VI and VII (Andersson *et al.*, 1973), as was discussed in Sect. 6.6.

6.9. Assembly of Adenoviruses

6.9.1. Mechanism of Assembly

During the late phase of the adenovirus infection, virus particles as well as empty capsids accumulate in the nucleus. The latter particles lack viral DNA. The amount of empty capsids formed varies between different serotypes. After infection with Ad 2 only minute amounts are found, whereas a large pool of empty capsids is found with Ad 3 and Ad 16 (Prage *et al.*, 1972; Wadell *et al.*, 1973).

Horwitz *et al.* (1969) reported that newly synthesized core polypeptides appear more rapidly in mature virus particles than the polypeptides of the outer capsid. However, their data did not allow a conclusion whether the virions are formed by injection of DNA cores into preformed capsids or by the assembly of capsomers around cores. Recently results have been obtained (Sundquist *et al.*, 1973*b*) which strongly suggest that the empty capsids are intermediates in the assembly of virions and that the most likely route is that DNA cores enter preformed capsids. This conclusion was based on several lines of evidence: (1) The increase in absolute amount of empty capsids coincided in time with the increase in the amount of virions. This indicates that the empty capsids are not created by breakdown of virions late in the infectious cycle. (2) Under conditions of continuous labeling with radioactive amino acids, labeled virions appeared with a lag period when compared to labeled empty capsids. (3) Pulse-chase experiments with labeled amino acids showed that radioactivity increased in the empty capsids faster than in the complete virions and without a detectable lag. The lag period before the appearance of labeled virions was 60–80 minutes. Later the radioactivity in the fraction containing empty capsids decreased, while the radioactivity in complete virions was still increasing. These results thus provide more direct evidence for a precursor–product relationship between the empty capsids and the mature virions and suggest that the DNA core complex is introduced into a preformed capsid. The experiments are not conclusive, however, since it is still possible that the DNA of intermediates may be released during purification of the particles.

Additional evidence for empty capsids as intermediates has been obtained through the analysis of the polypeptide patterns of these particles compared to the virions. The empty capsids lack the core proteins V and VII and also polypeptides VI, VIII, and X, but instead they contain polypeptides not present in the virion (Sundquist *et al.*, 1973*b*). Anderson *et al.* (1973) have recently reported results which suggest that two of the virion components are derived by cleavage of precursor polypeptides of higher molecular weight. In comparing their results with those of Sundquist *et al.* (1973*b*) one finds that two of the polypeptides found in abundance in the empty capsids (27K and 26K in Fig. 2) most probably correspond to the precursors suggested for polypeptide VI and VIII of the virion. In addition to these polypeptides, the empty capsids also contain other unidentified polypeptides in the molecular weight range from 45,000 to 100,000.

The virus serotypes which give rise to large amounts of empty capsids, like adenovirus type 3, also accumulate significant amounts of particles which band in CsCl at densities intermediate to those of empty capsids and virions (Prage *et al.*, 1972; Wadell *et al.*, 1973), suggesting that they are particles with varying amounts of DNA. These structures are here referred to as imcomplete particles. When they are analyzed for constituent polypeptides it is found that they also contain the precursor to polypeptide VI. The heavier incomplete particles contain less of the precursor than the lighter particles. One of the core proteins, polypeptide V, appears in particles containing the amount of DNA corresponding to half the adenovirus genome or more, and the other core protein, polypeptide VII, is not observed, except in particles containing the whole DNA complement (B. Edvardsson, unpublished). Whether the assembly of complete virions also involves exchange of proteins in a similar way as has been demonstrated for the *Salmonella typhimurium* phage P22 (King *et al.*, 1973) is not known.

It is noteworthy that the ratio of hexon to fiber polypeptide in the empty capsids is lower than in the virions, suggesting that some hexons in the facets of the icosahedron might be lacking in the empty capsids (Sundquist *et al.*, 1973*b*). This could reflect discontinuities in the capsid through which the DNA core is introduced. It has not been possible by electron microscopy to establish whether these discontinuities are located in the penton area or within the triangular facets of the virion. The final elucidation of the route of assembly probably will require studies with the recently isolated *ts* mutants of Ad 5. Interesting in this connection especially are the mutants of the serological class 1 (Russell *et al.*, 1972*a*) which make most of the virion proteins yet do not assemble properly. Another avenue to establish the route of assembly might be to use some of the defective assembly systems described below.

6.9.2. Defective Assembly

6.9.2a. Arginine Starvation

Synthesis of adenovirus has been shown to be strongly dependent on the concentration of arginine in the medium (Rouse and

Schlesinger, 1967; Russell and Becker, 1968). When arginine is omitted from the medium the virus yield is reduced by 3–4 orders of magnitude. This dramatic reduction is somewhat surprising when one considers that the components of the virus are still synthesized in amounts which would seem to be sufficient to give rise to normal virus production (Everitt *et al.*, 1971; Rouse and Schlesinger, 1972).

When arginine is added back to the infected cells the block is released and after a period of 4–5 hours the virions begin to accumulate and eventually an almost normal yield of virus is obtained. However, tenfold less of the capsid proteins made during the arginine starvation, compared with the amount synthesized and assembled after the release of the block (Everitt *et al.*, 1971), are utilized for the assembly of virions. The accumulation of virions after reversion is sensitive to inhibition of DNA, RNA, and protein synthesis, again emphasizing a dependence upon newly synthesized macromolecules.

When the condition of arginine starvation is introduced at different times after infection, almost normal yields are attained when starvation is introduced later than 14 hours after infection. This would imply that some factor(s), the synthesis of which is strongly dependent upon the presence of arginine in the medium, is made earlier than this time. If this factor is absent or present in insufficient amounts, the virus components cannot assemble efficiently. Furthermore, the virus components made in the absence of this factor appear to be defective since they do not participate in the virus assembly occurring after arginine is added back to the system. The factor(s) involved has not yet been identified. The synthesis of viral DNA is only slightly affected and late viral mRNAs appear to be synthesized normally (Raska *et al.*, 1972; Everitt, unpublished). Viral proteins including those rich in arginine are also synthesized (Everitt *et al.*, 1971; Rouse and Schlesinger, 1972). There is, however, a 2- to 4-fold relative suppression of the synthesis of the hexon polypeptide, but whether this has relevance for the defective assembly is not clear. This effect is similar to the effect of low concentrations of cycloheximide described above under control mechanisms.

Winters and Russell (1971) reported that viral DNA synthesized in arginine-deficient medium and present in extracts from these cells can assemble into virions when provided with extracts from infected cells maintained in a normal medium. This finding has not been confirmed, but this system would be extremely useful in the elucidation of the role of arginine in assembly and also in clarifying the route of virus assembly.

6.9.2b. Elevated Temperature

The synthesis of infectious virus particles is also reduced by two orders of magnitude if the infected cells are maintained in suspension culture at 42°C instead of at the optimal 37°C. Viral DNA and viral mRNA are synthesized at a faster rate at 42°C, but no virions are formed (Warocquier *et al.*, 1969; Okubo and Raskas, 1971). It appears that the viral polypeptides are formed in the right proportions but that the assembly of the polypeptides into capsomers is reduced to the same extent as the reduction in virus yield (Okubo and Raskas, 1971).

In addition, Perlman *et al.* (1972) have demonstrated that the relative amounts of different polypeptides synthesized at 40°C differ from those observed at 37°C. The hexon polypeptide appears to be synthesized at nearly normal rates, while the synthesis of the other capsid polypeptides is reduced. Normally the ratio of hexon to fiber polypeptide is about 4, but it may increase to 10 at 40°C. The results add further support to the idea that the translation of viral mRNA is under complex control. The route of assembly may thus involve several steps which may be controlled differently: control of rate of synthesis of individual polypeptides, assembly of capsid polypeptides into the respective capsomers, assembly of capsomers into empty capsids, and subsequently assembly of empty capsids into virions. The transport of newly synthesized polypeptides or capsomers into the nucleus may be another event which is at fault in defective assembly.

7. ABORTIVE INFECTIONS

Whether an adenovirus infection will be productive or abortive depends both on the serotype used and the type of cell involved and also on the state of growth of the cells. Table 4 shows the responses of different cells to infection with different adenovirus serotypes.

7.1. Abortive DNA Replication

Although hamster cells are permissive for Ad 2 and Ad 5, the replication of Ad 12 is restricted. The Ad 12 virus particles can penetrate baby hamster kidney cells (BHK21) and induce T-antigen synthesis. Thus the early genes appear to be expressed, at least in part (Strohl *et al.*, 1967; Raska and Strohl, 1972). However, the Ad 12

TABLE 4

Permissiveness for Human Adenovirus Infection of Cells from Different Species

Species	Cell	Serotype	Permissiveness[a]	Defective event
Human	KB or HeLa Primary fibroblasts or epithelial	All human	+	None
Monkey	AGMK Rhesus	Most human	−	Translation
Hamster	Primary BHK21 NIL-2	Ad 2	+	None
		Ad 5	+	None
		Ad 12	−	DNA replication
Rat	Primary	Ad 2	−	DNA replication?
		Ad 12	−	DNA replication?

[a] + denotes production of progeny virions; − denotes that production of virions is deficient.

genome is not replicated (Doerfler, 1968, 1969), and thus there is no late phase in these cells.

By following the fate of parental DNA, Doerfler (1968, 1969) obtained evidence that a fraction of Ad 12 DNA is integrated into the host genome after infection at high multiplicities. The interpretation of these experiments was challenged by Zur Hausen and Sokol (1969), who with a different hamster cell line, Nil-2, showed that a significant amount of the parental virus DNA was degraded and reutilized for synthesis of cellular DNA. Additional evidence for viral DNA integration was subsequently supplied (Doerfler, 1970), indicating that the incorporation of viral DNA in the host cell genome is independent of both host protein synthesis and host DNA synthesis. The differences in the results obtained in these investigations could be due to variations between the two cell lines used in the degradation and reutilization of the parental virus DNA. It is noteworthy that the DNA of Ad 12 cannot replicate, even in mixed infections with Ad 2 (Doerfler, 1969; Doerfler and Lundholm, 1970).

A small population of the Ad 12-infected BHK cells which survive the infection exhibit altered growth properties characteristic of transformation (Strohl *et al.*, 1970), but the majority of infected cells show extensive chromosome pulverization with concomitant degradation of the host cell DNA, and these cells finally die (Strohl, 1969*a,b*).

In BHK cells arrested in the G1 phase of the growth cycle, Ad 12 induces cellular DNA synthesis (Strohl, 1969*a,b*) with a concurrent activation of the enzymes involved in DNA replication. This occurs immediately after infection and prior to the lytic effect on the cells (Zimmerman *et al.*, 1970). The stimulation of DNA synthesis caused by Ad 12 under these conditions appears to be blocked by the addition of cyclic AMP, which also suppresses the synthesis of T antigen (Zimmerman and Raska, 1972). Thus the hamster cells offer an abortive system with regard to Ad 12 DNA replication, but other serotypes like Ad 2 and Ad 5 can utilize the same cell permissively.

7.2. Abortive Translation

Abortive infection has been observed also in monkey cells with several human adenoviruses (Rabson *et al.*, 1964*b*). The virus particles enter the cells and the early phase of the infection appears to proceed normally (Feldman *et al.*, 1966; Friedman *et al.*, 1970). In contrast to the situation in the abortive infection described above, in this case the viral genome does replicate (Rabson *et al.*, 1964*b*; Reich *et al.*, 1966), but still very few progeny virions are produced. Recently results have been presented suggesting that also the late-phase mRNA is produced and that most of the late sequences are present in the cytoplasm (Baum *et al.*, 1968; Fox and Baum, 1972; Lucas and Ginsberg, 1972*a*). Since capsid proteins cannot be detected (Friedman *et al.*, 1970; Baum *et al.*, 1972), however, the translation of the late mRNA appears to be impaired. The nonpermissiveness, therefore, might reside in the mechanism governing the switch from translation of host mRNA to translation of viral mRNA. The further elucidation of the mechanism of the block obviously requires more detailed investigations concerning the properties of the mRNA present in the cytoplasm under nonpermissive conditions.

It was discovered many years ago that the block in the replication of adenovirus in monkey cells can be overcome by coinfection with SV40 virus (O'Conor *et al.*, 1963; Rabson *et al.*, 1964*b*). Since then it has been shown that other viruses like the simian adenoviruses (Naegele and Rapp, 1967; Altstein and Dodonova, 1968), adeno–SV40 hybrid viruses (Rowe and Baum, 1965), and the unidentified agent MAC can also act as helper viruses (Butel and Rapp, 1967).

In the presence of SV40 as a helper virus, the normally abortive adenovirus infection in monkey cells is converted to an efficient lytic, replicative cycle, where the yield of the the adenovirus is increased by 2–3 orders of magnitude. Since the helper SV40 virus under these conditions is unable to replicate its own DNA, it appears likely that an early SV40 function aids the translation of late adenovirus mRNA (Friedman *et al.*, 1970).

7.3. Adeno–SV40 Hybrid Viruses

The adenovirus–SV40 hybrid viruses consist of all or part of the SV40 genome plus an intact or partially deleted adenovirus genome enclosed in an adenovirus capsid. Such hybrid viruses were originally isolated when adenovirus 7 vaccine strains were propagated in monkey cells. The stocks of adenoviruses used to produce vaccines were adapted to grow in rhesus monkey kidney cells for the Ad 1–5 and Ad 7 serotypes (Hartley *et al.*, 1956). The monkey cells were found to be contaminated with SV40 virus, and, when the adenoviruses were purified from the contaminant with the use of SV40 antiserum, they still gave rise to productive infection. The first isolated strain with these characteristics, designated E46^{+} or PARA virus (particles aiding replication of adenovirus), contained particles with both SV40 and adenovirus genetic material enclosed in an adenovirus capsid. The original PARA Ad 7 strain was found to contain two types of particles: wild-type Ad 7 and PARA particles consisting of a hybrid genome containing SV40 and Ad 7 (Rowe and Baum, 1964, 1965). The SV40 and adeno DNAs present in the PARA particles are covalently linked since they both band at the density of adenovirus DNA after sedimentation in alkaline CsCl (Baum *et al.*, 1966). Kelly and Rose (1971) showed that the hybrid DNA contains about 75% of the SV40 DNA inserted into the adeno DNA at 0.05 fractional length from one end of the hybrid DNA. The adeno 7 DNA was found to be deleted in approximately 10% of these hybrid particles (Kelly and Rose, 1971). The PARA Ad 7 plaque with two-hit kinetics on monkey cells, suggesting that both particles are necessary for successful infection. The progeny of such plaques on AGMK cells still consist of a mixture of Ad 7 and hybrid virus. On human cells on the other hand, the Ad 7 helper is selected for, and the hybrid is lost. In monkey cells, again, the T and U antigens of SV40 are still induced, but no SV40 capsid proteins are synthesized. The hybrid genome can be transferred to the capsid of a different adenovirus serotype (transcapsidation) when the cells are coinfected with PARA Ad 7 and an excess of another adenovirus distinct from Ad 7 (Rapp *et al.*, 1965; Rowe, 1965). The transcapsidants are insensitive to antisera against Ad 7, but induce Ad 7 T antigen in infected cells (Rowe and Pugh, 1966). Transcapsidation between PARA Ad 7 and adeno 1, 2, 5, 6, and 12 has been reported (Rowe, 1965; Rapp *et al.*, 1968).

Recently adeno–SV40 hybrids have been obtained also from a strain of Ad 2, propagated in monkey cells. The original strain (Ad 2^{++}) gave rise to wild-type Ad 2, complete SV40 virions, and a mixed population of hybrids between SV40 and Ad 2 (Lewis *et al.*, 1969). Stable hybrids have been isolated from this strain on AGMK cells and two such segregants, $Ad2^{+}$HEY and $Ad2^{+}$LEY, obviously contain the complete SV40 genome in covalent linkage to a defective adenovirus genome (Lewis and Rowe, 1970; Crumpacker *et al.*, 1970). These strains plaque with two-hit kinetics on AGMK cells with a requirement for Ad 2 helper functions. Lewis and co-workers (1969) isolated from these strains a second type of hybrid viruses which were nondefective (ND) on both human and monkey cells. The first strain, $Ad2^{+}ND_1$, plaques with one-hit kinetics on both cell types and yields almost equal titers on both cells (Lewis *et al.*, 1969). This strain contains only a small fragment of SV40 DNA, which remains stable during replication in human cell lines. The two genomes in $Ad2^{+}ND_1$ are covalently linked (Levin *et al.*, 1971) and the insertion corresponding to 18% of the complete SV40 DNA has been mapped at a fractional distance of 0.14 from one end of the adenovirus DNA. The Ad DNA apparently is deleted at the SV40 insertion and the size of the missing fragment is about 1.3×10^6 daltons (Kelly and Lewis, 1973). The only SV40 product found in cells infected with $Ad2^{+}ND_1$ is the U antigen, which ressembles the SV40 T antigen but is more heat stable and can be detected with antisera from hamsters carrying SV40-induced tumors (Lewis and Rowe, 1971). Four additional nondefective Ad 2–SV40 hybrids designated $Ad2^{+}ND_2$–$Ad2^{+}ND_5$ have been isolated (Lewis *et al.*, 1973), and the following general characteristics of these strains have been reported:

1. Each nondefective hybrid contains a piece of SV40 DNA and a deletion of adenovirus DNA.
2. In all the hybrids, SV40 DNA sequences begin at the same position in the Ad 2 DNA molecules, namely 14% of the distance from one end of the adenovirus DNA molecule (Kelly and Lewis, 1973). Recent sequence studies of the SV40 RNA transcribed by *E. coli* polymerase from these hybrids (Weissman, personal communication) has established that this region is 14% of the distance from the 5′ end of the light DNA strand of the hybrid after separation with ribopolymers (Patch *et al.*, 1972).
3. The SV40 DNA insertions are overlapping (Henry *et al.*, 1973) for the different nondefective hybrids, as shown in Table 5. In mapping the early SV40 functions (Kelly and Lewis,

TABLE 5

Properties and Characteristics of Nondefective Adeno–SV40 Hybrid Viruses

Strain designation	SV40 DNA segment: Size, daltons[a]	SV40 DNA segment: % of genome[b]	Deletion of Ad 2 genome, %	Permissive in AGMK[c]	SV40 antigens induced[d]
$Ad2^+ND_3$	4.8×10^4	7	5.3	−	None
$Ad2^+ND_1$	2.4×10^5	18	5.4	+	U
$Ad2^+ND_5$	5.3×10^5	28	7.1	−	None
$Ad2^+ND_2$	6.2×10^5	32	6.1	+	U, TSTA
$Ad2^+ND_4$	8.4×10^5	43	4.5	+	U, TSTA, T

[a] Amount of SV40 DNA estimated by hybridization of SV40 cRNA (RNA transcribed *in vitro* with *E. coli* polymerase) to hybrid virus DNA (Henry *et al.*, 1973). The nondefective adeno–SV40 hybrids are listed according to the size of the integrated SV40 fragment.

[b] Estimated by heteroduplex mapping (Kelly and Lewis, 1973) and refer to the size of the integrated fragment relative to intact SV40 DNA.

[c] Permissiveness refers to comparable yields of hybrid virus in human and African green monkey cells (AGMK).

[d] The terminology and the characteristics of the U and T antigens have been described (Lewis *et al.*, 1973). They are all early products of the SV40 virus infection. TSTA refers to SV40-specific transplantation antigen.

1973) the only anomaly is that $Ad2^+ND_5$, which contains all the SV40 DNA sequences in $Ad2^+ND_1$, nevertheless does not induce SV40-specific U antigen (Lewis *et al.*, 1973).

4. The site for initiation of RNA transcription on SV40 DNA with *E. coli* polymerase is present in the light strand of the DNA of all hybrid viruses (Zain *et al.*, 1973).

Adeno–SV40 hybrids have been described also for Ad 4 (Easton and Hiatt, 1965; Beardmore *et al.*, 1965), Ad 5 (Lewis *et al.*, 1966), and Ad 12 (Schell *et al.*, 1966).

The hybrid viruses, $Ad2^+ND_{1\text{-}5}$, have already proved useful for the physical characterization of the SV40 genome, and have led to a map of the early functions of the SV40 genome. Most certainly these hybrid viruses will serve as excellent tools also in the genetic analysis of the adenovirus genome.

8. CELL TRANSFORMATION

The human adenoviruses have been divided into highly oncogenic, weakly oncogenic, and nononcogenic serotypes, corresponding to the subgroups A, B, C, and D, respectively. This grouping was originally based on the frequency with which hamsters develop tumors after inoculation with virus. The tumor cytology appears, however, to be similar, irrespective of the types or the groups used to induce the tumors. The present discussion will primarily be concerned with the transformation observed in *in vitro* cell cultures. A general conclusion for human adenoviruses and several types from other species is that *in vitro* transformation has been induced most frequently in cells nonpermissive for virus replication.

8.1. Cell Transformation by Different Adenoviruses

The group A adenoviruses were first shown to transform newborn hamster kidney cells (McBride and Wiener, 1964). Freeman *et al.* (1967*a*) subsequently reported transformation of rat embryo fibroblasts with Ad 12. Both hamster and rat cells are nonpermissive for Ad 12 replication. Rat cells are also nonpermissive for group C viruses, and, accordingly, transformation is seen with Ad 2 and Ad 1 in rat embryo cells *in vitro* (Freeman *et al.*, 1967*b*; McAllister *et al.*, 1969*a*). Even the nononcogenic adenovirus group D can transform rat and hamster cells *in vitro* (McAllister *et al.*, 1969*b*). Further support for the hypothesis that transformation requires nonpermissive conditions has recently been obtained with *ts* mutants and fragmented adenovirus DNA. Williams (1973) could transform hamster cells with *ts* mutants of Ad 5, and Graham and Van der Eb (1973*b*) could transform rat cells with fragmented Ad 5 DNA with the same efficiency as observed for induction of the productive cycle with intact DNA. Fragmented DNA from simian adenovirus 7 may also induce tumor in hamsters (Mayne *et al.*, 1971).

Cell transformation with adenoviruses requires high multiplicities of infection, and the frequency of transformation is low. The transformed cells which can be selected because of their "infinite" growth potential have a characteristic morphology and Ad 12-transformed BHK21 cells can be distinguished from the same cells transformed by polyoma virus (Strohl *et al.*, 1967). The cells grow to higher saturation densities than untransformed cells; they grow in disoriented arrays and have an altered plasma membrane. Infectious virus cannot be detected in transformed cells, but the cells can be identified since they synthesize T antigen which can be detected with sera from tumor-bearing animals (Huebner *et al.*, 1964). In contrast to the papovaviruses, infectious virus has never been induced from adenovirus-transformed cells either by forming heterokaryons with permissive cells or by treating the cells with physical or chemical agents which can induce virion production for SV40 and polyoma virus (Gerber, 1966; Burns and Black, 1969; Fogel and Sachs, 1969, 1970). In spite of the failure to rescue virus production, several lines of evidence indicate, as discussed below, that at least parts of the adenovirus genome persist in transformed cells.

8.2. Viral DNA in Transformed Cells

Some evidence has been obtained that Ad 12 DNA may be integrated into host DNA in nonpermissive hamster cells (Doerfler, 1970), but there is no direct experimental evidence that the viral DNA is integrated into the host cell DNA of adenovirus-transformed cells. Viral DNA has, however, been detected in transformed cells utilizing the capacity of RNA transcribed *in vitro* (cRNA) from adenovirus DNA by *E. coli* RNA polymerase to hybridize to viral DNA. Such cRNA can specifically hybridize to DNA extracted from cells transformed with the homologous adenovirus (Green *et al.*, 1970). A hybridization technique was used which utilizes filter-bound DNA. When standardized with known amounts of viral DNA and DNA from transformed cells, this technique gave 14–37 copies of viral DNA in transformed Ad 2 rat cells, and 22 and 97 copies of viral DNA in hamster cells transformed by Ad 12 and Ad 7, respectively. With the same technique Doerfler (1970) found 5–60 adeno DNA equivalents in Ad 12 hamster cells after nonpermissive infection at 30–40 hours after infection. The assay with filter-bound DNA may give inaccurate results mainly because this method must be calibrated by reconstruction experiments in which some of the viral DNA may be lost from the filters (Haas *et al.*, 1972). However, when the rate of renaturation of ^{32}P-labeled viral DNA was followed in the presence and absence of DNA from transformed cells with a less ambiguous method involving liquid hybridization and hydroxyapatite chromatography to score for duplex DNA, Pettersson and Sambrook (1973) found one copy of viral DNA per diploid quantity of cell DNA in Ad 2-transformed rat cells. More detailed analysis of the Ad 2 DNA present in the transformed cells with restriction endonuclease fragments of Ad 2 DNA (Pettersson *et al.*, 1973) indicated that only 45% of the viral genome, mainly derived from the fragments A, C, D, and E were present in the transformed cells (Sambrook *et al.*, 1974). It has not yet been established that the same fragments of the adenovirus genome are present in other lines of Ad 2-transformed cells. The finding suggesting that the adeno genome is deleted in transformed cells may explain the failure to rescue adenovirus production from transformed cells.

In conclusion, adenovirus-transformed rat cells and possibly also other adenovirus-transformed cells may contain single copies and possibly only fractions of the viral DNA, probably integrated into the cellular genome. The latter conclusion is primarily based on results obtained from studies on transciption of viral genes in transformed cells.

8.3. Transcription

Since only single copies and possibly only fractions of single copies of the adenovirus genome are present in transformed cells, the viral transcripts would be expected to constitute only a minute fraction of the total RNA in the cells. In contrast, Fujinaga and Green (1966) found that as much as 2% of the mRNA in the cytoplasm or in the polysomes from Ad 12-transformed hamster cells was of viral origin. This probably infers that the viral DNA is preferentially transcribed in transformed cells. The sequences transcribed in Ad 2-transformed rat cells have been reported to correspond to 50% of the sequences transcribed early during the productive cycle which according to Green *et al.* (1970) and Fujinaga and Green (1970) correspond to 5–10% of the genome and, according to Tibbetts *et al.* (1974) would correspond to 20–30%. In Ad 7-transformed hamster cells approximately 50% of the adenovirus genome is expressed corresponding to all the sequences present early in productive infection (Green *et al.*, 1970). Although the morphology of cells transformed by different types are similar and their oncogenic potential once transformed is similar except for group C viruses (Gallimore, 1972), the viral RNA transcribed from cells transformed with viruses from subgroups A, B, and C adenoviruses is homologous with the DNA from viruses within the subgroup, but heterologous to the viral RNA transcribed in cells transformed with viruses from other subgroups (Fujinaga and Green 1967,*a*,*b*, 1968; Fujinaga *et al.*, 1969). Thus, in spite of the fact that the different human adenovirus subsgroups share between 10% and 30% of their DNA sequences, the sequences transcribed in transformed cells appear to be transcribed from portions of the adenovirus DNA which are subgroup-specific. This infers that if a "transforming gene" common to all adenoviruses exists, it has been exposed to considerable genetic drift. In view of recent findings with regard to the lack of homology between different histone mRNAs and different hemoglobin mRNAs, although the proteins have almost identical sequences, the mediator of transformation might still be almost identical at the protein level. The base composition of the viral mRNA in transformed cells shows that the sequences transcribed in transformed cells have a G+C content of 47–51% although the DNA from members of the different groups show a great variation in base composition (Fujinaga and Green, 1968).

In conclusion, it appears that only part of the genome is expressed, but is preferentially transcribed, in adenovirus transformed cells, and that the sequences of RNA transcribed in cells transformed by adenoviruses are different by hybridization between the three subgroups A, B, and C.

The original transcript containing viral RNA in transformed cells appears to be confined to HnRNA species similar in size to the HnRNA in uninfected cells (Green *et al.*, 1970; Wall *et al.*, 1973; Tseui *et al.*, 1972). By selecting large HnRNA molecules from Ad 2-transformed rat cells on adenovirus DNA and selecting for poly(A)-containing molecules, R. Wall (unpublished) provided evidence that the virus-specific sequences are located at the 3′ terminus of the HnRNA and that some of the sequences at the 5′ end are probably of cellular origin. HnRNA selected by hybridization to adenovirus DNA will hybridize to cellular DNA to the same extent as unselected HnRNA, which suggests that the integrated genomes are transcribed together with cellular DNA sequences (Tseui *et al.*, 1972; Wall *et al.*, 1973). In the cytoplasm of the transformed cells the viral RNA sequences are confined to species sedimenting at 10–25 S, and the viral mRNA pattern on SDS-polyacrylamide gels and sucrose gradients is very similar to the viral mRNA pattern observed in early productive infection (Wall *et al.*, 1973; Lindberg *et al.*, 1972). Since the size of the mRNA in the cytoplasm is distinctly smaller than the size of the largest HnRNA-containing virus-specific RNA in the nucleus, it appears that the viral mRNA in transformed cells is subjected to the same processing mechanism as HnRNA in uninfected cells (Wall *et al.*, 1973). As a complement to the productively infected cells, where integration might not occur, the transformed cells appear useful for the elucidation of the detailed mechanism of processing of mRNA in eukaryotic cells.

8.4. Phenotype of the Transformed Cells

Originally, Ad 12-transformed cells were noted as foci of epithelial-like cuboidal elements which tended to pile up on top of each other (McBride and Wiener, 1964). Similar findings were reported by others for both transformed rat and hamster cells (Pope and Rowe, 1964; Strohl *et al.*, 1966; Freeman *et al.*, 1967*a,b*; Kusano and Yamane, 1967; Rafajko, 1967). The frequency of unrestrained growth transformation, defined as the absence of a post-logarithmic resting phase which is observed in normal cells with a concurrent lack of cell cycle inhibition (Pontén, 1971), is extremely low and transformation only occurs unpredictably during the course of many months. The growth state of the cells at infection, the multiplicity of infection, as well as the composition of the medium influence the ability of the virus-transformed cells to survive (Schell and Schmidt, 1968; Schell *et al.*, 1968*a,b*). Under optimal conditions, colonies composed of tightly packed small polygonal cells can usually be identified 2–3 weeks post-inoculation. The cells have an increased content of DNA, which can be used for identification with Feulgen staining (Kusano and Yamane, 1967). Transformed cells further show an increased amount of membrane-bound mucopolysaccharides similar to what has been found after transformation by papovaviruses (Martinéz-Palomo and Brailovsky, 1968). One transforming unit corresponds to about 10^4 infectious units as measured by end-point titrations on human embryonic kidney cultures. It appears that normal levels of calcium are essential for primary transformation, but the establishment of cell lines require a reduction of the calcium concentration to 0.1 mM (Schell *et al.*, 1968*a*; Van der Noorda, 1968*a,b*). In order to assay the frequency of transformation Strohl *et al.* (1967) infected cloned established BHK21 cells with high multiplicity of purified Ad 12 and assayed for colony growth in soft agar; after three weeks 0.1–1% of the cells had formed visible colonies. McAllister and Macpherson (1968), on the other hand, reported a transformation frequency of 0.002% with Ad 12 in another established hamster cell line, Nil-2, but they failed to transform the BHK21. Human adenovirus belonging to subgroup C has been reported to cause unrestrained and infinite growth transformation of rat embryo cells (Freeman *et al.*, 1967*a,b*; Van der Noorda, 1968*b*, McAllister *et al.*, 1969*a*), and also in this case the transformation was reproducible only when the calcium content of the medium was low. A simian adenovirus, SA7, can cause transformation of hamster cells in the same way as human Ad 12 (Riggs and Lennette, 1967; Casto, 1968; Whitcutt and Gear, 1968; Alstein *et al.*, 1967). Simian adenovirus SA11 can also transform hamster cells (Casto, 1969). Bovine adenovirus type 3 has also been shown to be oncogenic and able to transform cells (Darbyshire, 1966; Gilden *et al.*, 1967; Tsukamoto and Sugino, 1972).

The adenoviruses obviously can induce specific cell morphology and specific growth characteristics in transformed cells, but it has not been unequivocally proven that the adenovirus is responsible for this effect. Transformation occurs only at low frequencies, but the unique morphology observed both *in vitro* and *in vivo* (Kusano and Yamane, 1967) has not been seen in spontaneously transformed rodent cells or in cells transformed by other viruses.

The adenovirus-induced products found in the transformed cells are the adenovirus-specific T antigens. They can be detected in cells transformed by human adenoviruses of all three groups (Huebner *et al.*, 1963; Pope and Rowe, 1964; Huebner, 1967), and in the productive infection this product belongs to the early phase (Hoggan *et al.*, 1965). There is no antigenic cross-reaction with T antigen found in papovavirus-transformed or -infected cells. There is also subgroup specificity with regard to the T antigens in that the T antigens from human adenovirus subgroups A, B, and C differ immunologically between the groups. In spite of numerous attempts to purify the T antigens (Tavitian *et al.*, 1967; Gilead and Ginsberg, 1968*a,b*; Tockstein *et al.*, 1968), these antigens have not yet been biochemically characterized. In addition to the T antigen, a tumor-specific transplantation antigen (TSTA), which appears to be responsible for transplantation rejection, appears on the surface of cells transformed by adenovirus (Sjögren *et al.*, 1967). Although surface changes have been reported for the productive infections (Salzberg and Raskas, 1972), it is not yet established that the TSTA antigen is present during productive infection. At present there is no direct evidence that either T antigen or TSTA is coded for by the viral genome.

9. ADENOVIRUS GENETICS

The first adenovirus mutants described were a number of cytocidal (*cyt*) mutants of Ad 12 (Takemori *et al.*, 1968). They were spontaneously encountered in frequencies of 0.01%, but UV irradiation increased the frequency about five-fold. The mutants can be distinguished from wild-type virus because they show an enhanced cytopathic effect and produce large clear plaques in comparison to wild-type Ad 12, which produces small fuzzy-edged plaques on human embryonic kidney cells. These *cyt* mutants are furthermore less oncogenic than wild type for newborn hamster, and they have lost the ability to transform hamster cells *in vitro* (Takemori *et al.*, 1968). True recombinants, which produce wild-type plaques and which have regained their oncogenicity, have been obtained (Takemori, 1972).

Isolation of temperature-sensitive (*ts*) mutants of adenoviruses was not reported until 1971, but at present there is a rapid accumulation of such *ts* mutants and a new nomenclature has recently been introduced (Ginsberg *et al.*, 1973). However, functional studies are only slowly being accumulated. Williams *et al.* (1971) and Ensinger and Ginsberg (1972) described a battery of *ts* mutants of Ad 5. Williams *et al.* (1971) isolated mutants after mutagenesis with nitrous acid, nitrosoguanidine, hydroxylamine, and bromodeoxyuridine. Mutation frequencies were in the range of 0.6–9.6%, and the mutants showed comparatively little leakiness. The reversion frequencies were of the order of 10^{-5}–10^{-6}. Complementation analysis has revealed at least 14 complementation groups (Williams and Ustacelebi, 1971; Wilkie *et al.*, 1973; Russell *et al.*, 1972*a*). In addition, recombination experiments appear to indicate that recombinants to wild type can only be obtained at frequencies in the range of 0.05–9.0% (Williams and Ustacelebi, 1971; Williams, personal communication). Ensinger and Ginsberg (1972) reported the isolation of 8 *ts* mutants of Ad 5 obtained in a similar way. Again, the reversion frequencies were low and these mutants were separated into three complementation groups. Recombination to wild type including reciprocal recombination could also be observed at frequencies of 0.1–15%. Temperature-sensitive mutants have also been described for Ad 12 by Lundholm and Doerfler (1971) and Shiroki *et al.* (1972). The latter group isolated 88 *ts* mutants. They found 13 complementation groups for the Ad 12 genome, which should be compared with the 14 groups for the Ad 5 genome (Williams and Ustacelebi, 1971). More recent data appear to indicate that there are at least 17 complementation groups for the Ad 5 genome (Williams, personal communication). The isolation of *ts* mutants of adenovirus type 31 (Suzuki and Shimojo, 1971) and of an avian adenovirus (celo) (Ishibashi, 1970) has also been reported. Ishibashi (1971) isolated 49 mutants of the avian celo virus which were grouped according to their capacity to transport virus antigen from the cytoplasm to the nucleus of infected cells at nonpermissive temperature. Several mutants which failed to carry out proper transport were encountered. However, there is probably a considerable variation in the sites of antigen accumulation with several different *ts* mutants, depending on the time of harvesting of the cells.

The Ad 12 *ts* mutants of Shiroki *et al.* (1972) and the Ad 5 *ts* mutants of Williams *et al.* (1971) have been most extensively mapped for functional defects, but this work has only begun. The 13 complementation groups of Ad 12 virus fall into four classes. The first contains the complementation groups A–D, where the mutants are unable to synthesize all major capsid proteins, hexon, penton base, and fiber at the nonpermissive temperature. The second class consists of the mutants in the 3 complementation groups E–G which are defective in production of 2 capsid polypeptides, whereas the third class comprises mutants in the complementation groups H–J, which are unable to produce either hexon (H), fiber (I), or penton base (J). The mutants in the fourth class, covering the remaining 3 complementation groups K–M, produce capsid proteins in normal amounts but are unable to assemble virions at the nonpermissive temperature. None of these Ad 12 mutants are defective in synthesis of viral DNA, and all the mutants induce T antigen and are oncogenic for hamsters. The Ad 5 mutants have also been classified into four serological classes (Russell *et al.*, 1972*a*) based on the identification of the defects with immunological tests for virion proteins. The largest class, class 4, comprising at least 50% of the mutants, shows a defect in the production of hexon antigen and this class can, with the aid of fluorescent antibodies, be subdivided into two subclasses. Class 4 A mutants exhibit abnormal synthesis of the hexon polypeptide and class 4 B mutants appear to have a defect in the transport of hexon antigen to the nucleus. Fifty percent of the mutants of Ensinger and Ginsberg (1972) were also defective in hexon synthesis. It is of interest that the serological class 4 consists of at least 6 different complementation groups, which strengthens the notion that hexon synthesis is a highly controlled event in virus reproduction. Although intracistronic complementation cannot be excluded, it appears that at least 4 and possibly 6 adenovirus genes are involved in controlling or specifying hexon antigen. The class 3 mutants comprising 3 complementation groups are all defective in fiber antigen production. The class 2 mutants only comprise one complementation group (Wilkie *et al.*, 1973) and they are defective in viral DNA synthesis at nonpermissive conditions, but they can still make the early form of the P antigen. The serological class 1 mutants, finally, comprising two complementation groups, correspond to the third complementation group of Ensinger and Ginsberg (1972). Both the DNA and the capsid proteins of the virions are made in normal amounts with these mutants but they fail to assemble intact virions. Two mutants of Ad 5 which complement each other fail to induce interferon in chick embryo cells (Ustacelebi and Williams, 1972). The results up till now only allow a crude classification of the *ts* mutants, which illustrates the limitations of immunological techniques for assessing the defectiveness of the mutants. A more careful analysis, relating the mutants to all the 22–25 virus-induced polypeptides late in infection (Anderson *et al.*, 1973), might specifically characterize the gene products affected in these mutants.

10. CONCLUSIONS AND PROSPECTS FOR THE FUTURE

This description of adenovirus reproduction has emphasized the presence of similarities between the expression of viral genes and the expression of host genes. Since the adenoviruses probably do not carry their own enzymes for the synthesis of either RNA or DNA, and since they do not appear to induce the synthesis of virus-specific polymerases after infection, it is implied that the virus relies almost entirely on the enzyme machinery of the host cell for expressing and replicating its genome. If this is true, the efficiency of the virus reproduction indicates that adenovirus has acquired functions by which it can when needed redirect the enzymes of the host cell to replicate preferentially viral DNA, to express mostly viral genes, and finally to take over almost the entire protein-synthesizing machinery of the host cell for synthesis of viral proteins. Examples of this kind are well known from studies in prokaryotic systems. The switch of these regulatory functions in favor of the virus would result in an interference with the synthesis of corresponding host cell products. A variety of such effects are seen in the interplay between adenovirus and its host cell, especially during the transition from the early to the late phase of the infection. An understanding of these regulatory mechanisms might inform us about the corresponding control mechanisms of the host cell or eukaryotic cells in general.

The ultimate result of the expression of the early genes is the triggering of viral DNA replication, and this in turn is crucial for the switch from early- to late-phase gene expression. The early viral product(s) necessary for the initiation of viral DNA synthesis has not been identified. It appears to be proteinaceous and synthesized comparatively late in the early phase. Furthermore, this proteinaceous product apparently does not need to be synthesized continuously since viral DNA replication, once it has begun, becomes insensitive to inhibition of protein synthesis. The isolation of *ts* mutants of adenovirus, which are blocked in viral DNA replication at nonpermissive temperatures, and the development of an *in vitro* system of isolated nuclei from infected cells, where viral DNA synthesis can be studied under a variety of conditions, are the two important advances, which soon might lead to the identification of the factor(s) involved in the replication of the viral DNA.

With respect to transcription and translation, several important questions remain to be answered especially with regard to the mechanisms controlling the switch from early to late gene transcription. It is known that if viral DNA replication is prevented, the mRNA of the late phase is not made, and, although all *ts* mutants have not as yet been fully characterized, there is no indication that viral DNA can be replicated without concurrent late mRNA production. This could suggest that the decision of whether or not late mRNA shall be made is inherent in the structure of the progeny DNA; or, expressed in another way, if the appropriate DNA template is provided through the replication of the viral DNA, the synthesis of late mRNA can occur.

Another important aspect pertains to the almost complete takeover by the virus of the protein-synthesizing machinery of the cell. Recent data seem to suggest that this effect is governed by control mechanisms functioning both in the nucleus and in the cytoplasm. First, there seems to be a virus-controlled function exerting its action at the post-transcriptional level, ensuring that viral mRNA is preferentially exported to the cytoplasm in the late phase. Second, there seems to be a mechanism operating at the translational level causing the replacement in the polysome fraction of the host mRNA by viral mRNA. Since the host mRNA is long-lived and no degradation of host mRNA has been observed, the latter effect may involve initiation of translation. It is possible that both of these effects are due to a property of the mRNA itself, but the evidence seems to favor the theory that they are caused by specific factors produced by the virus. Actually, a simple protein factor could be responsible for both these events, and a likely candidate for such a function is the extra polypeptide found on adenovirus mRNP as compared to mRNP from uninfected cells. A most interesting observation with respect to the control of the expression of the late adenovirus genes has been made on monkey cells, which normally are nonpermissive to adenovirus infection. These cells do allow the synthesis of viral DNA and also late viral mRNA. However, the late viral mRNA appearing in the cytoplasm cannot be used for translation in these cells unless at least a part of SV40 DNA is concurrently expressed. The SV40 DNA may be introduced as a separate helper virus or as a hybrid DNA integrated in the adenovirus DNA. The studies on nondefective adeno–SV40 hybrids suggest in fact that the early U antigen induced by SV40 DNA may be a regulatory protein (Table 5) enabling adenovirus translation. An attempt to identify the SV40 gene product in infection with adeno-SV40 hybrid virus (Ad $2^{+}ND_{1}$) has already been published (López-Revilla and Walter, 1973). Further experimentation on this problem should provide valuable information regarding the regulatory mechanisms of adenoviruses.

ACKNOWLEDGMENTS

The authors are indebted to several colleagues for supplying preprints and unpublished information for this review. Drs. B. Edvardsson, E. Everitt, K. Johansson, T. Persson, U. Pettersson, L. Prage, C. Tibbetts, B. Sundquist, and B. Vennström in our laboratory have contributed both experimentally and conceptually to this compilation. Drs. C. W. Anderson, W. Doerfler, H. S. Ginsberg, J. A. Rose, J. S. Sussenbach, R. Wall, and J. F. Williams have kindly provided information on unpublished work and work about to be published. Excellent secretarial aid was provided by Mrs. Berit Nordensved.

The investigations carried out in the author's laboratory and cited in this review were supported by grants from the Swedish Medical and Natural Science Research Councils and the Swedish Cancer Society. The survey of the literature pertaining to this review was completed in October, 1973.

11. REFERENCES

Altstein, A. D., and Dodonova, N. N., 1968, Interaction between human and simian adenoviruses in simian cells: Complementation, phenotypic mixing and formation of monkey cell "adapted" virions, *Virology* **35**, 248–254.

Altstein, A. D., Sárycheva, O. F., and Dodonova, N. N., 1967, Transforming activity of green monkey SA7 (C8) adenovirus in tissue culture, *Science* (*Wash., D.C.*) **158**, 1455–1456.

Anderson, C. W., Baum, S. G., and Gesteland, R. F., 1973, Processing of adenovirus 2-induced proteins, *J. Virol.* **12**, 241–252.

Bablanian, R., and Russell, W. C., 1974, Adenovirus polypeptide synthesis in the presence of non-replicating poliovirus. *J. Gen. Virol.* **24**, in press.

Baum, S. G., Reich, P. R., Huebner, R. J., Rowe, W. P., and Weissman, S. M., 1966, Biophysical evidence for linkage of adenovirus and SV40 DNAs in adenovirus 7–SV40 hybrid particles, *Proc. Natl. Acad. Sci. USA* **56**, 1509–1515.

Baum, S. G., Wiese, W. H., and Reich, P. R., 1968, Studies on the mechanism of enhancement of adenovirus 7 infection in African green monkey cells by simian virus 40: Formation of adenovirus-specific RNA, *Virology* **34**, 373–376.

Baum, S., Horwitz, M., and Maizel, J., Jr., 1972 Studies on the mechanism of enhancement of human adenovirus infection in monkey cells by simian virus 40, *J. Virol.* **10,** 211–219.

Beardmore, W. B., Havlick, M. J., Serafini, A., and McLean, I. W., Jr., 1965, Interrelationship of adenovirus (type 4) and papovavirus (SV-40) in monkey kidney cell cultures, *J. Immunol.* **95,** 422–435.

Béládi, I., 1972, Adenoviruses, *in* "Strains of Human Viruses" (M. Majer and S. A. Plotkin, eds., pp. 1–19, Karger, Basel.

Bellett, A. J. D., and Younghusband, H. B., 1972, Replication of the DNA of chick embryo lethal orphan virus, *J. Mol. Biol.* **72,** 691–709.

Bello, L. J., and Ginsberg, H. S., 1967, Inhibition of host protein synthesis in type 5 adenovirus-infected cells, *J. Virol.* **1,** 843–850.

Bello, L. J., and Ginsberg, H. S., 1969, Relationship between deoxyribonucleic acid-like ribonucleic acid synthesis and inhibition of host protein synthesis in type 5 adenovirus-infected KB cells, *J. Virol.* **3,** 106–113.

Bernhard, W., Granboulan, N., Barski, G., and Turner, P., 1961, Essais de cytochmie ultrastructurale. Digestion de virus sur coupes ultrafines, *C. R. Hebd. Seances Acad. Sci. Ser. D Sci. Nat.* **252,** 202–204.

Bhaduri, S., Raskas, H., and Green, M., 1972, A procedure for the preparation of milligram quantities of adenovirus messenger RNA, *J. Virol.* **10,** 1126–1129.

Bibrack, B., 1969, Untersuchungen über serologische einordnung von 9 in Bayern aus Schweinen isolierten Adenovirus-Stämmen, *Z. Vet. Med. B* **16,** 327–334.

Blobel, G., 1973, A protein of molecular weight 78,000 bound to the polyadenylate region of eukaryotic messenger RNAs, *Proc. Natl. Acad. Sci. USA* **70,** 924–928.

Boulanger, P., Torpier, G., and Biserte, G., 1970, Investigations on intranuclear paracrystalline inclusions induced by adenovirus 5 in KB cells, *J. Gen. Virol.* **6,** 329–332.

Boulanger, P., Torpier, G., and Rimsky, A., 1973, Crystallographic study of intranuclear adenovirus type 5 crystals, *Intervirology* **2,** 56–62.

Bresnick, E., and Rapp, F., 1968, Thymidine kinase activity in cells abortively and productively infected with human adenoviruses, *Virology* **34,** 799–802.

Burlingham, B., and Doerfler, W., 1971, Three size classes of adenovirus intranuclear deoxyribonucleic acid, *J. Virol.* **7,** 707–719.

Burlingham, B., and Doerfler, W., 1972, An endonuclease in cells infected with adenovirus and associated with adenovirions, *Virology* **48,** 1–13.

Burlingham, B., Doerfler, W., Pettersson, U., and Philipson, L., 1971, Adenovirus endonuclease: Association with the penton of adenovirus type 2, *J. Mol. Biol.* **60,** 45–64.

Burnett, J. P., and Harrington, J. A., 1968*a*, Simian adenovirus SA7 DNA: Chemical, physical and biological studies, *Proc. Natl. Acad. Sci. USA* **60,** 1023–1029.

Burnett, J. P., and Harrington, J. A., 1968*b*, Infectivity associated with simian adenovirus type SA7 DNA, *Nature* (*Lond.*) **220,** 1245–1246.

Burns, W. H., and Black, P. H., 1969, Analysis of SV40-induced transformation of hamster kidney tissue *in vitro*: VI. Characteristics of mitomycin C induction, *Virology* **39,** 625–634.

Butel, J. S., and Rapp, F., 1967, Complementation between a defective monkey cell-adapting component and human adenovirus in simian cells, *Virology* **31,** 573–584.

Caffier, H., and Green, M., 1971, Adenovirus proteins. III. Cellfree synthesis of adenovirus proteins in cytoplasmic extracts of KB cells, *Virology* **46,** 98–105.

Casto, B. C., 1968, Effects of ultraviolet irradiation on the transforming and plaque-forming capacities of simian adenovirus SA7, *J. Virol.* **2,** 641–642.

Casto, B. C., 1969, Transformation of hamster embryo cells and tumor induction in newborn hamsters by simian adenovirus SV11, *J. Virol.* **3,** 513–519.

Chambon, P., Gissinger, F., Mandel, J. L., Jr., Kedinger, C., Gniazdowski, M., and Meihlac, M., 1970, Purification and properties of calf thymus DNA-dependent RNA polymerase A and B, *Cold Spring Harbor Symp. Quant. Biol.* **35,** 693–707.

Chardonnet, Y., and Dales, S., 1970*a*, Early events in the interaction of adenovirus with HeLa cells. I. Penetration of type 5 and intracellular release of the DNA genome, *Virology* **40,** 462–477.

Chardonnet, Y., and Dales, S., 1970*b*, Early events in the interaction of adenovirus with HeLa cells. II. Comparative observation on the penetration of type 1, 5, 7, 12, *Virology* **40,** 478–485.

Chardonnet, Y., and Dales, S., 1972, Early events in the interaction of adenovirus with HeLa cells. II. Relationship between an ATPase activity in nuclear envelopes and transfer of core material: A hypothesis, *Virology* **48,** 342–359.

Chardonnet, Y., Gazzolo, L., and Pogo, B., 1972, Effect of α-aminitin on adenovirus 5 multiplication, *Virology* **48,** 300–304.

Cornick, G., Sigler, P. B., and Ginsberg, H. S., 1971, Characterization of crystals of adenovirus type 5 hexon, *J. Mol. Biol.* **57,** 397–401.

Cornick, G., Sigler, P. B., and Ginsberg, H. S., 1973, Mass of protein in the asymmetric unit of hexon crystals—a new method, *J. Mol. Biol.* **73,** 533–537.

Crowther, R. A., and Franklin, R. M., 1972, The structure of the groups of nine hexons from adenovirus, *J. Mol. Biol.* **68,** 181–184.

Crumpacker, C. S., Levin, M. J., Wiese, W. H., Lewis, A. M., Jr., and Rowe, W. P., 1970, The adenovirus type 2–simian virus 40 hybrid population: Evidence for a hybrid deoxyribonucleic acid molecule and the absence of adenovirus-encapsidated circular simian virus 40 deoxyribonculeic acid, *J. Virol.* **6,** 788–794.

Dales, S., 1962, An electron microscope study of the early association between two mammalin viruses and their hosts, *J. Cell. Biol.* **13,** 303–322.

Darbyshire, J. H., 1966, Oncogenicity of bovine adenovirus type 3 in hamsters, *Nature* (*Lond.*) **211,** 102.

Darnell, J. E., Wall, R., and Tushinski, R., 1971*a*, An adenylic acid-rich sequence in messenger RNA of HeLa cells and it possible relationship to reiterated sites in DNA, *Proc. Natl. Acad. Sci. USA* **68,** 1321–1325.

Darnell, J. E., Philipson, L., Wall, R., and Adesnik, M., 1971*b*, Polyadenylic acid sequences: Role in conversion of nuclear RNA into messenger RNA, *Science* (*Wash., D.C.*) **174,** 507–510.

Darnell, J. E., Wall, R., Adesnik, M., and Philipson, L., 1972, The formation of messenger RNA in HeLa cells by post-transcriptional modification of nuclear RNA, *in* "Molecular Genetics and Developmental Biology" (M. Sussman, ed.), pp. 201–225, Prentice-Hall, Englewwod Cliffs, N.J.

Doerfler, W., 1968, The fate of the DNA of adenovirus type 12 in baby hamster kidney cells, *Proc. Natl. Acad. Sci. USA* **60,** 636–646.

Doerfler, W., 1969, Non-productive infection of baby hamster kidney cells (BHK21) with adenovirus type 12, *Virology* **38,** 587–606.

Doerfler, W., 1970, Integration of the DNA of adenovirus type 12 into the DNA of baby hamster kidney cells, *J. Virol.* **6,** 652–666.

Doerfler, W., and Kleinschmidt, A. K., 1970, Denaturation pattern of the DNA of adenovirus type 2 as determined by electron microscopy, *J. Mol. Biol.* **50,** 579–593.

Doerfler, W., and Lundholm, U., 1970, Absence of replication of the DNA of adenovirus type 12 in BHK21 cells, *Virology* **40,** 754–756.

Doerfler, W., Hellman, W., and Kleinschmidt, A. K., 1972, The DNA of adenovirus type 12 and its denaturation pattern, *Virology* **47,** 507–512.

Doerfler, W., Lundholm, U., Rensing, U., and Philipson, L., 1973, Intracellular forms of adenovirus deoxyribonucleic acid. II. Isolation in dye-buoyant density gradients of a deoxyribonucleic acid–ribonucleic acid complex from KB cells infected with adenovirus type 2, *J. Virol.* **12,** 793–807.

Easton, J. M., and Hiatt, C. W., 1965, Possible incorporation of SV40 genome within capsid proteins of adenovirus 4, *Proc. Natl. Acad. Sci. USA* **54,** 1100–1104.

Edmonds, M., and Caramela, M. G., 1969, The isolation and characterization of adenosine monophosphate-rich polynucleotides synthesized by Ehrlich ascites cells, *J. Biol. Chem.* **244,** 1314–1324.

Edmonds, M., Vaughan, M. H., and Nakazoto, N., 1971, Polyadenylic acid sequences in the heterogeneous nuclear RNA and rapidly labeled polyribosomal RNA of HeLa cells: Possible evidence for a precursor relationship, *Proc. Natl. Acad. Sci. USA* **68,** 1336–1340.

Ellens, D. J., Sussenbach, J. S., and Jansz, H. S., 1974, Studies on the mechanism of replication of adenovirus DNA III. Electron microscopy of replicating DNA, *Virology,* in press.

Enders, J. F., Bell, J. A., Dingle, J. H., Francis, T., Jr., Hilleman, M. R., Huebner, R. J., and Payne, A. M.-M., 1956, Adenoviruses: group name proposed for new respiratory tract viruses, *Science* (*Wash., D.C.*) **124,** 119–120.

Ensinger, M. J., and Ginsberg, H. S., 1972, Selection and preliminary characterization of temperature-sensitive mutants of type 5 adenovirus, *J. Virol.* **10,** 328–339.

Epstein, J., 1959, Observation on the fine structure of type 5 adenovirus, *J. Biophys. Biochem. Cytol.* **6,** 523–526.

Epstein, M. A., Holt, S. J., and Powell, A. K., 1960, The fine structure and composition of type 5 adenovirus: An integrated electron microscopical and cytochemical study, *Brit. J. Exp. Pathol.* **41,** 567–576.

Everett, S. F., and Ginsberg, H. S., 1958, A toxin-like material separable from type 5 adenovirus particles, *Virology* **6,** 770–771.

Everitt, E., Sundquist, B., and Philipson, L., 1971, Mechanism of arginine requirement for adenovirus synthesis. I. Synthesis of structural proteins, *J. Virol.* **8,** 742–753.

Everitt, E., Sundquist, B., Pettersson, U., and Philipson, L., 1973, Structural proteins of adenoviruses. X. Isolation and topography of low-molecular-weight antigens from the virion of adenovirus type 2, *Virology* **52,** 130–147.

Feldman, L. A., and Rapp, F., 1966, Inhibition of adenovirus replication by 1-β-D-arabinofuranosylcytosine, *Proc. Soc. Exp. Biol. Med.* **122,** 243–247.

Feldman, L. A., Butel, J. S., and Rapp, F., 1966, Interaction of a simian papovavirus and adenovirus. I. Induction of adenovirus tumor antigen during abortive infection of simian cells, *J. Bacteriol.* **91,** 813–818.

Flanagan, J. F., and Ginsberg, H. S., 1962, Synthesis of virus-specific polymers in adenovirus-infected cells: Effect of 5-fluorodeoxyuridine, *J. Exp. Med.* **116,** 141–157.

Fogel, M., and Sachs, L., 1969, The activation of virus synthesis in polyoma-transformed cells, *Virology* **37,** 327–334.

Fogel, M., and Sachs, L., 1970, Induction of virus synthesis in polyoma-transformed cells by ultraviolet light and mitomycin C, *Virology* **40,** 174–177.

Fox, R., and Baum, S., 1972, Synthesis of viral ribonucleic acid during restricted adenovirus infection, *J. Virol.* **10**, 220–227.
Franklin, R. M., Pettersson, U., Akervall, K., Strandberg, B., and Philipson, L., 1971*a*, Structural proteins of adenoviruses. V. On the size and structure of the adenovirus type 2 hexon, *J. Mol. Biol.* **57**, 383–395.
Franklin, R. M., Harrison, S. C., Pettersson, U., Brändén, C. I., Werner, P. E., and Philipson, L., 1971*b*, Structural studies on the adenovirus hexon, *Cold Spring Harbor Symp. Quant. Biol.* **36**, 503–510.
Freeman, A. E., Black, P. H., Wolford, R., and Huebner, R. J., 1967*a*, Adenovirus type 12–rat embryo transformation system, *J. Virol.* **1**, 362–367.
Freeman, A. E., Black, P. H., Vanderpool, E. A., Henry, P. H., Austin, J. B., and Huebner, R. J., 1967*b*, Transformation of primary rat embryo cells by adenovirus type 2, *Proc. Natl. Acad. Sci. USA* **58**, 1205–1212.
Freeman, A. E., Vanderpool, E. A., Black, P. H., Turner, H. C., and Huebner, R. J., 1967*c*, Transformation of primary rat embryo cells by a weakly oncogenic adenovirus type 3, *Nature* (*Lond.*) **216**, 171–173.
Friedman, D. L., and Mueller, G. C., 1968, A nuclear system for DNA replication from synchronized HeLa cells, *Biochim. Biophys. Acta* **161**, 455–468.
Friedman, M. P., Lyons, M. J., and Ginsberg, H. S., 1970, Biochemical consequences of type 2 adenovirus and SV40 double infection of African green monkey kidney cells, *J. Virol.* **5**, 586–597.
Fujinaga, K., and Green, M., 1966, The mechanism of viral carcinogenesis by DNA mammalian viruses: Viral-specific RNA in polyribosomes of adenovirus tumor and transformed cells, *Proc. Natl. Acad. Sci. USA* **55**, 1567–1574.
Fujinaga, K., and Green, M., 1967*a*, Mechanism of viral carcinogenesis by deoxyribonucleic acid mammalian viruses: IV. Related virus-specific ribonucleic acids in tumor cells induced by "highly" oncogenic adenovirus type 12, 18 and 31, *J. Virol.* **1**, 576–582.
Fujinaga, K., and Green, M., 1967*b*, Mechanism of viral carcinogenesis by DNA mammalian viruses: II. Viral-specific RNA in tumor cells induced by "weakly" oncogenic human adenoviruses, *Proc. Natl. Acad. Sci. USA* **57**, 806–812.
Fujinaga, K., and Green, M., 1968, Mechanism of viral carcinogenesis by DNA mammalian viruses: V. Properties of purified viral-specific RNA from human adenovirus-induced tumor cells, *J. Mol. Biol.* **31**, 63–73.
Fujinaga, K., and Green, M., 1970, Mechanism of viral carcinogenesis by DNA mammalian viruses: VII. Viral genes transcribed in adenovirus type 2 infected and transformed cells, *Proc. Natl. Acad. Sci. USA* **65**, 375–382.
Fujinaga, K., Piña, M., and Green, M., 1969, The mechanism of viral carcinogenesis by DNA mammalian viruses: VI. A new class of virus-specific RNA molecules in cells transformed by group C human adenoviruses, *Proc. Natl. Acad. Sci. USA* **64**, 255–262.
Gallimore, P. H., 1972, Tumour production in immunosuppressed rats with cells transformed *in vitro* by adenovirus type 2, *J. Gen. Virol.* **16**, 99–102.
Garon, C. F., Berry, K., and Rose, J., 1972, A unique form of terminal redundancy in adenovirus DNA molecules, *Proc. Natl. Acad. Sci. USA* **69**, 2391–2395.
Garon, C. F., Berry, K. W., Hierholzer, J. C., and Rose, J. A., 1973, Mapping of base sequence heterologies between genomes from different adenovirus serotypes, *Virology* **54**, 414–426.
Gerber, P., 1966, Studies on the transfer of subviral infectivity from SV40-induced hamster tumor cells to indicator cells, *Virology* **28**, 501–509.

Ghabrial, S. A., and Lister, R. M., 1973, Anomalies in molecular weight determinations of tobacco rattle virus protein by SDS-polyacrylamide gel electrophoresis, *Virology* **51,** 485–488.

Gilden, R. V., Kern, J., Beddow, T. G., and Huebner, R. J., 1967, Bovine adenovirus type 3: Detection of specific tumor and T antigen, *Virology* **31,** 727–729.

Gilead, Z., and Ginsberg, H. S., 1965, Characterization of a tumorlike anitgen in type 12 and type 18 adenovirus-infected cells, *J. Bacteriol.* **90,** 120–125.

Gilead, Z., and Ginsberg, H. S., 1968*a*, Characterization of the tumorlike (T) antigen induced by type 12 adenovirus. I. Purification of the antigen from infected KB cells and a hamster cell line, *J. Virol.* **2,** 7–14.

Gilead, Z., and Ginsberg, H. S., 1968*b*, Characterization of a tumorlike (T) antigen induced by type 12 adenovirus. II. Physical and chemical properties, *J. Virol.* **2,** 15–20.

Ginsberg, H. S., 1969, Biochemistry of adenovirus infection, *in* "The Biochemistry of Viruses" (H. B. Levy, ed.), pp. 329–359, Marcel Dekker, New York.

Ginsberg, H. S., Pereira, H. G., Valentine, R. C., and Wilcox, W. C., 1966, A proposed terminology for the adenovirus antigens and virion morphological subunits, *Virology* **28,** 782–783.

Ginsberg, H. S., Bello, L. J., and Levine, A. J., 1967, Control of biosynthesis of host macromolecules in cells infected with adenovirus, *in* "The Molecular Biology of Viruses" (J. S. Colter and W. Paranchych, eds.), p. 547–572, Academic Press, New York.

Ginsberg, H. S., Williams, J. F., Doerfler, W. H., and Shimojo, H., 1973, Proposed nomenclature for mutants of adenoviruses, *J. Virol.* **12,** 663–664.

Giorno, R., and Kates, J. R., 1971, Mechanism of inhibition of vaccinia virus replication in adenovirus-infected HeLa cells, *J. Virol.* **7,** 208–213.

Girardi, A. J., Hilleman, M. R., and Zwickey, R. E., 1964, Tests in hamsters for oncogenic quality of ordinary viruses including adenovirus type 7, *Proc. Soc. Exp. Biol. Med.* **115,** 1141–1150.

Goodheart, C., 1971, DNA density of oncogenic and non-oncogenic simian adenoviruses, *Virology* **44,** 645–648.

Graham, F. L., and Van der Eb, A. J., 1973*a*, A new technique for the assay of infectivity of human adenovirus 5 DNA, *Virology* **52,** 456–467.

Graham, F. L., and Van der Eb, A. J., 1973*b*, Transformation of rat cells by DNA of human adenovirus 5, *Virology* **54,** 536–539.

Green, M., 1962*a*, Biochemical studies on adenovirus multiplication. III. Requirements of DNA synthesis, *Virology* **18,** 601–613.

Green, M., 1962*b*, Studies on the biosynthesis of viral DNA. IV. Isolation, purification and chemical analysis of adenovirus, *Cold Spring Harbor Symp. Quant. Biol.* **27,** 219–235.

Green, M., 1966, Biosynthetic modifications induced by DNA animal viruses, *Annu. Rev. Microbiol.* **20,** 189–222.

Green, M., 1970, Oncogenic viruses, *Annu. Rev. Biochem.* **39,** 701–756.

Green, M., Piña, M., and Kimes, R. C., 1967*a*, Biochemical studies on adenovirus multiplication: XII. Plaquing efficiencies of purified human adenoviruses, *Virology* **31,** 562–565.

Green, M., Piña, M., Kimes, R. C., Wensink, P. C., MacHattie, L. A., and Thomas, C. A., Jr., 1967*b*, Adenovirus DNA: I. Molecular weight and conformation, *Proc. Natl. Acad. Sci. USA* **57,** 1302–1309.

Green, M., Parsons, J. T., Piña, M., Fujinaga, K., Caffier, H., and Landgraf-Leurs, I., 1970, Transcription of adenovirus genes in productively infected and in transformed cells, *Cold Spring Harbor Symp. Quant. Biol.* **35,** 803–818.

Greenberg, J. R., 1972, High stability of messenger RNA in growing cultured cells, *Nature* (*Lond.*) **240,** 102–104.

Haas, M., Vogt, M., and Dulbecco, R., 1972, Loss of SV40 DNA–RNA hybrids from nitrocellulose membranes. Implications for the study of virus–host DNA interactions, *Proc. Natl. Acad. Sci. USA* **69,** 2160–2164.

Hartley, J. W., and Rowe, W. P., 1960, A new mouse virus apparently related to the adenovirus group, *Virology* **11,** 645–647.

Hartley, J. W., Huebner, R. J., and Rowe, W. P., 1956, Serial propagation of adenovirus (APC) in monkey kidney tissue cultures, *Proc. Soc. Exp. Biol. Med.* **92,** 667–669.

Henry, P. H., Schnipper, L. E., Samaha, R. J., Crumpacker, C. S., Lewis, A. M., Jr., and Levine, A. S., 1973, Studies on nondefective adenovirus 2–simian virus 40 hybrid viruses. VI. Characterization of the DNA from five nondefective hybrid viruses, *J. Virol.* **11,** 665–671.

Henshaw, E. C., 1968, Messenger RNA in rat liver polyribosomes: Evidence that it exists as ribonucleoprotein particles, *J. Mol. Biol.* **36,** 401–411.

Hilleman, M. R., 1966, Adenoviruses: History and future of a vaccine, *in* "Viruses Inducing Cancer" (W. J. Burdette, ed.), pp. 337–402, University of Utah Press, Salt Lake City, Utah.

Hilleman, M. R., and Werner, J. H., 1954, Recovery of new agent from patients with acute respiratory illness, *Proc. Soc. Exp. Biol. Med.* **85,** 183–188.

Hillis, W. D., and Goodman, R., 1969, Serological classification of chimpanzee adenoviruses by hemagglutination and hemagglutination inhibition, *J. Immunol.* **103,** 1089–1095.

Hodge, L. D., and Scharff, M. D., 1969, Effect of adenovirus on host cell DNA synthesis in synchronized cells, *Virology* **37,** 554–564.

Hoggan, M. D., Rowe, W. P., Black, P. H., and Huebner, R. J., 1965, Production of "tumor-specific" antigens by oncogenic viruses during acute cytolytic infections, *Proc. Natl. Acad. Sci. USA* **53,** 12–19.

Homburger, F., Merkow, L. P., and Slifkin, M., 1973, Oncogenic Adenoviruses, *in* "Progress in Experimental Tumor Research," Vol. 18, Karger, Basel.

Horne, R. W., Brenner, S., Waterson, A. P., and Wildy, P., 1959, The icosahedral form of an adenovirus, *J. Mol. Biol.* **1,** 84–86.

Horwitz, M., 1971, Intermediates in the synthesis of type 2 adenovirus deoxyribonucleic acid, *J. Virol.* **8,** 675–683.

Horwitz, M. S., Scharff, M. D., and Maizel, J. V., Jr., 1969, Synthesis and assembly of adenovirus 2: I. Polypeptide synthesis, assembly of capsomers and morphogenesis of the virion, *Virology* **39,** 682–694.

Horwitz, M. S., Maizel, J. V., Jr., and Scharff, M. D., 1970, Molecular weight of adenovirus type 2 hexon polypeptide, *J. Virol.* **6,** 569–571.

Horwitz, M. S., Brayton, C., and Baum, S. G., 1973, Synthesis of type 2 adenovirus DNA in the presence of cycloheximide, *J. Virol.* **11,** 544–551.

Huebner, R. J., 1967, Adenovirus-directed tumor and T antigens, *in* "Perspecitives in Virology" (M. Pollard, ed.), Vol. V, pp. 147–166, Academic Press, New York.

Huebner, R. J., Rowe, W. P., and Lane, W. T., 1962, Oncogenic effects in hamsters of human adenovirus type 12 and 18, *Proc. Natl. Acad. Sci. USA* **48,** 2051–2058.

Huebner, R. J., Rowe, W. P., Turner, H. C., and Lane, W. T., 1963, Specific adenovirus complement-fixing antigens in virus-free hamster and rat tumors, *Proc. Natl. Acad. Sci. USA* **50,** 379–389.

Huebner, R. J., Chanock, R. M., Rubin, B. A., and Casey, M. J., 1964, Induction by adenovirus type 7 of tumors in hamsters having the antigenic characteristics of SV40 virus, *Proc. Natl. Acad. Sci. USA* **52,** 1333–1340.

Huebner, R. J., Casey, M. J., Chanock, R. M., and Scheel, K., 1965, Tumors induced in hamsters by a strain of adenovirus type 3: Sharing of tumor antigens and "neoantigens" with those produced by adenovirus type 7 tumors, *Proc. Natl. Acad. Sci. USA* **54,** 381–388.

Hull, R. N., Johnson, I. S., Culbertson, C. G., Reimer, C. B., and Wright, H. F., 1965, Oncogenicity of the simian adenovirus, *Science* (*Wash., D.C.*) **150,** 1044–1046.

Isibashi, M., 1970, Retention of viral antigen in the cytoplasm of cells infected with temperature-sensitive mutants of an avian adenovirus, *Proc. Natl. Acad. Sci. USA* **65,** 304–309.

Isibashi, M., 1971, Temperature-sensitive conditional lethal of an avian adenovirus (CELO). I. Isolation and characterization. *Virology* **45,** 42–52.

Isibashi, M., and Maizel, J. V., 1972, Hexoseamine in fiber protein of type 2 adenovirus, *Abstr. 72nd Annu. Meet. Am. Soc. Microbiol.* **86,** 199.

Jacobson, M. F., Asso, J., and Baltimore, D., 1970, Further evidence on the formation of poliovirus proteins, *J. Mol. Biol.* **49,** 657–669.

Jelinek, W., Adesnik, M., Salditt, M., Sheiness, D., Wall, R., Molloy, G., Philipson, L., and Darnell, J. E., 1973, Further evidence on the nuclear origin and transfer to the cytoplasm of poly (A) sequences in mammalian cell RNA, *J. Mol. Biol.* **75,** 515–532.

Jörnvall, H., Ohlsson, H., and Philipson, L., 1974*a*, An acetylated N-terminus of adenovirus type 2 hexon protein, *Biochem. Biophys. Res. Comm.* **56,** 304–310.

Jörnvall, H., Pettersson, U., and Philipson, L., 1974*b*, Structural studies of adenovirus type 2 hexon protein, *Europ. J. Biochem.*, in press.

Kates, J., 1970, Transcription of the vaccinia virus genome and the occurrence of polyriboadenylic acid sequences in messenger RNA, *Cold Spring Harbor Symp. Quant. Biol.* **35,** 743–752.

Kelly, T., and Rose, J., 1971, Simian virus 40 integration site in an adenovirus 7–SV40 hybrid DNA molecule, *Proc. Natl. Acad. Sci. USA* **68,** 1037–1041.

Kelly, T. J., and Lewis, A. M., Jr., 1973, Use of non-defective adenovirus–simian virus 40 hybrids for mapping the simian virus 40, *J. Virol.* **12,** 643–652.

Kimes, R., and Green, M., 1970, Adenovirus DNA. II. Separation of molecular halves of adenovirus type 2, *J. Mol. Biol.* **50,** 203–205.

King, J., Lenk, E. V., and Botstein, D., 1973, Mechanism of head assembly and DNA encapsulation in Salmonella phage P22. II. Morphogenetic pathway. *J. Mol. Biol.* **80,** 697–732.

Kit, S., Piekarski, L. J., Dubbs, D. R., de Torres, R. A., and Anken, M., 1967, Enzyme induction in green monkey kidney cultures infected with simian adenovirus, *J. Virol.* **1,** 10–15.

Kjellén, L., Lagermalm, G., Svedmyr, A., and Thorsson, K. G., 1955, Crystalline-like patterns in the nuclei of cells infected with an animal virus, *Nature* (*Lond.*) **175,** 505–506.

Kline, L. K., Weissman, S. M., and Soll, D., 1972, Investigation on adenovirus-directed 4 S RNA, *Virology* **48,** 291–296.

Koczot, F. J., Carter, B. J., Garon, C. F., and Rose, J. A., 1973, Self-complementarity of terminal sequences within plus or minus strands of adenovirus-associated virus DNA, *Proc. Natl. Acad. Sci. USA* **70,** 215–219.

Köhler, K., 1965, Reinigung and Charakterisierung zweier Proteine des Adenovirus Type 2, *Z. Naturforsch.* **20b,** 747–752.

Korant, B. D., 1973, Cleavage of poliovirus-specific polypeptide aggregates, *J. Virol.* **12,** 556–563.

Kubinski, H., and Rose, J. A., 1967, Regions containing repeating base-pairs in DNA from some oncogenic and non-oncogenic animal viruses, *Proc. Natl. Acad. Sci. USA* **57,** 1720–1725.

Kumar, A., and Lindberg, U., 1972, Characterization of messenger ribonucleoprotein and messenger RNA from KB cells, *Proc. Natl. Acad. Sci. USA* **69,** 681–685.

Kusano, T., and Yamane, I., 1967, Transformation *in vitro* of the embryonal hamster brain cells by human adenovirus type 12, *Tohoku J. Exp. Med.* **92,** 141–150.

Lacy, S., Sr., and Green, M., 1964, Biochemical studies on adenovirus multiplication. VII. Homology between DNA's of tumorigenic and nontumorigenic human adenoviruses, *Proc. Natl. Acad. Sci. USA* **52,** 1053–1059.

Lacy, S. Sr., and Green, M., 1965, Adenovirus multiplication. Genetic relatedness of tumorigenic human adenovirus type 7, 12, 18, *Science (Wash., D.C.)* **150,** 1296–1298.

Lacy, S., Sr., and Green, M., 1967, The mechanism of viral carcinogenesis by DNA mammalian viruses: DNA–DNA homology relationship among the "weakly" oncogenic human adenoviruses, *J. Gen. Virol.* **1,** 413–418.

Landgraf-Leurs, M., and Green, M., 1971, Adenovirus DNA. III. Separation of the complementary strands of adenovirus types 2, 7, and 12 DNA molecules, *J. Mol. Biol.* **60,** 185–202.

Larsson, V. M., Girardi, A. J., Hilleman, M. R., and Zwickey, R. E., 1965, Studies of oncogenicity of adenovirus type 7 viruses in hamsters, *Proc. Soc. Exp. Biol. Med.* **118,** 15–24.

Laver, W. G., 1970, Isolation of an arginine-rich protein from particles of adenovirus type 2, *Virology* **41,** 488–500.

Laver, W. G., Suriano, J. R., and Green, M., 1967, Adenovirus proteins. II. N-terminal amino acid analysis, *J. Virol.* **1,** 723–728.

Laver, W. G., Pereira, H. G., Russell, W. C., and Valentine, R. C., 1968, Isolation of an internal component from adenovirus type 5, *J. Mol. Biol.* **37,** 379–386.

Laver, W. G., Wrigley, N. G., and Pereira, H. G., 1969, Removal of pentons from particles of adenovirus type 2, *Virology* **38,** 599–605.

Laver, W. G., Younghusband, H. B., and Wrigley, N. G., 1971, Purification and properties of chick embryo lethal orphan virus (an avian adenovirus), *Virology* **45,** 598–614.

Lawrence, W. C., and Ginsberg, H. S., 1967, Intracellular uncoating of type 5 adenovirus deoxyribonucleic acid, *J. Virol.* **1,** 851–867.

Lebleu, B., Marbaix, G., Huez, G., Temmerman, J., Burny, A., and Chantrenne, H., 1971, Characterization of the messenger ribonucleoprotein released from reticulocyte polyribosomes by EDTA treatment, *Eur. J. Biochem.* **19,** 264–269.

Ledinko, N., 1967, Stimulation of DNA synthesis and thymidine kinase activity in human embryonic kidney cells infected by adenovirus 2 or 12, *Cancer Res.* **27,** 1459–1469.

Ledinko, N., 1970, Transient stimulation of deoxyribonucleic acid-dependent ribonucleic acid polymerase and histone acetylation in human embryonic kidney cultures infected with adenovirus 2 or 12: Apparent induction of host ribonucleic acid synthesis, *J. Virol.* **6,** 58–68.

Ledinko, N., 1971, Inhibition by α-aminitin of adenovirus 12 replication in human embryonic kidney cells and of adenovirus transformation of hamster cells, *Nat. New Biol.* **233,** 247–248.

Ledinko, N., 1972, Nucleolar ribosomal precursor RNA and protein metabolism in human embryo kidney cultures infected with adenovirus 12, *Virology* **49,** 79–89.

Lee, Y., Mendecki, J., and Brawerman, G., 1971, A polynucleotide segment rich in adenylic acid in the rapidly labeled polyribosomal RNA component of mouse sarcoma 180 ascites cells, *Proc. Natl. Acad. Sci. USA* **68,** 1331–1335.

Levin, M. J., Crumpacker, C. S., Lewis, A. M., Jr., Oxman, M. N., Henry, P. H., and Rowe, W. P., 1971, Studies of non-defective Ad2–SV40 hybrid viruses. II. Relationship of adenovirus 2 and SV40 DNAs in the $Ad2^+ND_1$ genome, *J. Virol.* **7,** 343–351.

Levine, A. J., and Ginsberg, H. S., 1967, Mechanism by which fiber antigen inhibits multiplication of type 5 adenovirus, *J. Virol.* **1,** 747–757.

Levine, A. J., and Ginsberg, H. S., 1968, Role of adenovirus structural proteins in the cessation of host-cell biosynthetic functions, *J. Virol.* **2,** 430–439.

Lewis, A. M., Jr., and Rowe, W. P., 1970, Isolation of two plaque variants from the adenovirus type 2–simian virus 40 hybrid population which differ in their efficiency in yielding simian virus 40, *J. Virol.* **5,** 413–420.

Lewis, A. M., Jr., and Rowe, W. P., 1971, Studies on nondefective adenovirus-simian virus 40 hybrid viruses. I. A newly characterized simian virus 40 antigen induced by the $Ad2^+ND_1$ virus, *J. Virol.* **7,** 189–197.

Lewis, A. M., Jr., Baum, S. G., Prigge, K. O., and Rowe, W. P., 1966, Occurrence of adenovirus–SV40 hybrids among monkey kidney cell adapted strains of adenovirus, *Proc. Soc. Exp. Biol. Med.* **122,** 214–218.

Lewis, A. M., Jr., Levin, M. J., Wiese, W. H., Crumpacker, C. S., and Henry, P. H., 1969, A non-defective (competent) adenovirus–SV40 hybrid isolated from the Ad2–SV40 hybrid population, *Proc. Natl. Acad. Sci. USA* **63,** 1128–1135.

Lewis, A. M., Jr., Levine, A. S., Crumpacker, C. S., Levin, M. J., Samaha, R. J., and Henry, P. H., 1973, Studies of nondefective adenovirus 2–simian virus 40 hybrid viruses. V. Isolation of additional hybrids which differ in their simian virus 40-specific biological properties, *J. Virol.* **11,** 655–664.

Liem, I. T. F., Kron, I., and Wigand, R., 1971, Dodecon, das lösliche Hämagglutinin bei allen Adenoviren der Untergruppe II. *Arch. Ges. Virusforsch.* **33,** 177–181.

Lindberg, U., and Darnell, J. E., 1970, SV40-specific RNA in the nucleus and polyribosomes of transformed cells, *Proc. Natl. Acad. Sci. USA* **65,** 1089–1096.

Lindberg, U., and Persson, T., 1972, Isolation of mRNA from KB cells by affinity chromatography on polyuridylic acid covalently linked to Sepharose, *Eur. J. Biochem.* **31,** 246–254.

Lindberg, U., and Sundquist, B., 1974, Isolation of messenger ribonucleoproteins from mammalian cells, *J. Mol. Biol.* **86,** 1–18.

Lindberg, U., Persson, T., and Philipson, L., 1972, Isolation and characterization of adenovirus messenger RNA in productive infection, *J. Virol.* **10,** 909–919.

Lomberg-Holm, K., and Philipson, L., 1969, Early events of virus-cell interaction in an adenovirus system, *J. Virol.* **4,** 323–338.

López-Revilla, R., and Walter, G., 1973, Polypeptide specific for cells with adenovirus 2–SV40 hybrid $Ad2^+ND_1$, *Nat. New Biol.* **244,** 165–167.

Lucas, J. J., and Ginsberg, H. S., 1971, Synthesis of virus-specific ribonucleic acid in KB cells infected with type 2 adenovirus, *J. Virol.* **8,** 203 213, 1971.

Lucas, J. J., and Ginsberg, H. S., 1972*a*, Transcription and transport of virus-specific ribonucleic acids in African green monkey kidney cells abortively infected with type 2 adenovirus, *J. Virol.* **10,** 1109–1118.

Lucas, J. J., and Ginsberg, H. S., 1972*b*, Identification of double-stranded virus-specific ribonucleic acid in KB cells infected with type 2 adenovirus, *Biochem. Biophys. Res. Commun.* **49,** 39–44.

Lundholm, U., and Doerfler, W., 1971, Temperature-sensitive mutants of adenovirus type 12, *Virology* **45,** 827–829.

Lwoff, A., Horne, R., and Tournier, P., 1962, A system of viruses, *Cold Spring Harbor Symp. Quant. Biol.* **27,** 51–55.

McAllister, R. M., and Macpherson, I., 1968, Transformation of a hamster cell line by adenovirus type 12, *J. Gen. Virol.* **2,** 99–106.

McAllister, R. M., Nicolson, M. O., Lewis, A. M., Jr., Macpherson, I., and Huebner, R. J., 1969*a*, Transformation of rat embryo cells by adenovirus type 1, *J. Gen. Virol.* **4,** 29–36.

McAllister, R. M., Nicolson, M. O., Reed, G., Kern, J., Gilden, R. V., and Huebner, R. J., 1969*b*, Transformation of rodent cells by adenovirus 19 and other group D adenoviruses, *J. Natl. Cancer Inst.* **43,** 917–923.

McBride, W. D., and Wiener, A., 1964, *In vitro* transformation of hamster kidney cells by human adenovirus type 12, *Proc. Soc. Exp. Biol. Med.* **115,** 870–874.

McFerran, J. B., Nelson, R., McCracken, J. M., and Ross, J. G., 1969, Viruses isolated from sheep, *Nature* (*Lond.*) **221,** 194–195.

McFerran, J. B., Clarke, J. K., and Connor, T. J., 1972, Serological classification of avian adenoviruses, *Arch. Ges. Virusforsch.* **39,** 132–139.

McGuire, P. M., Swart, C., and Hodge, L. D., 1972, Adenovirus messenger RNA in mammalian cells: Failure of polyribosome association in the absence of nuclear cleavage, *Proc. Natl. Acad. Sci. USA* **69,** 1578–1582.

Maden, B. E. H., Vaughan, M. H., Warner, J. R., and Darnell, J. E., 1969, Effect of valine deprivation on ribosome formation in HeLa cells, *J. Mol. Biol.* **45,** 265–275.

Magnusson, G., Pigiet, V., Winnacker, E.-L., Abrams, R., and Reichard, P., 1973, RNA-linked short DNA fragments during polyoma replication, *Proc. Natl. Acad. Sci. USA* **70,** 412–415.

Maizel, J., Jr., White, D., and Scharff, M., 1968*a*, The polypeptides of adenovirus. I. Evidence of multiple protein components in the virion and a comparison of types 2, 7 and 12, *Virology* **36,** 115–125.

Maizel, J. V., Jr., White, D. O., and Scharff, M. D., 1968*b*, The polypeptides of adenovirus. II. Soluble proteins, cores, top components and the structure of the virion, *Virology* **36,** 126–136.

Mak, S., 1971, Defective virions in human adenovirus type 12, *J. Virol.* **7,** 426–433.

Mak, S., and Green, M., 1968, Biochemical studies on adenovirus multiplication. XIII. Synthesis of virus-specific ribonucleic acid during infection with human adenovirus type 12, *J. Virol.* **2,** 1055–1063.

Martinez-Palomo, A., and Brailovsky, C., 1968, Surface layer in tumor cells transformed by adeno-12 and SV40 viruses, *Virology* **34,** 379–382.

Marusyk, R., Norrby, E., and Lundquist, U., 1970, Biophysical comparison of two canine adenoviruses, *J. Virol.* **5,** 507–512.

Marusyk, R., Norrby, E., and Marusyk, H., 1972, The relationship of adenovirus-induced paracrystalline structures to the virus core protein(s), *J. Gen. Virol.* **14,** 261–170.

Mautner, V., and Pereira, H. G., 1971, Crystallization of a second adenovirus protein (the fibre), *Nature* (*Lond.*) **230,** 456–457.

Mayne, N., Burnett, J. P., and Butler, L. K., 1971, Tumour induction by simian adenovirus SA7 DNA fragments, *Nat. New Biol.* **232,** 182–183.

Mendecki, J., Lee, S., and Brawerman, G., 1972, Characteristics of the polyadenylic acid segment associated with messenger ribonucleic acid in mouse sarcoma 180 ascites cells, *Biochemistry* **11,** 792–798.

Mohanty, S. B., 1971, Comparative study of bovine adenoviruses, *Am. J. Vet. Res.* **32,** 1899–1905.

Molloy, G. R., Sporn, M. B., Kelley, D. E., and Perry, R. P., 1972, Localization of polyadenylic acid sequences in messenger ribonucleic acid of mammalian cells, *Biochemistry* **11,** 3256–3260.

Morel, C., Kayibanda, B., and Scherrer, K., 1971, Proteins associated with globin messenger RNA in avian erythroblasts: Isolation and comparison with the proteins bound to nuclear messenger-like RNA, *FEBS* (*Fed. Eur. Biochem. Soc.*) *Lett.* **18,** 84–88, 1971.

Morel, C., Gander, E. S., Herzberg, M., Dubochet, J., and Scherrer, K., 1973, The duck-globin messenger-ribonucleoprotein complex. Resistance to high ionic strength, particle gel electrophoresis, composition and visualisation by dark-field electron microscopy, *Eur. J. Biochem.* **36,** 455–464.

Morgan, C., Goodman, G., Rose, H., Howe, C., and Huang, J., 1957, Electron microscopic and histochemical studies of an unusual crystalline protein occurring in cells infected by type 5 adenovirus, *J. Biophys. Biochem. Cytol.* **3,** 505–508.

Morgan, C., Rosenkranz, H. S., and Mednis, B., 1969, Structure and development of viruses as observed in the electron microscope. X. Entry and uncoating of adenoviruses, *J. Virol.* **4,** 777–796.

Mulder, C., Sharp, P. A., Delius, H., and Pettersson, U., 1974, Specific fragmentation of DNA of adenovirus serotypes 3, 5, 7 and 12 and adeno-SV40 hybrid virus ad2$^+$ND$_1$ by restriction endonuclease *Eco*RI, *J. Virol.*, in press.

Murray, R. E., and Green, M., 1973, Adenovirus DNA. IV. Topology of adenovirus genomes, *J. Mol. Biol.* **74,** 735–738.

Naegele, R. F., and Rapp, F., 1967, Enhancement of the replication of human adenoviruses in simian cells by simian adenovirus SV15, *J. Virol.* **1,** 838–840.

Nicolson, M. O., and McAllister, R. M., 1972, Infectivity of human adenovirus-1 DNA, *Virology* **48,** 14–21.

Norrby, E., 1966, The relationship between the soluble antigens and the virion of adenovirus type 3: I. Morphological characteristics, *Virology* **28,** 236–248.

Norrby, E., 1968, Biological significance of structural adenovirus components, *Curr. Top. Microbiol. Immunol.* **43,** 1–43.

Norrby, E., 1969*a*, The relationship between soluble antigens and the virion of adenovirus type 3. IV. Immunological characteristics, *Virology* **37,** 565–576.

Norrby, E., 1969*b*, The structural and functional diversity of adenovirus capsid components, *J. Gen. Virol.* **5,** 221–236.

Norrby, E., 1969*c*, Capsid mosaic of intermediate strains of human adenoviruses, *J. Virol.* **4,** 657–662.

Norrby, E., and Skaaret, P., 1967, The relationship between soluble antigens and the virion of adenovirus type 3. III. Immunological identification of fiber antigen and isolated vertex capsomer antigen, *Virology* **32,** 489–502.

O'Conor, G. T., Rabson, A. S., Berezesky, I. K., and Paul, F. J., 1963, Mixed infection with simian virus 40 and adenovirus 12. *J. Natl. Cancer Inst.* **31**, 903–917.

Ogino, T., and Takahashi, M., 1969, Altered properties of thymidine kinase induced in hamster kidney cells by adenovirus type 5 and 12, *Biken J.* **12**, 17–23.

Ohe, K., 1972, Virus-coded origin of a low-molecular-weight RNA from KB cells infected with adenovirus 2, *Virology* **47**, 726–733.

Ohe, K., and Weissman, S. M., 1970, Nucleotide sequence of an RNA from cells infected with adenovirus type 2, *Science* (*Wash., D.C.*) **167**, 879–881.

Ohe, K., Weissman, S. M., and Cooke, N. R., 1969, Studies on the origin of a low-molecular-weight ribonucleic acid from human cells infected with adenoviruses, *J. Biol. Chem.* **244**, 5320–5332.

Okazaki, R., Okazaki, T., Sakabe, K., Sugimoto, K., and Sugino, A., 1968, Mechanism of DNA chain growth. I. Possible discontinuity and unusual secondary structure of newly synthesized chains, *Proc. Natl. Acad. Sci. USA* **59**, 598–605.

Okubo, C. K., and Raskas, H. J., 1971, Thermosensitive events in the replication of adenovirus type 2 at 42°, *J. Virol.* **46**, 175–182.

Parsons, J. T., and Green, M., 1971, Biochemical studies on adenovirus multiplication. XVIII. Resolution of early virus specific RNA species in adeno 2-infected and -transformed cells, *Virology* **45**, 154–162.

Patch, C. T., Lewis, A. M., Jr., and Levine, A. S., 1972, Evidence for a transcription control region of SV40 in the adenovirus 2–SV40 hybrid Ad2^{+}ND$_1$, *Proc. Natl. Acad. Sci. USA* **69**, 3375–3379.

Pearson, G. D., and Hanawalt, P. C., 1971, Isolation of DNA replication complexes from uninfected and adenovirus infected HeLa cells, *J. Mol. Biol.* **62**, 65–80.

Pereira, H. G., 1958, A protein factor responsible for the early cytopathic effect of adenoviruses, *Virology* **6**, 601–611.

Pereira, H. G., and Skehel, J. J., 1971, Spontaneous and tryptic degradation of virus particles and structural components of adenoviruses, *J. Gen. Virol.* **12**, 13–24.

Pereira, H. G., Huebner, R. J., Ginsberg, H. S., and Van der Veen, J., 1963, A short description of the adenovirus group, *Virology* **20**, 613–620.

Pereira, H. G., Valentine, R. C., and Russell, W. C., 1968, Crystallization of an adenovirus protein (the hexon), *Nature* (*Lond.*) **219**, 946–947.

Pereira, M. S., Pereira, H. G., and Clarke, S. K., 1965, Human adenovirus type 31: a new serotype with oncogenic properties, *Lancet* **1**, 21–23.

Perlman, S., Hirsch, M., and Penman, S. Utilization of messenger in adenovirus-2-infected cells at normal and elevated temperatures, *Nat. New Biol.* **238**, 143–144.

Perry, R. P., and Kelley, D. E., 1968, Messenger RNA-protein complexes and newly synthesized ribosomal subunits: Analysis of free particles and components of ribosomes, *J. Mol. Biol.* **35**, 37–59.

Pettersson, U., 1971, Structural proteins of adenoviruses. VI. On the antigenic determinants of the hexon, *Virology* **43**, 123–136.

Pettersson, U., 1973*a*, The adenoviruses, in "Molecular Biology of Tumor Viruses" (J. Tooze, ed.), pp. 420–469, Cold Spring Harbor Laboratory, Cold Spring Harbor, N.Y.

Pettersson, U., 1973*b*, Some unusual properties of replicating adenovirus type 2 DNA, *J. Mol. Biol.* **81**, 521–527.

Pettersson, U., and Höglund, S., 1969, Structural proteins of adenoviruses. III. Purification and characterization of the adenovirus type 2 penton antigen, *Virology* **39**, 90–106.

Pettersson, U., and Sambrook, J., 1973, Amount of viral DNA in the genome of cells transformed by adenovirus type 2, *J. Mol. Biol.* **73,** 125–130.

Pettersson, U., Philipson, L., and Höglund, S., 1967, Structural proteins of adenoviruses. I. Purification and characterization of adenovirus type 2 hexon antigen, *Virology* **33,** 575–590.

Pettersson, U., Philipson, L., and Höglund, S., 1968, Structural proteins of adenoviruses. II. Purification and characterization of adenovirus type 2 fiber antigen, *Virology* **35,** 204–215.

Pettersson, U., Mulder, C., Delius, H., and Sharp, P., 1973, Cleavage of adenovirus type 2 DNA into six unique fragments with endonuclease R · RI, *Proc. Natl. Acad. Sci. USA* **70,** 200–204.

Philipson, L., 1967, Attachment and eclipse of adenovirus, *J. Virol.* **1,** 868–875.

Philipson, L., and Pettersson, U., 1973, Structure and function of virion proteins of adenoviruses, *Progr. Exp. Tumor Virus Res.* **18,** 1–55.

Philipson, L., Lonberg-Holm, K., and Pettersson, U., 1968, Virus receptor interaction in an adenovirus system, *J. Virol.* **2,** 1064–1075.

Philipson, L., Wall, R., Glickman, G., and Darnell, J. E., 1971, Addition of polyadenylate sequences to virus-specific RNA during adenovirus replication, *Proc. Natl. Acad. Sci. USA* **68,** 2806–2809.

Philipson, L., Lindberg, U., Persson, T., and Vennström, B., 1973, Transcription and processing of adenovirus RNA in productive infection, *in* "Advances in the Biosciences," Vol. 11 (G. Raspé, ed.), pp. 167–183, Pergamon Press, Vieweg.

Piña, M., and Green, M., 1965, Biochemical studies on adenovirus multiplication: IX. Chemical and base composition analysis of 28 human adenoviruses, *Proc. Natl. Acad. Sci. USA* **54,** 547–551.

Piña, M., and Green, M., 1968, Base composition of the DNA of oncogenic simian adenovirus SA7 and homology with human adenovirus DNAs, *Virology* **36,** 321–323.

Piña, M., and Green, M., 1969, Biochemical studies on adenovirus multiplication: XIV. Macromolecule and enzyme synthesis in cells replicating oncogenic and nononcogenic human adenovirus, *Virology* **38,** 573–586.

Polasa, H. and Green, M., 1965, Biochemical studies on adenovirus multiplication. VIII. Analysis of protein synthesis, *Virology* **25,** 68–79.

Pontén, J., 1971, "Spontaneous and Virus-Induced Transformation in Cell Culture," Springer, Vienna.

Pope, J. H., and Rowe, W. P., 1964, Immunofluorescent studies of adenovirus 12 tumors and of cells transformed or infected by adenoviruses, *J. Exp. Med.* **120,** 577–588.

Prage, L., and Pettersson, U., 1971, Structural proteins of adenoviruses. VII. Purification and properties of an arginine-rich core protein from adenovirus type 2 and type 3, *Virology* **45,** 364–373.

Prage, L., Pettersson, U., and Philipson, L., 1968, Internal basic proteins in adenovirus, *Virology* **36,** 508–511.

Prage, L., Pettersson, U., Höglund, S., Lonberg-Holm, K., and Philipson, L., 1970, Structural proteins of adenoviruses. IV. Sequential degradation of the adenovirus type 2 virion, *Virology* **42,** 341–358.

Prage, L., Höglund, S., and Philipson, L., 1972, Structural proteins of adenoviruses. VIII. Characterization of incomplete particles of adenovirus type 3, *Virology* **49,** 745–757.

Price, R., and Penman, S., 1972*a*, Transcription of the adenovirus genome by an α-aminitine-sensitive ribonucleic acid polymerase in HeLa cells, *J. Virol.* **9,** 621–626.
Price, R., and Penman, S., 1972*b*, A distinct RNA polymerase activity, synthesizing 5.5S, 5S and 4S RNA in nuclei from adenovirus 2-infected HeLa cells, *J. Mol. Biol.* **70,** 435–450.
Rabson, A. S., Kirschstein, R. L., and Paul, F. J., 1964*a*, Tumours produced by adenovirus 12 in mastomys and mice, *J. Natl. Cancer Inst.* **32,** 77–87.
Rabson, A. S., O'Conor, G. T., Berezesky, I. K., and Paul, F. J., 1964*b*, Enhancement of adenovirus growth in African green monkey kidney cell cultures by SV40, *Proc. Soc. Exp. Biol. Med.* **116,** 187–190.
Rafajko, R. R., 1967, Routine establishment of serial lines of hamster embryo cells transformed by adenovirus type 12, *J. Natl. Cancer Inst.* **38,** 581–591.
Rapp, F., Butel, J. S., and Melnick, J. L., 1965, SV40–adenovirus "hybrid" populations: Transfer of SV40 determinants from one type of adenovirus to another, *Proc. Natl. Acad. Sci. USA* **54,** 717–724.
Rapp, F., Jerkofsky, M., Melnick, J. L., and Levy, B., 1968, Variation in the oncogenic potential of human adenoviruses carrying a defective SV40 genome (PARA), *J. Exp. Med.* **127,** 77–90.
Raska, K., and Strohl, W. A., 1972, The response of BHK21 cells to infection with type 12 adenovirus. VI. Synthesis of virus-specific RNA, *Virology* **47,** 734–742.
Raska, K., Frohwirth, D., and Schlesinger, R. W., 1970, Transfer ribonucleic acid in KB cells infected with adenovirus type 2, *J. Virol.* **5,** 464–469.
Raska, K., Prage, L., and Schlesinger, R. W., 1972, The effects of arginine starvation on macromolecular synthesis in infection with type 2 adenovirus. II. Synthesis of virus-specific RNA and DNA, *Virology* **48,** 472–484.
Raskas, H. J., Thomas, D. C., and Green, M., 1970, Biochemical studies on adenovirus multiplication. XVII. Ribosome synthesis in uninfected and infected KB cells, *Virology* **40,** 893–902.
Raskas, H. J., and Okubo, C. K., 1971, Transcription of viral RNA in KB cells infected with adenovirus type 2, *J. Cell Biol.* **49,** 438–449.
Reeves, W. C., Scrivani, R. P., Pugh, W. E., and Rowe, W. P., 1967, Recovery of an adenovirus from a feral rodent peromuscus maniculatus, *Proc. Soc. Exp. Biol. Med.* **124,** 1173–1175.
Reich, P. R., Baum, S. G., Rose, J. A., Rowe, W. P., and Weissman, S. M., 1966, Nucleic acid homology studies of adenovirus type 7–SV40 interactions, *Proc. Natl. Acad. Sci. USA* **55,** 336–341.
Riggs, J. L., and Lennette, E. H., 1967, *In vitro* transformation of newborn hamster kidney cells by simian adenoviruses, *Proc. Soc. Exp. Biol. Med.* **126,** 802–806.
Roeder, R. G., and Rutter, W. J., 1970, Specific nucleolar and nucleoplasmic RNA polymerases, *Proc. Natl. Acad. Sci. USA* **65,** 675–682.
Rose, J. A., Reich, P. R., and Weissman, S. M., 1965, RNA production in adenovirus-infected KB cells, *Virology* **27,** 571–579.
Rosen, L., 1960, Hemagglutination-inhibition technique for typing adenovirus, *Am. J. Hyg.* **71,** 120–128.
Rouse, H. C., and Schlesinger, R. W., 1967, An arginine-dependent step in the maturation of type 2 adenovirus, *Virology* **33,** 513–522.
Rouse, H. C., and Schlesinger, R. W., 1972, The effects of arginine starvation on macromolecular synthesis in infection with type 2 adenovirus. I. Synthesis and utilization of structural proteins, *Virology* **48,** 463–471.

Rowe, W. P., 1965, Studies of adenovirus-SV40 hybrid viruses: III. Transfer of SV40 genes between adenovirus types, *Proc. Natl. Acad. Sci. USA* **54,** 711-716.

Rowe, W. P., and Baum, S. G., 1964, Evidence for a possible genetic hybrid between adenovirus type 7 and SV40 viruses, *Proc. Natl. Acad. Sci. USA* **52,** 1340-1347.

Rowe, W. P., and Baum, S. G., 1965, Studies of adenovirus-SV40 hybrid viruses. II. Defectiveness of the hybrid particles, *J. Exp. Med.* **122,** 955-966.

Rowe, W. P., and Pugh, W. E., 1966, Studies of an adenovirus-SV40 hybrid virus: V. Evidence for linkage between adenovirus and SV40 genetic materials, *Proc. Nat. Acad. Sci.* **55,** 1126-1132.

Rowe, W. P., Huebner, R. J., Gillmore, L. K., Parrott, R. H., and Ward, T. G., 1953, Isolation of a cytogenic agent from human adenoids undergoing spontaneous degeneration in tissue culture, *Proc. Soc. Exp. Biol. Med.* **84,** 570-573.

Rowe, W. P., Hartley, J. W., Roizmann, B., and Levey, H. B., 1958, Characterization of a factor formed in the course of adenovirus infection of tissue cultures causing detachment of cells from glass, *J. Exp. Med.* **108,** 713-729.

Russell, W. C., and Becker, Y., 1968, A maturation factor for adenovirus, *Virology* **35,** 18-27.

Russell, W. C., and Knight, B., 1967, Evidence for a new antigen within the adenovirus capsid, *J. Gen. Virol.* **1,** 523-528.

Russell, W. C., and Skehel, J. J., 1972, The polypeptides of adenovirus-infected cells, *J. Gen. Virol.* **15,** 45-57.

Russell, W. C., Valentine, R. C., and Pereira, H. G.,1967*a*, The effect of heat on the anatomy of the adenovirus, *J. Gen. Virol.* **1,** 509-522.

Russell, W. C., Hayashi, K., Sanderson, P. J., and Pereira, H. G., 1967*b*, Adenovirus antigens—A study of their properties and sequential development in infection, *J. Gen. Virol.* **1,** 495-507.

Russell, W. C., McIntosh, K., and Skehel, J. J., 1971, The preparation and properties of adenovirus cores, *J. Gen. Virol.* **11,** 35-46.

Russell, W. C., Newman, C., and Williams, J. F., 1972*a*, Characterization of temperature-sensitive mutants of adenovirus type 5—Serology, *J. Gen. Virol.* **17,** 265-279.

Russell, W. C., Skehel, J. J., Machado, R., and Pereira, H. G., 1972*b*, Phosphorylated polypeptides in adenovirus-infected cells, *Virology* **50,** 931-934.

Salzberg, S., and Raskas, H. J., 1972, Surface changes of human cells productively infected with human adenoviruses, *Virology* **48,** 631-637.

Samarina, O. P., Lukanidin, E. M., Molnar, J., and Georgiev, G. P., 1968, Structural organization of nuclear complexes containing DNA-like RNA, *J. Mol. Biol.* **33,** 251-263.

Sambrook, J., Sharp, P. A., and Keller, W., 1972, Transcription of simian virus 40. I. Separation of the strands of SV40 DNA and hybridization of the separated strands to RNA extracted from lytically infected and transformed cells, *J. Mol. Biol.* **70,** 57-71.

Sambrook, J., Sharp, P. A., Ozanne, B., and Pettersson, U., 1974, Studies on transcription of simian virus 40 and adenovirus type 2. In: "Control of Transcription" (B. B. Biswas, R. K. Mandal, A. Stevens, and W. E. Cohn, eds.). Plenum Press, New York, pp. 167-179.

Sarma, P. S., Huebner, R. J., and Lane, W. T., 1965, Inductions of tumors in hamsters with an avian adenovirus (CELO), *Science* (*Wash., D.C.*) **149,** 1108.

Schell, K., and Schmidt, M., 1968, Adenovirus transformation of hamster embryo cells. I. Assay conditions. *Arch. Ges. Virusforsch.* **24,** 332–341.

Schell, K., Lane, W. T., Casey, M. J., and Huebner, R. J., 1966, Potentiation of oncogenicity of adenovirus type 12 grown in African green monkey kidney cell cultures preinfected with SV40 virus: Persistence of both T antigens in the tumors and evidence for possible hybridization, *Proc. Natl. Acad. Sci. USA* **55,** 81–88.

Schell, K., Maryak, J., and Schmidt, M., 1968*a*, Adenovirus transformation of hamster embryo cells. III. Maintenance conditions, *Arch. Ges. Virusforsch.* **24,** 352–360.

Schell, K., Maryak, J., Young, J., and Schmidt, M., 1968*b*, Adenovirus transformation of hamster embryo cells. II. Inoculation conditions, *Arch. Ges. Virusforsch.* **24,** 342–351.

Schlesinger, R. W., 1969, Adenoviruses: The nature of the virion and of controlling factors in productive or abortive infection and tumorigenesis, *Adv. Virus Res.* **14,** 1–61.

Schmidt, N. J., King, C. J., and Lennette, E. H., 1965, Hemagglutination and hemagglutination-inhibition with adenovirus type 12, *Proc. Soc. Exp. Biol. Med.* **118,** 208–211.

Schochetman, G., and Perry, R. P., 1972, Characterization of the messenger RNA released from L cell polyribosomes as a result of temperature shock, *J. Mol. Biol.* **63,** 577–590.

Sharp, P. A., Pettersson, U., and Sambrook, J., 1974, Viral DNA in transformed cells. I. A study of the sequences of adenovirus DNA in a line of transformed rat cells using specific fragments of the viral genome, *J. Mol. Biol.*, in press.

Shimojo, H., Yamamoto, H., and Abe, C., 1967, Differentiation of adenovirus 12 antigens in cultured cells, *Virology* **31,** 748–752.

Shiroki, K., Irisawa, J., and Shimojo, H., 1972, Isolation and preliminary characterization of temperature-sensitive mutants of adenovirus 12, *Virology* **49,** 1–11.

Singer, R. H., and Penman, S., 1972, Stability of HeLa cell mRNA in actinomycin, *Nature (Lond.)* **240,** 100–102.

Sjögren, H. O., Minowada, J., and Ankerst, J., 1967, Specific transplantation antigens of mouse sarcomas induced by adenovirus type 12, *J. Exp. Med.* **125,** 689–701.

Smith, K. O., Gehle, W. D., and Trousdale, M. D., 1965, Architecture of the adenovirus capsid, *J. Bacteriol.* **90,** 254–261.

Soeiro, R., Vaughan, M. H., Warner, J. R., and Darnell, J. E., 1968, The turnover of nuclear DNA-like RNA in HeLa cells, *J. Cell Biol.* **39,** 112–118.

Sohier, R., Chardonnet, Y., and Prunieras, M., 1965, Adenoviruses: Status of current knowledge. *Progr. Med. Virol.* **7,** 253–325.

Spirin, A. S., and Nemer, M., 1965, Messenger RNA in early sea-urchin embryos: Cytoplasmic particles, *Science (Wash., D.C.)* **150,** 214–217.

Stasny, J. T., Neurath, A. R., and Rubin, B. A., 1968, Effect of formamide on the capsid morphology of adenovirus types 4 and 7, *J. Virol.* **2,** 1429–1442.

Strohl, W. A., 1969*a*, The response of BHK21 cells to infection with type 12 adenovirus: I. Cell killing and T antigen synthesis as correlated viral genome functions, *Virology* **39,** 642–652.

Strohl, W. A., 1969*b*, The response of BHK21 cells to infection with type 12 adenovirus: II. Relationship of virus-stimulated DNA synthesis to other viral functions, *Virology* **39,** 653–665.

Strohl, W. A., Rouse, H. C., and Schlesinger, W. R., 1966, Properties of cells derived from adenovirus-induced hamster tumors by long-term *in vitro* cultivation. II. Nature of the restricted response to type 2 adenovirus, *Virology* **28,** 645–658.

Strohl, W. A., Rabson, A. S., and Rouse, H., 1967, Adenovirus tumorigenesis: Role of the viral genome in determining tumor morphology, *Science* (*Wash., D.C.*) **156,** 1631–1633.

Strohl, W. A., Rouse, H., Teets, K., and Schlesinger, R. W., 1970, The response of BHK21 cells to infection with type 12 adenovirus. III. Transformation and restricted replication of superinfecting type 2 adenovirus, *Arch. Ges. Virusforsch.* **31,** 93–112.

Sugino, A., Hirose, S., and Okazaki, R., 1972, RNA-linked nascent DNA fragments in *Escherichia coli*, *Proc. Natl. Acad. Sci. USA* **69,** 1863–1867.

Sundquist, B., Pettersson, U., and Philipson, L., 1973*a*, Structural proteins of adenoviruses. IX. Molecular weight and subunit composition of the adenovirus type 2 fiber, *Virology* **51,** 252–256.

Sundquist, B., Everitt, E., Philipson, L., and Höglund, S., 1973*b*, Assembly of adenoviruses, *J. Virol.* **11,** 449–459.

Sussenbach, J. S., 1967, Early events in the infection process of adenovirus type 5 in HeLa cells, *Virology* **33,** 567–574.

Sussenbach, J. S., 1971, On the fate of adenovirus DNA in KB cells, *Virology* **46,** 969–972.

Sussenbach, J. S., and Van der Vliet, P., 1972, Characterization of adenovirus DNA in cells infected with adenovirus type 12, *Virology*, **49,** 224–229.

Sussenbach, J. S., and Van der Vliet, P. C., 1973, Studies on the mechanism of replication of adenovirus DNA. 1. The effect of hydroxyurea, *Virology* **54,** 299–303.

Sussenbach, J. S., Van der Vliet, P. C., Ellens, D. J., and Jansz, H. S., 1972, Linear intermediates in the replication of adenovirus DNA, *Nat. New Biol.* **239,** 47–49.

Sussenbach, J. S., Ellens, D. J., and Jansz, H. S., 1973, Studies on the mechanism of replication of adenovirus DNA. 2. The nature of single-stranded DNA in replicative intermediates, *J. Virol.* **12,** 1131–1138.

Suzuki, E., and Shimojo, H., 1971, A temperature-sensitive mutant of adenovirus 31, defective in viral deoxyribonucleic acid, *Virology* **43,** 488–494.

Takahashi, M., Veda, S., and Ogino, T., 1966, Enhancement of thymidine kinase activity of human embryonic kidney cells and newborn hamster kidney cells by infection with human adenovirus types 5 and 12, *Virology* **30,** 742–743.

Takemori, N., 1972, Genetic studies with tumorgenic adenoviruses. III. Recombination in adenovirus type 12, *Virology* **47,** 157–167.

Takemori, N., Riggs, J. L., and Aldrich, C., 1968, Genetic studies with tumorigenic adenoviruses. I. Isolation of cytocidal (*cyt*) mutants of adenovirus type 12, *Virology* **36,** 575–586.

Tavitian, A., Peries, J., Chuat, J., and Boiron, M., 1967, Estimation of the molecular weight of adenovirus 12 tumor CF antigen by rate-zonal centrifugation, *Virology* **31,** 719–721.

Tavitian, A., Uretsky, S. C., and Acs, G., 1968, Selective inhibition of ribosomal RNA synthesis in mammalian cells, *Biochim. Biophys. Acta* **157,** 33–42.

Tejg-Jensen, B., Furugren, B., Lindqvist, I., and Philipson, L., 1972, Röntgenkleinwinkelstreuung an Hexon aus Adenovirus Typ 2, *Monatshefte Chem.* **103,** 1730–1736.

Thomas, D. C., and Green, M., 1966, Biochemical studies on adenovirus multiplication: XI. Evidence of a cytoplasmic site for the synthesis of viral-coded proteins, *Proc. Natl. Acad. Sci. USA* **56,** 243–146.

Thomas, D. C., and Green, M., 1969, Biochemical studies on adenovirus multiplication: XV. Transcription of the adenovirus type 2 genome during productive infection, *Virology* **39,** 205–210.

Tibbetts, C., and Pettersson, U., 1974, Complementary strand-specific sequences from unique fragments of adenovirus type 2 DNA for hybridization-mapping experiments, *J. Mol. Biol.*, in press.

Tibbetts, C., Johansson, K., and Philipson, L., 1973, Hydroxylapatite chromatography and formamide denaturation of adenovirus DNA, *J. Virol.* **12,** 218–225.

Tibbetts, C., Pettersson, U., Johansson, K., and Philipson, L., 1974*b*, Relationship of messenger ribonucleic acid from productively infected cells to the complementary strands of adenovirus type 2 deoxyribonucleic acid, *J. Virol.* **13,** 370–377.

Tockstein, G., Plasa, H., Piña, M., and Green, M., 1968, A simple purification procedure for adenovirus type 12 T and tumor antigens and some of their properties, *Virology* **36,** 377–386.

Tonegawa, S., Walter, G., Bernardini, A., and Dulbecco, R., 1970, Transcription of the SV40 genome in transformed cells and during lytic infection, *Cold Spring Harbor Symp. Quant. Biol.* **35,** 823–832.

Trentin, J. J., Yabe, Y., and Taylor, G., 1962, The quest for human cancer viruses, *Science* (*Wash., D.C.*) **137,** 835–841.

Tseui, D., Fujinaga, K., and Green, M., 1972, The mechanism of viral carcinogenesis by DNA mammalian viruses: RNA transcripts containing viral and highly reiterated cellular base sequences in adenovirus-transformed cells, *Proc. Natl. Acad. Sci. USA* **69,** 427–430.

Tsukamoto, K., and Sugino, Y., 1972, Nonproductive infection and induction of cellular deoxyribonucleic acid synthesis by bovine adenovirus type 3 in a contact-inhibited mouse cell line, *J. Virol.* **9,** 465–473.

Ustacelebi, S., and Williams, J. F., 1972, Temperature-sensitive mutants of adenovirus defective in interferon induction at non-permissive temperature, *Nature* (*Lond.*) **235,** 52–53.

Valentine, R. C., and Pereira, H. G., 1965, Antigens and structure of the adenovirus, *J. Mol. Biol.* **13,** 13–20.

Van der Eb, A. J., 1973, Intermediates in type 5 adenovirus DNA replication, *Virology* **51,** 11–23.

Van der Eb, A. J., Kestern, L. W., and Van Bruggen, E. F. J., 1969, Structural properties of adenovirus DNA's, *Biochim. Biophys. Acta* **182,** 530–541.

Van der Nordaa, J., 1968*a*, Transformation of rat kidney cells by adenovirus type 12, *J. Gen. Virol.* **2,** 269–272.

Van der Nordaa, J., 1969*b*, Transformation of rat cells by adenovirus types 1, 2 and 3, *J. Gen. Virol. 3*, 303–304.

Van der Vliet, P. C., and Levine, A. J., 1973, DNA-binding proteins specific for cells infected by adenovirus, *Nature New Biol.* **246,** 170–174.

Van der Vliet, P. C., and Sussenbach, J. S., 1972, The mechanism of adenovirus-DNA-synthesis in isolated nuclei, *Eur. J. Biochem.* **30,** 548–592.

Velicer, L., and Ginsberg, H. S., 1968, Cytoplasmic synthesis of type 5 adenovirus capsid proteins, *Proc. Natl. Acad. Sci. USA* **61,** 1264–1271.

Velicer, L., and Ginsberg, H. S., 1970, Synthesis, transport and morphogenesis of type 5 adenovirus capsid proteins, *J. Virol.* **5,** 338–352.

Wadell, G., 1970, Structural and biological properties of capsid components of human adenoviruses, Ph.D. Thesis, Karolinska Institutet, Stockholm.

Wadell, G., and Norrby, E., 1969, Immunological and other biological characteristics of pentons of human adenoviruses, *J. Virol.* **4,** 671–680.

Wadell, G., Hammarskjöld, M.-L., and Varsanyi, T., 1973, Incomplete virus particles of adenovirus type 16, *J. Gen. Virol.* **20,** 287–303.

Wall, R., Philipson, L., and Darnell, J. E., 1972, Processing of adenovirus-specific nuclear RNA during virus replication, *Virology* **50,** 27–34.

Wall, R., Weber, J., Gage, Z., and Darnell, J. E., 1973, Production of viral mRNA in adenovirus transformed cells by the post-transcriptional processing of heterogeneous nuclear RNA containing viral and cell sequences, *J. Virol.* **11,** 953–960.

Wallace, R. D., and Kates, J., 1972, On the state of the adenovirus 2 DNA in the nucleus and its mode of transcription. Studies with viral protein complexes and isolated nuclei, *J. Virol.* **9,** 627–635.

Warocquier, R., Samaille, J., and Green, M., 1969, Biochemical studies on adenovirus multiplication. XVI. Transcription of the adenovirus genome during abortive infection at elevated temperatures, *J. Virol.* **4,** 423–428.

Watson, J. D., 1972, Origin of concatemeric T7 DNA, *Nat. New Biol.* **239,** 197–201.

Whitcutt, J. M., and Gear, H. S., 1968, Transformation of newborn hamster cells with simian adenovirus SA7, *Int. J. Cancer* **3,** 566–571.

White, D. O., Scharff, M. D., and Maizel, J. V., Jr., 1969, The polypeptides of adenoviruses. III. Synthesis in infected cells, *Virology* **38,** 395–406.

Wickner, W., Brutlag, D., Schekman, R., and Kornberg, A., 1972, RNA-synthesis initiates *in vitro* conversion of M13 DNA to its replicative form, *Proc. Natl. Acad. Sci. USA* **69,** 965–969.

Wigand, R., 1969, Adenovirus types 20, 25, and 28: A typical members of group II, *J. Gen. Virol.* **6,** 325–328.

Wilcox, W. C., and Ginsberg, H. S., 1963, Protein synthesis in type 5 adenovirus infected cells. Effect of p-fluorophenyl alanine on synthesis of protein nucleo acids and infectious virus, *Virology* **20,** 269–280.

Wilcox, W. C., Ginsberg, H. S., and Anderson, T. F., 1963, Structure of type 5 adenovirus. II. Fine structure of virus subunits. Morphologic relationship of structural subunits to virus-specific antigens from infected cells, *J. Exp. Med.* **118,** 307–314.

Wilhelm, J. M., and Ginsberg, H. S., 1972, Synthesis in vitro of type 5 adenovirus capsid proteins, *J. Virol.* **9,** 973–980.

Wilkie, N. M., Ustacelebi, S., and Williams, J. F., 1973, Preliminary characteristics of temperature-sensitive mutants of adenovirus type 5. Nucleic acid synthesis, *Virology* **51,** 499–503.

Willems, M., Penman, M., and Penman, S., 1969, The regulation of RNA synthesis and processing in the nucleous during inhibition of protein synthesis, *J. Cell Biol.* **41,** 177–187.

Williams, J. F., 1973, Oncogenic transformation of hamster embryo cells in vitro by adenovirus type 5, *Nature* (*Lond.*) **243,** 162–163.

Williams, J. F., and Ustacelebi, S., 1971, Complementation and recombination with temperature-sensitive mutants of adenovirus type 5, *J. Gen. Virol.* **13,** 345–348.

Williams, J. F., Gharpure, M., Ustacelebi, S., and McDonald, S., 1971, Isolation of temperature-sensitive mutants of adenovirus type 5, *J. Gen. Virol.* **11,** 95–102.

Wills, E. J., Russell, W. C., and Williams, J. F., 1973, Adenovirus-induced crystals: studies with temperature-sensitive mutants, *J. Gen. Virol.* **20,** 407–412.

Wilner, B. I., 1969, "A Classification of the Major Groups of Human and Other Animal Viruses," pp. 120–132, Burgess, Minneapolis, Minn.

Winnacker, E.-L., Magnusson, G., and Reichard, P., 1972, Replication of polyoma DNA in isolated nuclei. I. Characterization of the system from mouse fibroblast 3T6 cells, *J. Mol. Biol.* **72,** 523–537.

Winters, W. D., and Russell, W. C., 1971, Studies on assembly of adenovirus *in vitro*, *J. Gen. Virol.* **10,** 181–194.

Wolfson, J., and Dressler, D., 1972, Adenovirus 2-DNA contains an inverted terminal repetition, *Proc. Natl. Acad. Sci. USA* **69,** 3054–3057.

Yabe, Y., Samper, L., Bryan, E., Taylor, G., and Trentin, J. J., 1964, Oncogenic effect of human adenovirus type 12 in mice, *Science* (*Wash., D.C.*) **143,** 46–47.

Zain, B. S., Dhar, R., Weissman, S. M., Lebowitz, P., and Lewis, A. M., Jr., 1973, Preferred site for initiation of RNA transcription by *Escherichia coli* RNA polymerase within the simian virus 40 DNA segment of the nondefective adenovirus–simian virus 40 hybrid viruses Ad2+ND_1 and Ad2+ND_3, *J. Virol.* **11,** 682–693.

Zimmerman, J.E., and Raska, K., 1972, Inhibition of adenovirus type 12 induced DNA synthesis in G1-arrested BHK21 cells by dibutyryl adenosine cyclic 3′, 5′-monophosphate, *Nat. New Biol.* **239,** 145–147.

Zimmerman, J., Raska, K., and Strohl, W. A., 1970, The response of BHK21 cells to infection with type 12 adenovirus. IV. Activation of DNA-synthesizing apparatus, *Virology* **42,** 1147–1159.

Zur Hausen, H., and Sokol, F., 1969, Fate of adenovirus type 12 genomes in non-permissive cells, *J. Virol.* **4,** 256–271.

Zylber, E. A., and Penman, S., 1971, Products of RNA polymerases in HeLa cell nuclei, *Proc. Natl. Acad. Sci. USA* **68,** 2861–2865.

CHAPTER 4

The Replication of Herpesviruses

Bernard Roizman* and Deirdre Furlong

Department of Microbiology
The University of Chicago
Chicago, Illinois 60637

1. OBJECTIVES AND SCOPE

1.1 Objectives

This chapter deals with the structure and replication of herpesviruses in susceptible cells. Having stated its scope, we shall take license to say what what we think of it.

Reviews such as this generally have an objective. A plausible, but not credible objective of writing a review in a rapidly expanding scientific field is to bring order. But what is order? A recital of established "facts"? The expert all too frequently has an inflexible view of his universe and for him another expert's order is anathema. The beginner might be more susceptible to expert incantations, but alas, even the most fluent summation of credible data frequently makes dull reading, and too much order, like too little entropy in chemical reactions, is not a suitable environment on which to nurture the urge to discover. We cannot claim either good reading or too much order. Our excuse, if one need be offered, is the veritable flood of information on herpesviruses which has erupted from just about every corner of the world in recent years. Whereas an earlier review from this laboratory (Roizman, 1969) took six weeks to write, this one took over eight months, even though it is

* Also in the Department of Biophysics

much more limited in scope. With deep regret we left out the many scores of papers on abortive infections, latency, and the mushrooming field of herpesvirus genetics, all of which we hope will be covered in a subsequent volume of this series. Also, once again we feel compelled to make the trite apology for the failure to do justice, or injustice, as the case may be, to the many hundreds of papers published on herpesviruses each year, or to the many thousand papers published on herpesviruses since the first of the members of the family was experimentally transmitted to a heterologous host more than half a century ago (Grüter, 1924).

1.2. The Herpesviruses

The reader interested in the derivation of the name "herpes" and other historical tidbits should look up Beswick (1962), Roizman (1969, 1971*a*), and Nahmias and Dowdle (1968).

Herpesviruses are defined as having DNA as their genetic material, an icosahedral capsid containing 162 capsomers, and an envelope; no one has ever reported counting all 162 capsomers. The most frequent identification of a virus as a putative herpesvirus or, more frequently, as a herpes-type virus, is based on the presence of particles approximately 100 nm in diameter budding through the inner nuclear membrane. Based on this identification, herpesviruses have been reported in nearly every eukaryotic species examined in detail, from fungi (Kazama and Schornstein, 1972) to man. The list includes oysters (Farley *et al.*, 1972), fish (Wolf and Darlington, 1971), frogs, birds, and a variety of domestic and wild animals. Are they all herpesviruses? Must enveloped DNA viruses budding through the nuclear membrane necessarily be related? What should be criteria for relatedness in the face of (1) the fact that there is apparently little DNA–DNA homology among herpesviruses other than herpes simplex 1 and 2 and the possibility that the common antigens reported at length might well be host or serum contaminants? From the point of view of this paper, the questions are academic, particularly since of the 2 or 3 score herpesviruses reported to date only a handful, i.e., herpes simplex (HSV), pseudorabies, the Epstein-Barr virus (EBV), equine herpesviruses, and some cytomegaloviruses, have been studied in some detail.

Although we do not wish to discuss any classification of herpesviruses, some comments concerning the nomenclature of herpesviruses should be made. The extent of variation in the current naming of herpesviruses is exemplified by herpesviruses infecting man which are named after the host (herpesvirus hominis), after the disease (varicella-zoster) or histological manifestation (cytomegaloviruses) with which they are associated, or after the names of their discoverers (Epstein-Barr virus). This is obviously not a very satisfactory state of affairs, but permanent alternatives are not readily acceptible since sufficient information is as yet unavailable concerning the phylogenetic relationships of the herpesviruses to one another to permit the assembly of a binomial nomenclature related to an hierarchical system of classification. Until the time when such a nomenclature can be instituted, the ICNV Herpesvirus Study Group proposed (Roizman *et al.*, 1973*b*) a provisional system for the labeling of herpesviruses based on the following rules: (1) the label for each herpesvirus would be in an anglicized form, (2) each herpesvirus would be named after the taxonomic unit, the family, to which its primary natural host belongs, and (3) the herpesviruses within each group would be given arabic numbers. The number would not be preceded by the word "type." New herpesviruses would receive the next available numbers.

To facilitate reading of this chapter, the common names now in use rather than the proposed new designation will be used throughout the text. Table 1 however lists both the new and the common names, the few abbreviations used in the text, as well as the pertinent properties of viral DNAs reported to date. As a further concession to the observation that herpesvirus strains passaged in the laboratory differ from isolates passaged a limited number of times in cell culture (Heine *et al.*, 1974), we included wherever desirable the name of the virus strain in parentheses after the virus designation or its abbreviation.

2. THE HERPESVIRION

2.1. Architectural Components

The large virion consists of four major architectural components, the centrally located core being surrounded by three concentric structures: the capsid, the tegument, and the envelope (Fig. 1).

2.1.1. The Core

In thin sections of extracellular virions or virions in infected cells, the core has the appearance of an electron-dense ring surrounding an electron-translucent center (Fig. 1E) or of an electron-dense bar (Fig. 2A, B). These two images are compatible with the hypothesis that the core is an electron-dense toroid with a less-dense plug or spool filling the hole (Fig. 1H). Indeed, many, although not all, pictures of the core could be accounted for in terms of the angle of the plane in which the section is cut (Fig. 1D). Two lines of evidence have led to the con-

TABLE 1

Provisional Labels, Common Names, and Properties of the DNA of Herpesviruses

Provisional label[a]	Common name	Abbreviation	G + C, moles %	Buoyant density in CsCl	Reference	Mol. wt. $\times 10^{-6}$	Reference	Comments
Human herpesvirus 1	Herpes simplex type 1	HSV-1	66	1.725	Graham *et al.* (1972)	68	Russell and Crawford (1964)	Sedimentation constant
	Herpesvirus hominis 1		67	1.726	Kieff *et al.* (1971)		Becker *et al.* (1968)	Length and cosedimentation with poxvirus DNA in sucrose density gradients
			68	1.727	Goodheart *et al.* (1968), Russell and			
			68	1.727	Crawford (1964)			
						88 ± 14	Graham *et al.* (1972)	Sedimentation constant
						99 ± 5	Kieff *et al.* (1971)	Cosedimentation with T4 in sucrose density gradient
						95 ± 1	Frenkel and Roizman (1971)	Reassociation kinetics
						85	Mosmann and Hudson (1973)	Cosedimentation with T4 in sucrose density gradient
Human herpesvirus 2	Herpes simplex type 2	HSV-2	68	1.727	Ludwig *et al.* (1971*a*)	99 ± 5	Kieff *et al.* (1971)	Cosedimentation with T4 in sucrose density gradient
			68	1.727	Graham *et al.* (1972)			
	Herpesvirus hominis 2		69	1.728	Kieff *et al.* (1971)			
			70	1.729	Goodheart *et al.* (1968)			
Human herpesvirus 3	Varicella-zoster virus	—	46	1.705	Ludwig *et al.* (1972*a*)			
Human herpesvirus 4	Epstein-Barr virus	EBV	59	1.718	Wagner *et al.* (1970), Schulte-Holthausen and Zur Hausen (1970), Nonoyama and Pagano (1971), Jehn *et al.* (1972)			
			61	1.720	Weinberg and Becker (1969)			
			63	1.723	Ludwig *et al.* (1971*b*)			

Human herpesvirus 5	Cytomegalovirus		57	1.717	Plummer *et al.* (1969)	32	Crawford and Lee (1964)	Bandwidth in CsCl
			56–57	1.717	Ludwig (1972)			
			59	1.718	Crawford and Lee (1964)			
Ceropithecid herpesvirus 1	B Virus							
Ceropithecid herpesvirus 2	SA 6		51	1.710	Goodheart and Plummer (1974)			
Ceropithecid herpesvirus 3	SA 8							
Cebid herpesvirus 1	Herpesvirus tamarinus; herpesvirus platyrrhini marmoset herpesvirus		67	1.726	Goodheart and Plummer (1974)			
Cebid herpesvirus 2	Herpesvirus saimiri		50	1.709	B. Fleckenstein (Personal communication)			
Cebid herpesvirus 3	Spider-monkey herpesvirus		72	1.731	Goodheart and Plummer (1974)			
Callitrichid herpesvirus 1	Marmoset herpesvirus							
Tupaiid herpesvirus 1	Tree-shrew herpesvirus		66	1.725	Ludwig (1972)			
Canine herpesvirus 1	Canine herpesvirus		33	1.692	Plummer *et al.* (1969)			
Feline herpesvirus 1	Feline rhinotracheitis virus		46	1.705	Plummer *et al.* (1969)			
Equid herpesvirus 1	Equine abortion virus, equine rhinopneumonitis virus	EAV	55	1.714	Russell and Crawford (1964)	84	Russell and Crawford (1964)	Sedimentation constant
			55	1.714	Ludwig (1972)	84–94	Soehner *et al.* (1965)	Sedimentation constant
			57	1.716	Soehner *et al.* (1965)			
			57	1.716				
Equid herpesvirus 2	Slowly growing, cytomegalo-type viruses		58	1.717	Plummer *et al.* (1969, 1973)			

TABLE 1—*continued*

Provisional label[a]	Common names	Abbreviation	G + C, moles %	Buoyant density in CsCl	Reference	Mol. wt. $\times 10^{-6}$	Reference	Comments
Equid herpesvirus 3	Coital-exanthema virus		66	1.725	Ludwig *et al.* (1971*a*)			
Bovid herpesvirus 1	Infectious bovine rhinotracheitis virus		71	1.730	Russell and Crawford (1964)	54	Russell and Crawford (1964)	Sedimentation constant
			71	1.730	Ludwig (1972)			
			72	1.731	Plummer *et al.* (1969)			
Bovid herpesvirus 2	Bovine mammalitis virus		64	1.723	Martin *et al.* (1966)	82	Martin *et al.* (1966)	Band in CsCl
Bovid herpesvirus 3	Wildebeest herpesvirus, malignant catarrhal fever virus							
Bovid herpesvirus 4	Herpesvirus from sheep pulmonary adenomatosis							
Pig herpesvirus 1	Pseudorabies virus		72	1.731	Plummer *et al.* (1969)	35	Kaplan and Ben-Porat (1964)	Bandwidth in CsCl
			72	1.731	Ludwig (1972)	16, 22	Graham *et al.* (1972)	Sedimentation constant
			74	1.733	Russell and Crawford (1964)	68	Russell and Crawford (1964)	Sedimentation constant
			73	1.732	Kaplan and Ben-Porat (1964)			
Pig herpesvirus 2	Inclusion-body rhinitis virus, pig cytomegalovirus							
Murid herpesvirus 1	Mouse cytomegalovirus from Mus		59	1.718	Mosmann and Hudson (1973)	132	Mosmann and Hudson (1973)	Cosedimentation with T4 in sucrose density gradients
Murid herpesvirus 2	Rat cytomegalovirus from Rattus							

Sciruid herpesvirus 1	Cytomegalovirus from European ground squirrel							
Caviid herpesvirus 1	Guinea-pig cytomegalovirus		57	1.716	Nayak (1971)			
Lagomorph herpesvirus 1	Rabbit herpesvirus							
Phasianid herpesvirus 1	Infectious laryngotracheitis virus		45	1.704	Plummer *et al.* (1969)			
Phasianid herpesvirus 2	Marek's disease virus	MDV	46	1.705	Lee *et al.* (1971)	103	Lee *et al.* (1971)	Cosedimentation with T4 in sucrose density gradients
Turkey herpesvirus 1	Turkey herpesvirus		46	1.706	Lee *et al.* (1972)			
Anatid herpesvirus 1	Duck-plague herpesvirus							
Pigeon herpesvirus 1	Pigeon herpesvirus							
Cormorant herpesvirus 1	Cormorant herpesvirus							
Iguana herpesvirus 1	Iguana herpesvirus							
Ranid herpesvirus 1	Lucke virus		44–45	1.703	Wagner *et al.* (1970), Collard *et al.* (1973)			
Ranid herpesvirus 2	Frog virus 4		56	1.716	Gravell (1971)			
Catfish herpesvirus 1	Catfish herpesvirus		56	1.715	Goodheart and Plummer (1974)			

[a] Recommended by Herpesvirus Study Group. See Roizman *et al.* (1973*b*).

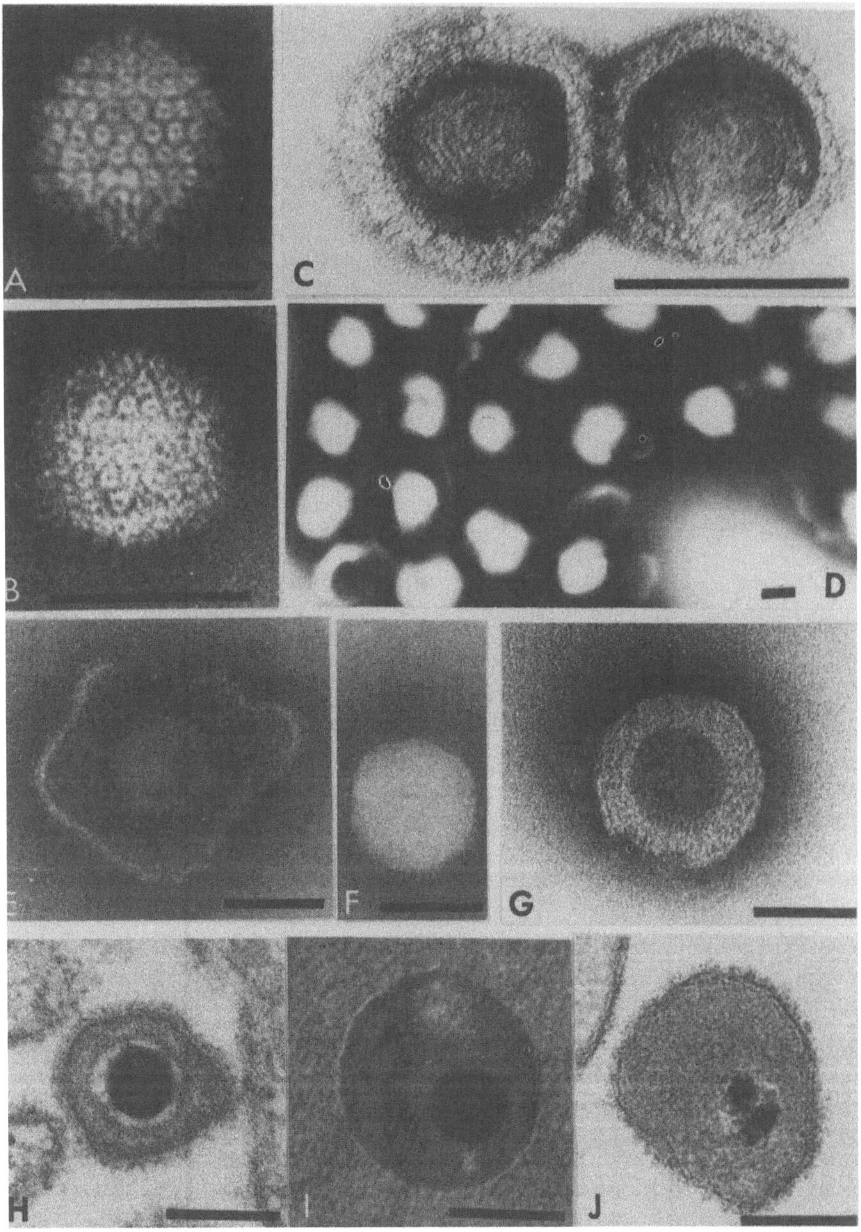

Fig. 1. The structure of the herpesvirus capsid and virion. (A, B, C) Negatively stained preparations of HSV-1 capsids. (A and B) Capsids were stained with phosphotungstic acid and show 3- and 2-fold symmetry, respectively. (C) Capsids penetrated by uranyl acetate, bringing into relief threadlike structures showing periodic striations on the surface of the core. (D) Negative-stain preparation of purified HSV-1 virions. Intact virions are impermeable to negative stain and appear as white blobs with tails. The irregular shape and tailing probably result from stretching of the envelope during centrifugation. (E) Ruptured HSV-1 virion penetrated by negative stain.

clusion that the core contains DNA. First, numerous investigators (Epstein, 1962*a*; Zambernard and Vatter, 1966; Chopra *et al.*, 1970; Cook and Stevens, 1970; Trung and Lardemer, 1972) have shown that the central region of the virion is sensitive to DNase but not to RNase or to proteolytic enzymes, though there is little of the virion left to be recognized after proteolysis. Although enzymatic digestion showed that the DNA is in the core, it was not immediately clear whether the DNA was contained in the electron-dense area, the electron-translucent area, or both areas. Recent studies (Furlong *et al.*, 1972) have taken advantage of a technique developed by Bernhard (1969) to show that the DNA is contained in the electron-dense region making up the toroid (Fig. 2). The technique is based on the observation that EDTA treatment of thin sections, fixed with glutaraldehyde only, and stained with uranyl acetate, selectively removed uranyl ions bound to DNA. The appearance of the plug was unaltered by this treatment and nothing is known of its structure. Strandberg and Carmichael (1965) reported that the core of the canine herpesvirion had a helical appearance. Herpes simplex capsids negatively stained with uranyl acetate showed coiled threadlike structures on the surface of the core (Fig. 2C). The threadlike structures were 4.0–5.0 nm wide and showed some indications of a periodic superstructure (coils?) along the thread. We suspect, but have no proof, that these structures are the spooled DNA.

2.1.2. The Capsid

In thin sections the capsid appears as a moderately electron-dense hexagon or ring separated from the core by an electron-translucent shell. The outer dimensions of the capsid reported to date (Wolf and Darlington, 1971; Stackpole and Mizell, 1968; Strandberg and Carmichael, 1965; Wildy *et al.*, 1960; Nazerian *et al.*, 1971; Abodeely *et al.*, 1970; McCombs *et al.*, 1971; Nayak, 1971; Nazerian and Witter, 1970) vary considerably, but in general fall between 85 and 110 nm. It is not clear whether the variability of dimensions reported to date reflects artifacts of manipulation or inherent structural differences among herpesviruses. The morphologic subunits of the capsid, the

Note the outlines of the capsomers and of the membrane with spikes on its outer surface and tegument on its inner surface. (F) Intact HSV-1 virion impermeable to negative stain shown at same magnification as E and G. (G) HSV-1 virion showing loss of membrane from one side. Note tegument adhering to capsid. (H, I, J) Thin sections of HSV-1, MDV, and Tree shrew herpesvirus virions, respectively. Note extensive tegument in I and J and densely staining membrane with spikes on the outer surface in H and J. Bar in this and all subsequent figures = 100 nm.

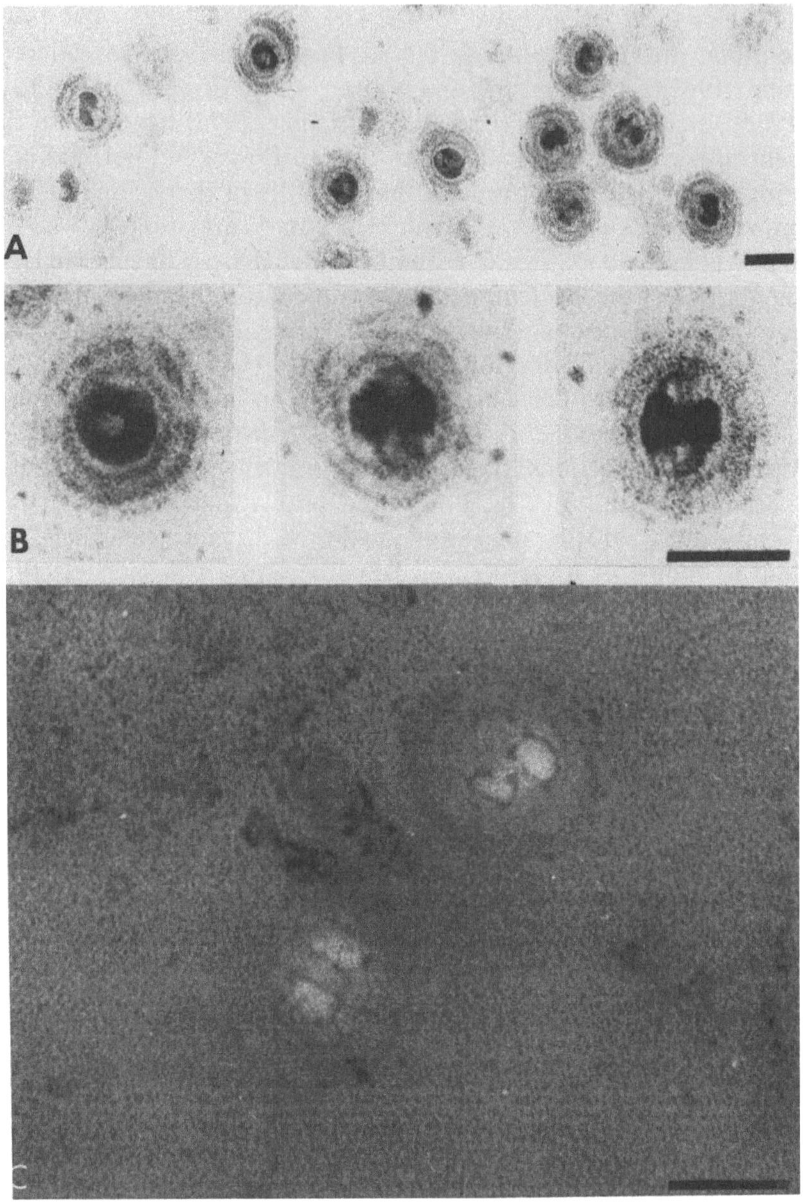

Fig. 2. The structure of the core of the herpesvirion. (A) Electron micrographs of thin sections of HSV-1 virions showing the core cut at various angles. (B) Three selected views of the core shown at high magnification. The material shown in A and B was fixed with glutaraldehyde, and post-fixed with osmium tetroxide. Sections were stained with uranyl acetate and lead citrate. All thin sections shown in this chapter were fixed and stained in this way unless otherwise indicated. (C) Electron micrograph of glutaraldehyde-fixed thin sections of

capsomers, are readily visible in negatively stained preparations (Fig. 1A,B). The capsomer arrangement shows 2-, 3-, and 5-fold symmetry. From the number of capsomers (five) along the side of the triangular face and the axis of symmetry, Wildy *et al.*, (1960) concluded that the capsid is made of 162 capsomers arranged in the form of an icosadeltahedron. The number of capsomers has not been confirmed by actual count. Icosadeltahedral arrangement requires pentameric as well as hexameric capsomers. Hexameric capsomers are commonly seen in negatively stained preparations. Pentagonal capsomers have been reported (Wildy *et al.*, 1960), but pictures showing clearly resolved pentamers and hexamers have not been published. One explanation is that herpesvirion pentamers may be unstable under conditions of negative staining. It would be expected that the predicted 12 pentamers and the 150 hexamers would differ with respect to structural subunits. However, the composition and molar ratios of the capsid proteins (Sect. 2.6.3) do not readily permit assignment of proteins to pentameric and hexameric structures. The hexameric capsomer appears to be 12.5 nm long. The end projecting outside the capsid has a diameter of 8.0–9.0 nm (Wildy *et al.*, 1960). A hole, 4.0 nm in diameter, runs through the long axis of the capsomer. The negative stain completely fills the capsomers broken off the capsid during the staining processing in a manner which suggests that the canal runs all the way through the hexamer. However, in the intact capsid, the hole fills only partly with negative stain, suggesting that in the intact capsid the hole is blocked at the proximal end. The electron-translucent space between the core and the capsid, the pericore, is not empty. The core remains in the center of the capsid even after prolonged centrifugation and pelleting (Gibson and Roizman, 1972). If the space were empty, it would be predicted that the core would be displaced toward one side. In some negatively stained capsids this space is filled with stain (Toplin and Shidlovsky, 1966). Also, in capsids treated with the nonionic detergent NP-40 for 30 minutes, the core loses its shape and can be seen lying next to the capsid which itself does not appear to be morphologically altered by the treatment (Abodeely *et al.*, 1970). Cook and Stevens (1970) report that after DNase digestion of thin sections, there is some dense material adhering to the inside of the capsid. The nature of the material in the pericore regions is unclear. Comparison of the protein composition of empty and full capsids suggests that, if this ma-

---

virions in the cytoplasm of HSV-1-infected cell. The sections were treated with EDTA to selectively remove the uranyl binding to DNA according to the procedure of Bernhard (1969).

terial is protein, it is probably a structural component of the capsid, arranged differently in empty capsids and in capsids containing DNA.

2.1.3. The Tegument

The tegument is defined as the structure located between the capsid and the envelope. A structure corresponding to the tegument was recognized by numerous authors (Morgan *et al.*, 1968; Wildy *et al.*, 1960; Schwartz and Roizman, 1969*a*) as a fibrous coat around the capsid. In some publications (Shipkey *et al.*, 1967; Heine *et al.*, 1971; Roizman, 1969) it was designated as an inner membrane, a misnomer since the tegument does not have a trilaminar unit-membrane structure. In thin section it appears as a layer of amorphous material (Fig. 2B,C), and in negatively stained virions it has a fibrous appearance (Fig. 1E,F). Although the amount of this material is variable, when the layer is complete the material is slightly denser than either the adjacent capsid or inner lamella of the envelope. This is true of unstained material as well as of sections stained with uranyl acetate or lead citrate, or both. The material comprising the tegument seems to be present in most virions, but the amount seen is variable from virion to virion, even in the same cell. Fong *et al.* (1973) recently reported that the tegument is absent in guinea pig herpes virions in the perinuclear space but is present in virions in cytoplasmic vacuoles or in the extracellular space in infected cultured guinea pig cells. The width of the layer varies considerably among the herpesviruses (Kazama and Schoenstein, 1973; Nayak, 1971; Strandberg and Carmichael, 1965); for example, in herpes simplex the layer is frequently interrupted (Nii *et al.*, 1968*a*) and in virions from Lucké adenocarcinoma the layer is generally complete (Stackpole and Mizell, 1968). In virions from Marek's disease virus (Nazerian and Witter, 1970) or from the herpesvirus of tree shrews (McCombs *et al.*, 1971), there is enough of this material to extend the size of enveloped particles to 250 or 260 nm in diameter (Fig. 1I,J). The extent of this layer may be determined by the virus strain. Herpes simplex grown in tree shrew fibroblasts does not show enlarged virions. However, the tree shrew herpesvirus produces enlarged virions in both human embryonic fibroblasts and rabbit kidney cells, indicating that the structure of the tegument is controlled at least in part by the virus.

2.1.4. The Envelope

The outermost structure of the herpesvirion is the envelope (Siegert and Falke, 1966; Falke *et al.*, 1959; Morgan *et al.*, 1954; Chopra *et al.*, 1970; Darlington and Moss, 1969; McCombs *et al.*, 1971) consisting of a trilaminar membrane with spikes projecting from its outer surface (Fong *et al.*, 1973; McCraken and Clarke, 1971; Wildy *et al.*, 1960). The envelope is impermeable to negative stain (Fig. 1D,F), but becomes permeable if damaged (Watson, 1968) (Fig. 1E,F). The virion envelope seems to be especially fragile (Cook and Stevens, 1970). Envelopes of virions are seen to be broken in thin sections of cells where the cellular membranes are well preserved. Manipulation of the virion results in striking alterations of the envelope, and "tailing" frequently results (Fong *et al.*, 1973; Nayak, 1971; Spear and Roizman, 1972) (Fig. 1D). The efficiency of envelopment as well as the stability of the envelope varies considerably among the herpesviruses (Nazerian and Witter, 1970; Schwartz and Roizman, 1969*b*; Underwood, 1972; Witter *et al.*, 1972; Gershon *et al.*, 1973). In the case of cytomegalovirus infection, intact enveloped particles are rarely seen.

2.2. Purification and Fractionation of Herpesviruses

2.2.1. General Considerations

Analyses of chemical and some biologic properties of herpesviruses requires purified virions. For many years the purification of herpesviruses presented difficulties which were surmounted in part only in recent years. Before discussing the purification of herpesviruses, it seems worthwhile to discuss the problems involved.

(1) The source of virus for purification is either the infected cell in which the bulk of the virus accumulates or the extracellular fluid into which a fraction of the virus is released as a consequence of autolysis of the cell and of active transport of the virus to the cell surface (Sect. 3). By definition, pure preparations are free of cellular and extracellular contaminants, both host and viral. Neither the fact that the virus bands in a gradient nor the observation that all its proteins label post-infection is evidence that the preparation is pure. Purity is never absolute; it must be defined with respect to a specific contaminant, i.e., host proteins, host DNA, or even viral proteins.

(2) Homogeneity is an important but usually ill-defined property of herpesvirus preparations. The problem stems from two considerations. First, the virion is structurally as well as functionally labile. Many preparations of "purified virus" which had been analyzed in great detail with respect to protein and amino acid composition consisted of undetermined mixtures of enveloped and unenveloped virus or of empty and full capsids. The second problem is more com-

plex and less amenable to solution. Under rare, and usually well-defined, conditions the particle-to-infectivity ratio approaches a limit of 10 or less. Usually, however, the particle-to-infectivity ratio for large batches is at least one order of magnitude higher. The question, obviously, is whether the composition of purified, homogeneous particles in such preparations reflects the composition of the infectious particles. This is a general problem encountered with all biologically active macromolecules but which acquires very special significance for viruses.

2.2.2. Virus Purification

The major contaminants encountered in the course of purification of the virus and its components are (1) host DNA, (2) host proteins and particulate structures such as ribosomes and mitochondria, and (3) membrane vesicles carrying both viral and cellular constituents. In general, purification of virions, subviral structures, and viral DNA with respect to host DNA presents no serious problem. The general procedures for purification of virions and subviral particles with respect to host proteins yield preparations free of host DNA. From the point of view of maximum recovery, viral DNA is best prepared from nucleocapsids (Kieff *et al.*, 1971). These can be obtained in two ways: following lysis of infected cell nuclei with deoxycholate and partial digestion of host DNA with DNase, or following NP-40, deoxycholate, and urea treatment of cytoplasmic or of extracellular virions (Gibson and Roizman, 1972). Consistently good preparations were obtained by banding the DNA released from purified nucleocapsids with sodium dodecyl sulfate (SDS) and sarkosyl in sucrose density gradients prepared in buffer at neutral *p*H (Kieff *et al.*, 1971).

The purification of herpesvirions presents several problems resulting from the presence of the envelope and the site of maturation and accumulation of the virus. Briefly, as discussed in Sect. 4, herpesviruses acquire an envelope predominantly from the inner lamella of the nuclear membrane, and accumulate in the perinuclear space and in the cisternae of the endoplasmic reticulum. Virus release from infected cells is inefficient and acquires momentum only after the infected cell disintegrates, spilling both virus and debris. In our hands, extracellular virions from cells in culture frequently contained a damaged envelope (permeable to negative stain), and it was not always a satisfactory source of virus for analysis. In addition, separation of intracellular virus from cellular organelles and particularly from membrane vesicles generated during the disruption of the cell presents special problems. The difficulty is compounded by the fact that the membranes of infected cells carry virus-specific proteins. Another problem is that infected cells contain not only virions, but also naked nucleocapsids in ratios which differ greatly from one cell line to another.

The procedure for the preparation of purified virions which has been most successful to date consists of three steps (Spear and Roizman, 1972). The particular object of this purification procedure is to separate enveloped nucleocapsids in the cytoplasm from cellular organelles and from membrane vesicles generated during cell disruption. Of the various contaminants, membrane vesicles are the most prominent and the most difficult to deal with. The first step consists of disruption of the cell and the separation of the cytoplasm from nuclei. Careful preparation of cytoplasm is essential and isotonic sucrose solutions are used to stabilize the nuclei and prevent leakage of nucleocapsids and nucleoproteins which tend to aggregate membranous structures. In general, if the nuclei are broken, the purification scheme will not be effective and the preparation is best discarded. The second step consists of rate zonal centrifugation of the cytoplasmic extract on Dextran 10 gradients. The centrifugation effectively separates virions from soluble proteins and most cellular membrane vesicles, which remain near the top of the gradient, and from aggregates of virions, cytoplasmic organelles, and large debris, which pellet. Dextran specifically is used here because of its low osmolarity; presumably, intact membrane vesicles stay distended rather than collapse and sediment more slowly than the virion which has a high density core. After centrifugation, virions are found in a diffuse, light-scattering band just above the middle of the tube. This band contains partially purified, largely enveloped virus and will be referred to in the text as "Dextran 10 virus." The purpose of the third step in the purification procedure is to separate the virions from the remaining contaminants, i.e., membrane fragments and small amounts of unenveloped nucleocapsids. There are several variations of the third step. The procedure which gives the purest virus is flotation of the partially purified virus through a discontinuous sucrose gradient. Membrane-containing contaminants float to a lower density and a higher position in the tube than the virions, whereas nucleocapsids, because of their high density, will pellet. Although this procedure yields almost exclusively enveloped virus, free of membrane contaminants, many of the particles exhibit damaged envelopes. The second variation of the last step is to centrifuge Dextran

10 virus through a potassium tartrate gradient instead of the discontinuous sucrose gradient. In this way the normal morphology of the virion is better maintained. A third procedure is to reband the Dextran gradient. It should be pointed out that the first two alternatives yield equivalent results in terms of protein recovered in the final product. The third procedure results in a reduction in the extent of contamination of virions with nucleocapsids but probably does not significantly lower the contamination of the preparation with membrane vesicles.

2.2.3. Monitoring the Purification Procedure

The most common techniques for monitoring the purification of herpesvirons are the determination of the extent of purification of some specific activity (infectivity, a capsid protein, etc.) relative to total protein mass, or the determination of the extent of purification of virus from either an artificial mixture of virus and labeled host debris or from infected cells which had been labeled with radioactive amino acids before infection. These techniques will obviously reflect the extent of purity of the virus relative to host proteins but not relative to other viral products, and in particular they are not absolute. There are no satisfactory techniques for monitoring the presence of traces of unique host DNA sequences, particularly if they are covalently linked to viral DNA.

The second comment stems from the fact that purity is not synonymous with homogeneity. Electron microscopy is useful in monitoring qualitative but not quantitative aspects of homogeneity. In general, intact virions are not permeable to negative stain and, hence, the presence of substantial amounts of particles penetrated by negative stain indicates that the procedures employed damage the envelope. The presence of collapsed membrane vesicles either empty or flooded with negative stain is a good indiction of contamination by cellular membranes. Their absence is less meaningful. The usefulness of negative stain in determining the presence of empty and full nucleocapsids is questionable since negative stain permeates both. Negative stain does not differentiate between the B and C forms of nucleocapsids, even though C capsids in our hands almost invariably contain substantial amounts of glycoproteins and yet lack a major protein of the B capsid.

Monitoring homogeneity of capsids presents numerous problems arising from the fact that, although at least 3 types of capsids are readily separated from infected cells, the three forms which differ in protein composition cannot be readily differentiated by examination following negative staining. Forms A and B correspond to full and empty capsids and are prepared from infected cell nuclei by the technique described by Gibson and Roizman (1972). Form C is derived by stripping the envelope off the virions. Forms A and B are readily differentiated by differences in the sedimentation rates in sucrose density gradients. Detergent treatment followed by fractionation of whole infected cells separates A capsids from B and C capsids, but not B capsids from C capsids.

2.3. Herpesvirus DNA

Although it has been long suspected that herpesviruses contain DNA, definitive evidence did not emerge until some 12 years ago (Russell, 1962; Ben-Porat and Kaplan, 1962). The literature on herpesvirus DNA has steadily increased in recent years, with some notable and striking observations. We shall consider the herpesvirus DNA from the points of view of composition, structure, and genetic relatedness.

2.3.1. Composition

Cumulative catalogues of the base composition of herpesviruses are shown in Table 1 (Roizman, 1969; Bachenheimer *et al.*, 1972; Plummer *et al.*, 1969; Goodheart and Plummer, 1974; Ludwig, 1972). There is basic agreement that herpesvirus DNA varies in base composition from 33 to 74 G+C (guanine plus cytosine) moles percent and three comments should be made in this connection. First, most studies have been done by equilibrium centrifugation of the DNA in CsCl solutions in a preparative ultracentrifuge, and, therefore, the obtained bouyant densities should be considered as approximate but not absolute values. Densities obtained by centrifugation in the analytical centrifuge with known standards, and suitably corrected for the conditions of centrifugation, are more reliable. However, many determinations made in the analytical centrifuge use DNA extracted from whole infected cells rather than DNA extracted from purified virus. This procedure will not generate much error in the case of DNAs of high G+C content from viruses producing large yields in the infected cells. However, these data are less convincing for the DNAs whose density approaches host DNA and for poorly growing viruses.

The second comment concerns the interpretation of the density measurements, and several recent findings significantly bear on this point. First, the presence of labeled markers, and in particular ^{14}C, may significantly increase the density of the DNA due to the difference in atomic weights of the substituted isotopes. Indeed, Cassai and Bachenheimer (1973) readily demonstrated differences in the buoyant density of HSV-1 and HSV-2 due to incorporation of ^{3}H- or ^{14}C-thymidine. Second, isopycnic centrifugation of some herpesvirus DNAs yields more than one band differing in density (Plummer *et al.*, 1969). Mosmann and Hudson (1973) reported that the double band observed for murine cytomegalovirus DNA is a consequence of extensive heterogeneity in the distribution of A-T and G-C base pairs in the DNA molecule which on fragmentation forms 2 populations differing in base composition.

Third, determinations of base composition on the basis of buoyant density assumes the absence of significant amounts of unusual bases which might throw such calculations off. In several instances (Kieff *et al.*, 1971; Graham *et al.*, 1972) the determinations of the base composition from buoyant-density determinations (Fig. 3) have been confirmed by determinations based on melting profiles of the DNA (Fig. 4).

The final comment concerns the significance of the observation that the base composition of herpesviruses exceeds by a wide margin the range of any other family of DNA viruses known. It has been suggested that because of the redundancy in the genetic code, the same

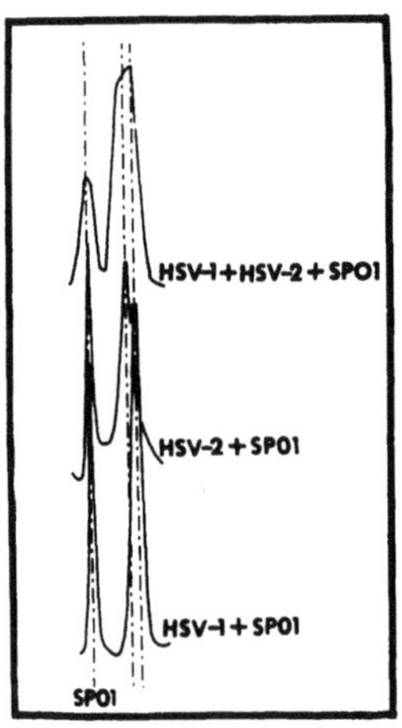

Fig. 3. Buoyant-density determinations of HSV-1 (F1) and HSV-2 (G) DNAs. The buoyant densities of these DNAs were determined by cocentrifugation with SPO1 DNA or with both SPO1 and micrococcal DNA in the Spinco Model E centrifuge. The UV-absorption photographs were scanned with a Joyce Loebl microdensitometer. Data from Kieff *et al.* (1971).

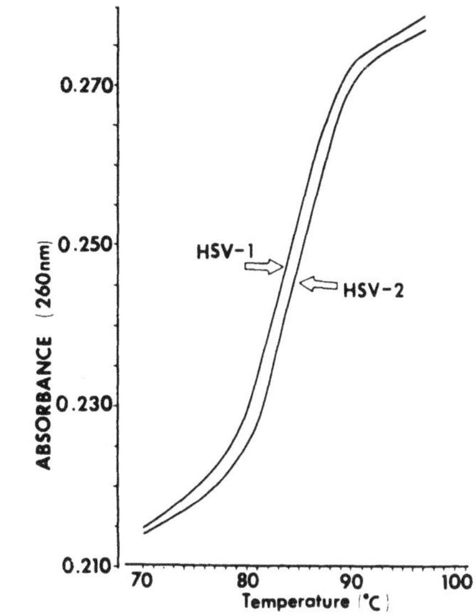

Fig. 4. UV-absorbance–thermal-denaturation profile of HSV-1 (F1) and HSV-2 (G) DNAs in 0.015 M sodium chloride and 0.015 M sodium citrate, measured in a Gilford Model 2000 spectrophotometer equipped with an automatic temperature and absorbance recorder. Data from Kieff *et al.* (1971).

protein could be readily specified by DNAs of both high and low G+C content (Goodheart, 1970). This does not of course explain the range. Differences in the base composition of the host DNAs also do not explain the range, particularly since there is no great variation in the base composition among animal cells and also because the base composition of human herpesviruses alone varies from 45 to 69 G+C moles percent. The main question facing us is whether herpesviruses have unique properties which permit base substitutions to fit unusual requirements of the cells in which they are grown. It is noteworthy that Cassai (unpublished data) found significant differences in the base composition of 2 HSV-1 strains, i.e., between the DNAs of the F strain which is maintained at a maximum of 4 passages in HEp-2 cells from its isolation, and the Schooler strain which has been passaged an indeterminate number of times in numerous cell lines. HSV-1 and HSV-2 DNAs which are closely related (see below) differ in their base composition by 2 G+C moles percent, corresponding to the substitution of 3500 adenosines or thymidines in HSV-1 or by cytosine or guanosine in HSV-2. Although in nature both HSV-1 and HSV-2 cause occasional generalized infections, the most frequent habitats of these two viruses are the face, including oral mucosa, and the urogenital tract, respectively. Do the base composition of herpesviruses reflect some peculiar characteristic of the differentiated cell in which they grow? The answer is unavailable, but potentially interesting.

2.3.2. Structure, Size, and Conformation

It is now generally agreed that herpesvirus DNAs are linear, double-stranded molecules, the strands of which fragment on heat or alkaline denaturation. Each of the points requires further amplification.

The evidence for the linearity of herpes simplex DNA is based on electron microscopic data (Becker *et al.*, 1968). The published sizes of the DNAs of herpesvirus as catalogued and annotated in Table 1 vary considerably, so much, in fact, that a few points should be made. In principle, there are several techniques for the determination of the absolute molecular weight of DNA molecules greater than 50×10^6, i.e., from enumeration of ^{32}P disintegrations on autoradiograms, band width in equilibrium sedimentation, measurements of the length of DNA spread by the Kleinschmidt technique, electrophoretic mobility, reassociation kinetics, and by a tedious indirect technique based on phosphate analysis and the molecular weight determination of the whole virion from equilibrium sedimentation and sedimentation diffusion analyses. However, it is unlikely that the last technique (Freifelder, 1970) will find wide application for the DNAs of enveloped viruses. Probably the best procedures are those involving comparisons by sedimentation of DNA molecules of unknown size with known DNA molecules of approximately the same size or those involving comparisons of DNA molecules of unknown size with known molecules of approximately the same base composition by gel electrophoresis, reassociation kinetics, or electron microscopy. Application of these techniques have not yielded uniform molecular weights for herpesvirus DNAs. Thus, HSV-1 DNA has been estimated to be 85 and 100×10^6 by cosedimentation with T4 DNA (Mosmann and Hudson, 1973, and Kieff *et al.*, 1971, respectively); 100×10^6 by electron microscopy (Becker *et al.*, 1968); 85×10^6 by band width (Wagner, personal communication); and 85 and 95×10^6 by reassociation kinetics (Wagner, personal communication, and Frenkel and Roizman, 1971, respectively). HSV-2 does not appear to differ in sedimentation properties from HSV-1 (Kieff *et al.*, 1971); on the other hand, Lee *et al.* (1971) reported that Marek's disease herpesvirus DNA sediments faster than HSV-1 DNA and has a molecular weight of $103 \pm 5 \times 10^6$. Mosmann and Hudson (1973) reported the largest molecular weight for any herpesvirus, i.e., 132×10^6 for murine cytomegalovirus DNA. It should be pointed out that even in this instance, it is not clear whether the differences between the size of the DNAs of MDV and HSV or for murine cytomegalovirus and HSV are real or whether they reflect differences in conformation due to widely different base composition (47 and 59 G+C moles percent for MDV DNA and murine cytomegalovirus, respectively, compared to 67–69 G+C moles percent for the DNAs of HSV-1 and HSV-2, respectively). Parenthetically, it is likely that not all T4 DNAs sediment at the same rates.

An interesting feature of herpesvirus DNAs is their fragmentation upon denaturation with alkali. This was first observed during studies on the DNAs of MDV (Lee *et al.*, 1971) and HSV (Kieff *et al.*, 1971) and subsequently extended to EBV DNA (Nonoyama and Pagano, 1972*a*). The observations consist chiefly of the following: HSV-1 DNA denatured with alkali and sedimented on alkaline sucrose density gradients formed several bands (Fig. 5). Statistical analysis of the

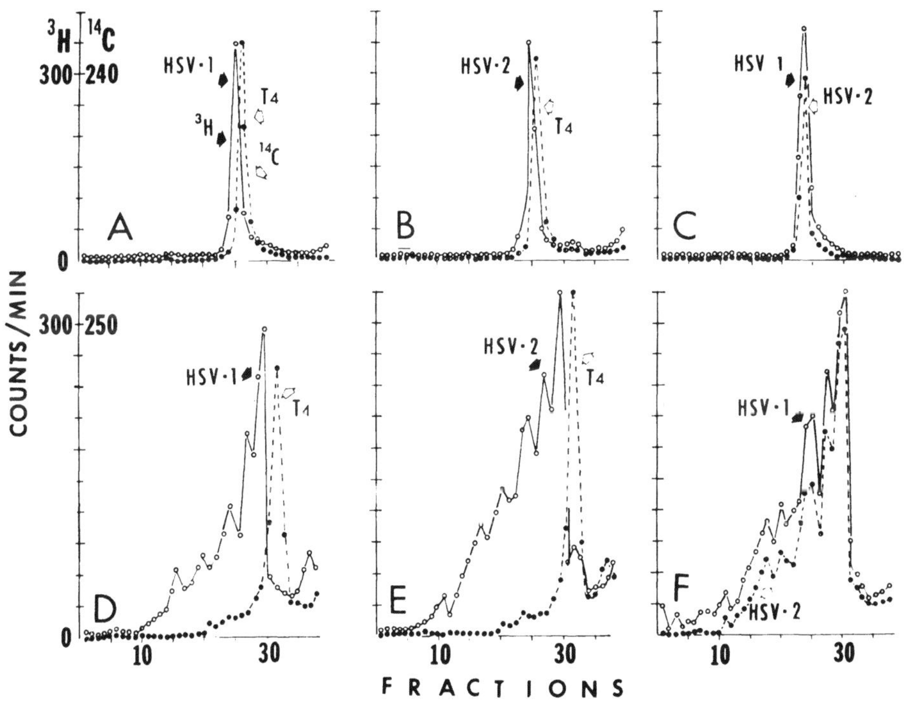

Fig. 5. Zone sedimentation of HSV DNA in neutral and alkaline sucrose density gradients. HSV-1 DNA and HSV-2 DNA were labeled with ^{3}H- or ^{14}C-thymidine and cosedimented with each other (C, F) and with T4 DNA (A, B, D, E). The DNAs were centrifuged for 3.5 hours in a SW41 rotor at 40,000 rpm and 20°C. (- - - -) ^{14}C-labeled DNA; (——); ^{3}H-labeled DNA. Direction of sedimentation is to the right. Details of the preparation and denaturation of the DNA, and of the neutral and alkaline sucrose density gradients are given in Kieff *et al.* (1971).

distances of sedimentation bands from numerous gradients revealed 7 clusters with mean sedimentation values ranging from 36 to 68 S and corresponding to DNA fragments of 9.8, 15.1, 21.1, 26.6, 32.9, 39.0, and 48 × 10^6 daltons (Frenkel and Roizman, 1972*a*) (Fig. 5). The most rapidly sedimenting band had a sedimentation constant and molecular weight corresponding to that of an intact strand. The reassociation of the DNA from the band containing the intact strand (Fig. 7), or from those containing fragments, was considerably slower and less complete than that of alkali-denatured, unfractionated DNA, indicating that each of the bands preferentially contained sequences from one strand (Frenkel and Roizman, 1972*a*). The data permitted reconstruction of six possible types of DNA duplexes differing in the position of the strand interruptions (Fig. 6). Two questions arise in connection with these data. The first question concerns the underlying cause of the fragmentation of the DNA in alkali. The second, and perhaps more readily answerable, question is whether advantage can be taken of the fragments of the DNA to sequence the genetic information along the viral DNA.

With respect to the first question, the fragmentation of the DNA could result from single-strand nicks, gaps, ribonucleotides and, possibly, cross-linking agents or endonucleases bound at specific sites. These are not mutually exclusive alternatives and, in fact, the cause of the fragmentation has not been completely resolved. What has been excluded with reasonable rigor is the presence of nonspecific nucleases in the virion (Kieff *et al.*, 1971; Bachenheimer *et al.*, 1972). As indicated in the Sect. 3.6., newly made DNA sediments as small fragments in alkaline gradients. Fragments approximating the size of intact DNA strands appear only after prolonged labeling. It is conceivable, therefore, that maturation of newly made DNA requires ligation and/or repair and that the fragmentation of the DNA in alkaline solu-

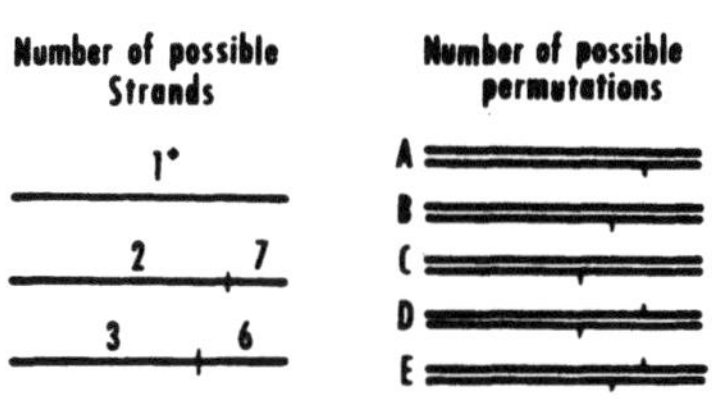

Fig. 6. Diagrammatic reconstruction of single strands and duplexes of HSV-1 (F1) DNA from the intact strand (band 1) and the various fragments contained in bands 2–7. The permutations shown describe types of duplexes without assigning polarity to the fragments. The number of permutations becomes 9 with the assigned polarity since molecules of the types D, E, and F could occur in two forms, depending on whether the fragmented strands have the same base sequences as the intact strand or the complementary sequences. Data from Frenkel and Roizman (1972*a*).

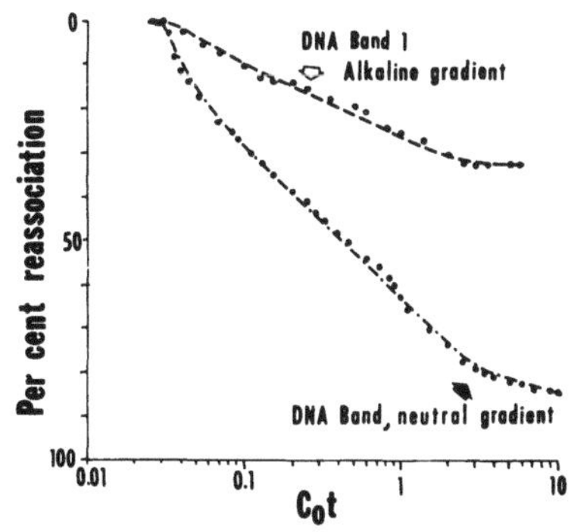

Fig. 7. Comparison of the self-reassociation of the DNAs in alkaline density gradient band 1 (intact strand) and the DNA banded in neutral sucrose density gradient. The DNAs were sonically disrupted in alkaline solutions to fragments of 5 S. The hybridization tests were done at 75°C in 0.3 M phosphate and were monitored in a Gilford recording spectrophotometer. The data were adjusted for 0.12 M phosphate according to the relationship given by Britten and Smith (1970) to enable comparison with the renaturation studies published previously (Frenkel and Roizman, 1971). Data from Frenkel and Roizman (1972*a*).

tions is a consequence of incomplete repair (Frenkel and Roizman, 1972*a*). Gordin *et al.* (1973) reported that HSV-1 DNA "denatured" with formamide consists of intact strands. The data, however, are not very convincing. The conclusion is based on the fact that in neutral sucrose density gradients the formamide-denatured DNA sediments "twice as fast" as native DNA. In fact, the formamide-denatured DNA forms a very broad band occupying nearly half the gradient. Moreover, the conditions for denaturation of the DNA (95% formamide, 0.1 × SSC, 37°C, 20 minutes) do not completely denature the DNA (S. Wadsworth and B. Roizman, unpublished data), and, lastly, the expected sedimentation rate of intact single-stranded DNA should be more than twice the sedimentation rate of native DNA. Regrettably, no markers of known size were included.

With respect to the second question, attempts to improve the separation of the intact strand and fragments by electrophoresis in polyacrylamide gels have been only partially successful. By comparison with T5 DNA, which forms discrete bands readily separated from each other, the HSV-1 DNA bands are less discrete, and, moreover, considerable amounts of DNA migrate in the space between the bands, indicating that the interruptions occur predominantly but not exclusively at unique sites. DNA fragments produced by restriction endonucleases *Hin* III and *Eco* R1 (Roizman *et al.*, 1974) appear to be much more promising for determinations of the topology of genetic information in HSV DNA.

Recent studies (Sheldrick and Berthelot, 1974; Wadsworth and Roizman, unpublished studies) indicate that the HSV-1 and HSV-2 DNA have both terminal and internal redundancy. The terminal redundancy demonstrated by the endonuclease digestion amounts to 2% of the

DNA molecule. The internal redundancy demonstrable after heat renaturation of intact strands resulted in the formation of barbells which consisted of double-stranded regions formed by the ends with adjacent internal regions of the DNA and single-stranded loops at both ends of the bar.

2.3.3. Studies on the Reassociation Kinetics of Herpesvirus DNA

The questions we are concerned with here are what is the length of the unique sequences in the DNA and does herpesvirus DNA have measurable amounts of repetitive sequences. The design of the experiment to yield the answers to these questions is to shear the DNA to fragments of uniform size, denature them, and monitor the reassociation by changes in the UV absorbance, changes in the susceptibility to digestion by single-strand-specific nucleases, or by separation of reassociated from nonreassociated DNA on hydroxyapatite columns. In principle, reassociation of DNA consisting of unique sequences should follow second-order kinetics, with a single rate constant according to the relationship $D_t/D_0 = 1/(1 + KC_0t)$, where D_t/D_0 is the fraction of DNA remaining single-stranded, K is the hybridization rate constant at the particular salt concentration and fragment size, t is time, and C_0 is concentration of DNA in mole/liter. A plot of D_0/D_t *vs.* t should yield a straight line with an intercept of 1. The reassociation of HSV-1 DNA, as monitored both by absorbance (Fig. 8) and by susceptibility to single-strand-specific nucleases, followed second order kinetics with a single rate constant. The calculated kinetic complexity (i.e., the sum of unique sequences) of the DNA according to Wetmur and Davidson (1968) was $95 \pm 1 \times 10^6$ daltons (Frenkel and Roizman, 1971), indicating that one molecule of $95–99 \times 10^6$ daltons in molecular mass contains all of the genetic information necessary for the production of infectious progeny in permissive cells.

The reassociation of HSV-2 DNA has been monitored with enzymes only. In initial studies, two preparations of HSV-2, isolated continents apart, appeared to differ from that of HSV-1 in that the intercept of the line relating D_0/D_t to t is substantially greater than 1. The data would suggest that HSV-2 DNA stocks used in these experiments have rapidly reassociating sequences amounting to more than 10% of the DNA (Frenkel *et al.*, 1972; Roizman and Frenkel, 1973). However, more recent studies with DNA obtained from plaque purified virus free of defective virus yielded intercepts of 1 (N. Frenkel and B. Roizman, unpublished data).

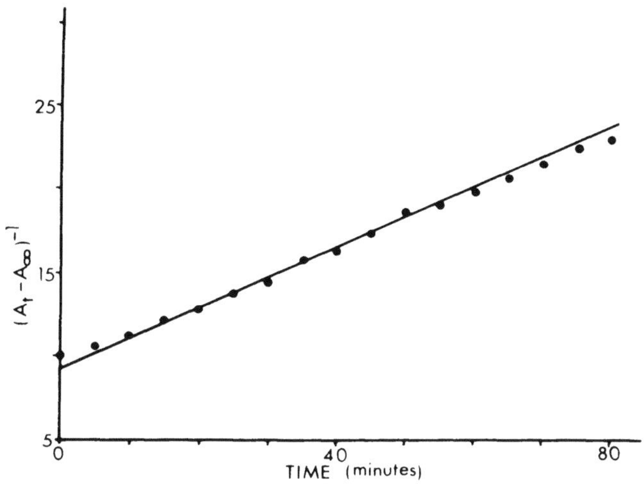

Fig. 8. Second-order rate plot for the renaturations of HSV-1 DNA in 1 x SSC at 70°C. The slope was calculated by the least square regression method to be (3.03 ± 0.04) × 10^{-3}/second, at the 95% confidence level. The value of k_2 is obtained from the following relationship $1/(A_t - A_\infty) = 2.72 \times 10^{-4}\, k_2 t + 1/0.27\, A_\infty$, where A_t is the 260 nm absorbance at time t, A_∞ is the absorbance of native DNA, k_2 is the second-order rate constant, and t is the time of renaturation. This equation was derived from the relationship given by Wetmur and Davidson (1968) with the correction for the effect of the base composition on the hyperchromicity. Here, k_2 derived from the slope 1 × SSC is 11.1 ± 0.1 liter/mole/second. The kinetic complexity (i.e., the size of the unique DNA), N_d, was calculated from the relationship given by Wetmur and Davidson (1968): $N_d = [5.5 \times 10^8 (s_{20,w}^{pH\,13})^{1.25}]/k_2$, where $s_{20,w}^{pH\,13}$ of the DNA fragments was determined by the alkaline band sedimentation to be 9.1 S, and 92 ± 1 liter/mole/second is the corrected value of k_2 for 1.0 monovalent cation concentration. The kinetic complexity for HSV-1 genome was found to be (95 ± 1) × 10^6 daltons. Data from Frenkel and Roizman (1971).

2.3.4. Genetic Relatedness among Herpesviruses

The genetic relatedness among viruses is generally measured by hybridization of DNA or RNA of one virus to excess DNA of another, either free in solution or fixed to filters. Demonstration of "relatedness" is generally simple; quantitation of the relatedness is not. The problem arises from two considerations. First, the reassociation rate constant is greatly affected by the extent of matching of base pairs, and even minor mismatching greatly favors homologous rather than heterologous reassociation (Southern, 1971; Sutton and McCallum, 1971). Ideally, therefore, one DNA in amounts so small that it cannot reassociate with itself during the allotted time should be hybridized to an excess amount of a second DNA fixed so that it is available for hybridization to the first DNA but not to itself. In our experience, immobilization of the second DNA to filters is not satisfactory since, for reasons poorly understood, the amount of DNA available for hybridization varies considerably from filter to filter, even though the total amounts of DNA fixed to the filters are identical. The second consideration arises from the use of RNA in place of DNA. If the transcription is asymmetric the use of RNA eliminates the problem of reassociation of at least one of the DNAs to itself. However, the RNA must be shown to contain all of the sequences present in the DNA; all of the sequences must be present in equimolar amounts; and, in addition, the kinetics of hybridization of RNA to the homologous DNA should be known. There have been numerous publications dealing with the relatedness among herpesviruses which might be summarized as follows:

(1) Quantitative hybridization tests involving two-phase systems (DNA in solution exposed to DNA bound to filters) revealed no significant hybridization between EBV DNA and HSV DNA (Zur Hausen and Schulte-Holthausen, 1970); between EBV DNA and DNA extracted from MDV (Zur Hausen *et al.*, 1970); between frog virus 4 (a herpesvirus) and the Lucké herpesvirus purified from a tumor (Gravell, 1971); between MDV and HSV-1 or -2 (Bachenheimer *et al.*, 1972); between varicella zoster and HSV-1 or -2 (Ludwig, 1972); between equine coital exanthema virus and HSV-1 (Ludwig, 1972); and between infectious bovine rhinotracheitis virus and pseudorabies (Ludwig, 1972). The relatedness of pseudorabies to HSV-1 and -2 was described as less than 10% (Ludwig, 1972).

(2) Two quantitative studies have been published. In the first, Ludwig *et al.* (1972*b*) measured hybridization of HSV-1 and HSV-2 DNA in solution to homologous and heterologous DNA on filters. In spite of the way the points were connected, the hybridizations did not reach completion. The authors concluded that 68% of the sequences of HSV-1 and HSV-2 were homologous. Evaluation of these data is difficult, chiefly because the size of the DNA in solution is not given. Large fragments containing both homologous and heterologous sequences would hybridize as homologous sequences. In the second study, Kieff *et al.* (1972) measured the relatedness in two ways. First, mixtures of labeled HSV-1 and HSV-2 DNA in solution were hybridized to DNA on filters and the initial rates of hybridization were measured. In the second experiment, mixtures of labeled HSV-1 and HSV-2 DNA were added to a 50-fold excess of unlabeled DNA in so-

lution and allowed to reassociate until the homologous reassociation was nearly complete. The reassociated and nonreassociated DNAs were then separated on a hydroxyapatite column and the thermal elution of the DNAs was measured (Fig. 9). The study concluded that 47% of the HSV-1 and HSV-2 DNA were homologous with 85% matching of base pairs. In the latter study, the DNA fragments were approximately 400 nucleotides long. A striking feature of the reassociation was the absence of both completely matched and very poorly matched heteroduplexes even though the temperature of hybridization should have permitted formation of both. This is probably a minimal estimate chiefly because the mismatching of base pairs favors homologous reassociation and poorly matched sequences would have little or no chance to hybridize. Based on these considerations, it seems reasonable to conclude that at least 47% of the sequences of HSV-1 and HSV-2 are homologous with minimal mismatching of base pairs, and that the remaining 53% either have no measurable homology or that the matching of base pairs is too poor to have been detected.

(3) Two studies have employed DNA–RNA hybridization tests to measure relatedness. In one, Bronson *et al.* (1972) hybridized labeled RNA extracted from HSV-1- and HSV-2-infected cells at 16 and 18 hours post-infection, respectively, to DNA on filters. The authors reported homology amounting to 40% for HSV-2 RNA hybridized to HSV-1 DNA and 14% for HSV-1 RNA hybridized to HSV-2 DNA. In this test, HSV-1 and HSV-2 RNA hybridized to infectious bovine tracheitis DNA with efficiencies of 9 and 4%, respectively; to pseudorabies DNA with efficiencies of 2 and 6%, respectively; and to *Escherichia coli* DNA with efficiencies of 3 and 3%, respectively. The data are difficult to interpret; as stated at the beginning of this discussion, it is not clear just which RNA sequences have been labeled and are present in the hybridization mixture. In the second study (Frenkel *et al.*, 1973), excess unlabeled RNA extracted from HSV-1-infected cells at 8 hours post-infection was hybridized in solution with amounts of labeled HSV-1 and HSV-2 DNA too small to allow self-renaturation under conditions used. The hybridization was monitored by susceptibility to single-strand-specific nucleases, as discussed in more detail in Sect. 3.3. The 8-hour RNA contained viral RNA sequences complementary to 48% of viral DNA, i.e., to 96% of the genome, assuming transcription is asymmetric. The data showed that HSV-1 RNA hybridized to only 24% of the HSV-2 DNA, and, if the same assumption holds, the RNA was complementary to 48% of the HSV-2 genome. These data reinforce the conclusions of the

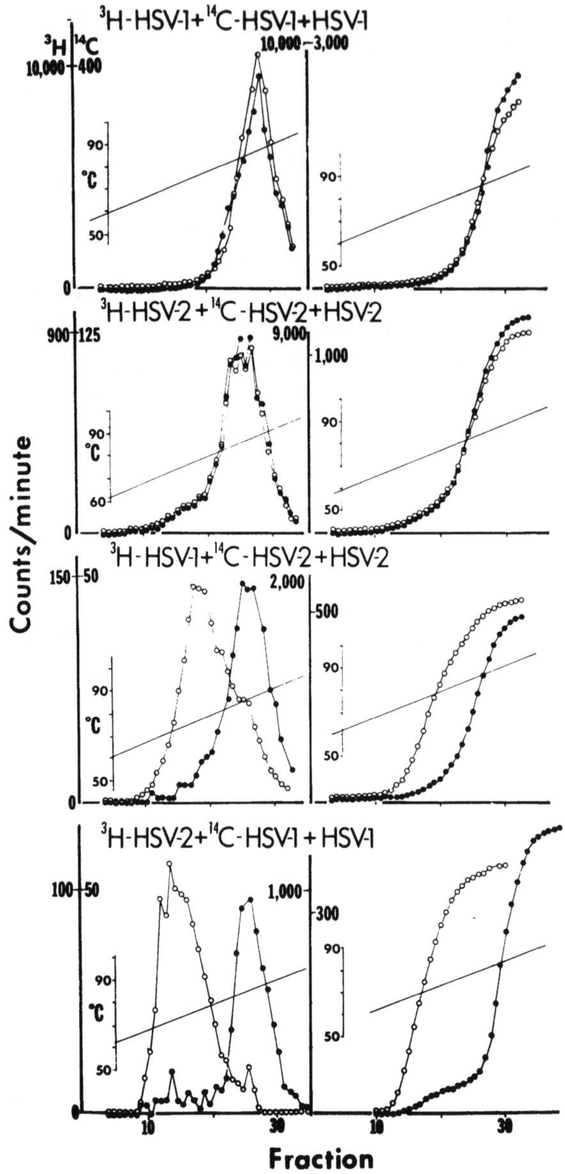

Fig. 9. Thermal elution of HSV-1 (F1) and HSV-2 (G) homo- and heterohybrid DNA from hydroxyapatite columns. Fragmented labeled HSV-1 or HSV-2 DNA were hybridized with a large excess of either HSV-1 or HSV-2 DNA. The hybrid was then absorbed to the hydroxyapatite. The DNA absorbed to the column (homo- and heterohybrid DNA) was then eluted with 0.14 M phosphate buffer containing 8 M urea and a thermal gradient (0.25°C/min). The fractions were assayed for radioactivity. Note that the heterohybrid elutes at a lower temperature than the homohybrid. Left panels: cpm/fraction. Right panels: cumulative elution of DNA as a function of temperature. (O——O) ³H-labeled DNA; (●——●) ¹⁴C-labeled DNA. Data from Kieff *et al.* (1972).

DNA–DNA hybridization studies of Kieff *et al.* (1972) that 47–48% of the HSV-1 and HSV-2 DNA sequences are homologous with good matching of base pairs.

2.4. Herpesvirus Proteins

2.4.1. General Considerations

In the past several years numerous papers have reported on the number, characteristics, and topology of the proteins in the virion. Although they undoubtedly represent an enormous amount of work, their significance is limited for one or more of 3 reasons, i.e., (1) the purity of the preparation was not adequately documented, (2) the preparation consisted of naked and enveloped nucleocapsids and hence was inhomogeneous, or (3) the method chosen to separate and display the viral proteins lacked adequate resolving power. The first two points have already been discussed in the section on virus purification (Sect. 2.2). With respect to the third point, the most convenient and most rapid separation of the proteins is by electrophoresis in SDS-polyacrylamide gel. Two aspects of this technique should be considered further. First, the technique of Davis (1964) involved electrophoresis of proteins soluble in aqueous solutions in polyacrylamide gels without SDS, and the separation obtained in this fashion is due to net charge as well as to the secondary and tertiary structure of the protein. Summers *et al.* (1965) significantly contributed to this technique by denaturing the proteins with SDS prior to their electrophoresis and by incorporating SDS into the buffer and gels. This made it possible to separate proteins insoluble in aqueous solutions and to render the separation dependent on size only, i.e., independent of net charge and of the secondary and tertiary structure in their native conformation. However, the changes introduced in the original techniques at the same time reduced the resolving power of the gels. Laemmli (1970) and Dimmock and Watson (1969) independently modified the techniques of Davis (1964) to the extent of adding SDS, and thereby retained much if not all of its resolving power. The second comment relates to the visualization of the proteins in the gel. Fundamentally two techniques are available. The first is to stain the whole gel with a dye such as Coomassie brilliant blue or fast green, which show absorbances following Beer's law and binds to proteins in proportion to their concentration. This technique displays all the proteins present in the gel. The second technique is most commonly used to differentiate between proteins made at different times either pre- or post-infection by radiolabeling. Assays of radiolabeled proteins by slicing gels and counting the amount of radioactivity in each slice yield a display whose resolving power is roughly equivalent to the resolving power of the gels of Summers *et al.* (1965). A less-sensitive method, but one which retains the resolving power of the Laemmli (1970) and Dimmock and Watson (1969) gels, is to subject the gels to autoradiography (Fig. 10). Both the stained gels and the autoradiograms may then be scanned with nearly perfect superimposition of bands. The autoradiographic method is most successful with ^{14}C-, ^{35}S-(methionine), or ^{32}P-labeled proteins, and generally is not satisfactory with ^{3}H-labeled material.

2.4.2. The Virion Proteins

At the time of writing, analyses of herpesvirus protein preparations conforming at least in part with the requirements set forth in the preceding section are available only for several strains of HSV-1 passaged a limited number of times in cell culture and for HSV-1 strains with a history of numerous passages in cell culture. To some extent the current designations are based in part on historical accident and should be amplified as follows:

(1) Initial studies of the number of species of virion polypeptides were based on electrophoretic separations on polyacrylamide gels cross-linked with methylene(bis)acrylamide (Spear and Roizman, 1972). These studies revealed 24 bands numbered consecutively as virion proteins (VP1–24), from the slowest (VP1) to the most rapidly migrating band (VP24). Some of the bands, such as VP1 and 2, were resolved only on gels prepared from low concentrations of acrylamide. It soon became apparent, however, that the 24 bands represented a minimum number of polypeptides. Thus, in subsequent studies it emerged that band 19 consisted of at least 2 polypeptides—a nonglycosylated polypeptide, designated VP19C, contained in the capsid and a glycosylated polypeptide, designated VP19E, contained in the virion, presumably in its envelope. Another polypeptide was also noted in region between VP6 and 7. However, the major source of disaffection with this technique emerged from attempts to determine which of the polypeptides were glycosylated. Although stained gels and autoradiograms of ^{14}C-labeled polypeptides showed discrete bands, the autoradiograms of electrophoretically separated polypeptides labeled with ^{14}C-glucosamine showed 3 wide bands covering roughly the regions occupied by VP7–8, 12–14, and 17–19, respectively (Spear and Roizman, 1972). The slowest band was skewed slightly toward VP8. In

view of the wide bands formed by the ^{14}C-glucosamine-labeled polypeptides, it was impossible to ascertain precisely whether only a few or all of the polypeptides in those regions were glycosylated (Fig. 11).

(2) In attempts to obtain better resolution of the polypeptides contained in the virion, the polypeptides were subjected to electrophoresis on polyacrylamide gels cross-linked with N,N′-diallyltartardiamide (DATD) according to a slight modification of the procedure described by Anker (1970). In DATD gels, virion polypeptides formed at least 33 bands (Fig. 10). The major differences were: (1) VP7 and 8 formed 3 bands. Of these, the more highly glycosylated band (VP8) migrated more slowly in DATD cross-linked gels than the glycosylated VP7. (2) The band seen between VP6 and VP7 in methylene(bis)acrylamide gels comigrated with VP8 and was seen only in electropherograms of virion polypeptides obtained from HSV-1 strains lacking VP8 (Sect. 2.8). (3) A considerable gain in resolution was observed in the region of polypeptides VP9 and VP18.

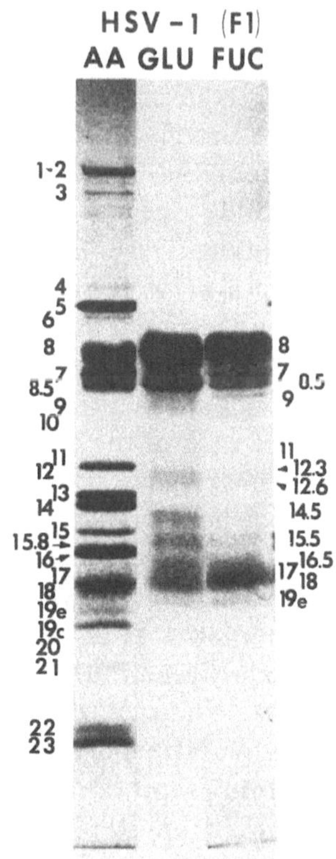

Fig. 10. Autoradiograms of polyacrylamide gels containing electrophoretically separated HSV-1 (F1) polypeptides labeled with ^{14}C-amino acids (left), ^{14}C-glucosamine (middle), or with ^{14}C-fucose (right). The polypeptides labeled with ^{14}C-amino acids are numbered on the left, the glycosylated polypeptides on the right. Although the three viral preparations were subjected to electrophoresis on the same gel slab, because of the differences in the intensities of the autoradiographic image, the autoradiographic image of each gel was printed at different exposure time. Data from Heine *et al.* (1974).

Specifically DATD cross-linked gels separated several polypeptides which contained little or no glucosamine or fucose from highly glycosylated polypeptides. Moreover, VP16 was found to contain at least 2 polypeptides migrating close together. (4) Polypeptides VP19E and VP19C, which could not be resolved in methylene(bis)acrylamide gel, were readily resolved in DATD cross-linked gels (Figs. 10, 11).

(3) The finding of these new bands raised several questions concerning both the relationship of the polypeptides to each other and their nomenclature. The problem arises from the fact that if all polypeptides currently resolved consist of a unique amino acid sequence, it would follow that the HSV-1 virion consists of at least 33 polypeptides. Although there is no formal objection to any number of polypeptides, provided it does not exceed the potential maximum information content of viral DNA, we must consider the possibility that the virion contains inhomogeneously modified polypeptides. The modifications could be the result of cleavages of precursor polypeptides, glycosylation, phosphorylation, etc. Although rapid post-translational cleavages have not been seen (Honess and Roizman, 1973), some polypeptides appeared to undergo modification resulting in an altered electrophoretic mobility. It is conceivable that both precursor and cleavage products are present in the virion or that the DATA cross-linked gels discriminate better than the methylenebis(acrylamide) gels between the glycosylated polypeptide and its nonglycosylated precursor. From this point of view, and particularly since the number of polypeptide species is at least 33 and could be still higher, the polypeptides were not renumbered, but rather, the newly resolved polypeptides were designated with decimal numbers intermediate between the previously recognized polypeptides migrating more slowly and more rapidly, respectively.

It should be stressed that the fundamental issue raised here, i.e., whether each of the polypeptide species currently detected in the virion is a primary gene product, remains unresolved. The proposed alternative that both precursor and product are present in the virion are not wholly satisfactory for they raise the question why the modifications, i.e., cleavage, glysocylation, etc., are not carried to completion. Since the alternative is as yet undocumented, a discussion of this question is presently premature.

(4) The size of virion proteins may be estimated from their mobilities on SDS-polyacrylamide gels since, in general, the distance of migration is proportional to the logarithm of the molecular weight (Shapiro *et al.*, 1967). Injudicious application of this relationship has led to erroneous estimates, and two points are worth making. First, the

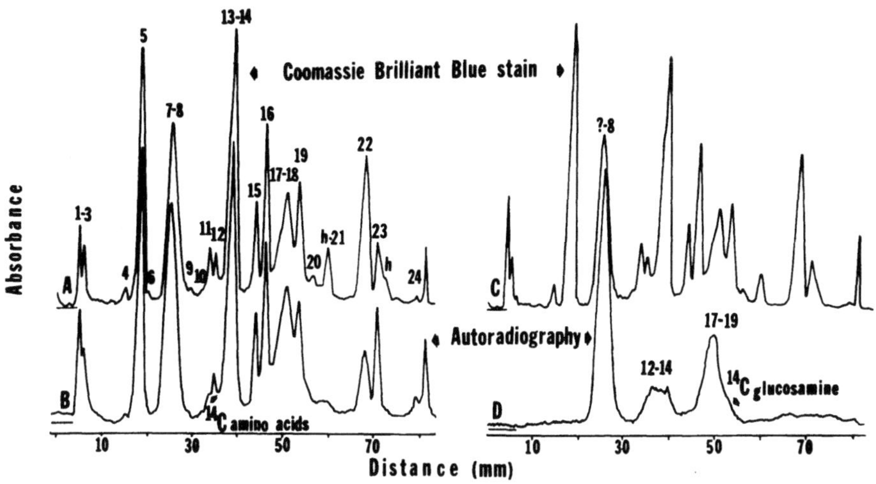

Fig. 11. Electropherogram of the proteins and glycoproteins of enveloped nucleocapsids of HSV-1. HSV-1 enveloped nucleocapsids were purified from the cytoplasm of infected cells by banding in a Dextran 10 gradient followed by flotation in a discontinuous sucrose gradient. The infected cells were incubated in medium containing ^{14}C-labeled amino acids or ^{14}C-glucosamine from 5 to 24 hours post-infection, i.e., after the complete shutdown of all host protein synthesis. The solubilized viral proteins were subjected to electrophoresis in discontinuous acrylamide gels cross-linked with methylene(bis)acrylamide and containing sodium dodecyl sulfate. After electrophoresis, the gels were stained with Coomassie brilliant blue stain, scanned with a Gilford spectrophotometer, dried, and then placed in contact with X-ray film. The autoradiographic image on the X-ray films was scanned on a Joyce-Loebl spectrophotometer. The electrophoresis was from left to right. The numbered protein bands were labeled only after infection. They remain in constant ratio during purification. The protein bands designated with the letter *h* were identified as cellular by the fact that these bands were labeled only when the cells were incubated in radioactive medium prior to infection or when labeled, uninfected cell cytoplasm was mixed with unlabeled, infected cell cytoplasm prior to purification. The cellular proteins are probably not virus constituents as they are separable from the virus in the final stages of purification. The line under the letters A–D indicates the base line for the absorbance measurements. The unnumbered band at the extreme right consists of material migrating with the buffer front. (A) Absorbance of stained bands of purified viral proteins labeled with ^{14}C-amino acids. (B) Autoradiographic tracing of the same gel. (C) Absorbance of stained bands of purified viral proteins labeled with ^{14}C-glucosamine. (D) Autoradiographic tracing of the same gel. In this and all following electropherograms, the position of the base line is indicated by a short segment at the extreme left of the profile. Data from Spear and Roizman (1972).

relationship of the distance of migration to the logarithm of the molecular weight is not always linear, and serious deviations may occur particularly in the region of the gel containing proteins greater than 100,000 daltons in molecular weight (Weber and Osborn, 1969). A simple way to both check the linearity of the relationship and to establish the molecular weights of the individual polypeptides is to coelectrophorese them with virion polypeptide proteins of known molecular weights which span the molecular weights of the viral proteins. As shown in Fig. 12, the relationship was linear for only one of the gels tested; the molecular weights of viral proteins as determined directly from this gel and extrapolated from the others are generally in agreement and indicate that the viral polypeptides range in molecular mass from 25,000 to 275,000 daltons. The second point concerns glycoproteins specifically; it has been reported that at least one glycoprotein (Bretcher, 1971) migrates anomalously in gels, in that its mobility is dependent on the concentration of polyacrylamide in the gels and differs from that predicted for its molecular weight estimated by another technique. The fact that the migration of most herpesvirion proteins in the gels cross-linked with methylene(bis)acrylamide is independent of the concentration of the polyacrylamide does not obviate the possibility that the migration of one or more of the glycoproteins

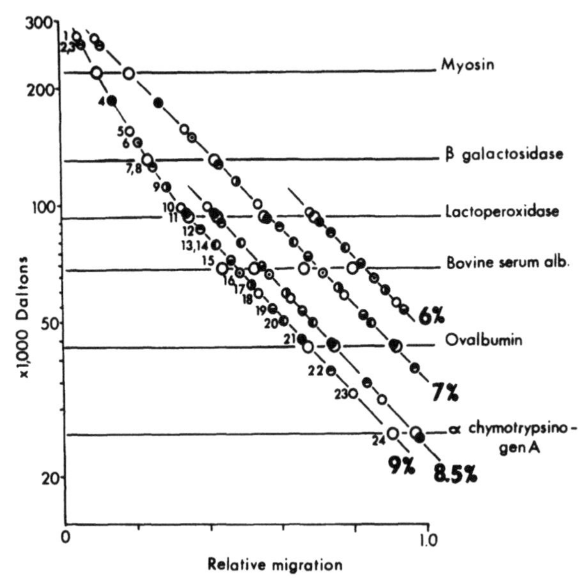

Fig. 12. The molecular weights of HSV-1 virion proteins graphically determined by their migration rates relative to known protein standards in 6, 7, 8.5, and 9% polyacrylamide gels. The large open circles represent the protein standards of known molecular weight. The small circles represent viral proteins as numbered on the left.

does not accurately reflect its molecular weight. In point of fact, whereas glycoprotein 8 migrates more rapidly than glycoprotein 7 in gels cross-linked with methylene(bis)acrylamide, it migrates more slowly than VP7 in gels cross-linked with DATD. Under the circumstances, the estimation of size based on their migration might well be wrong, reflecting the size of nonglycosylated proteins migrating at the same rates.

It should be noted that viral polypeptides differ widely in size and, moreover, a surprisingly large number of polypeptides appear to be greater than 100,000 daltons in molecular mass. Attempts to dissociate the high-molecular-weight polypeptides by reduction and alkylation in the presence of 6 M urea were not successful (Spear and Roizman, 1972), suggesting that they are, in fact, single polypeptides and not aggregates. '

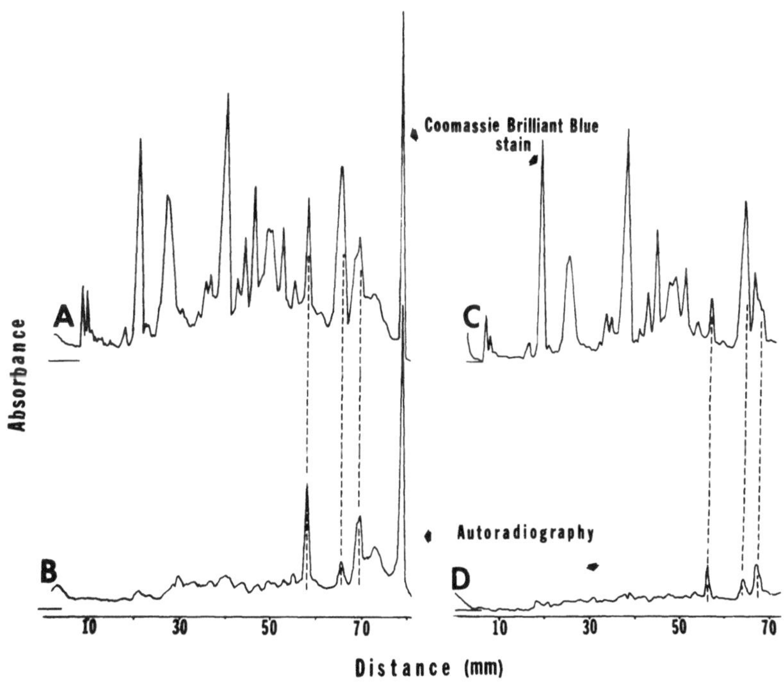

Fig. 13. Electrophorograms of partially purified (A and B) and purified (C and D) HSV-1 (strain F1) virus prepared from HEp-2 cells which had been incubated with ^{14}C-labeled amino acids for 48 hours prior to infection. The proteins were analyzed in cylinders of 8.5% acrylamide. Absorbance profiles of Coomassie brilliant blue-stained gels and of autoradiograms prepared from the same gels are shown. The dashed lines mark the positions of the most prominent cellular proteins. Data from Spear and Roizman (1972).

(5) As far as we can tell, host proteins are not structural components of the virion. This conclusion is based on analysis of the virion proteins in the course of purification. Figure 13 compares the amounts of residual host proteins labeled for 4–8 hours before infection in partially purified (Dextran 10 fraction) and in highly purified virus. The striking observation is that the amount of host proteins does not bear a constant ratio to the amount of viral proteins. In fact, host proteins appear to be associated with fragments of infected cell membrane which are separated only with difficulty from the virion. The point is of interest in relation to the observation that HSV-1 virions are readily agglutinated by sera made against host cells (Watson and Wildy, 1963), and it suggests that the antigens reacting with these sera are nonprotein constituents, either free or linked to viral proteins.

2.4.3. Capsid Proteins

As pointed out already, three kinds of capsids of HSV-1 and HSV-2 have been isolated, and they have been designated forms A, B, and C, corresonding to empty and full capsids extracted from nuclei prior to envelopment and full capsids derived by stripping the envelope off virions. The electropherograms of the proteins of the three forms are shown in Fig. 14, along with those of the virion. Analysis of the A, B, and C capsids showed the following: First, form B and form C

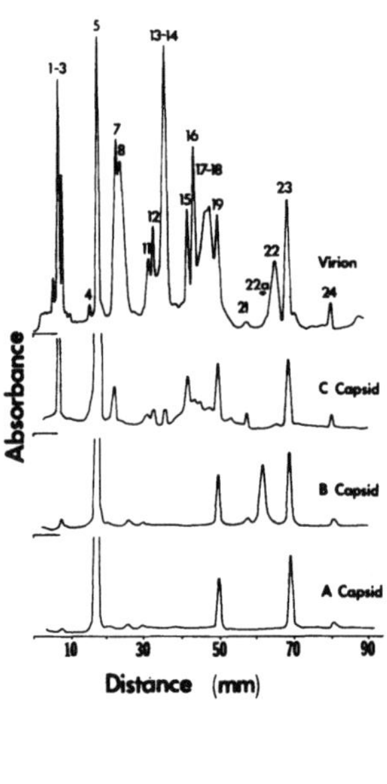

Fig. 14. Electrophorograms of the protein components of enveloped HSVs isolated by the procedure of Spear and Roizman (1972). C capsids, derived from enveloped virions by detergent treatment, as described by Gibson and Roizman (1972); B capsids, containing viral DNA and recovered from the nuclei of infected cells; and A capsids, recovered from the nuclei of infected cells and containing no DNA. The absorbance profiles of the virion, A capsids, and B capsids were made from autoradiograms. The C-capsid profile was made from a gel stained with Coomassie brilliant blue stain. Details of the techniques used to isolate, solubilize, and electrophorese these virus forms have been presented elsewhere (Gibson and Roizman, 1972). Data from Gibson and Roizman (1973).

both contain DNA of 100 × 10^6 molecular weight. Preparations of A capsids have only 10% of the DNA content of B capsids, and they band at a lower density. This DNA is also of 100 × 10^6 molecular weight, indicating that one in ten A capsids contains DNA. Why these "full" A capsids band with the empty ones is not entirely clear. The extent of contamination of B capsids with empty nucleocapsids is also not known. Examination of negatively stained preparations can be deceiving, but both negatively stained preparations and thin sections of pelleted material suggest that fewer than 1 B capsid in 10 or 15 lacks a core. Second, the three types of capsids differ in protein composition. All 3 forms contain proteins 5, 19, 23, and 24. In addition, B capsids contain protein 21 and a protein (22a) which is absent from the virion. The C capsids contain proteins 5, 19, 21, 23, and 24, in addition to proteins 1–3 and trace amounts of some glycoproteins. As documented elsewhere (Spear and Roizman, 1972; Gibson and Roizman, 1972) removal of the envelope by nonionic detergents with or without urea and deoxycholate does not completely remove all traces of glycoproteins.

2.4.4. Chemical Composition of Virion Proteins

Very little is known concerning the chemical composition of virion proteins, but what little there is deserves attention. Dreesman *et al.* (1972) reported the amino acid composition of HSV capsids obtained by treatment of cell extracts with nonionic detergents, then with 1% deoxycholate and 1% Tween 40, and finally by centrifugation in a sucrose density gradient containing 2 M urea. Dreesman *et al.* compared their data, incidentally showing only minor differences between HSV-1 and HSV-2, with the amino acid composition of equine abortion virions published by O'Callaghan *et al.* (1972) though the substantive difference between the types of materials being compared is not mentioned. Unfortunately, the nature of the material being analyzed, i.e., the extent of contamination of C capsids with A and B capsids and with residual glycoproteins, is not described. Thus the significance of these data is limited.

Several proteins of both capsid and envelope appear to be phosphorylated (Gibson and Roizman, 1974). The major phosphorylated proteins appear to be proteins 6, 12, 13, 14, 16, and especially 22.

The composition of polysaccharide chains in the glycoproteins is not known. Six constituents, i.e., glucosamine, galactosamine, fucose, mannose, sialic acid, and galactose, have so far been identified, but this list is incomplete.

2.4.5. Quantitation of Viral Proteins

In principle, there are several techniques for the quantitation of viral proteins. The simplest method, but also one fraught with many dangers, is based on two determinations; first, the overall ratio of protein to DNA, and, second, the ratio of the proteins to each other. The first determination relies on the assumption that only one molecule of DNA is present per virion, which is probably correct (Frenkel and Roizman, 1971; Furlong *et al.*, 1972), but its accuracy is largely dependent on the homogeneity of the preparation. The second determination, if made on the basis of electrophoretic separation in polyacrylamide gels, assumes that each band contains only one polypeptide, and its accuracy is further dependent on measurements of protein content which are indirect and subject to error. Briefly, two methods of quantitation of viral proteins in gels are available. The first is based on measurements of dyes such as Coomassie brilliant blue or fast (FCF) green bound to the proteins in the bands. The inherent assumption is that the affinity of the dye is uniform and proportional to overall amino acid content. This assumption is not entirely defensible in the frame of reference of herpesvirus proteins. The problem is exacerbated by the fact that some stained proteins destain more rapidly than others (Fig. 15) (Gibson and Roizman, 1974) possibly because chemical changes in the protein during the destaining process alter its net charge. The second method is based on the quantitation of proteins on the basis of their radioactivity following labeling of the infected cells with a mixture of amino acids of uniform specific activities. The errors in this determination arise from two sources. In the first source, a protein may appear to be in a lower concentration than it really is if it contains a disproportionately high amount of an amino acid not present in the mixture. The second source of error arises from the fact that structural proteins are not made uniformly at the same time. A specific example is the HSV-1 (F) proteins 1, 13–14, and 22, which are labeled more intensively in cells labeled continuously beginning 2 hours post-infection than in cells labeled continuously beginning 5 hours post-infection (Fig. 16) (Gibson and Roizman, 1974), although their relative amounts of proteins, as judged by the absorbance of the stain, were the same in both preparations.

With these qualifications in mind, the available data may be summarized as follows:

(1) Lampert *et al.* (1969) reported the dry mass of the virion, the empty and full nucleocapsids, and the core. The determinations were based on the principle that in the electron microscope an object's mass—in this instance the dry weight—is inversely proportional to its

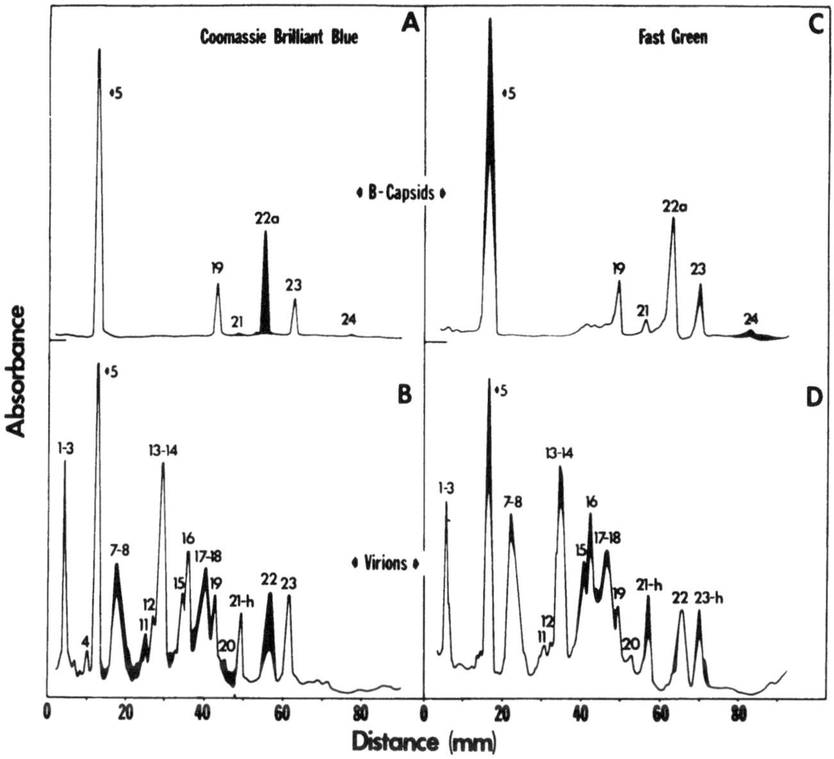

Fig. 15. Differential loss of Coomassie brilliant blue (CBB) and fast green (FG) stains from HSV B-capsid and virion proteins. HSV-2 B capsids and HSV-1 virions were isolated, solubilized, subjected to electrophoresis, and stained, wither with CBB or FG, as described by Gibson and Roizman (1972). Shown here are tracings comparing (A) B-capsid proteins stained with CBB and scanned after 2 days and 5 days of destaining (absorbance of peak 5 was 2.50 and 1.00 OD units, respectively); (B) virion proteins stained with CBB and scanned after 2 days and 6 days of destaining (absorbance of peak 5 was 1.30 and 0.50 OD units, respectively); (C) B capsid proteins stained with FG and scanned first after destaining at *p*H 3.0 and again after destaining at *p*H 6.0 (absorbance of peak 5 was 1.00 and 0.45 OD units, respectively); and (D) virion proteins stained with FG and scanned after destaining at *p*H 3.0 and again after destaining at *p*H 6.0 (absorbance of peak 5 was 0.30 and 0.15 OD units, respectively). The shaded portion in each panel represents the amount of stain lost during the second period of destaining. Two host proteins (Spear and Roizman, 1972) are designated by the letter *h*. Electrophoresis was from left to right in this and subsequent figures, and the position of the base line in the top panels is indicated by a short line at the extreme left of the profile. Data from Gibson and Roizman (1974).

permeability to the electron beam. The mass can then be determined by measuring the amount of visible light transmitted through the electron micrograph in reference to some known standard, provided that certain operating conditions of the microscope and of the photographic processing are fulfilled. Lampert *et al.* reported that the

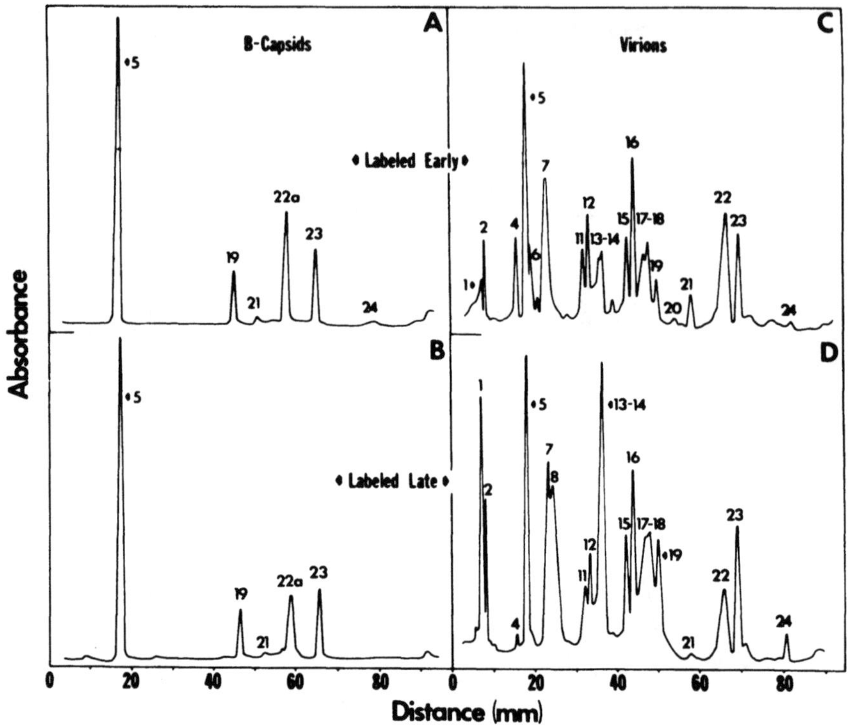

Fig. 16. Electrophorograms of HSV B-capsid and virion proteins labeled with ^{14}C-amino acids at different times during the reproductive cycle. HSV-2 B capsids were isolated from cells labeled with ^{14}C-amino acids (1 μCi/ml) beginning at 2 or 5 hours after infection. HSV-1 virions were isolated from cells labeled with ^{14}C-amino acids (1 μCi/ml) during the interval 2–4 hours after infection, and from cells labeled during the interval 8–20 hours after infection. These preparations were solubilized; subjected to electrophoresis, and autoradiograms made from each. Absorbance profiles of the autoradiograms are shown as follows: (A) B capsids labeled beginning at 2 hours after infection; (B) B capsids labeled beginning 5 hours after infection; (C) virions iabeled 2–4 hours after infection; and (D) virions labeled 8 to 20 hours after infection. The absorbance of peak 5 in these tracings was (A) 0.60; (B) 1.30; (C) 1.40, and (D) 2.00 OD units. Data from Gibson and Roizman (1972).

dry mass of virions, full nucleocapsids, empty nucleocapsids, and cores were $13.33 \pm 2.56 \times 10^{-16}$ g, $7.55 \pm 1.11 \times 10^{-16}$ g, $5.22 \pm 1.10 \times 10^{-16}$ g, and $2.07 \pm 0.95 \times 10^{-16}$ g, respectively. Based on these determinations, the average mass ratios of the protein to DNA in the virion, nucleocapsid, and the core are 8.1, 4.6, and 1.25, respectively.

(2) Analyses of total protein and DNA in five independently purified preparations of HSV-1 (F1) virions yielded a ratio of 10.73 ± 0.96 (mean $\pm$ one standard deviation) (Heine *et al.*, 1974). Based on the estimated molecular weight of 10^8 for the DNA and assum-

ing one DNA molecule per virion (see Sects. 2.1 and 2.3), it was calculated that the virion contained 19.4×10^{-16} g of protein. This datum was in good agreement with calculations based on enumeration of particles and measurement of protein content of purified preparations (Honess and Roizman, unpublished data) although the error in the latter determinations was somewhat higher. The amounts of each polypeptide in the virion were then calculated based on the molecular weight of the virion polypeptides, the estimates of the fraction of total mass constituted by each polypeptides from either autoradiograms of electrophoretically separated polypeptides or scans of Coomassie brilliant blue-stained gels, and on the calculated amount of virion protein per virion. The estimated number of copies of each polypeptide species is shown in Table 2.

2.5. Polyamines

The search for polyamines in the herpesvirion stemmed from two considerations. First, the condensation of viral DNA into a tight coil to fit within the toroidal space it occupies within the nucleocapsid would necessarily require neutralization of its electronegativity. Calculations based on the dimension of the toroidal core and of the known length of the herpesvirus DNA molecule (Furlong *et al.*, 1972) indicate that the core is barely adequate to accommodate the tightly coiled DNA. The hypothesis that basic proteins neutralize the DNA is not supported by electron microscopic data, and, in fact, presents steric problems. Polyamines, on the other hand, according to the model of Liquori *et al.* (1967), can bind in the narrow groove of the DNA and thus neutralize the phosphate charges without presenting similar steric problems. Parenthetically, in contradistinction to the report by Olshevsky and Becker (1970), who reported an arginine-rich peptide (VII), neither the capsid protein 22a nor the other major B-capsid protein contain an unusually high percentage of basic residues (Gibson and Roizman, unpublished data). The minor proteins (21 and 24) have not been similarly analyzed, but together they represent only about 3% of the total capsid protein (Gibson and Roizman, 1972). It is apparent from those data, then, that the herpesvirion does not contain core proteins of the type present in adenoviruses (Laver, 1970; Prage and Pettersson, 1971) or in oncornaviruses (Fleissner, 1971), which contain nearly 20% arginine. The second consideration stems from the observation that herpesvirus replication is strongly dependent on the amino acid arginine (Tankersley, 1964; Roizman *et al.*, 1967; Inglis, 1968; Mark and Kaplan, 1971, 1972). Since arginine can serve as a precursor to

TABLE 2

Designation and Known Properties of Virion Polypeptides of HSV-1 (F1) and HSV-1 (F5)

Virion proteins	Mol. wt.[a] × 10^{-3}	Molecules[b] / Particles	Glycosylation: Glucosamine	Glycosylation: Fucose
1	275	150 (a)	–	–
2	275			
3	260	40 (a)	–	–
4	177	64 (a)	–	–
5	157	810 (a)	–	–
6	149	80 (a)	–	–
6.5	130	40 (c)	–	–
7	126	450 (a)	+?	+
8	129	590 (a)	+	+
8.5	119	400 (a)	+	+
9	115	86 (c)	+	+
10	100	100 (c)	–?	–?
11	94	360 (c)	+	+
12	91	280 (c)	–	–
12.3	88	—	+	+
12.6	86	—	+	+
13	82	710 (a)	–	–
14	78	950 (a)	–	–
14.5	76	—	+	+
15	73	460 (a)	–	–
15.5	70	—	+	+
15.8	69	—	+?	+?
16	68	—	–	–
17	62	680 (a)	+	+
18	59	1020 (a)	+	+
19E	57	660 (a)	+	+
19C	55	810 (a)	–	–
20	51	350 (a)	+	+
21	47	—	–	–
22	39	1470 (a)	–	–
23	36	1720 (a)	–	–
24	25	—	–	–

[a] Molecular weight estimates as reported by Spear and Roizman (1972) for polypeptides 1–3 and 24 and by Honess and Roizman (1974) for polypeptides 4–23.

[b] Data from Heine *et. al.* (1974). Estimates are based on analyses of autoradiograms of numerous polyacrylamide gels containing electrophoretically separated virion polypeptides (a) or of absorbance scans of gels stained with Coomassie brilliant blue (c). Some of the polypeptides are present in amounts too small to permit accurate estimates; others (VP 24) comigrated in these gels with polypeptides smaller than 25,000 in molecular weight and could not be estimated accurately.

polyamines and since a number of workers have proposed that polyamines may serve to neutralize DNA-phosphate, the question arose whether the observed effect of arginine starvation, namely absence of capsid assembly, might in fact be due to an effect on polyamine metabolism rather than on protein synthesis. The evidence for the presence of polyamines in the virion (Gibson and Roizman, 1971, 1973) is based on the following:

(1) Highly purified preparations of herpes simplex virions were found to contain the polyamines spermidine and spermine in a nearly constant molar ratio of 1.6 + 0.2. On the basis of polyamine-to-viral DNA ratios in these preparations, it was calculated that each virion contains approximately 40,000 molecules of spermine and about 70,000 of spermidine. The basic groups in spermine are adequate to neutralize over half of the viral DNA-phosphate groups.

(2) Polyamines seem to be specific structural components of the virion rather than nonspecifically-bound host molecules. This conclusion is based on the results of an experiment in which an infected-cell lysate containing unlabeled virions was mixed with a similarly prepared lysate of uninfected cells containing labeled polyamines. Virions isolated from the mixture contained polyamines with specific activities which were less than 10% of those in the initial mixture, indicating that there had been little exchange between unlabeled polyamines bound to the virions and labeled extraneous polyamines.

(3) Lastly, it was shown that disruption of the viral envelope with a nonionic detergent and urea removed up to 95% of the viral spermidine, but left the amount of spermine essentially unchanged and in a nearly constant ratio to the viral DNA-phosphate. Although spermine was not shown to be specifically localized in the core, it is likely but remains unproven that it functions in this virus in a manner similar to that proposed for putrescine and spermidine found in the bacteriophage T4—namely, to neutralize DNA-phosphate (Ames *et al.*, 1958; Ames and Dubin, 1960). Since spermidine was removed in parallel with the envelope constituents following detergent treatment, it was suggested that this polyamine is specifically associated with the envelope (Table 3).

2.6. Lipids

Very little is known about the structural lipids of the herpes virion. The problem arises from the fact that the purity of the virus analyzed for lipid content is uncertain. The available data may be summarized as follows: (1) The infectivity of the mature virion is destroyed by phospholipase C (Spring and Roizman, 1968) suggesting that the li-

TABLE 3

Characterization of Polyamine Binding to the Virion and Subviral Components[a]

Material analyzed	DNA,[b] μg	Polyamine,[c] nmole		Ratio	Polyamine-N[d]/DNA-P	
		Sp	Spd	Sp/Spd	Sp	Spd
Purified virions	20.3	7.5	13.0	0.6	0.50	0.60
Purified virions plus NP-40	25.4	9.5	10.5	0.9	0.50	0.40
Purified virions plus NP-40 and urea	17.8	6.0	0.4	15.0	0.45	0.02

[a] From Gibson and Roizman (1971). Details of the experimental procedures were published elsewhere (Gibson and Roizman, 1971). Briefly, a large pool of enveloped virions was prepared according to Spear and Roizman (1972), and divided into three equal aliquots. The first was pelleted and extracted with perchloric acid; the second was exposed to NP-40 (0.5% vol/vol final concentration) at 0°C for 15 minutes prior to pelleting and acid extraction; and the third was similarly treated with NP-40 but pelleted through a barrier of 10% wt/wt sucrose–0.01 M sodium phosphate (*p*H 7.1)–2 M urea.

[b] DNA determinations were made on the cold PCA-insoluble pellets by the diphenylamine technique. Herring sperm DNA was used as a standard.

[c] Polyamines were extracted from virions with perchloric acid, dansylated (1-dimethyl-amino-naphthalene-5-sulfonyl chloride, obtained from Pierce Chemical Co., Rockford, Ill.), and separated by thin-layer chromatography on silica gel G plates developed in ethylacetate-cyclohexane, 2:3, essentially by the method of Seiler and Weichmann (1967) as modified and described by Dion and Herbst (1967). More specific details of the experimental procedures have been previously published (Gibson and Roizman, 1971). Sp, spermine; Spd, spermidine.

[d] Calculations were based on an average nucleotide weight of 322, and on 3 nitrogen atoms for spermidine and 4 nitrogen atoms for spermine.

pids essential for infectivity are contained in the envelope. (2) Phospholipids made prior to infection cosediment with virions at least partially purified from extracellular fluid (Asher *et al.*, 1969; Ben-Porat and Kaplan, 1971). (3) The distribution of radiophosphorus in phospholipids of the virus was reported to be similar to that in phospholipids of nuclei of infected cells (Ben-Porat and Kaplan, 1971), but the significance of the finding is not clear since the nuclei were prepared with the aid of detergents.

2.7. Topology of Structural Components in the Herpesvirion

Attempts to localize the virion proteins were made by Becker and Olshevsky (1972), who claimed that of 7 major proteins, out of a maximum of 13 detected, 1 was located in the outer envelope, 2 in the inner envelope, 3 in the capsid, and 1 in the core. The localization of peptide VII in the core was based on the evidence (Olshevsky and Becker, 1970) that it is arginine-rich. Similarly, the conclusion that glycoprotein V was in close proximity to the capsid was deduced from its association with the capsid presumably after NP-40 treatment of the virion. However, since the molecular weights of the 13 peptides, determined in another study (Olshevsky and Becker, 1970), cannot be reconciled with those reported by Spear and Roizman (1972), the placement of the 7 peptides is difficult to compare with those of subsequent studies by Gibson and Roizman (1972). The present data on the topology of the structural components of the herpes virions are to a large extent preliminary and are summarized diagrammatically in Fig. 17. The assignments are based on the following: (1) As discussed in the preceding section, the DNA is now firmly localized in the core by a variety of cytochemical tests. (2) The core probably contains proteins in addition to the DNA. At the moment the only protein present in the B capsid containing DNA but not in the empty A capsid and which might be the "core" protein is No. 21. It is conceivable that in the capsids containing DNA one or more of the proteins present in the A capsid becomes a core protein (3) We have operationally defined as "capsid" proteins those found in the A capsid, i.e., 5, 19C, 23, and 24. By exclusion, the proteins external to the capsid and not in the envelope are defined as being in the tegument. This would include proteins 1–4, 6, 6.5, 20, and 22. All of these proteins are absent from either A or B capsids and, with exception of proteins 1–3, also from the C capsid, which is derived by stripping the virion of its envelope (Gibson and Roizman, 1972, 1974). At the same time, analysis of cellular membranes (Heine *et al.*, 1972; Heine and Roizman, unpublished data) indicates that these proteins do not bind to isolated membranes. As will be discussed later in the text (see Sect. 3.5), protein 22, found in the virion, is probably a processed form of protein 22a, which is present in the B capsids but not in the virion (Gibson and Roizman, 1972, 1974). We suspect that protein 22a, which binds tenaciously to the B capsid, is cleaved and that the resulting polypeptide (No. 22) is either trapped in the virion or is bound very loosely to the capsid and becomes detached during treatment of the virion with detergents (Gibson and Roizman, 1972). (4) Additional data on the localization of tegument proteins emerged from attempts to determine the topology of the envelope glycoproteins in HSV-1 (F) virions. Two series of experiments were done. In the first (Gibson and Roizman, unpublished data), the removal of envelope and tegument proteins by NP-40 and by NP-40 with deoxycholate and urea were

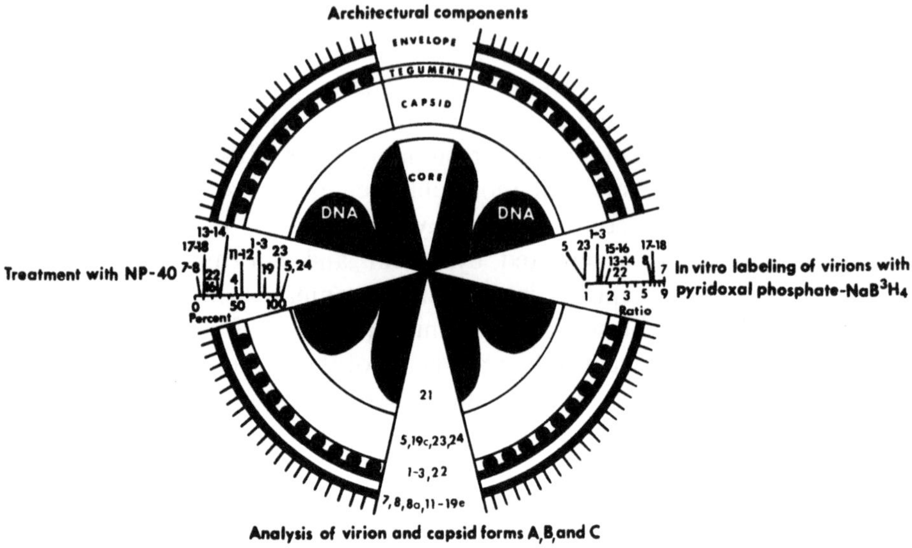

Fig. 17. Schematic diagram of the HSV-1 (strain F) herpesvirion and its architectural components drawn to scale according to thin-section measurements. Top segment contains the designations of the architectural components. Bottom segments shows the polypeptides present in the various components of the virion as determined from the polypeptide composition of the intact virion (Spear and Roixman, 1972), empty (A) and full (B) capsids (Gibson and Roizman, 1972), and polypeptides incorporated into membranes (Heine *et al.*, 1972). Left segment shows one possible order of polypeptides in the envelope and tegument of the virion based on the relative efficiency of stripping from the virion by the nonionic detergent NP-40. The polypeptides number is plotted in terms of the percentage remaining attached to the virion after extraction with detergent. The data are from Gibson and Roizman (unpublished experiments). Right segment shows another possible order of polypeptides in the envelope and tegument of the virion based on the relative accessibility of the polypeptides to labeling by tritiated borohydride reduction of the Schiff's base formed between the polypeptide and pyridoxal phosphate. The polypeptide number is plotted as a function of the ratio of ^{3}H-external-label donated by tritiated borohydrate borohydrate and ^{14}C-label derived from ^{14}C-amino acids mixture added to the culture medium during viral biosynthesis. The experiment predicts that the higher the ratio, the more external is the polypeptide. Data from Heine, Honess, and Roizman, unpublished experiments.

measured. Based on the amounts of residual proteins remaining attached to the capsid, a relative measure of tenacity of binding was determined as shown in the left sector of the diagram shown in Fig. 17. The rationale of this experiment, admittedly poor, was that the relative recovery of proteins may reflect proximity and binding to the capsid. In the second experiment (Heine, Honess, and Roizman, unpublished data), virions were externally labeled by tritiated borohydride reduction of Schiff's bases formed between pyridoxal phosphate and the protein amino group on the surface of the virions. The rationale of this experiment was that within a short labeling interval, the extent of external labeling should be inversely proportional to the distance from the surface of the virions. The ratios of extrinsic label (^{3}H) to the ^{14}C-amino acid-labeled proteins is shown in the right segment of the diagram in Fig. 17. The interesting finding is that both techniques suggest that VP7, 8, 17, and 18 are either extrinsic to the envelope (spikes) or are localized on the surface of the hydrophobic layer of the envelope, in contrast to the proteins 1–3 and some of the other polypeptides, which are both more tightly bound to the capsid and are labeled poorly with tritiated borohydride.

2.8. Variability of Virion Proteins among Wild and Laboratory Strains

In the past several years various observations on virus development (Schwartz and Roizman, 1969*a*) and virion proteins (Roizman *et al.*, 1970*b*; Roizman, 1971*b*) suggested that laboratory strains of HSV-1 may differ from "wild" strains with respect to some features of replication and protein composition. Differences in the development of laboratory and wild strain human cytomegalovirus were also reported (Kanich and Craighead, 1972). In these instances "wild" strains were defined as isolates passaged a limited number of times in cultured human cells in contrast to the laboratory strains passaged numerous times in one or more host cells. Unfortunately at that time neither the techniques for purification of the virus nor the separation of viral proteins were adequate to determine precisely the differences in the composition of virion proteins of wild and laboratory strains. In a more recent study (Heine *et al.*, 1974), two wild strains of HSV-1 were compared with several laboratory strains (Fig. 18). No differences were detected among the virion proteins of the two wild strains. On the other hand, the electrophoretic profiles of virion proteins of the laboratory strains differed both qualitatively and quantitatively from each other and from those of the wild strains. The major qualitative differences were the absence of some glycoproteins and the presence of new glycoproteins (Fig. 18). This is probably not an isolated incident but rather a more generalized phenomenon of considerable potential significance. Since the initial isolates of the laboratory strains were not available, it was not possible to differentiate between the hypothesis that the laboratory strains resembled in composition the parent isolate, and

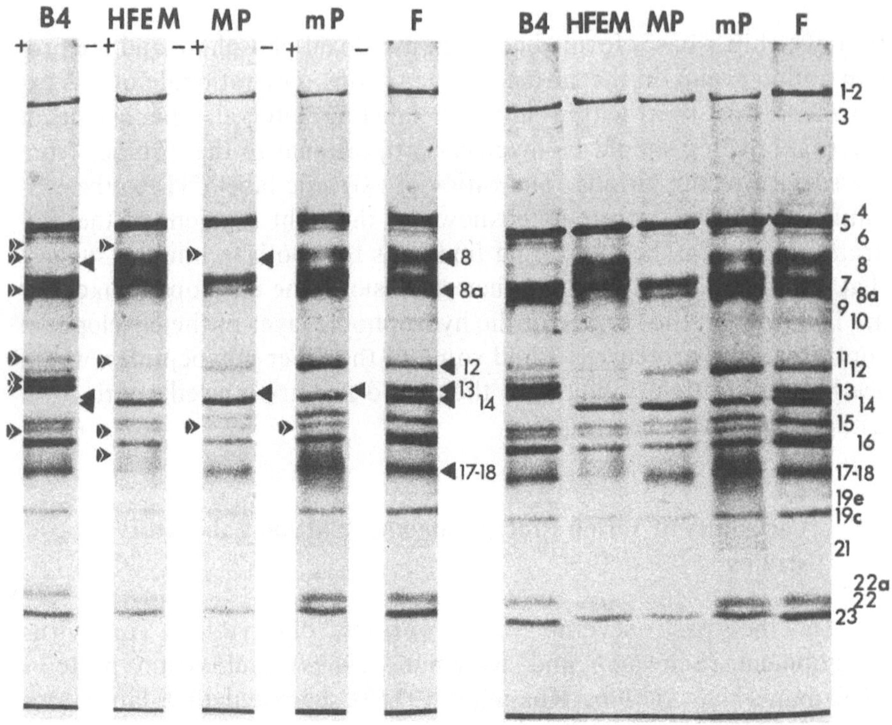

Fig. 18. Autoradiograms of a polyacrylamide gel slab containing electrophoretically separated polypeptides of HSV-1 (F1), HAV-1 (mP), HSV-1 (MP), HSV-1 (HFEM) and HSV-1 (13VB4) virions. The entire autoradiogram is shown on the right together with the current designations of polypeptides of HSV-1 (F1), to facilitate comparisons, the autoradiograms of each gel are also shown on the left. The arrows on the (−) side of each gel pinpoint the position of HSV-1 (F1) polypeptides absent from the various HSV-1 strains tested. The arrows on the (+) side indicate the polypeptides for which there is no electrophoretic analogue in the HSV-1 (F1) virions. The arrows against the gel containing the electrophoretically separated HSV-1 (F1) polypeptides pinpoint all the polypeptides whose electrophoretic analogue is absent in one or more of the HSV-1 strains tested. Data from Heine *et al.*, 1974.

the alternative hypothesis that they resembled mutants selected for some advantageous property in the cells in which they were passaged. The first hypothesis implies that wild strains of HSV-1 vary both in time and geographically. While there are no data that would substantiate or refute this hypothesis, the alternative has, in fact, some substance. HSV-1 (MP) arose spontaneously in a culture infected with HSV-1 (mP) and has a distinct advantage in that it spreads from infected to uninfected cells more rapidly than the parent strain (Hoggan and Roizman, 1959*a*). Another example is the selection of spontaneous mutants HSV-1

(MPdk$^+$Lp) and HSV-1 (MPdk$^+$sp) from dog kidney cells abortively infected with HSV-1 (MP) (Roizman and Aurelian, 1965). The mutants had the advantage of being able to grow in these cells. In this instance the mutants differed from the parent HSV-1 (MP) with respect to immunologic specificity, buoyant density in CsCl solution, and stability at 40°C, all consistent with altered structure.

The rather extensive differences among the electrophoretic profiles of the laboratory strains both *vis-à-vis* each other and in reference to those of the wild strains suggest that noncapsid polypeptides may vary considerably. Although we are not certain that the variation is not detrimental, it is clear that it is not lethal since these viruses multiply and produce good yields and are stable with regard to laboratory manipulation. Concurrent with or related to the variation in the noncapsid protein content are two other properties of the virus. Thus, the virion may exhibit altered immunologic specificity, as has been clearly demonstrated for HSV-1 (MP) which is neutralized equally well by both anti-HSV-1 and -HSV-2 sera (Ejercito *et al.*, 1968; Plummer *et al.*, 1970; Aurelian *et al.*, 1970), whereas HSV-1 (mP) is more readily neutralized by anti-HSV-1 than by anti-HSV-2 sera. Moreover, cells infected with the laboratory strain differ from the HSV-1 (F) physiologically as evident from electron microscopic studies (Schwartz and Roizman, 1969*a*), and from the nature of the social interaction (i.e., tight clumps, loose clumps, polykaryocytosis) of infected cells among themselves (Ejercito *et al.*, 1968). The implications are as follows. Viruses used as antigenic reagents in seroepidemiologic studies must be shown to be identical with the strains circulating in the population being analyzed since misleading data may otherwise result. Also, biochemical studies and reports should include sufficient information on the virus strain being used to assess the validity of comparisons of data generated in different laboratories.

2.9. The Function of Structural Components in Relation to Particle Infectivity

2.9.1. The Problems

We are concerned here with two questions: First, what is the minimum amount and type of material necessary to infect cells and to produce infectious progeny; and second, what is the function of all of the structural components which, in accordance with the answer to the first question, are not absolutely necessary for infection of the cell and for biosynthesis of infectious progeny?

2.9.2. The Minimal Requirements for Infection of the Cell and Production of Infectious Progeny

Lando and Ryhiner (1969) and, subsequently in more detail, Sheldrick *et al.* (1973) reported that herpes simplex virus DNA is infectious. The data reported by Sheldrick *et al.* (1973) are particularly interesting in that they conclusively demonstrated that both native, double-stranded DNA with a sedimentation constant of 56 S and alkali-denatured, single-strand-specific nuclease-sensitive DNA with a sedimentation constant (*p*H 13) of 67 S are infectious. The specific infectivity of the DNA was 3–5 $\times$ 10^6 molecules and 40 $\times$ 10^6 molecules per infectious unit for native DNA (100 $\times$ 10^6 molecular weight) and denatured (50 $\times$ 10^6), respectively. With respect to the specific issue raised here, the low specific infectivity of viral DNA makes it extremely difficult to exclude the possibility that 1 in 10^6 or more molecules of DNA still contained undenatured protein attached to it. The specific infectivity of the DNA is to some extent dependent on the presence of "competent" cells. The competence in this instance is not mere uptake of DNA, but rather some other function not required if the cell is exposed to the virion instead of DNA. We may speculate that this function may be the transcription of viral DNA (see Sect. 3.4 10) but it could also be a more trivial function, e.g., the absence of enzymes capable of degrading the DNA. However, analyses of the properties and sedimentation of infectious, native, and denatured viral DNAs preclude the possibility of large amounts of proteins remaining attached to the viral DNA. It thus seens reasonable to exclude therefore all envelope proteins and the bulk of the capsid proteins not in direct juxtaposition to viral DNAs as being essential for viral expression in infected cells.

2.9.3. The Role of the Capsid and Envelope in Infectivity

The role of the envelope in the infectivity of herpesviruses was probably the dominant theme of herpesvirus research a decade ago. Like many other subjects it was never resolved, but simply faded away. The pertinent data may be summarized as follows:

(1) Although both virions and nucleocapsids are present in infected cells, only virions are present in the extracellular fluid. The nucleocapsids are either not released or are unstable in the extracellular fluid (Nii *et al.*, 1968*a*). It would seem that the enveloped nucleocapsid is selected as best suited for the extracellular envi-

ronment, and there is general agreement that it represents the mature infectious form of the virus, but the question nevertheless arises whether the naked nucleocapsid is capable of infecting cells. (ii) The specific infectivity of the naked nucleocapsid is not exactly known though it probably is not as low as that of enveloped particles (4–100 particles/PFU for HSV-1), nor as low as that of native or denatured DNA. The problem arises from methodological considerations. The cleanest fractionation of enveloped and unenveloped nucleocapsids is either by isopycnic centrifugation in CsCl solution or by stripping the envelope with nonionic detergents (Smith and Dukes, 1964; Stein *et al.*, 1970; Rubenstein *et al.*, 1972). Prolonged centrifugation in CsCl solutions leads to disaggregation of nucelocapsids (Spring and Roizman, 1968), and under these conditions loss of infectivity could be due to a number of different causes unrelated to the intrinsic infectivity of the nucleocapsids. In a similar vein, nucleocapsids prepared by detergent treatment of virions are also suspect because minute amounts of detergent trapped in the particle might inhibit virus multiplication. It is perhaps interesting to note that Abodeely *et al.* (1970) found that, on exposure of cells to enveloped and de-enveloped nucleocapsids, the nucleocapsids from enveloped particles were ultimately present free in the cytoplasm whereas those of de-enveloped particles were confined to cytoplasmic vesicles. Experimental approaches to this problem by either relating particle count to infectivity or by some method of fractionation have yielded other kinds of data, but the techniques employed are less reliable. Thus, Watson *et al.* (1964) reported that in some preparations the number of plaque forming units exceeded the number of enveloped particles. This finding led to the conclusion that probably both kinds of particles were infectious. Fractionation of enveloped and unenveloped virus on the basis of the cellular compartment in which they accumulate has also led to the conclusion that the "nuclear" or "nonenveloped" particles are infectious. Thus, Watson *et al.* (1964), Siminoff and Menefee (1966), and Spring and Roizman (1968) showed that "nuclear" virus is also infectious, and Hochberg and Becker (1968) concluded that the nuclear virus is as infectious as cytoplasmic virus. The major difficulty in interpreting these reports is the impossibility of unequivocal equating the unenveloped with the nuclear virus. Indeed, analysis of the nuclear virus by Spring and Roizman (1968) showed that it differs in stability and hydrodynamic properties from cytoplasmic, fully enveloped virus. However, the infectivity of both nuclear and cytoplasmic particles is sensitive to phospholipase C, and one justifiable conclusion is that intact envelopes are not required for infectivity.

Although we cannot determine the specific infectivity of the various nucleocapsids, two lines of evidence suggest that the specific function of the envelope is to stabilize the nucleocapsid and to facilitate entry of the virus into cells. As indicated earlier in the text, Spring and Roizman (1968) reported that the infectivity of fully enveloped cytoplasmic particles is far more stable than that of partially enveloped or unenveloped nuclear particles. Coupled with the observation by Nii *et al.* (1968*a*) that naked particles are absent from extracellular fluid, the data suggest that one of the functions of the envelope is to protect the capsid from the deleterious effects of the extracellular environment. The role of the envelope in initiation of infection has been proven by several authors who noted that enveloped particles more readily absorbed to cells than naked ones (Holmes and Watson, 1961; Siegert and Falke, 1966; Dales and Silversberg, 1969; Abodeely *et al.*, 1970).

3. BIOSYNTHESIS OF THE VIRUS IN PERMISSIVE CELLS

3.1. The Reproductive Cycle and Its Outcome

3.1.1. Definition

The outcome of animal cell infection with wild herpesvirus strains varies from one cell line to another and is genetically determined by both the virus and the cell. It is convenient at this point to define the genetic predisposition of the host and resulting infection. A cell may be defined as nonpermissive if one or more sets of essential viral functions is not expressed and no progeny results; the infection itself is abortive. On the other extreme, cells which allow full expression of all essential viral functions are permissive and the infection is productive. Intermediate between these two are cell lines which restrict, in comparison with permissive cells, the yield of infectious progeny, but since some infectious progeny is made, all essential viral functions are expressed and the infection is not, strictly speaking, abortive. Restrictive cells are sometimes difficult to differentiate from either nonpermissive or permissive cells, and, in fact, it could be argued that all cells are more or less restrictive. This section will deal with the reproductive cycle of virus in permissive cells although in many instances the cells are actually at least partially restrictive.

Parenthetically, permissiveness defines a host phenotype under specified optimal conditions for virus multiplication. One does not render the host nonpermissive by merely elevating the temperature, by adding cytosine arabinoside, actinomycin D, or some other inhibitor of virus multiplication, as has been claimed, even though the infection is obviously aborted.

3.1.2. General Characteristics of the Reproductive Cycle

The general characterization of the reproductive cycle of a number of herpesviruses has been reviewed elsewhere (Kaplan, 1969; Roizman, 1969). Much more is known about the "fast" herpesviruses, i.e., HSV, pseudorabies, EAV, etc., than about EBV and the "slow" cytomegaloviruses. For the purpose of this discussion it seems useful to provide some dimensions of time and quantity, particularly for the "fast" herpesviruses which have been analyzed in greater detail. We are concerned here with three parameters, i.e., (1) the duration of the eclipse terminated by the appearance of new virus, (2) duration of the reproductive cycle, and (3) the yield of virus per cell.

(1) The duration of the eclipse varies from 3 to 8 hours for most herpesviruses. It is affected by the temperature of incubation (Farnham and Newton, 1959; Hoggan and Roizman, 1959*b*; Smith, 1963), by the multiplicity of infection, and by prior infection of cells with another mutant (Roizman, 1963*a*, 1965*a*). Once a minimum eclipse period is attained (3 hours for pseudorabies in rabbit kidney cells, 5 hours for HSV in HEp-2 cells) it cannot be shortened by increasing the multiplicity of infection (Kaplan and Vatter, 1959; Roizman, 1963*a*).

(2) The duration of the reproductive cycle of herpesvirus varies considerably. The cycles of herpes simplex in HEp-2 cells and that of EAV in LM cells last from 13 to 19 hours, depending on the multiplicity of infection and temperature of incubation (Roizman *et al.* 1963; O'Callaghan *et al.*, 1968*a*; Kaplan and Vatter, 1959). At 37°C and 50 PFU of HSV per cell, the cycle lasts 17 hours. The cycle of pseudorabies in rabbit kidney cells is somewhat shorter.

(3) The yield of virus from infected cells increases exponentially from the end of the eclipse phase until almost the end of the reproductive cycle. Under optimal conditions the yield of HSV is 10,000 to probably no more than 100,000 virions per HEp-2 cell. The best preparation of Watson *et al.* (1964) contained about 10 virions/PFU; routine preparations of virus contain on the order of 100–1000 virions/PFU (Smith, 1963). Cells grown and maintained in monolayer cultures usually yield more virus than those suspended after infection (Roizman and Spear, 1968). The relative amounts of enveloped and naked nucleocapsids in herpesvirus-infected cells vary not only with the conditions of infection but also with the host species in which the virus is grown. High-titer virus is more readily obtained in rapidly growing cells maintained after infection in an enriched medium (Roizman and

Spear, 1968) at the *p*H (Roizman, 1965*b*) and temperature (Farnham and Newton, 1959; Hoggan and Roizman, 1959*b*; Smith, 1963; Roizman, 1965*a*) optimal for that particular strain.

3.2. Entry of Virus into the Cell

The principal events and requirements for attachment of herpesvirions to cells, rate of penetration, and uncoating are known, as is evidence that viral DNA finds its way into the nucleus. The available data on these points have been summarized (Roizman, 1969).

In evaluating the data accrued in recent years, one major point of uncertainty still exists. Briefly, electron microscopic studies indicate that entry of the virus into the cell occurs by two mechanisms (Abodeelly *et al.*, 1970; Iwasaki *et al.*, 1973). The earliest described is by phagocytosis of particles from the extracellular fluid. For infection to take place, the vacuolar contents would have to be released into the cytoplasmic sap. More recently it has been shown that the envelope of extracellular virions fuses with plasma membranes and that as a consequence of fusion the virion contents are released directly into the cytoplasm (Abodeely *et al.*, 1970; Iwasaki *et al.*, 1973; Morgan *et al.*, 1968; Hummeler *et al.*, 1969).

Although it has been suggested that (1) capsids found in vacuoles are a more advanced form of degradation than those entering by fusion of the virion and plasma membranes and that (2) fusion is more likely to occur rapidly during initial exposure of virus to cells whereas phagocytosis occurs throughout exposure (Hummeler *et al.*, 1969), there are no fullproof markers which would allow differentiation between capsids taken into cytoplasmic sap from vacuoles and those entering the cytoplasm directly as a consequence of fusion. Electron microscopic techniques do not readily lend themselves to differentiating which type of entry, if either, is uniquely required for virus multiplication. The fate of the capsids taken into the cytoplasmic sap is unclear, and, in any event, it is not uniform (Miyamoto and Morgan, 1971). Hochberg and Becker (1968) as well as Sydiskis (personal communication) found that viral DNA becomes at least in part accessible to nucleases, indicating at least partial dissociation of the DNA from the capsid. Electron microscopic studies do not contradict these conclusions. Specifically, it has been reported that, with passage of time after exposure of virus to cells, more and more capsids either lacking cores or containing particles showing decreased amounts of electron-dense core material accumulate at or near the nuclear membranes (Morgan *et al.*, 1968; Iwasaki, 1973). These observations suggest that release of the DNA occurs through some port in the capsid and does not require its dissolution. The mechanism of this putative release of DNA from capsids is not clear. Miyamoto and Morgan (1971) reported that pretreatment of cells with puromycin or with actinomycin D did not prevent entry and deenvelopment, but it prevented the release of core material since capsids with cores accumulated in the cytoplasm. The fate of the capsids which have lost their core is uncertain (Iwasaki, 1973).

Fundamentally, there are three sources of possible error in the results available from electron microscopic studies. Briefly (1) efficient examination of cells by electron microscopy requires that cells be infected at high multiplicity to ensure that at least some of the thin sections contain virus. Some of the observations made to date could be the consequence of artifacts introduced by high multiplicity of infection. (2) None of the studies reported to date was done with purified, homogeneous preparation. For example, Miyamoto and Morgan (1971) observed two distinct morphologic types of particles in the inoculum of HSV-1 (Miyama). These particles seemed to behave differently after uncoating. One type, designated as "light capsids," was characterized by a clearly defined capsid and an electron-translucent shell between the capsid and the core. After entry into the cell light capsids were seen minus the core at or near the nuclear membrane, where they remained seemingly intact for several hours. The second type of particle, designated as "dense capsids," was characterized by an ill-defined capsid and by the presence of fine granular material in the region between the core and the capsid. Upon entry into the cell, the dense capsids appeared to disintegrate, while still quite near the plasma membrane. Although the two types of particles were present in nearly equal quantity, they have not been separated and nothing is known concerning the specific infectivity of each. (3) In some studies the infected cells yielding the virus stock used for these experiments were in fact sonicated (Morgan *et al.*, 1968). Does a naked nucleocapsid or a damaged virus particle penetrate in the same way as an intact virion?

There are no data on the mechanism of entry into the nucleus of the viral nucleic acid, but dense intranuclear clumps of material have been seen close to nuclear pores soon after loss of cores by cytoplasmic capsids (Miyamoto and Morgan, 1971). A shuttle service by the nuclear pores suggests itself. Except for the margination of the chromatin, electron microscopy gives no information about the progress of the infection until the appearance of granular material in the cytoplasmic sap and especially near the nuclear pores, where they assume an hourglass shape. Among the unresolved problems are (1) the nature of the enzymes involved in uncoating of the virus, particularly since new synthesis of viral protein is not required for uncoating and since highly purified virus particles exhibit proteolytic activity; (2) the mechanism of transport of viral material from the periphery of the cytoplasm toward the nucleus, where it functions; (3) the mechanism of release of core material from the capsid; and (4) the mechanism of transport of viral DNA from the cytoplasm to the nucleus. In addition, the interference with virus adsorption by sulfated polyanions has been well documented (Vaheri and Penttinen, 1962; Vaheri and Cantell, 1963; Takemoto and Fabish, 1964; Nahmias and Kibrick, 1964; Tytell and Neuman, 1963; Nahmias *et al.*, 1964; Benda, 1966; Hadhazy *et al.*, 1966), but extensive clinical trials with these agents have not been attempted. Since future emphasis is likely to be on the potential risk of malignancy following infection, it would not be too surprising if the search for nontoxic molecules capable of interfering with absorption of virus to cells were to be renewed.

3.3. Viral RNA Synthesis and Metabolism

3.3.1. Introduction

After entry and uncoating, transcription of viral DNA is the first major event in the reproductive cycle of herpesviruses required for virus multiplication (Roizman, 1963*b*; Sauer *et al.*, 1966; Sauer and Munk, 1966; Flanagan, 1967). The events with which we are concerned, in a logical order, are: the requirements for transcription; the nature of the products; the control of transcription; the post-transcriptional modification of viral RNA; and the transport, function, and finally, degradation of the viral RNA. However, since all of these events necessitate detection and quantitation of viral RNA in the frame of reference of the amount of DNA sequences transcribed in the infected cells, it is convenient to begin the analysis of viral RNA metabolism with a discussion of detection and quantitation of viral RNA followed by an analysis of the transcriptional process.

3.3.2. Detection and Quantitation of Viral RNA

Differentiation between viral and cellular RNA is based solely on sequence recognition in hybridization tests with viral DNA. The choice of technique is based on the fact that the total amount of virus-specific RNA in relation to cell RNA is small (Frenkel and Roizman, 1972*b*; Roizman and Frenkel, 1973; Silverstein *et al.*, 1973). Moreover, host RNA synthesis continues after infection, albeit at a reduced rate, and therefore neither the total RNA nor the fraction labeled after infection is indicative of the amount of viral RNA sequences present. Initially at least, the main questions are (1) the extent of transcription of viral DNA in the course of a productive infection by HSV in cell culture, (2) the number of classes of viral RNA present in the cell in different molar concentrations, (3) the fraction of total viral DNA serving as template for the different abundance classes of RNA (4) the fraction of total infected cell RNA which is virus-specific, and lastly (5) the changes in the concentration of RNA species as a function of the time post-infection.

Logistically, the hybridization tests designed to answer these questions fall into two categories. Until rather recently all hybridization tests involving herpesviruses were done in a two-phase system consisting of labeled RNA in liquid exposed to viral DNA on filters. More recently, a one-phase system consisting of labeled viral DNA and excess unlabeled viral RNA was introduced (Frenkel and Roizman, 1972*b*). In practice, the hybridization of labeled RNA to filters containing viral DNA is a simple technique, permitting rapid separation of hybrid from unhybridized RNA (Gillespie and Spiegelman, 1965). Moreover, fixation of denatured DNA to filters prevents its reassociation. The technique suffers from several shortcomings and at best provides a rather qualitative answer. The problems specifically are as follows: (1) A characteristic of herpesvirus DNA on filters is that the relative amounts of its sequences available for hybridization to RNA are rather small. (2) The bonds holding the DNA to filters are unstable, and DNA completely saturated with RNA tends to fall off the filter. (3) Analysis of the rates of hybridization would not readily permit detection of two or more classes of RNA complementary to different regions of DNA and differing widely in molar concentrations. Thus, two classes, A and B, differing 20-fold or more in molar concentrations would not be readily detected even if the scarce class were complementary to 50% of the total transcribed DNA. (4) Quantitative comparisons of two or more preparations of RNA labeled at different times in the productive cycle must necessarily correct for the nucleotide pools during the labeling intervals or make unwarranted assumptions; in any event, analyses of RNA labeled late in infection are complicated by the possible presence of RNA made before the label was added but sharing the same sequences.

In the one-phase hybridization system, the separation of hybridized and unhybridized reagents is technically more complex in that centrifugation, hydroxyapatite chromatography, or digestion with single-strand-specific nucleases is required. However, the results are readily amenable to analysis which is not complicated by any of the shortcomings listed above. Specifically, if the concentration of labeled

DNA is so low that no significant DNA reassociation occurs and if the concentration of unlabeled viral RNA is in excess, the hybridization of RNA to DNA is described by the relationship (Frenkel and Roizman, 1972*b*)

$$D_t/D_o = \alpha_1 e^{-k_1 R_1 t} + \cdots + \alpha_n e^{-k_n R_n t} + 1 - (\alpha_1 + \cdots + \alpha_n)$$

where D_t/D_o is the fraction of DNA remaining single-stranded, subscripts $1, \ldots, n$ refer to classes of RNA differing in molar concentrations, $R_1, \ldots, R_n$ represent RNA concentrations in moles per liter, and $\alpha_1, \ldots, \alpha_n$ are corresponding fractions of total DNA serving as templates for the classes of RNA differing in abundance. Assuming the base composition of the various RNA species does not vary appreciably, the numerical value of the rate constant $k_1, \ldots, k_n$ appearing in the different terms of the equation should be about the same and may be referred to as k. The rate constant k for herpesvirus DNA–RNA hybridizations has not been determined and was in all of the studies reported to date assumed to be the same as the rate constant for herpesvirus DNA–DNA hybridizations, defined by $dD/dt = {}^-kD^2$, where D is the molar concentration of herpes simplex 1 DNA, i.e., (g/liter)/(99 $\times$ 10^6 g/mole). The assumption that rate constants for DNA–DNA and DNA–RNA hybridization are the same is probably incorrect; Melli *et al.* (1971), on the basis of hybridization of excess *E. coli* DNA to complementary RNA prepared *in vitro*, calculated that the rate constants differ by a factor of 0.43. In principle, the numerical value of the hybridization rate constant is required for calculation of the absolute values of viral RNA concentrations in the cell and, therefore, all calculated values actually may be 2.3 times lower than the true concentrations if the measurements of Melli *et al.* (1971) are applicable here. The numerical value is not required for the calculation of the ratios of molar concentrations of the RNA species or for the determination of the size of the template from which they are transcribed. In general, the sensitivity of the technique in resolving the number of classes of RNA differing in molar concentration hinges on the number and precision of the measurements of D_t/D_o at various times in the course of the hybridization, but probably does not exceed 3 or 4.

3.3.3. The Transcription of Viral RNA

Evidence for transcriptional controls in herpesvirus-infected cells has recently been reported by a number of investigators (Huang *et al.*, 1971; Wagner, 1972; Wagner *et al.*, 1972; Frenkel and Roizman, 1972*b*; Roizman and Frenkel, 1973; Frenkel *et al.*, 1973; Rakusanova *et al.*, 1971). Most of these studies employed the two-phase hybridization involving labeled RNA in solution and DNA on filters and, therefore, as discussed earlier in the text, they dealt largely if not exclusively with abundant RNA species and their templates. The most extensive analyses employing excess unlabeled RNA and labeled viral DNA were done on HSV-1- and HSV-2-infected HEp-2 cells (Frenkel and Roizman, 1972*b*; Frenkel *et al.*, 1973) and will be the ones discussed here in detail.

The analyses of viral RNA transcripts in HSV-1-(F)- and HSV-2-(G)-infected HEp-2 cells were done on RNA extracted at 2 and 8 hours post-infection. The time intervals were selected on the basis of observations that viral DNA synthesis begins at about 3 hours post-infection (Roizman and Roane, 1964; Roizman, 1969) and that infectious progeny appears at 6 hours and is present at appreciable levels at 8 hours post-infection (Roizman, 1969). The 2-hour sample should therefore contain the bulk of the RNA made before viral DNA synthesis begins, whereas the 8-hour RNA should contain all of the species made late in infection and required for the synthesis of infectious progeny.

A representative hybridization is shown in Fig. 19. The analyses of the data are shown in Table 4. The results of the experiments performed to date may be summarized as follows:

(1) Late in infection, the RNA present in infected cells is complementary to 48–50% of viral DNA. In no hybridization test was this figure exceeded. The data imply that only half of the DNA, or the equivalent of 1 strand, is transcribed. There are no data on whether all transcripts arise from one strand or whether both strands are transcribed in part.

(2) Two points should be made in connection with the amounts of DNA transcribed early in infection. First, the difference between the total amounts of HSV-1 DNA hybridizing with homologous RNA early (44%) and late (48–50%) might be due to low concentration of viral RNA early in infection. The second point concerns the difference between the amount of DNA transcribed early in HSV-1-(44%) and HSV-2-(21%) infected cells. The significance of the difference is not entirely clear; among the possible hypotheses which might explain our data are the following: (1) there is a real difference in the transcriptional process of HSV-1 and HSV-2 DNA; (2) HSV-2 DNA is transcribed to the same extent as HSV-1 DNA, but the transcripts of 23% of HSV-2 DNA (the difference between 44 and 21%) are present in concentrations too small to be detected; (3) the transcripts of the 23% of HSV-2 DNA are made but are rapidly degraded; or (4) the infection proceeds more slowly in HSV-2- than in HSV-1-infected cells, and therefore the

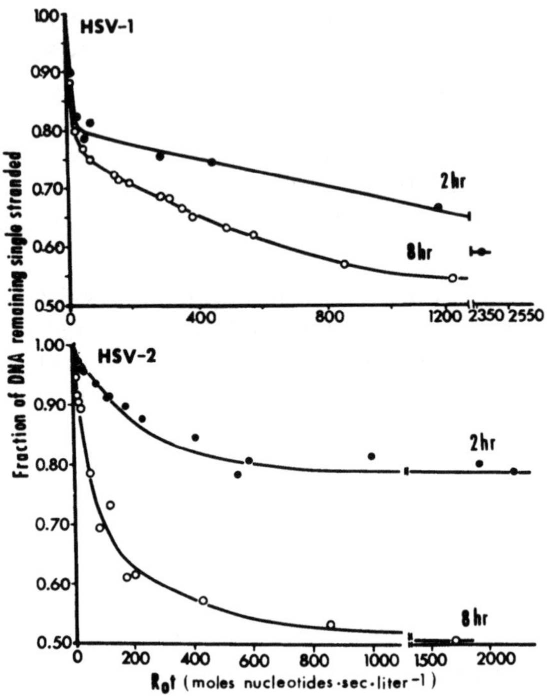

Fig. 19. Single-phase hybridization of labeled HSV-1 DNA with excess RNA extracted from HEp-2 cells at 2 and 8 hours post-infection. The hybridization tests were carried out in 0.07 M NaCl, 0.05 M Tris buffer. After hybridization the residual single-stranded viral DNA was digested with single-strand-specific nucleases. These completely digested single-stranded DNA but did not attack duplexes. Details of the procedures are given in Frenkel and Roizman (1972) and Frenkel *et al.* (1973). Since variable amounts of RNA were hybridized for different time intervals as described in the text, the data are presented as a plot of the fraction of single-stranded DNA as a function of the input concentration of RNA (R_o) in moles nucleotides/liter times the time of hybridization (t). Experimentally determined points are shown as circles. The line is a computer plot calculated according to equation given in Sect. 3.3.2 and fitted by nonlinear least squares method (Fig. 20). The computer analysis involves the determination of parameters α_n and R_n in the equation in which the number of classes of RNA (n) is assumed to be 1, 2, 3, 4, etc. The fit for $n = 2$ is better than that for $n = 1$. When n is greater than 2, the results are meaningless. Thus, for the hybridization of HSV-1 RNA extracted at 2 hours post-infection, for $n = 3$, α_1 is less than 0.001 whereas α_2 and α_3 retain the same values as α_1 and α_2 in the case of $n = 2$. Data from Frenkel *et al.* (1973).

transient period of limited transcription observed within the first 2 hours of HSV-2 infection might occur earlier in HSV-1 infection. In our experiments HSV-1-infected cells may have been examined too late to detect the limited early transcription observed in HSV-2 infection. This possibility is suggested by the two-phase hybridization tests reported by

TABLE 4

Analysis of the Transcripts of HSV-1 and HSV-2 DNAs in Productively Infected HEp-2 Cells at 2 and 8 Hours Post-Infection[a]

Virus	Time of extraction, hr post-infection	Fraction of viral DNA transcribed			RNA abundance		
		α_1	α_2	Total	R_1,[b] $\frac{\text{nmoles}}{\text{liter}}$	R_2,[b] $\frac{\text{nmoles}}{\text{liter}}$	R_1/R_2
HSV-1	2	0.14	0.30	0.44	7.9	0.058	136.2
	8	0.19	0.28	0.48	7.1	0.183	40.3
HSV-2	2	0.21	0.00	0.21	0.51	0.0	—
	8	0.31	0.19	0.50	2.23	0.28	7.96

[a] Data from Frenkel *et al.* (1973).
[b] Moles of viral RNA fragments of $0.99 \times 10^8 \cdot \alpha_n$ daltons in molecular weight.

Wagner *et al.* (1972). We cannot differentiate among these hypotheses at present.

(3) Analyses of the kinetics of the hybridization of RNA to viral DNA revealed the existence of two classes of viral RNA, differing in molar concentrations late in HSV-1- and HSV-2-infected cells and early in HSV-1-infected cells. Virus-specific RNA present early in infection of HEp-2 cells with HSV-2 formed only one abundance class. The size of the template from which each class is derived, the molar concentrations of the RNA in each class, and the molar ratios of the abundant to scarce class are summarized in Table 4. Several points should be made in connection with these data. First, the calculation of the number of classes is based on fitting to the experimental points a series of curves by nonlinear least squares regressions of the parameters $\alpha_1, \ldots, \alpha_n$ and $R_1, \ldots, R_n$ for $n = 1, 2, 3$, etc. (Fig. 20). In all but one case, i.e., RNA present early in cells infected with HSV-2, the fit for $n = 2$ was better than those for $n = 1, 3, 4$, etc. Although further determinations of homogeneity of the RNA in each class were not done, the existence of two classes differing in molar abundance was confirmed by their actual separation as described later in the text. The second point concerns the question of whether the nucleotide sequences in the most abundant class at 2 hours post-infection are those in the most abundant class present in 8-hour infected cells. To answer this question, a series of abundance competition tests were done. This type of analysis is designed to answer the general question whether any class of RNA with a given abundance contains the same nucleotide sequences as any other class of RNA with the same or different abun-

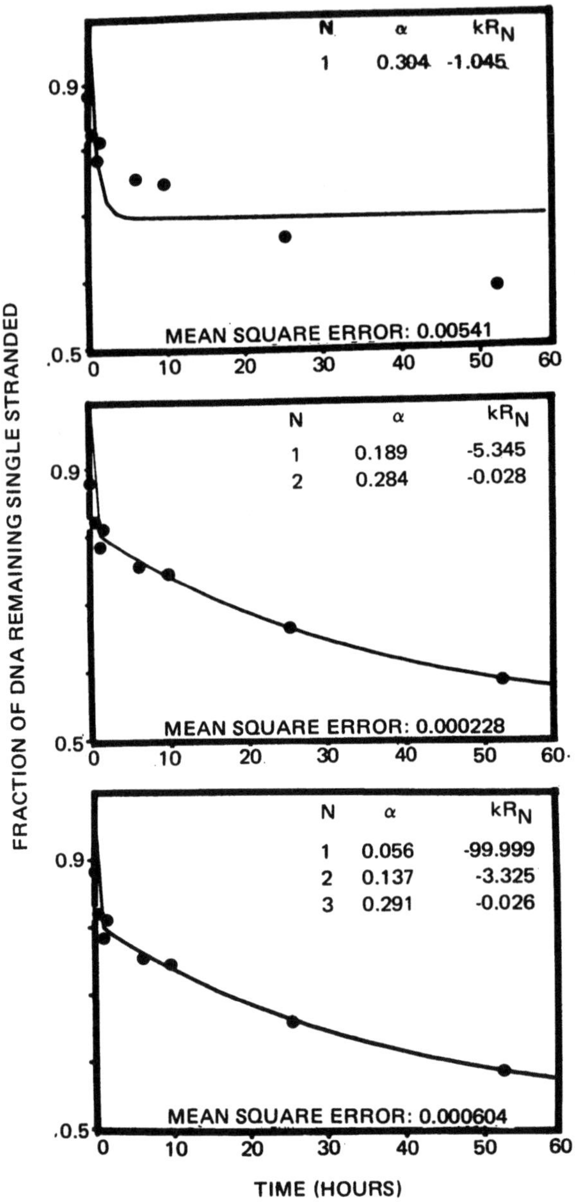

Fig. 20. Computer-aided analysis of the population of viral RNA transcripts made in HEp-2 cells productively infected with HSV-1 (F1). The filled circles are the actual amounts of labeled viral DNA remaining single-stranded after hybridization with excess unlabeled RNA from infected cells. Solid lines were fitted by least squares regression with the aid of the computer according to equation in Sect. 3.3.2 for $n = 1$ (top panel), $n = 2$ (middle panel), and $n = 3$ (bottom panel). Note that the line does not fit the experimental points for $n = 1$. The fit for $n = 3$ is almost as good as for $n = 2$, but to obtain the fit for $n = 3$ it must be assumed that the cells contains a class of viral RNA derived from 5.6% of DNA ($\alpha = 0.056$) and present in a concentration 30 times more abundant than the next most abundant RNA class.

dance. As shown in Table 5, the test consists of 3 sets of hybridizations. In the control set the 2- and 8-hour RNA were each hybridized to denatured DNA at a concentration (R_o) and time (t) such that all of the single-stranded DNA fragments complementary to the most abundant RNA were driven into DNA–RNA hybrids. In the abun-

TABLE 5

Abundance Competition between Most Abundant RNA 2 Hours and 8 Hours After Infection of HEp-2 Cells with HSV-2[a]

Competition experiment	RNA source	R_0t,[b] moles nucleotide·sec/liter	Observed DNA in hybrid, %	Predicted DNA in hybrid, %	
				Same[c]	Different[d]
1	2-hr	71.6	18.8		
	8-hr	72.5	24.8		
	8-hr	144.9	28.8		
	8-hr	72.5			
	+		22.6	24.8–28.8	43.6
	2-hr	71.6			
2	2-hr	71.6	18.8		
	8-hr	289.9	32.6		
	8-hr	362.4	35.6		
	8-hr	289.9			
	+		35.6	32.6–35.6	51.4
	2-hr	71.6			

[a] Data from Frenkel and Roizman (1972).
[b] The 2-hr RNA, 8-hr RNA, or both were incubated with labeled DNA to R_0t specified.
[c] All abundant sequences in 2-hour RNA are present in competing 8-hr RNA.
[d] Most abundant species in 2-hour RNA different from those in competing 8-hr RNA.

dance competition set, the 8- and 2-hour RNA were mixed. However, the concentration of each RNA species was adjusted so that for a common time of incubation, t_c, the product R_0t_c was exactly the same as the product R_0t for each of the RNAs in the control set. The data indicate that the sequences in the most abundant 2-hour RNA were a subset of the sequences present in the most abundant RNA at 8 hours post-infection.

The data presented above are significant from two points of view. First, the most abundant class comprises 99.3 to 93.5% of the early and late viral RNA sequences, respectively, yet it arises from at most 24% of the DNA. By contrast, the scarce species which comprise at most 0.7% to 6.5% of the RNA early and late in infection, respectively, arise from 27% of the DNA. It is useful to point out that conventional hybridization tests of labeled RNA to DNA on filters would barely detect the scarce species at their highest concentration. Appreciation of the fact that these scarce species arise from more than half of the viral genome is due to the hybridization techniques used in these studies. The second point concerns the fact that the technique of hybridizing unlabeled RNA to labeled DNA measures the amount and abundance of viral RNA accumulating in the cell at different times after infection. It does not directly reveal the time of synthesis of the various species or their turnover rates. This type of analysis requires some form of labeling of the viral RNA. In general the technique measures stable RNA species which do not arise by symmetric transcription. The technique is applicable to cytoplasmic RNA in which symmetric transcripts are absent and only with reservations to nuclear RNA which contains both symmetric and nonsymmetric transcripts (Kozak and Roizman, manuscript in preparation).

3.3.4. The Requirements for Synthesis and Site of Synthesis of RNA

We have so far established the extent of transcription of viral DNA and the relative amounts of viral RNA made in human cells

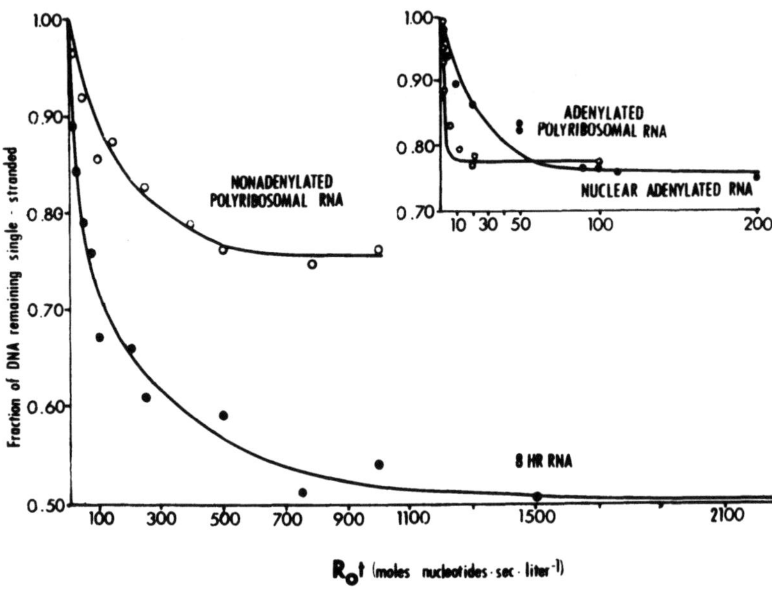

Fig. 21. Hybridization of labeled HSV-1 DNA with excess RNA extracted from the nucleus and polyribosomes of infected cells. The lines are computer plots calculated according to the equation in Sect. 3.3.2 and fitted by nonlinear least squares regression method. The computer analysis involved the determination of parameters α_n and R_n in the equation in which the number of classes of RNA (n) was assumed to be 1, 2, 3, 4 etc. The fit for $n = 2$ was better than $n = 3$, etc. for 8-hour whole-cell RNA and nonadenylated polyribosomal RNA, whereas $n = 1$ gave a better fit than $n = 2$, 3, etc. for nuclear (●) and polyribosomal (O) adenylated RNA (insert). Data from Silverstein *et al.* (1973).

productively infected with HSV-1 and HSV-2. It is convenient at this point to consider the requirements for and site of synthesis of viral RNA. The data may be summarized briefly as follows:

(1) The synthesis of viral RNA takes place in the nucleus. This conclusion is based on observations (Wagner and Roizman, 1969*a*; Roizman *et al.*, 1970*a*) that virus-specific labeled RNA appears in the nucleus, and specifically in the nuclear particulate fraction, without a lag, but that a significant (10–15 minute) lag exists between the time of labeling and the appearance of virus-specific sequences in the cytoplasm of the infected cell. Parenthetically, Miyamoto and Morgan (1971) suggested that transcription of viral DNA be initiated in the cytoplasm. This suggestion was made to explain observations on uncoating of nucleocapsids in the cytoplasm. All of the available evidence is that viral DNA is transcribed in the nucleus, and, although there is no direct evidence on this point, it is likely that the transcription is also initiated in the nucleus.

(2) At least 44–45% of the DNAs of HSV-1 and HSV-2 are transcribed early in the absence of protein synthesis (Table 6) and, by extension, in the absence of DNA synthesis. It seems clear that the enzyme transcribing viral DNA is either brought into the cell by the virus or is a host enzyme which has the competence to transcribe at

TABLE 6

The Effect of Cycloheximide on the Extent of Transcription of Herpes Simplex Virus 1 and 2 DNAs Early in Infection[a]

Virus	Treatment	Total DNA transcribed, %	R_0t[b]
HSV-1	None	44	2350
	Cycloheximide	44	500–2500[c]
HSV-2	None	21	800–2000[c]
	Cycloheximide	45	3020

[a] The cycloheximide in concentrations of 100 μg/ml of medium was present during and after exposure of cells to virus. The hybridization procedures were as described by Frenkel *et al.* (1973). Data from Frenkel *et al.* (1973).

[b] The product of RNA concentration (R_0) in moles nucleotide/sec and time of hybridization to which the hybridizations were carried out.

[c] No further hybridization occurred in this range.

least 90% of the total transcribable DNA. We cannot differentiate as yet between these two hypotheses. However, if the infectious DNA isolated by Sheldrick *et al.* (1973) is fully deproteinated, it would follow that the viral DNA is at least able to utilize a host polymerase. The 24% increase in the amount of HSV-2 DNA transcribed in 2-hour-infected HEp-2 cells in the presence of cycloheximide requires an explanation. Among the various hypotheses which could account for these data are the following: (1) transcription of 24% of the DNA is normally prevented by a protein made early after infection, (2) degradation of RNA transcripts from 24% of viral DNA is blocked or reduced in the presence of cycloheximide, and (3) the transcription of the DNA is accelerated in some unknown fashion in the presence of the cycloheximide. It should be pointed out that Rakusanova *et al.* (1971) reported that in pseudorabies-infected cells the RNA made in the presence of cycloheximide was complementary to only 25% of the template transcribed "early" in the absence of the drug. It is conceivable that the requirements for the transcription of pseudorabies DNA are different from those observed for HSV-infected cells. However, Rakusanova *et al.* were measuring the rates of saturation of viral DNA on filters with labeled RNA extracted from treated and untreated infected cells and they observed that RNA made in the presence of cycloheximide required more DNA before it became depleted than that made in the absence of the drug. An alternative explanation of their data is that the concentration of viral RNA in treated infected cells was higher than in untreated cells.

3.3.5. The Processing of Viral RNA

For the purposes of this discussion it is convenient to define processing of RNA as any change in the structure or in the site of accumulation of viral RNA. Based on this definition there appear to be 4 kinds of processing, i.e., (1) post-transcriptional cleavage of RNA, (2) selective adenylation of RNA species, (3) transport of RNA from nucleus to cytoplasm, and (4) cytoplasmic modifications and degradation of RNA.

3.3.5a. Post-Transcriptional Cleavage

The evidence for post-transcriptional cleavage is inferential, based on studies of the size of virus-specific RNA in nuclei and polyribosomes of infected cells (Wagner and Roizman, 1969*b*; Roizman *et al.*, 1970*a*). The data can be summarized as follows: Virus-specific nuclear RNA species range in sedimentation constant from <10 to >60 S. The distribution reflects the size of the RNA and not its conformation since RNA denatured with dimethyl sulfoxide gave a similar distribution. Polyribosomal RNA, on the other hand, ranged in size from < 10 to about 35 S. In hybridization competition tests, unlabeled polyribosomal RNA competed 75–80% of the labeled nuclear RNA fractionated according to size ranges of 10–28 S, 28–50 S, 50–60 S, and >60 S. These data suggested that at least 75–80% of the viral sequences present in the high-molecular-weight nuclear RNA find their way into smaller nuclear RNA species as well as into the smaller polyribosomal mRNA, and that at least a fraction of viral RNA is made as a high-molecular-weight precursor which is cleaved before appearing in polyribosomes. However, the data are incomplete, some experimental uncertainties remain, and few firm conclusions can be made at this time. Specifically, (1) the high-molecular-weight RNA present in the nucleus has not been chased completely into smaller, polyribosomal-size, viral RNA. The smallest pulse-labeling interval tested, 12 minutes, labeled not only high-molecular-weight RNA but also fragments in the size range of polyribosomal RNA. Although this may be unlikely, the possibility still exists that the high-molecular-weight RNA represents aberrant transcription and that the sequences present in it do not find their way into polyribosomes. (2) As pointed out, the hybridization competition tests were done in a two-phase system. In consequence, the hybridizations involved abundant RNA species, leaving unanswered the question of whether the high-molecular-weight RNA consisted of both abundant and scarce or of abundant sequences only. (3) Perhaps more significant is the obvious conclusion that if the high-molecular-weight RNA serves as a precursor to viral mRNA, the nucleus must contain enzymes recognizing specific sequences in viral RNA. The finding of such an enzyme would materially strengthen the hypothesis that the high-molecular-weight RNA functions as a precursor.

3.3.5b. Adenylation

Adenylation of HSV and pseudorabies RNA has been reported by Bachenheimer and Roizman (1972) and by Rakusanova *et al.* (1972) respectively. (1) Polyadenylic [poly(A)] sequences comprise 15–20% of total HSV mRNA present in polyribosomes at 4.5 hours post-in-

TABLE 7

Nonviral Origin of Poly(A) Sequences Attached to Viral RNA[a]

Determination	^{3}H-Adenosine, cpm	Amount of hybrid, %
DNA–RNA hybrid eluted with 0.3 M PB	8304	100
Poly(A) bound to filter after digestion of hybrid with ribonuclease in 0.26 M PB	1320	16
DNA–RNA hybrid not bound to filter after digestion with ribonuclease	5520	66
Trichloroacetic acid-precipitable cpm after denaturation of hybrid and digestion with ribonuclease	149	2

[a] Sonically treated, denatured HSV-1 DNA was mixed with ^{3}H-adenosine, ^{14}C-uridine-labeled RNA prepared from the cytoplasm of HSV-1-infected cells and adjusted to a final concentration of 0.044 M PB (stock solution of phosphate buffer consisted of equal volumes of 2.4 M Na_2HPO_4 and 2.4 M NaH_2PO_4). The hybridization mixture was incubated for 25 minutes at 79°C, after which time it was diluted 10-fold and loaded onto a 1-ml hydroxyapatite column. Under these conditions, only the most abundant RNA species will hybridize to DNA. Single-stranded material, including unreassociated DNA and unhybridized RNA, was eluted with 0.18 M PB, and the DNA–RNA hybrids and reassociated DNA were eluted at 0.3 M PB. The column fractions containing the hybrids were pooled, adjusted to 0.26 M PB (equivalent to 0.39 M monovalent cation) and digested with pancreatic ribonuclease (200 μg/ml) and T1 ribonuclease (5 units/ml) for 30 minutes at 37°C. The digest was extracted with phenol as described in the footnote to Table 1, and the aqueous phase was precipitated with an equal volume of 15% trichloroacetic acid. The precipitate was washed with 80% ethanol, resuspended in 0.5 M KCl buffer, and filtered slowly through a cellulose nitrate filter presoaked in the same buffer. The filter was dried and counted, while a fraction of the filtrate was precipitated with trichloroacetic acid, washed with ethanol, and resuspended in TSM buffer. The hybrids were denatured and digested with deoxyribonuclease and ribonuclease, and the digest was assayed for trichloroacetic acid-precipitable radioactivity.

fection and consist of chains ranging up to 160 nucleotides long (Bachenheimer and Roizman, 1972). (2) Several lines of evidence indicate that adenylation takes place in the nucleus. For example, the appearance of ^{3}H-adenosine in poly(A)-rich nuclear RNA lagged behind the appearance of the label in total nuclear RNA. In agreement with the expectation that the adenylated RNAs entering the polyribosomes immediately after a short pulse of ^{3}H-adenosine should contain the bulk of the label in the poly(A) chain but that on prolonged labeling the label should be randomized, the proportion of ^{3}H-adenosine in the poly (A) moiety of adenylated polyribosomal RNA decreased with time after addition of the label (Bachenheimer and Roizman, 1972). (3) Addition of poly(A) appears to be a post-transcriptional event. The strongest evidence in support of this conclusion is the observation that the poly(A) chains covalently linked to viral RNA are nearly quantitatively released from viral DNA–RNA hybrids by ribonuclease digestion in 0.26 M phosphate buffer (equivalent to 0.39 M monovalent cation) (Table 7). Lastly (4), Extensive adenylation is restricted to a specific set of transcripts comprising the abundant viral RNA species. Thus, Silverstein *et al.* (1973) showed that adenylated nuclear and polyribosomal RNA extracted from HEp-2 cells 8 hours post-infection with HSV-1 formed one abundance class and were complementary to 24 and 22% of DNA, respectively (Fig. 21 and Table 8). Hybridization competition tests showed that nuclear and polyribosomal adenylated RNA competed quantitatively with each other and with abundant RNA species extracted from 8-hour-infected cells (Table 9). By contrast, the nonadenylated polyribosomal RNA formed two classes, i.e., abundant and scarce, complementary to 6 and 21% of viral DNA, respectively (Table 8). In abundance competition tests (Table 9) the nonadenylated polyribosomal RNA was found to compete in part with the adenylated polyribosomal RNA. Thus, of the 27% of total viral DNA sequences complementary to nonadenylated RNA, 6% of the total are also complementary to adenylated species. Among the hypotheses which could explain these data are that some transcripts are present in both highly adenylated and poorly adenylated form or that

TABLE 8

Herpes Simplex Virus 1 and 2 RNA Transcripts, the Size of the DNA Template from Which They Arise, and Their Abundance[a]

RNA fraction tested	Fraction of DNA transcribed			RNA abundance, nmoles/liter		Viral RNA in the fraction,[b] wt/wt %
	α_1	α_2	Total	R_1	R_2	
Whole cell	0.22	0.27	0.49	4.71	0.357	2.6
Adenylated nuclear	0.24	—	0.24	5.36	—	3.2
Adenylated polyribosomal	0.22	—	0.22	69.8	—	37.1
Nonadenylated polyribosomal	0.06	0.21	0.27	6.29	0.382	1.3

[a] α_1, α_2 are the fraction of the viral DNA serving as templates for RNA present in abundancies R_1 and R_2, respectively. R_1 and R_2 expressed in moles of viral RNA fragments of $0.99 \times 10^8 \cdot \alpha_n$ daltons in molecular weight. Data from Silverstein *et al.* (1973).

[b] Calculated by dividing R_n (moles/liter)$\cdot \alpha_n \cdot 0.99 \times 10^8$ g/mole by the total RNA concentration in the hybridization mixture expressed in g/liter.

TABLE 9

Abundance Competition Tests between Adenylated Nuclear RNA, Adenylated Polyribosomal RNAs, Nonadenylated Polyribosomal RNA, and Whole-Cell Abundant RNA Extracted from Infected Cells 8 Hours Post-Infection[a]

Competition experiment	RNA source	R_0t, moles nucleotides · sec/liter	Observed DNA in hybrid, %	Predicted DNA in hybrid, %: Identical templates	Predicted DNA in hybrid, %: Nonoverlapping templates
1. *Adenylated polyribosomal vs. adenylated nuclear 8-hour RNA*					
	Adenylated polyribosomal	20	22		
	Adenylated nuclear	90	24		
	Mixture: adenylated nuclear + adenylated polyribosomal	20 90	23	25	46
2. *Adenylated polyribosomal vs. abundant RNA from 8-hour-infected cells*					
	Adenylated polyribosomal	25	23		
	8-hour-infected cells	80	28		
	8-hour-infected cells	105	32		
	Mixture: adenylated polyribosomal + 8-hour-infected cells	25 80	30	32	50
3. *Adenylated nuclear vs. abundant RNA from 8-hour-infected cells*					
	Adenylated nuclear	101.1	26		
	8-hour-infected cells	80	28		
	8-hour-infected cells	180	35		
	Mixture: Adenylated nuclear + 8-hour-infected cells	101 80	30	35	50
4. *Adenylated vs. nonadenylated polyribosomal RNA*					
	Adenylated polyribosomal	25	23		
	Nonadenylated polyribosomal	780	26		
	Mixture: adenylated polyribosomal + nonadenylated polyribosomal	25 780	43	26	49

[a] Adenylated nuclear and polyribosomal RNA and the nonadenylated polyribosomal RNA were prepared from nuclei and cytoplasm of 8-hour-infected cells as described in (Silverstein *et al.*, 1973). The RNA extracted from 8-hour-infected cells and used in experiments 2 and 3 was not fractionated further. The R_0t values chosen for adenylated nuclear and polyribosomal RNA and nonadenylated polyribosomal RNA yield a maximum amount of hybridization detected with the species (see Fig. 21). Data from Silverstein *et al.* (1973).

the nonadenylated RNA fraction is contaminated by fragments arising from sheared adenylated RNA.

Parenthetically, if the assumption is made that the DNA–RNA hybridization rate constant is the same as the DNA–DNA reassociation rate constant, virus-specific adenylated RNA sequences may be calculated to constitute 37% of total adenylated sequences in the polyribosomes (Table 8). If the DNA–RNA hybridization rate constant is 0.43 of that of DNA–DNA reassociation rate constant, as calculated by Melli *et al.* (1971), it would follow that at least 86.4% of all adenylated RNAs in polyribosomes are virus-specific. This figure does not include the poly(A) moiety of the adenylated mRNA, which does not hybridize with viral DNA and comprises 15–20% of the molecule.

The finding that only a specific set of transcripts is adenylated raises several questions with regard to (1) the signal which identifies which transcripts are to become adenylated, (2) the relationship of the adenylation of mRNA and its relative abundance, and (3) the function of the poly(A) moiety of the mRNA. The answers to these question are not known. It seems probable that the signal for adenylation is contained in viral DNA and that its appearance in the 3′ terminus of the transcript calls for its adenylation. Likewise, it is conceivable, although definitive evidence is still lacking, that adenylated RNA has a longer half life and that the difference between abundant and scarce RNA is determined, wholly or in part, more by the presence of poly(A) than by differential rates of transcription.

3.3.6. Transport of Viral RNA

The transport of viral RNA sequences from nucleus to cytoplasm has not been studied extensively. The available information is limited to two observations. First, following a short pulse with radioactive uridine, there is a 10- to 20-minute lag between the appearance of radioactively labeled HSV-1 RNA in the nucleus and in polyribosomes (Roizman *et al.*, 1970*a*; Wagner and Roizman, 1969*b*). Second, whereas the viral RNA sequences in the cytoplasm and polyribosomes are identical, the nucleus contains sequences which are not present in polyribosomes or free in the cytoplasm. Thus in the course of productive infection only 42–43% of DNA is represented in polyribosomes. Transcripts of the remaining regions of transcribed DNA are retained in the nucleus and connot be detected in the cytoplasm. A more striking difference between RNA sequences present in cytoplasm and nuclei emerged in cells treated with cycloheximide for several hours from the time of infection. In these cells, the viral RNA sequences contained in the nuclei arose from transcription of 50% of viral DNA. In contrast, the cytoplasm contained sequences arising from only 10% of the DNA (Roizman *et al.*, 1974; Kozak and Roizman, 1974). In accord with this finding, only a small subset of viral polypeptides, i.e., the α group (see Sect. 3.4.7) was made in the infected cells immediately after withdrawal of cycloheximide. Attempts to chase the RNA sequences selectively accumulating in the nucleus into cytoplasm failed. In fact, *de novo* RNA synthesis was required for the synthesis of viral polypeptides other than those in the α group (Roizman *et al.*, 1974; Honess and Roizman, 1974). The precise difference between the transcripts from the same regions of the DNA which at one point in infection are retained in the nucleus and at another found in association with polyribosomes is unknown.

3.3.7. Cytoplasmic Modification of Viral RNA

Studies by Bachenheimer and Roizman (unpublished data) indicate that the poly(A) chains attached to the mRNA decrease in size The decrease in size is gradual, extending over several hours, and also occurs in the presence of inhibitors of both RNA and protein synthesis. The significance of these findings remains unclear. It is not known whether complete degradation of the poly(A) moiety occurs; whether the removal of all poly(A) results in rapid degradation of the remaining mRNA molecule; by which mechanism the adenosine residues are removed; and whether the amounts of gene product made is related to the extent of adenylation of mRNA. Parenthetically, it is not likely that the relationship between peptide chain completion and adenosine removal bear a one-to-one relationship in the light of the fact that the de-adenylation is much slower than the calculated average rate of protein synthesis.

3.3.8. Genetic Relatedness of Sequences Specifying Abundant and Scarce RNA in HSV-1- and HSV-2-Infected Cells

In general, the genetic relatedness of two viruses may be determined by DNA–DNA reassociation (described in Sect. 2.3). DNA–RNA hybridizations may be used to obtain similar data and, more relevant here, they may be used to determine the genetic relatedness of specific sets of templates transcribed in the course of infection. This type of analyses has been done with HSV-1 and HSV-2

(Frenkel *et al.*, 1973; Roizman and Frenkel, 1973). Briefly, excess unlabeled RNA extracted at 8 hours post-infection of HEp-2 cells with HSV-1 was hybridized with labeled HSV-1 and HSV-2 DNAs. The basis for these experiments was as follows: First, HSV-1 RNA transcribed off DNA sequences common to both HSV-1 and HSV-2 DNAs must necessarily have the same molar ratios *vis-à-vis* the complementary regions of both HSV-1 and HSV-2 DNA. Second the thermal elution profiles of sheared HSV-1 and DNA–HSV-2 DNA heterohybrids off hydroxyapatite columns showed a very uniform matching of base pairs even though the T_m of the heterohybrids was 10°C lower than that of homohybrids (Kieff *et al.*, 1972). This finding suggested that the matching of base pairs is relatively uniform for all DNA–DNA heterohybrids and that the hybridization rate constant for all of the DNA–RNA heterohybrids should be uniform. The hybridization tests showed that 8-hour HSV-1 RNA was complementary to 48% of homologous DNA, but to only 24% of heterologous DNA, i.e., in good agreement with the results of the DNA–DNA reassociations described in Sect. 2.3. Analysis of the distribution of HSV-1 RNA sequences complementary to HSV-2 DNA indicated that 13% of HSV-2 DNA is complementary to a corresponding amount of HSV-1 DNA serving as template for RNA of high abundance. Correspondingly, 11% of HSV-2 DNA is complementary to HSV-1 DNA serving as template for RNA of low abundance. However, the HSV-1 DNA sequences shared in common by HSV-1 and HSV-2 DNA represent 71% of the HSV-1 templates specifying the adenylated RNA present in high abundance and only 39% of the templates specifying the scarce, nonadenylated RNA. The data indicate that in the course of evolution of HSV-1 and HSV-2 from a common ancestor the sequences specifying abundant RNA diverged less than those specifying the scarce RNA.

3.3.9. Viral Transfer RNAs

In a series of papers published since 1965, Subak-Sharpe and coworkers (Subak-Sharpe and Hay, 1965; Subak-Sharpe *et al.*, 1966; Hay *et al.*, 1966, 1967; Subak-Sharpe, 1966) presented data in support of the hypothesis that herpesviruses specify arginyl- and seryl-tRNA. The hypothesis was based on the assumption that in the course of evolution the amount of tRNA for each of the amino acids would become adjusted so that it would be in direct proportion to the frequency of the codon in the mRNA. The argument was advanced that, since the G+C content of herpesvirus DNA is considerably greater than that of the mammalian cell DNA, it would be expected that the number of codons containing G and/or C would be greater in viral mRNA than in cellular mRNA. Consequently, cells infected with DNA viruses rich in G+C should develop a shortage of tRNAs with recognition sites complementary to codons containing G+C. The argument was refined on the basis of nearest-neighbor analysis, and the putative scarce tRNA species were identified as those containing complementary recognition sites for codons containing the doublet CpG (CpGpX and XpCpG). The authors' finger of suspicion pointed squarely to arginyl- and seryl-tRNA.

In the first paper, Subak-Sharpe and Hay (1965) reported as evidence for the synthesis of new tRNA the coincidence of two activities in one fraction obtained by chromatography of RNA from infected cells on a methylated albumen-kieselguhr column. The fraction contained newly synthesized RNA complementary to viral DNA and RNA capable of accepting activated amino acids. The paper did not show whether the RNA complementary to viral DNA was the same one accepting the amino acids, thus leaving open the possibility that small fragments of labeled viral mRNA could have eluted together with unlabeled host transfer RNA synthesized before infection. Two subsequent experiments seemed more convincing. In the first, Subak-Sharpe *et al.* (1966) showed partial separation of arginyl-tRNA from infected and uninfected cells. In the second, they digested with ribonuclease T_1 artificial mixtures of ^{3}H-labeled amino-acyl-tRNA from uninfected cells and ^{14}C-labeled amino-acyl-tRNA from infected cells. The fragments were then chromatographed. Ribonuclease T_1 hydrolyzes RNA at guanine-phosphate bonds exclusively. If the putative virus-specific tRNAs were to differ from the corresponding host tRNAs with respect to the position of the first guanine base following the terminal triplet pCpCpA, it would be expected that after digestion with T_1 at least one set of fragments attached to the amino-acyl group of the tRNA from infected cells should differ from those of uninfected cells. The results did in fact show a difference between arginyl-acyl-oligonucleotides from infected and uninfected cells.

This series of experiments gave rise to considerable interest, motivated in part by the possibility that tRNAs and corresponding enzymes might explain the rapidity and ease with which herpesviruses seem to inhibit completely host protein synthesis without affecting their own. However, attempts to demonstrate more rigorously the existence of arginyl- and seryl-tRNAs failed. Morris *et al.* (1970) could not differentiate by chromatography on reverse-phase Freon columns arginyl- and seryl-tRNAs from infected and uninfected cells. In addition, highly purified arginyl- and seryl-tRNAs from infected cells failed to hybridize to viral DNA. Moreover, RNA of high molecular weight (>50,000) precluded the hybridization of 4 S RNA with viral DNA in competition tests, indicating that 4 S virus-specific RNAs were cleavage products of larger RNA sequences. A subsequent paper from Subak-Sharpe's laboratory (Bell *et al.*, 1971) retracted the original findings, blaming impurities in labeled amino acid preparations for the earlier results. The possibility that herpes simplex viruses specify substantial amounts of arginyl- and seryl-tRNAs seems dead. However the possibility still exists that herpesviruses do specify some tRNA species. Fisher and Fisher (1973) noted, for example, that phenylalanyl-tRNA from HSV-2-infected HEp-2 cells had 110 times the amino acid acceptor activity of the uninfected preparation. It is too early to tell whether the increase is due to tRNA specified by the virus.

3.3.10. Interim Conclusions

Herpesvirus RNA metabolism is characterized by three salient features. (1) We have no evidence that a temporal control of transcription of viral DNA is operative since the DNA sequences transcribed in the absence of viral DNA and protein synthesis cannot be differentiated from those transcribed throughout productive infection. This conclusion is in accord with independent observations that both structural membrane proteins (Roizman, 1972) and capsids (O'Callaghan *et al.*, 1968*a*; Rosenkranz *et al.*, 1968: Falke *et al.*, 1972) were found in herpesvirus-infected cells treated with inhibitors of DNA synthesis. (2) A transcript abundance control is apparent from analyses of viral RNA sequences accumulating in the infected cell. The mechanism by which this control is effected is not clear; it could be operative at a post-transcriptional level and could involve both the extent of adenylation and the longevity of the transcript. (3) Transport of RNA sequences from the nucleus to the cytoplasm also appear to be regulated. This is apparent from the observation that nuclei retain viral RNA sequences which are not present in the cytoplasm. It seems fair to add that the landscape of viral RNA metabolism is dotted by many uncertainties and questions. Thus, little is known concerning the enzyme involved in the transcription of the DNA, details of transcription and processing, or the transport and longevity of viral RNA.

3.4. Viral Protein Synthesis and Metabolism

3.4.1. General Characteristics and Requirements

Protein synthesis in infected cells, as measured by either amino acid incorporation into peptides (Sydiskis and Roizman 1966; Fujiwara and Kaplan, 1967; Honess and Roizman, 1973) or by the amount of polyribosomes present in the infected cells (Sydiskis and Roizman, 1966, 1967), takes place exclusively in the cytoplasm and follows a period of initial decline coinciding with disaggregation of polyribosomes specifying host proteins, a period of increased rates of synthesis coinciding with appearance of new, more rapidly sedimenting viral polyribosomes, and, lastly, a period of slow irreversible decline. At least a portion of the polypeptides made on the cytoplasmic polyribosomes find their way into the nucleus (Fujiwara and Kaplan, 1967; Olshevsky *et al.*, 1967; Spear and Roizman, 1968; Ben-Porat *et al.*, 1969). No special requirement for viral protein synthesis in permissive cells is known. However, a special, and perhaps unique, feature of herpesvirus-infected cells, in contrast to cells infected with some of the small DNA viruses, is the synthesis of structural proteins and at least partial assembly (empty capsids) in cells treated with inhibitors of DNA synthesis (O'Callaghan *et al.*, 1968*a*; Rosenkranz *et al.*, 1968; Roizman, 1972).

3.4.2. Enumeration of Proteins Made in Infected Cells

The purpose of this and the following sections is to analyze the available data on the synthesis and function of the proteins specified by herpesviruses. The major problems which arise in these analyses stem from enumeration of viral proteins and will be dealt with in this section; those that arise from the recognition of viral proteins will be dealt with in a subsequent section.

Enumeration of proteins made in the infected cell presents problems arising chiefly from the number of polypeptides which could potentially be specified by herpesviruses. Assuming a molecular weight of 10^8, asymmetric transcription, and total utilization of transcripts for protein synthesis, it may be readily calculated that HSV-1 is capable of specifying the sequence of some 55,000 amino acids. The structural polypeptides vary in molecular weight from 25,000 to 280,000. If the size distribution of nonstructural proteins is the same, we might reasonably expect 50–60 polypeptides. Thus, early attempts to analyze viral proteins both in this laboratory (Spear and Roizman, 1968) and in others (Courtney *et al.*, 1971; Kaplan *et al.*, 1970) were doomed to failure for technical reasons. These studies utilized the technique of polyacrylamide gel electrophoresis designed around the continuous buffer system (Summers *et al.*, 1965), and they relied on detection of the separated proteins by counting radioactivity in the slices of the gels.

These techniques could not resolve the large number of proteins made in the herpesvirus-infected cell. Further, the procedures then used for the purification of herpesvirus particles were inadequate, as discussed earlier in the text, and hence no meaningful classification of proteins as either structural or nonstructural could be made.

Application of high-resolution polyacrylamide gel electrophoresis technique led Honess and Roizman (1973) to the recognition of at least 49 polypeptides made in the infected cells. These range in molecular weight from 280,000 to 15,000 and have been numbered 1–49 (Table 10, columns 1 and 2).

3.4.3. Specificity of Proteins Made in Infected Cells

The source of genetic information for the polypeptides made in infected cells presents a second major problem in studies of the synthesis of viral proteins. Ideally, conclusive evidence that a protein is specified by a virus is its synthesis in an *in vitro* protein-synthesizing system directed by a viral mRNA. In view of the complexity of the herpesvirus genome, such data may not be available for the foreseeable future. The experimental approaches utilized in analyses of the source of genetic information for the polypeptides detected in HSV-1-infected cells are as follows:

(1) Experimental demonstration that the structure and function of specific proteins are genetically determined by the virus. This type of evidence has been provided for some of the structural proteins of the virus (Schaffer *et al.*, 1971, 1973) and for several enzyme activities (thymidine kinase, deoxycytidine kinase, etc., see Sect. 3.5) which become elevated following infection. In this particular instance, comparisons were made between an HSV-1 strain passaged a limited number of times in HEp-2 cells [HSV-1 (F)] with an HSV-1 strain passaged numerous times in cell culture [HSV-1 (mP)] and with a type 2 strain [HSV-2 (G)]. The intratype comparison showed that polypeptides synthesized in HSV-1 (F)- and HSV-1 (mP)-infected cells were largely congruent in both number and electrophoretic mobility, but differed in the relative amounts of a small number of polypeptides which were made within the same time interval. Thus, significantly more of ICP 1-2 10, 26, 27, and 31 was present in HSV-1 (F)-infected cells than in HSV-1 (mP)-infected cells. Conversely, HSV-1 (mP)-infected cells accumulated more of ICP 4, 21, 22, 23, 24, 42, 43, and 48 than did HSV-1 (F)-infected cells. In the intertype comparison, the general appearance of the gels containing electrophoretically separated

TABLE 10

Enumeration, Classification, and Evidence for Virus Specificity of HSV-1 (F1)-Infected Cell Polypeptides (ICP)[a]

1	2[b]	3	4[c]	5[d]	6[e]		7[f]		8[g]
					Precipitation by antiserum				
ICP	Polypeptide mol. wt. × 10^{-3}	Relationship to VP	Classification of ICP and stability to prolonged chase	Specific stimulation and kinetics of synthesis	Available for precipitation	Precipitated	F ICP different from G ICP	F ICP different from mP ICP	Criteria for virus specificity satisfied, columns
1–2	275	VP1–2	S	+A	0			F > mP	5, 7
3	260	VP3	S	+	0				5
4	184	VP4	S	+C	0			mP > F	5, 7
5	155	VP5	S	+A	+	+	+		5, 6, 7
6	146	VP6	S	+C	+	+	+		5, 6, 7
7	140	—	NS	+	+	+			5, 6
8	135	VP6.5	S	+C	+	+			5, 6
9	130	VP7	U	+B	+	+			5, 6
10	126	[VP8, 8.5	S	+B	+	+	+	F > mP	5, 6, 7
11	120	(broad virion	U	+	+	+	+		5, 6, 7
12	117	glycoprotein	U	+B	+	+			5, 6
13		bands)	U	+	0				5
14	112	VP9]	U	+	0				5
15			S	+	0				5
16		—	NS	+	+	+	+		5, 6, 7
17	103	—	NS(D)	+A	0				5
18	98	VP10	S	+E	0		+		5, 7
19	93	VP11	S	+C	+	+	+		5, 6, 7
20	87	VP12	S	+C	0		+	mP > F	5, 7
21	78	VP14	S	+A	+	+	+	mP > F	5, 6, 7

22	76	—	NS	+	+	−		mP > F	5, 7
23	74	—	NS(D)	+F	+	−		mP > F	5, 7
24	72	—	NS(I)	+	+	−		mF > F	5, 7
25	71	VP15	S	+C	+	−			5
26	65	VP16	S	+B	+	+		F > mP	5, 6, 7
27	63	—	NS	+F	+	+	+	F > mP	5, 6, 7
28	61	VP17	S	+F	+	−	+		5, 7
29	59		U	+F	+	+	+		5, 6, 7
30	57	VP18	S	+	+	−			5
31	53	VP19	S	+A	+	−	+	F > mP	5, 7
32	50	VP20	S	+		−			5
33	48	—	NS	±	0				0
34	44	VP21	S	−E	0				0
35	43	—	NS	−E	+	−			0
36	42	—	NS	+D	+	+			5, 6
37	40	—	NS	+F	+	+			5, 6
38	37	—	NS	+D	+	+			5, 6
39	36	VP22	S	+D	+	+			5, 6
40	33	VP23	S	+F	+	+	+		5, 6, 7
41	32	—	NS	+F	+	−	+		5, 7
42	31	—	U	−E	0			mP > F	7
43	30	—	U	−E	trace	−		mP > F	7
44	27		U	+	0		+		5, 7
45	25	VP24	S	+	0		+		5, 7
46	23		U	+	0				5
47		—	NS	+					5
48		—	NS	+				mP > F	5, 7
49		—	NS	+					5

[a] Data from Honess and Roizman (1973).

[b] Molecular weights of ICP obtained by interpolation of previously determined values for virion polypeptides of Spear and Roizman (1972).

[c] S = structural, NS = nonstructural, U = unassigned. See text for criteria applied in the classification. In addition to the 24 polypeptides described by Spear and Roizman (1972), these analyses took cognizance of additional polypeptides found in purified virions (Heine *et al.*, 1974).

(*Continued on page 308*)

Footnotes to Table 10

Among them were VP6.5, a minor virion polypeptide which coelectrophoresed with ICP8, another was a glycoprotein designated VP8.5. Due to poor resolution of VP7, 8, and 8.5, the mobility analogue of VP8.5 among the ICPs is uncertain and all of the ICPs included within this mobility range are therefore classified as S (ICP 10 which coelectrophoreses with VP8) or as U. (D) = components declining or disappearing on prolonged chase, (I) = components increasing or appearing on prolonged chase.

[d] + = amino acid incorporation into components migrating in these regions was higher in infected than in uninfected cells. − = amino acid incorporation into components migrating in these regions in infected cells was comparable to, or lower than, that observed in the same regions of uninfected cells. See Fig. 3 and related text. Letters A–F refer to the patterns of synthesis as shown in Fig. 29. Absence of a letter designation indicates the pattern of synthesis is uncertain.

[e] In the left-hand column (available for precipitation) 0 and + indicate, respectively, undetectable amounts and the presence of a particular ICP in the supernatant fluid. In the right-hand column, + indicates precipitated and −, not precipitated, by antiserum. See Fig. 22 and related text.

[f] In the comparison of HSV-1 (F1) and HSV-2 (G) (left-hand column), + indicates any HSV-1 (F1) ICP for which there was no exact counterpart in mobility or amount in HSV-2 (G)-infected cells. Comparisons of HSV-1 (F1) and HSV-1 (mP) (right-hand column) indicate significant quantitative differences between ICP with identical mobilities produced in cells infected with the two viruses, i.e., F > mP refers to a polypeptide produced in larger amounts in F- than in mP-infected cells.

[g] Numbers refer to the columns summarizing those criteria for virus specificity which were satisfied by a particular ICP of HSV-1 (F1). For example, ICP 5 satisfied the criterion of induction (column 5), possessed viral antigenic specificity (column 6), and its mobility was affected by virus-type-specific variation (column 7). Symbol 0 indicates that none of these criteria were fulfilled.

HSV-1 (F)- and HSV-2 (G)-infected cell polypeptides were similar, but there were differences in both mobility and amount of polypeptides migrating in corresponding regions of the polyacrylamide gel. Thus, HSV-2 (G)-infected cells lacked exact counterparts in mobility or amounts to HSV-1 (F) polypeptides, 5, 6, 10, 11, 16, 18, 19, 21, 27, 28, 29, 40, 41, 44, and 45.

(2) Demonstration that proteins with the same immunologic specificity are not present in the uninfected cells. The most impressive studies utilizing this technique are those of Watson *et al.* (1966), Watson and Wildy (1969), and Honess and Watson (1974), which used antisera prepared in rabbits against freeze-dried, infected rabbit kidney cells grown in rabbit serum. The assumption that the rabbit makes antibodies against the virus-specified proteins and not against modified host proteins could at least in part be checked by testing the reactivity of these sera with extracts of infected cells of other animal species on the basis of the added assumption that antisera against modified proteins of species A should not react with modified proteins of species B, and that the structure and immunologic specificity of proteins specified by the virus should be independent of the species in which it was made. Analysis of infected-cell extracts by this technique utilizing a serum produced by Honess and Watson (1974) showed (Fig. 22) that of the "soluble" polypeptides available for immunoprecipitation, polypeptides 5, 6, 7, 8, 9–12, 19, 26, 32, 36, 38, and 39 were precipitated efficiently; 16, 21, 28, 29, 37, and 40 were precipitated less efficiently, whereas, 22–24 and 41 did not precipitate at all. It should be emphasized that although precipitation by such antisera is additive evidence for virus specificity, the failure to precipitate a particular polypeptide is not evidence that it is specified by the host.

(3) Demonstration that the appearance and function of protein require genetic expression of the virus. The line of evidence involves the demonstration that the protein or its activity is absent from uninfected cells, that its appearance in the infected cell requires new synthesis and, lastly, that its synthesis is regulated. Thus, differences between the electrophoretic mobilities of infected- and uninfected-cell proteins, e.g., failure of infected-cell proteins to be synthesized in cells treated with inhibitors and mock-infected with inactivated virus, etc., could serve to bolster the hypothesis that specific proteins are specified by the virus. In practice, direct comparisons of the electrophoretic mobilities of infected- and uninfected-cell proteins is not very useful. The problem arises from the fact that uninfected-cell polypeptides are so numerous that even in high-resolution gels nearly every infected-cell polypeptide comigrates with some uninfected-cell polypeptide. A more promising

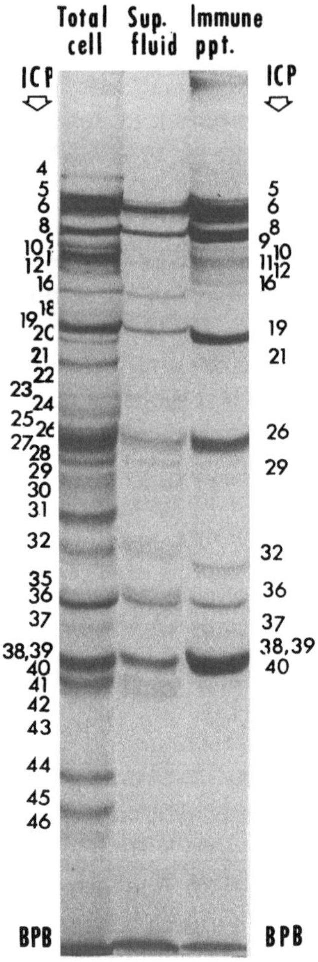

Fig. 22. Autoradiograms of 8.5% polyacrylamide gels containing electrophoretically separated (Total cell) total HSV-1 (C)-infected HEp-2 cell lysate labeled with ^{14}C-amino acids 4.25–6.5 hours post-infection, (Sup. fluid) the supernatant fluid obtained after high-speed sedimentation of the lysate, and (Immune ppt.) an immune precipitate formed by the addition of antiserum specific for virus antigens to the supernatant fluid. Numbers to the left of the three gels refer to infected-cell polypeptide in the total cell lysate, those to the right indicate the infected-cell polypeptide precipitated by the virus-specific antiserum. The antiserum was prepared by Honess and Watson in rabbits immunized with a lyophilized preparation of RK-13 cells infected with HSV-1. Data from Honess and Roizman (1973).

approach is to examine the rates of synthesis of uninfected- and infected-cell polypeptides based on the observation that the rate of synthesis of host polypeptides is generally lower than that of viral polypeptides and decreases in infected cells. Viral polypeptides on the whole tend to show increasing rates of synthesis, at least early in infection.

A summary of the analyses is furnished in Table 10. It can be seen that evidence for virus specificity, either kinetic, immunologic, or genetic is available for 47 of the 49 polypeptides.

3.4.4. Differentiation of the Structural and Nonstructural Polypeptides

Current differentiation between structural and nonstructural polypeptides is based on electrophoretic mobility in polyacrylamide gels (Honess and Roizman, 1973) (Fig. 23). Based on this technique, of the 47 polypeptides analyzed to date 23 were found to comigrate with virion proteins in parallel gels, 15 migrated sufficiently differently to be classified as nonstructural, and 9 could not be classified as either structural or nonstructural (Table 10). In general, classification of polypeptides on the basis of migration alone is fraught with problems arising from post-translational modification of polypeptides, i.e., chiefly glycosylation which tends to change the electrophoretic mobility of the polypeptides. The problem is also complicated by the fact that heterogeneity in the extent of glycosylation of the polypeptides may be reflected in the rather broad bands formed by glycoproteins, whereas the nonglycosylated polypeptides tend to make much narrower bands. The unclassified polypeptides either form individual bands coinciding with broad glycoprotein bands, or can not be identified with certainty as precursors of specific glycoproteins. Ultimately, precise

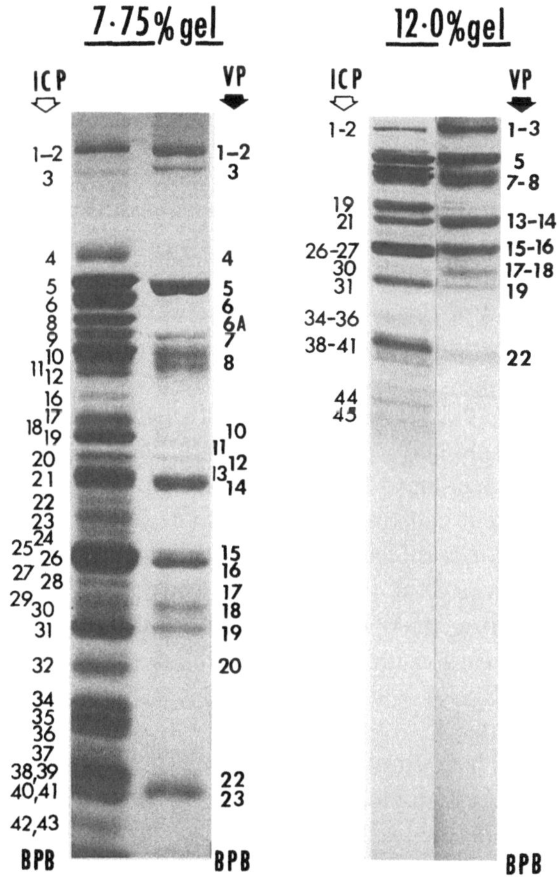

Fig. 23. Autoradiographic comparisons of electrophoretically separated virion polypeptides labeled with ^{14}C-amino acids (purified from cells and labeled 3–20 hours post-infection) and polypeptides of HSV-1 (C) infected cells labeled with ^{14}C-amino acids 3.5–8.7 hours post-infection. Solubilized labeled virions and labeled infected cells were subjected to electrophoresis in parallel on 7.75% and 12% slab gels. These were then processed for autoradiography. The figures are reproductions of photographs of the original autoradiogram and, therefore, suffer from loss of resolution. The darker area of the top of the 7.75% gel is due to uneven illumination during photography. The 12% gel was printed with different exposure periods for each of the two samples. Data from Honess and Roizman (1973).

identification of a glycosylated protein and its nonglycosylated precursors will have to be done by tryptic peptide analysis.

3.4.5. Post-Translational Modification of Viral Proteins

At least 3 kinds of post-translational modifications may alter the mobility of virus polypeptides in polyacrylamide gels, i.e., (1) rapid post-translational cleavages such as those seen in picornavirus-infected cells (Summers and Maizel, 1968; Jacobson *et al.*, 1970; Summers *et al.*, 1972), (2) slow post-translational cleavages occurring during intracellular translocation or during assembly of the virion, and (3) conjugation or addition of prosthetic groups, i.e., glycosylation, phosphorylation, amidation, acetylation, methylation, etc. It is convenient to discuss each type of post-translational modification separately.

3.4.5a. Rapid Post-Translational Cleavages

Two series of experiments indicate that the polypeptides synthesized in HSV-1 (F)-infected cells do not undergo rapid post-translational cleavages of the kind seen in picornavirus-infected cells (Honess and Roizman, 1973). (1) In the first series, infected-cell lysates prepared immediately after a 15-minute labeling period and after a 40-minute chase were subjected to electrophoresis on polyacrylamide gels. No differences between the relative amounts and electrophoretic mobilities of the labeled proteins were found. (2) The second series of experiments was based on the assumption that if rapid post-translational cleavages did occur, they could be blocked by the inhibitors of proteolytic enzymes, tolyl-sulfonyl-lysyl chloromethyl ketone (TLCK) and toly-sulfonyl-phenylalanyl chloromethyl ketone (TPCK), which have been used to prevent the cleavage of the poliovirus polyproteins in different cell lines (Summers *et al.*, 1972). Controls included lysates of poliovirus-infected HEp-2 cells pulse-labeled in the presence and absence of the inhibitors. As shown in Fig. 24, in contrast to poliovirus proteins, HSV-1 polypeptides made in the presence and absence of the inhibitors could not be differentiated with respect to electrophoretic mobility or with respect to the accumulation of high-molecular-weight proteins.

These findings are significant from one point of view. Although it will become clear that some polypeptides are in fact "modified" subsequent to their synthesis, the absence of a rapid post-translational cleavage conforms with the operational definition that herpesvirus polypeptides are specified by monocistronic messages.

3.4.5b. Modification of Proteins during Translocation from Cytoplasm to the Nucleus

Little is known concerning the requirements and potential modification of proteins for transport from the cytoplasm to the nucleus. However, it has now been well documented that in infected cells all proteins are made in the cytoplasm and that at least some are transported into the nucleus (Olshevsky *et al.*, 1967; Spear and Roizman, 1968; Ben-Porat *et al.*, 1969). Radiolabeling experiments suggest that the equilibration between the cytoplasmic and nuclear pools of translocated polypeptides is rather slow (Spear and Roizman, 1968). If the movement of polypeptides from cytoplasm to nucleus is indeed slow, the observation by Roizman *et al.* (1967) and by Ross *et al.* (1968) of compartmentalization of herpesvirus antigens is significant, and it implies that if the same proteins are present in both nucleus and cytoplasm, they must display different immunologic reactivities in the two compartments. Fujiwara and Kaplan (1967) also reported that in pulse-labeled cells infected with pseudorabies virus, radioactive peptides first appeared in the cytoplasm; they were then chased into the nucleus. Nuclear labeled peptides reacted with antiviral antibody whereas cytoplasmic labeled peptides did not. Mark and Kaplan (1971) reported that translocation of proteins from cytoplasm to nucleus requires arginine and is reduced or does not occur in deprived cells. Again, these experiments need confirmation and extension to determine whether arginine is required for the translocation of both structural and nonstructural proteins and whether arginine is required for translocation *per se* or for aggregation into subviral structures.

3.4.5c. Phosphorylation and Glycosylation of the Viral Proteins

Studies by Gibson and Roizman (1974) indicate that there is an apparent stimulation of phosphorylation of proteins in infected cells in both the nucleus and cytoplasm of cells. The cytoplasmic fractions of infected cells contained 4 major bands of ^{32}P radioactivity, 3 of which had electrophoretic mobilities indistinguishable from VP6, 12, and 16. The fourth migrated in the gel a little slower than protein 16 and was phosphorylated to about the same extent. The band corresponding in

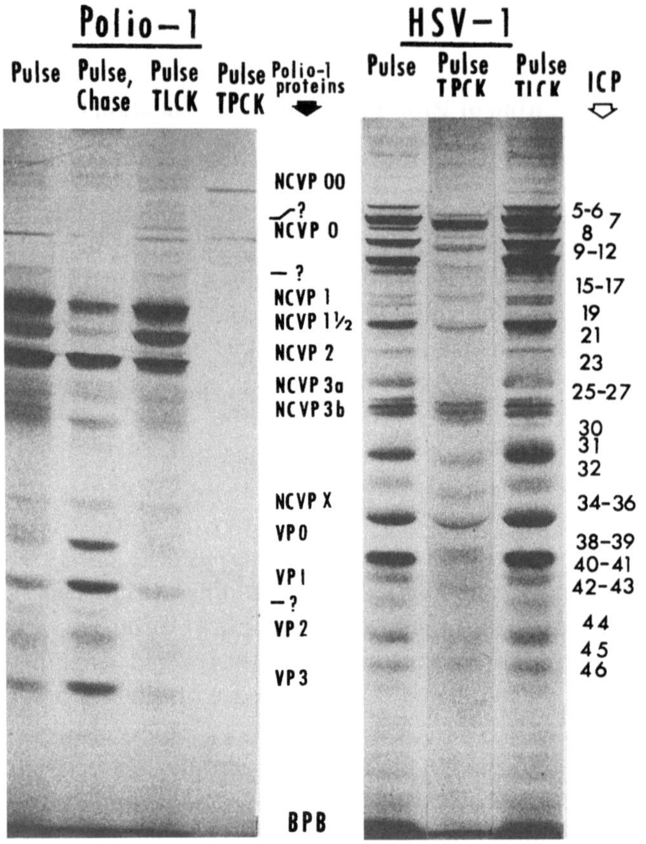

Fig. 24. Autoradiogram of a 9.0% polyacrylamide slab gel containing electrophoretically separated polypeptides synthesized in HSV-1 (F)- and poliovirus-1-infected HEp-2 cells labeled with ^{14}C-amino acids in the presence and absence of specific inhibitors of trypsins (TLCK) and chymotrypsin (TPCK) proteases. Samples on the left are the proteins synthesized in poliovirus-infected cells, pulsed for 15 minutes in the absence of inhibitors (Pulse), pulsed for 15 minutes and chased for 30 minutes in the absence of inhibitors (Pulse, Chase), and pulsed for 15 minutes in the presence of 10^{-4} M TLCK (Pulse TLCK) and TPCK (Pulse TPCK). Samples on the right are of HSV-1 (F) ICP synthesized in the absence (Pulse) or presence of inhibitors (TPCK, TLCK). The polio-1 proteins are designated according to Jacobson *et al.* (1970). The numbers to the right of HSV-1 (F) samples refer to selected HSV-1 (F)-infected cell polypeptides. The original autoradiogram was cut for the preparation of this figure, and the exposure interval employed for the printing of the autoradiograms of HSV-1 (F) Pulse-TPCK sample was longer than that employed for other samples. The molecular-weight estimates of poliovirus proteins agreed well with those of virion proteins subjected to electrophoresis in peripheral chambers of the same gel slab, but they were removed to simplify this figure. Data from Honess and Roizman (1973).

electrophoretic mobility to VP12 migrates only slightly slower than a cell protein which is phosphorylated in the uninfected cells. Nuclci obtained by NP-40 treatment of infected cells contained all 4 major cytoplasmic ^{32}P-labeled polypeptides plus 2 additional polypeptides corresponding in electrophoretic mobility to a nonstructural protein and to protein 22a. It is noteworthy that the relative amounts of ^{32}P bound to the electrophoretically separated structural proteins in cell lysates do not, in general, correspond to the relative amounts of ^{32}P bound to virion proteins. The major point that can be made about these data is that phosphorylation of virus-specified proteins is not random, but selective, and that the structural proteins are incorporated into subviral structures after phosphorylation.

Information concerning the glycosylation of viral polypeptides is minimal. (1)The glycosylation of viral proteins occurs in the absence of any detectable glycosylation of host proteins. It can be inferred from this that while we have no information on the stability of polysaccharide chains on host glycoproteins, there is no net exchange of sugars from host glycoproteins during infection (Heine *et al.*, 1972; Roizman and Heine, 1972). (2) Nothing is known concerning the nature of the glycosyl transferases which glycosylate viral proteins. In one study (Spear and Roizman, 1970) it was noted that the relative amounts of glycosylation of the viral glycoproteins was host dependent. In retrospect, it is not entirely clear whether the differences were due to a more rapid rate of polysaccharide chain elongation or to a different sequence of sugars in the polysaccharide chains. Parenthetically, little is known concerning the composition and sequence of the sugars in the polysaccharide chains. What is known is that virion glycopeptides show a size distribution similar to that of glycopeptides obtained from viral glycoproteins inserted into the plasma membranes of infected cells and to that of glycopeptides obtained from the plasma membranes of uninfected hosts. (3) Viral glycoproteins present in the infected cell partition with membranes (Spear and Roizman, 1970). Pulse-chase experiments done in the same study indicate that the protein moiety binds to membranes after synthesis and is glycosylated *in situ*. Moreover, glycosylation of HSV-1 proteins is accompanied by a decreased mobility in polyacrylamide gel, indicating a substantial change in size and conformation. (4) The requirements for glycosylation are not fully understood. It is apparent that glycosylation can take place in the presence of inhibitors of protein synthesis, albeit at a reduced rate. Subsequent attempts (Honess and Roizman, in preparation) to elucidate the continued glycosylation of proteins in infected cells have revealed additional puzzles but offered no documented explanation to account for this observation. The problem arises chiefly from the expectation that when protein synthesis is blocked, glycosylation would continue until the pool of unglycosylated proteins is exhausted. The results did not agree with this expectation. Specifically, incorporation of both glucosamine and fucose continues at the same rate for a brief interval, then decreases and remains relatively level at about half the rate of that of untreated cells for at least 4 hours. Analyses of cell lysates of infected treated and untreated cells indicate that the same proteins are glycosylated in the presence and absence of protein-synthesis inhibitors. On the other hand, the glycopeptides labeled in the presence of inhibitors of protein synthesis are smaller than those in untreated infected cells, and, moreover, the ratio of fucose to glucosamine in the small glycopeptides obtained from treated infected cells is significantly higher than in untreated infected cells, suggesting that fucose incorporation into large fragments is inhibited in the absence of protein synthesis. We still have no explanation for the continued glycosylation of the infected cell polypeptides in the treated cells for many hours at a reduced rate.

3.4.6. Kinetics of Synthesis of Viral Proteins

Analysis of the kinetics of synthesis of the polypeptides made in HSV-1 (F)-infected cells was recently attempted by Honess and Roizman (1973). These analyses, as demonstrated in the preceding section, showed that all polypeptides made in the infected cells are primary products of translation and presumably not secondary, or cleavage, products of some polyproteins. The assumption, discussed in Sect. 2.4.5, is that the relative amount of labeled amino acids incorporated into polypeptides accurately represents the relative amounts of protein actually synthesized during the labeling interval. In general, estimates of relative amounts of proteins based on staining with Coomassie brilliant blue and autoradiography of labeled polypeptides agree very well for structural proteins (Gibson and Roizman, 1972, 1974), but this has not yet been demonstrated for nonstructural proteins.

The analyses of the kinetics of synthesis consisted of two series of experiments. In the first series, the rate of incorporation of labeled amino acids during the reproductive cycle was measured at different times during the reproductive cycle to describe the overall pattern of protein synthesis. As shown in Fig. 25, the overall rate of synthesis declines at first rather precipitously, reaching a minimum at 2–3 hours post-infection, corresponding to the disappearance of host polyribosomes reported earlier (Sydiskis and Roizman, 1966, 1967).

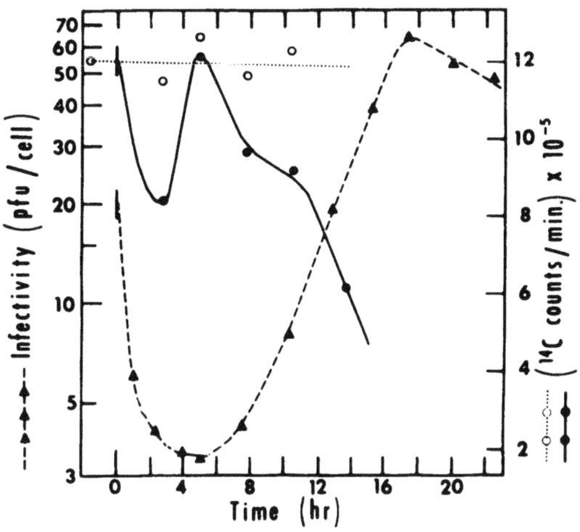

Fig. 25. Accumulation of infectious HSV-1 virus (▲——▲——▲) and changes in the rate of amino acid incorporation into acid-insoluble material in infected (-●-●-●-) and mock-infected HEp-2 cells (O-O-O-O). Amino acid incorporation was measured as acid-insoluble radioactivity at the end of 2-hour labeling intervals during infection or mock infection and plotted at the midpoint of the interval. Points for infected cultures are the mean of duplicate determinations (2 × 10^6 cells/determination); those for mock infected cells are results of a single determination. Accumulation of infectious virus was measured by plaque titrations of duplicate samples of homogenates containing 2 × 10^6 infected cells for each time interval. Data from Honess and Roizman (1973).

The rate of protein synthesis then increases and reaches a maximum comparable to that of mock-infected cultures at 5–6 hours post-infection, corresponding to the time at which viral polyribosomes reach maximum levels in the infected cell (Sydiskis and Roizman, 1966, 1967). Thereafter, the rate of synthesis of protein decreases at a nonuniform rate. To determine the kinetics of synthesis of the individual polypeptides, in the second series of experiments the cells were pulse-labeled during 4 intervals, i.e., 2.5–2.9; 3.75–4.0; 5.0–5.25, and 11.00–11.5 hours post-infection, respectively. These time intervals coincide with the end of the initial decline due to disappearance of host protein synthesis, the period just before and at the maximum rates of viral protein synthesis, period of the terminal decline in protein synthesis.

The autoradiograms of the infected-cell lysates separated in polyacrylamide gels are shown in Fig. 26, along with the virion polypeptides used as molecular weight markers. The data were analyzed as follows: First, the relative molar rates of synthesis of each polypeptide were calculated with the aid of a Gilford scanner attached

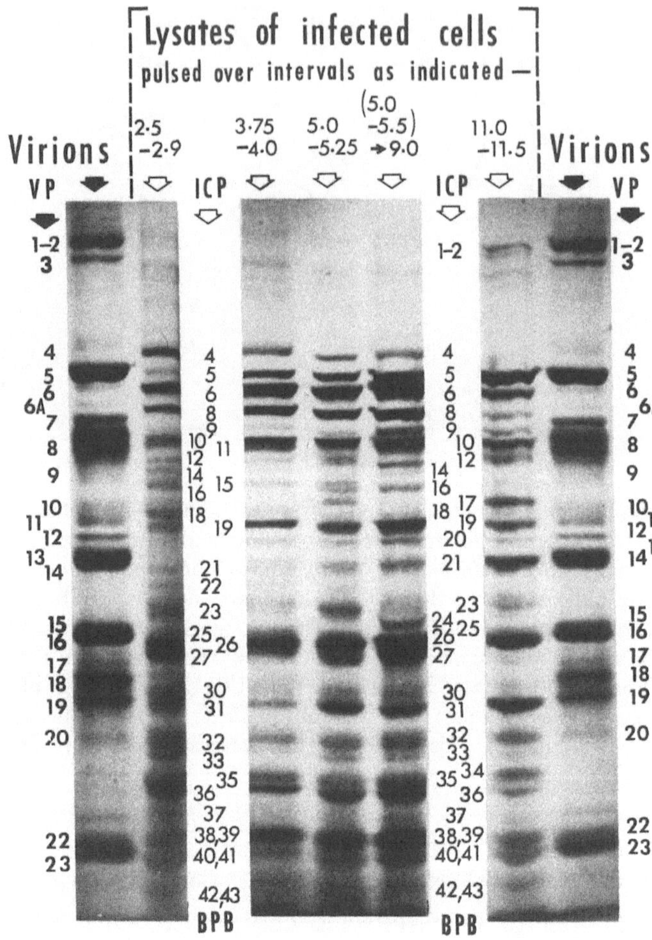

Fig. 26. Autoradiogram of a part of a 7.75% polyacrylamide gel slab containing electrophoretically separated proteins pulse-labeled in HSV-1 (F)-infected HEp-2 cells during intervals shown on top of each sample. The legend on top of the third gel from the right indicates that cells were labeled 5.0–5.5 hours post-infection and then incubated in medium lacking labeled precursors until 9 hours post-infection. Peripheral samples are virion proteins, identified by large number to the extreme right and left of the figure. The autoradiogram was cut in the preparation of this figure to permit convenient location for numbering the infected-cell polypeptides (smaller numbers between sample positions 2 and 3 and 5 and 6 of this figure). Data from Honess and Roizman (1973).

to a computer. The function of the computer, as illustrated in Fig. 27, was to expand the absorbance profile of the autoradiogram and to integrate the absorbance of each band. Some of the lysates were subjected to electrophoresis on several gels differing in polyacrylamide concentration in order to achieve maximum separation of the indi-

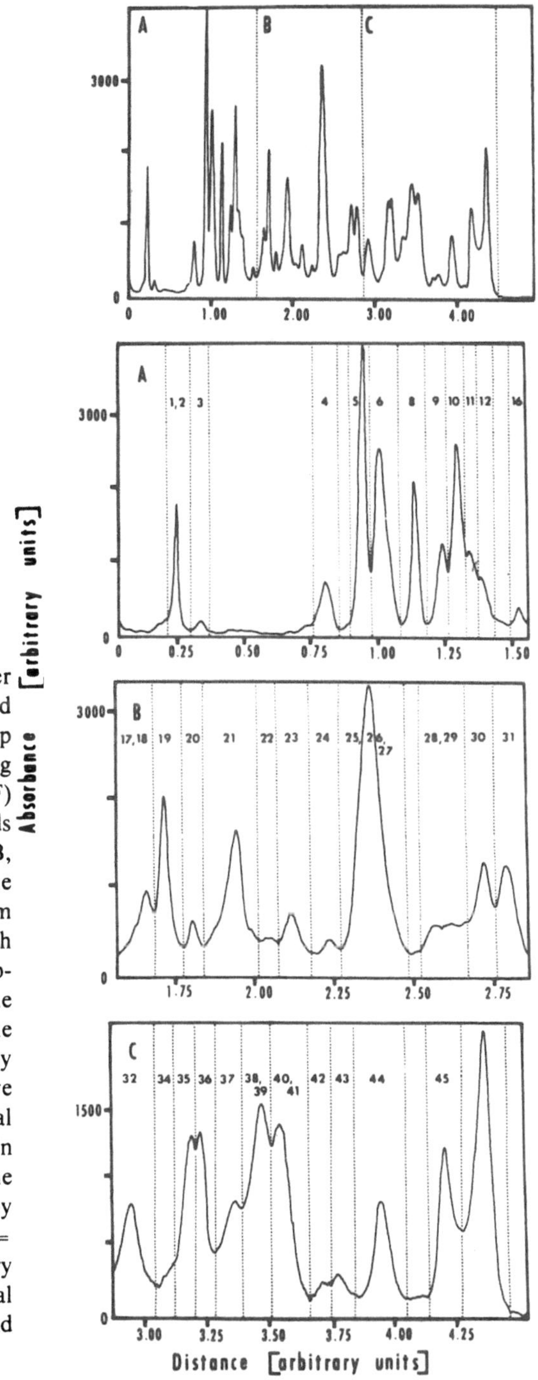

Fig. 27. Illustration of computer planimetry of the proteins separated in polyacrylamide gels. The top panel shows the absorbance tracing of an autoradiogram of HSV-1 (F) ICP labeled with ^{14}C-amino acids 3.5–8.7 hours post-infection. A, B, and C indicate those regions of the profile shown in the expanded form in the lower three panels. In each panel the areas bounded by the abscissa, absorbance profile, and the broken lines parallel to the absorbance axis were calculated by the computer. The numbers above the profile refer to the numerical designation of the ICP contained in each band. The arbitrary units on the ordinate are related to absorbance by the equation 1.0 absorbance unit = 1138 arbitrary units. One arbitrary unit of distance (abscissae) was equal to 2.0 cm. Data from Honess and Roizman (1973).

vidual bands. The absorbance of each band, corrected for the relative rate of incorporation of labeled amino acids into proteins calculated from the data shown in Fig. 26, was then divided by the molecular weight of the polypeptides in each band, yielding the relative molar rate of synthesis of the polypeptides. These results are shown in two ways. Figure 28 is a bar diagram showing the molar rates of synthesis of the various proteins plotted as a function of their molecular weight. Figure 29 shows the manner in which the relative molar rates of synthesis of polypeptides varied throughout infection.

(1) Analysis of the kinetics of synthesis of the various polypeptides shown in Fig. 29 indicated the existence of a temporal control of protein synthesis. The data show that the polypeptides synthesized in the infected cells may be grouped into at least 5 groups (A–E). Group A, B, C, and D proteins all show an initial increase in molar rate of synthesis followed by either continued increase (A), leveling off (B), or decline (C and D) in rate synthesis. Polypeptides in group E show continuous decline only; this would be expected for host proteins and also for "early early" polypeptides specified by the virus. Group F is probably an artifact and may comprise bands containing at least one polypeptide belonging to group E and one polypeptide belonging to group C or D. By varying the concentration of the polyacrylamide gels, some of the bands were actually shown to contain at least two polypeptides, differing in kinetics of their synthesis. The striking aspects of the grouping of the polypeptides according to the kinetics of their synthesis is that groups A and C contain only structural polypeptides, whereas groups B and E contain largely, but not exclusively, nonstructural polypeptides. Another interesting observation is that group A polypeptides, which might be designated as "late" structural proteins, are efficiently incorporated into virions (i.e., the intracellular pools of unassembled polypeptides are relatively small) whereas the "early" structural polypeptides (group C) are, with few exceptions, produced in very much larger amounts than are actually incorporated into the virion.

(2) Analysis of the relative molar rates of synthesis of the polypeptides made in the infected cells shows the superimposition of an abundance control over the temporal control. This is evident from the fact that even the polypeptides comprising the kinetic groups A, B, C, or D are not synthesized at the same relative molar rates. The maximum difference among the relative molar rates of synthesis of the various polypeptides is 130-fold (polypeptides 1–2 *versus* polypeptide

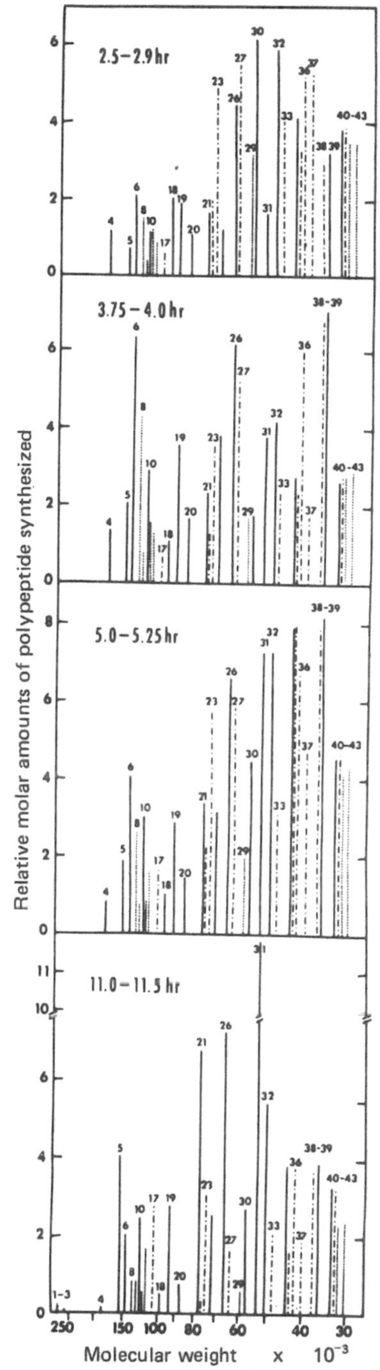

Fig. 28. Relative molar amounts of infected-cell polypeptides synthesized during short labeling intervals computed from data shown in Fig. 26 according to the procedure described in the text. The relative molar amounts of polypeptides classified as structural, undefined, and nonstructural are represented by solid lines, dotted lines, and dashed lines respectively. The polypeptides are identified by number (above bars) and the molecular-weight estimate (abscissa). Data from Honess and Roizman (1973).

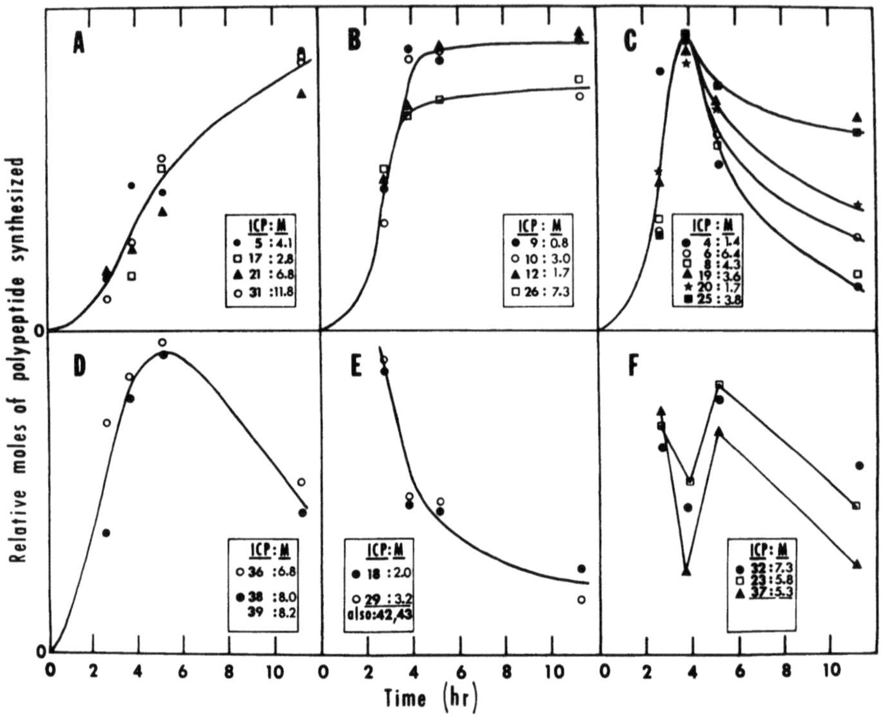

Fig. 29. Patterns of synthesis of infected-cell polypeptides in HSV-1 (F)-infected cells. The ICP were grouped according to the pattern of synthesis and designated A–F. Since the relative molar rates of synthesis in infected-cell polypeptides comprising each group are not the same, the plots were normalized. However, the maximum relative molar rate of synthesis observed during the reproductive cycle is shown under column M in the boxed legend next to the number and symbolic representation of the infected-cell polypeptides. Data from Honess and Roizman (1973).

31 at 11.0–11.5 hours post-infection). However, the bulk of the polypeptides vary no more than 8-fold in their molar rates of synthesis. The structural, nonstructural, and unclassified polypeptides cannot be segregated into nonoverlapping classes on the basis of their molar rates of synthesis. The range and average molar rates of synthesis of these three groups of polypeptides are about the same.

3.4.7. Coordinate Regulation of Viral Protein Synthesis

Evidence for coordinated regulation of viral protein synthesis emerged from studies (Honess and Roizman, 1974) designed to analyze the order of translation of viral messages based on the observation that RNA homologous to nearly 44% of HSV-1 DNA is transcribed and accumulates in cells treated with cycloheximide from the time of infection (Frenkel *et al.*, 1973). These studies led to the subdivision of the polypeptides specified by the virus into 3 coordinate groups, designated α, β, and γ. The interrelationships among these groups in the context of the experimental design may be summarized as follows:

(1) Polypeptides made immediately after removal of cycloheximide or of puromycin which was present during and after infection or added within 1 hour post-infection belong to the α group. These polypeptides comprise one structural (VP4) and several nonstructural polypeptides, ICP 27 and ICP 0 (recognized for the first time in this study). The longer the treatment with inhibitors, the higher the initial rate of synthesis of the α polypeptides, suggesting that the mRNA for these polypeptides accumulates during exposure to the inhibitors. On continued incubation of cells following removal of inhibitors, the rate of synthesis of α polypeptides decreases, coinciding with the appearance of β polypeptides. Addition of actinomycin D immediately after the removal of cycloheximide prevents both the appearance of α polypeptides and the decline in the rate of synthesis of α polypeptides, indicating that the synthesis α polypeptides requires new RNA synthesis and that the decline in synthesis of α polypeptides is causally related to the appearance of β polypeptides.

(2) The β polypeptides comprise most of the nonstructural and a few minor structural polypeptides belonging to the kinetic classes B, C, and D described above. The β polypeptides are made immediately after removal of the drugs only if these are added to the medium 2–5 hours post-infection. The rate of synthesis of β polypeptides exceeds that observed in control cultures provided the drug is added 3.5–5 hours post-infection and removed at 7 hours or later. On continued incubation of cells following removal of inhibitors, the rate of synthesis of β polypeptides decreases, and this coincides with the appearance of γ polypeptides. Addition of actinomycin D at the time of maximal rates of synthesis of β polypeptides precludes both the decline in β polypeptide synthesis and the appearance of γ polypeptides. The data, therefore, indicate that the decline in β polypeptide synthesis is due to appearance of γ polypeptides. In this instance, however, it is not clear that the synthesized polypeptides require new RNA synthesis since γ polypeptide synthesis is not stable in the presence of actinomycin D.

(3) The γ polypeptides comprise the major structural polypeptides belonging to the kinetic class A described above. Characteristic features of γ polypeptide synthesis are that it has a short half life in the presence of actinomycin D and that its rate of synthesis cannot be increased by any manipulation of the infected cells with inhibitors of protein synthesis.

(4) Inhibitors of DNA synthesis (hydroxyurea and cytosine arabinoside) added at the time of infection do not affect the rates of synthesis of α and β polypeptides, but they grossly reduce the rate of synthesis of the γ polypeptides. Combined with the short half life of γ mRNA in the presence of actinomycin D, the data suggest that γ mRNA arises by transcription of both parental and progeny DNA.

3.4.8. The Relationship between the Transcriptional and Translational Controls

The studies on the synthesis of viral proteins described in the preceding sections brought forth evidence of two kinds of regulations. First, analyses of the temporal patterns of viral protein synthesis (Honess and Roizman, 1973) show that viral polypeptides form several coordinately regulated groups. Second, analyses of viral polypeptides made immediately after removal of inhibitors of protein synthesis (Honess and Roizman, 1974) show that the synthesis of coordinately regulated polypeptides is ordered sequentially. These findings, together with the available data on viral RNA metabolism (Sect. 3.3), carry several implications: (1) The initiation of synthesis of at least α polypeptides is determined by the availability of competent viral RNA in the cytoplasm. On the other hand, it is difficult to explain the shut-off of α polypeptide by a lack of flow of new transcripts specifying this polypeptide since ongoing α polypeptide synthesis is stable if β polypeptide synthesis is blocked. These data suggest that the shut-off of α protein synthesis is regulated at some post-transcriptional level, possibly at the level of translation. (2) Both production of functional RNA molecules and cessation of function of these molecules in a coordinate manner implies that both the DNA templates giving rise to these transcripts and the transcripts themselves must carry cognate signals which initially serve to specify proper transcription and which subsequently specify that a particular set of transcripts is no longer to be translated. (3) The molar rates of synthesis within each coordinately regulated polypeptide group vary as much as 10-fold or more—certainly more than the error of the method employed in these measurements. This observation suggests that a polypeptide abundance control is superimposed on top of everything else. We cannot at the moment explain this phenomenon; conceivably the answer may come from comparisons of purified mRNAs from strains which produce "normal" amounts of polypeptides with those of strains which cause specific polypeptides to be overproduced.

Particular attention should also be given to the observation that in infected cells treated with cycloheximide from the time of infection, transcripts arising from 50% of the DNA accumulate in the nucleus but only 10% of the DNA is represented in cytoplasmic transcripts (Sect. 3.3.6) and only α polypeptides are made after withdrawal of the inhibitor (Sect. 3.4.8). Since the transcripts made in the presence and absence of protein synthesis represent the same DNA sequenced (Kozak and Roizman, 1974; Roizman *et al.*, 1974), the data imply that we should differentiate between a viral RNA transcript, which may or may not be functional, and viral mRNA, which by definition is functional, even though they may arise from the same DNA template. The data also suggest that only viral RNA molecules competent to function as mRNA are transported into the cytoplasm.

3.5. Enzymes Specified by Herpesviruses

3.5.1. General Considerations

By anology with other viruses, the biosynthesis of herpesviruses should require numerous enzymes to accomplish the transcription of the DNA, translation of RNA, post-translational modification of proteins, biosynthesis of deoxynucleotide precursors and of the progeny DNA, etc. Of the various enzymes which may or should be involved in biosynthesis of virus, only a few have been investigated in some detail, and little or nothing is known about the others.

The enzymes analyzed to date fall into 3 categories, i.e., (1) enzymes whose activities following infection become elevated relative to those of uninfected cell, (2) enzymes reported to be structural components of the herpesvirion, and (3) enzymes which predictably should be involved in viral biosynthesis but which do not fall into either of the first 2 categories. However, it is convenient to list the enzymes according to function rather than by their categories since placement in the categories does not yet reflect the source of genetic information for the enzyme. Thus, elevation of enzyme activity could result from enzyme stabilization and not necessarily from new enzyme synthesis. Similarly, the presence of enzyme activity in virions could be the consequence of contamination with trace amounts of host proteins, undectable except for their enzymatic activity.

3.5.2. DNA-Dependent RNA Polymerase

The source of genetic information necessary for the transcription of herpesvirus DNA, i.e., whether the virus initially uses a host enzyme or whether it carries its own enzymes, is not known. In fact, the possibility that the virus brings into the cell a small mRNA molecule has not been unequivocally excluded. The report by Sheldrick (1973) that fully deproteinized DNA is infectious suggests that viral DNA is able to use a host polymerase, but this information does not prove that this occurs normally. Keir (1968) reported as much as a twofold stimulation in the activity of this enzyme following infection. Since the amount of enzyme protein was not measured, it is not clear whether the stimulation reflects synthesis of new enzyme, enzyme stabilization, or availability of template. Of particular interest is a report by Mannini-Palenzona *et al.* (1971) that α-amanitin, a toxin of the toadstool *Amanita phalloides*, inhibits the multiplication of herpes simplex virus but not of vaccinia virus in chick embryo fibroblasts. α-Amanitin inhibits the activity of the extranucleolar enzyme (polymerase II) but not of polymerase I nor of the polymerase incorporated into the vaccinia core. α-Amanitin does not readily permeate into human (KB) cells, and hence its effect on virus multiplication in human cells in which the virus naturally multiplies is not known. It seems reasonable to conclude that HSV can utilize an enzyme with properties of host polymerase II at some point of its replication in a host (chick cells) which it does not naturally infect; it is not clear that utilization of this enzyme is required in the natural host or whether this is the enzyme utilized throughout the replicative cycle of the virus.

3.5.3. Thymidine (TdR) and Deoxycytidine (CdR) Kinases

TdR and possibly CdR kinases are "scavenger" enzymes, catalyzing the conversion of TdR and CdR to TdR-MP and CdR-MP, respectively, in cells grown *in vitro*. In BHK21 cells, activity increases as much as twentyfold from 2 to 8 hours after infection with HSV (Klemperer *et al.*, 1967); thereafter activity falls off. An increase in activity was also reported in rabbit kidney cells infected with pseudorabies virus (Hamada *et al.*, 1966). Stimulation of CdR kinase activity in BHK cells infected with HSV was also reported (Hay *et al.*, 1971), but the function of the enzyme is not known since the product of the enzyme (CdR-MP) is not incorporated into DNA. In recent years a considerable amount of information has accumulated on the source of genetic information and properties of these enzymes. These may be summarized as follows:

(1) In a rather elegant series of studies Kit and Dubbs (Kit and Dubbs 1963*a,b*; Dubbs and Kit, 1964) showed that TdR kinase activity

is not essential for the growth of cells or for HSV in cells in culture. First, they obtained a BUdR-resistant strain of mouse fibroblasts lacking TdR kinase activity by growing the cells in media containing the analogue. TdR kinase activity was induced in these cells by HSV and vaccinia viruses, but prevented by puromycin and actinomycin D. Subsequently, they obtained TdR kinase-less mutants of HSV by growing wild strains in BUdR-resistant cells in the presence of analogues.

(2) Several lines of evidence from many laboratories clearly established that the TdR kinases specified by different herpesviruses are readily differentiated among themselves and from host enzymes on the basis of their immunologic specificity. Thus Klemperer *et al.*, (1967) showed that the thymidine kinase produced in cells infected with HSV was inhibited by antiserum prepared against rabbit kidney strain 13 (RK-13) cells infected with the same virus; the enzyme activity in uninfected cells was unaffected. Subsequently, Buchan and Watson (1969) reported that thymidine kinase induced by pseudorabies virus in RK-13 cells was neutralized by homologous rabbit anti-infected-cell serum but not by serum produced against HSV-infected cells. In the same vein, the enzyme induced by HSV in these cells was neutralized by homologous but not by heterologous antibody. The report by Buchan and Watson (1969) refuted the suggestion of Hamada *et al.* (1966) that the TdR kinase activity in pseudorabies-infected cells was the result of allosteric alterations in the enzyme caused by environmental changes within the host cell. In more recent studies, Thouless (1972) reported that the thymidine kinases of HSV-1 and HSV-2 differ in their immunologic specificities. However, the study revealed several paradoxes which need clarification. First, common antigenic determinants were demonstrable in enzyme neutralization tests but not by the precipitin test. Second, some antisera to HSV-1 enzyme stabilized the TdR kinase specified by HSV-2 with only partial or no concommitant reduction in activity. The stabilized enzyme was still capable of being neutralized by homologous antiserum, albeit at a reduced rate. One set of hypotheses which could conceivably explain these observations but which has not been considered is that the active TdR enzyme is an oligomer which carries two sets of antigenic sites, i.e., a type-specific antigenic site near the active center of the enzyme and a common antigenic site formed by the junction of monomers at some distance from the active center. The stabilization of type 2 TdR kinase by anti-HSV-1 serum could be visualized as cross-linking of the monomers in the oligomeric form.

(3) Hay *et al.* (1971) postulated that both TdR and CdR kinase

activities may be functions of the same polypeptides, but at different sites. This conclusion is based on the observation that HSV mutants selected for resistance to cytosine arabinoside, or to BUdR are deficient in both CdR and TdR kinase activities, as is the BUdR-resistant mutant of Dubbs and Kit (1964). This conclusion is most interesting, not because the enzymes differ in their sensitivity to heat and to feedback inhibition, but rather because there is a 1.5-hour interval between the appearance of CdR and TdR kinase activities in the infected cell. In principle, it would be useful to analyze mutants deficient in this enzyme which are selected by a procedure which does not involve either of the drugs. Assuming that TdR and CdR kinase are truly two activities of the same gene product, an alternative plausible explanation of the finding is that a single polypeptide chain has the activity of CdR kinase in monomeric form and of TdR kinase in a multimeric form. If this were the case, it could be predicted on the basis of immunologic studies described in the preceding section that the neutralization of CdR kinase activity would be considerably more type-specific than that of TdR kinase activity.

(4) Compared with the enzyme from uninfected cells, the TdR kinase from infected BHK21 cells was reported to have a low *p*H optimum, a low K_m, to be relatively stable at 40°C, and to be insensitive to inhibition by deoxythymidine triphosphate (Klemperer *et al.*, 1967). On the other hand, TdR kinase activity of African green monkey kidney cells (BSC-1) could not be differentiated from that of the same cells infected with HSV with respect to thermal stability and optimal temperature (Prusoff *et al.*, 1965). Ogino and Rapp (1971) and Thouless and Skinner (1971) reported that HSV-2 TdR kinase is more thermolabile than that of HSV-1 and gives nonlinear dilution responses. Munyon *et al.* (1972) reported that the average mobility in polyacrylamide gels of HSV-1 TdR kinase was slower than that of the host enzyme by almost a factor of 2. It seems pertinent to note here that many of the biochemical characterizations have been done on enzymes isolated from cells infected with strains passaged for many years in cell culture. Some of these viral strains differ significantly from wild strains, suggesting that mutants were not only accumulated but selected (Heine *et al.* 1974). Although differences in the TdR kinase properties of wild and laboratory strains have not been reported, they cannot be *a priori* excluded.

(5) Of special interest are a series of studies on the properties of the TdR kinase-less mutants of HSV. Not necessarily in chronological order, the data may be summarized as follows. Dubbs and Kit (1965)

reported on the properties of 3 kinase-less mutants, i.e., B2006, B2010, and B2015. B2006 produced extremely low levels of TdR kinase independent of temperaure of incubation. B2010 and B2015 produced no detectable TdR kinase when grown at 37°C irrespective of the temperature at which the enzyme was tested. At 31°C these mutants produced one-tenth the level of TdR kinase detected in extracts of cells tested at 31°C. The activity was independent of the temperature at which it was tested, and from autoradiographic assays it appears to have been distributed throughout most cells. Buchan *et al.* (1970*a*) found that extracts of cells infected with mutant B2006 did not block the serum neutralization of wild-type TdR kinase, and they concluded that mutant B2006 did not produce nonfunctional cross-reacting antigen of TdR kinase. The picture became more complex when it was found (Munyon and Kit, 1965) that in cells doubly infected with wild-type and the TdR kinase-less mutant the enzyme activity was reduced 3- to 4-fold. Especially interesting was the observation that the TdR kinase-less mutant did not affect the production of thymidine kinase by vaccinia virus. On the other hand, thymidine kinase-less vaccinia virus was a potent inhibitor of herpesvirus TdR kinase in cells doubly infected with TdR kinase-less vaccinia and a wild strain of HSV. Munyon and Kit (1965) excluded the possibility that vaccinia produces an inhibitor of TdR kinase activity or a repressor specific for the TdR kinase gene, but they left open the possibilities that TdR kinase might be an oligomeric protein which in doubly infected cells consists of functional and nonfunctional subunits, or that the parent and mutant viruses might compete for some cellular structure necessary for the expression of TdR kinase. Buchan *et al.* (1970*b*) confirmed the observation of Munyon and Kit (1965) but failed again to find a cross-reacting antigen or a viral antigen present in cells infected with B2006 mutant but absent in cells infected with the wild type. They concluded that it was unlikely that the inhibition of thymidine kinase production in cells infected simultaneously with the mutant and wild-type viruses can be explained by the interaction of functional and nonfunctional subunits. The data are puzzling; the hypothesis that the parental and mutant viruses compete for a common cellular site such as "ribosomes" or "polymerase" as suggested by Munyon and Kit (1965) and by Buchan *et al.* (1970*b*) is not very appealing; it simply calls the black box by another name. Superficially, at least, the more appropriate line of investigation is to determine whether in cells infected with the TdR kinase-less mutants the translation of the TdR kinase mRNA is prematurely terminated, producing a product of limited immunogenicity but still capable of competing with the intact polypeptide in the assembly of the putative oligomeric enzyme. This hypothesis would explain the reduction in viral TdR kinase activity in cells infected with both parent and mutant viruses but not the inhibition of host enzyme activity in cells infected with the mutant virus alone (Buchan *et al.*, 1970*a*) unless the host and viral enzymes are capable of interacting.

3.5.4. Ribonucleotide Reductase

One of the functions of this enzyme is to catalyze the conversion of cytidylate to deoxycytidylate. In mammalian cells, dTTP acts as an allosteric inhibitor of this enzyme (Reichard *et al.*, 1960, 1961), and the synchronization of mammalian cells by the addition and subsequent removal of excess thymidine is thought by many to be the result of depletion of dCTP from the cellular pool as a consequence of inhibition of this enzyme. Cohen (1972) observed that excess thymidine blocked, as expected, the synthesis of host DNA in uninfected KB cells, but it did not affect the synthesis of either viral or cellular DNA in infected cells. Based on the reasoning that blocked infected cells have a source of dCTP which is unavailable in blocked uninfected cells, Cohen tested the ribonucleotide reductase activity of infected and uninfected cells. The results showed that the activity of this enzyme was increased twofold at 3 hours post-infection and that the enzyme from infected cells retained 60% activity even in the presence of 2mM dTTP, i.e., considerably more than required to inhibit the enzyme extracted from uninfected cells. There is as yet no genetic or immunochemical evidence that the enzyme is specified by the virus.

3.5.5. TdR-MP, CdR-MP, AdR-MP, and GdR-MP Kinases

Hamada *et al.* (1966) reported that TdR-MP kinase activity increased in rabbit kidney cells infected with pseudorabies virus whereas the activity of AdR-MP, GdR-MP, and CdR-MP kinases remained unaltered. The same laboratory previously reported (Nohara and Kaplan, 1963) that TdR-MP kinase from pseudorabies-infected rabbit kidney cells was more stable at 37°C than the corresponding enzyme from uninfected cells. Prusoff *et al.* (1965) observed a similar increase in TdR-MP kinase in African green monkey kidney cells infected with herpes simplex, but they were unable to differentiate between the properties of the enzyme in extracts of infected and uninfected cells. AdR-MP kinase from infected and uninfected cells could not be differentiated in neutralization tests by antisera prepared against infected and uninfected cell extracts (Hamada *et al.* 1966).

3.5.6. TdR-MP Synthetase

Frearson *et al.* (1965) reported that TdR-MP synthetase activity did not increase in mouse fibroblasts, HeLa, and rabbit kidney cells infected with HSV.

3.5.7. CdR-MP Deaminase

An increase in CdR-MP deaminase in BHK21 cells infected with HSV was reported by Keir (1968). The enzyme catalyzes one of the reactions concerned with *de novo* synthesis of TdR-MP. No additional information is available.

3.5.8. Deoxyribonuclease

An increase in DNase activity (measured at *p*H 7.3) in BHK21-C14 cells infected with herpes simplex virus was reported by Keir and Gold (1963). The increase in enzyme activity leveled off 7–9 hours after infection (Keir and Gold, 1963; Russell *et al.*, 1964). Subsequently, McAuslan *et al.* (1965) and Sauer *et al.* (1966) reported an increase in activity of an "alkaline" DNase in infected monkey kidney cells and L cells, respectively. An increase in "acid" DNase has been observed in HSV-infected HeLa and L cells (Newton, 1964) but not in monkey kidney cells (McAuslan *et al.*, 1965) or in KB cells (Flanagan, 1966).

The DNase studied by Kier and Gold (1963) and subsequently by Morrison and Keir (1966, 1967, 1968*a,b*) appeared to differ from the host DNase with respect to several properties. The induced DNase was readily inactivated at 45°C, adsorbed to DEAE-cellulose, preferred Mg^{2+} to Mn^{2+}, and required 50–60 mM Na^+ or K^+, whereas uninfected cell enzyme was stable at 45°C, did not adsorb to DEAE-cellulose, did not differentiate between Mg^{2+} and Mn^{2+}, and was inhibited by Na^+ or K^+ at concentrations greater than 15 mM. Purified induced DNase emerges in the void volume during gel filtration through Sephadex G-

200 whereas the uninfected-cell enzyme was retarded. The enzymes extracted from BHK21 and HEp-2 cells infected with HSV or with pseudorabies were DNA exonucleases capable of degrading both native and denatured DNA to deoxynucleoside-5′-monophosphates whereas the enzyme extracted from uninfected cells was a DNA endonuclease effective primarily against denatured DNA. Rabbit antisera prepared against extracts of allotypic rabbit kidney cells infected with HSV neutralized the DNA endonuclease from infected cells but not the endonuclease extracted from the uninfected cells (Morrison and Keir, 1967; Keir, 1968).

3.5.9. DNA Polymerase (DNA Nucleotidyl Transferase)

A comprehensive discussion of DNA polymerases in herpesvirus-infected cells was published by Keir (1968). Briefly, Keir and Gold (1963) reported that in BHK21 cells infected with HSV, the DNA polymerase activity of nuclei and of the mitochondria-microsome fraction increased two- to sixfold between 2 and 5.5 hours after infection. Actinomycin D prevented the increase in activity (Keir, 1968). The polymerases extracted from infected and from uninfected cells differed with respect to immunologic specificity, heat stability (both with and without primer), primer requirements, and sensitivity to iodoacetamide and *p*-mercuribenzoate (Keir, 1965; Keir *et al.* 1966*a,b*; Shedden *et al.*, 1966). The enzyme assay developed by Keir and associates took advantage of the observation that the enzyme induced after infection was optimally active at 0.2 M NH_4^+ whereas the uninfected cell polymerase, endogenous DNase, and virus-induced DNase were all virtually inactive at that concentration of NH_4^+ (Keir *et al.*, 1966*b*; Keir, 1968). Sephadex G-200 gel only slightly retarded DNA polymerase (Keir, 1968). The DNA polymerase activity in bovine kidney cells infected with bovine rhinotracheitis virus was also more heat stable than the enzyme activity of uninfected cells (Stevens and Jackson, 1967). Hay *et al.* (1971) reported that the DNA polymerase activities in HSV-1- and HSV-2-infected cells differed in heat stability and NH_4^+ optimum.

It is of interest to note that Hamada *et al.* (1966) could not differentiate in neutralization tests between DNA polymerase extracted from pseudorabies-infected and uninfected rabbit kidney cells. The results of Hamada *et al.* are of interest; the rabbit in not a natural host for pseudorabies virus and this makes it even more improbable that virus and host polymerase are immunologically related. The enzyme neutralization tests were done with the serum of a rooster injected with sonicates of infected stationary (arrested by contact or density inhibition) cultures of rabbit kidney cells. The serum prepared against the uninfected extract lacked neutralizing activity. The effect of the serum made against the infected cell extract on the template was not tested.

3.5.10. Uridine Kinase

There is 1.5- to 2-fold stimulation of uridine kinase in rabbit kidney cells infected with some strains of herpes simplex but not in the case of others (Dundaroff and Falke, 1972). Cycloheximide and actinomycin D did not preclude the stimulation of the activity of this enzyme, but the concentrations tested were too low to inhibit even the stimulation of TdR kinase activity.

3.5.11. Choline Kinase

Dundaroff and Falke (1972) reported a 1.5- to 2-fold increase in the activity of choline kinase in rabbit kidney cells infected with four HSV-2 strains and with a few of the numerous HSV-1 strains tested. The enzyme reached maximal activity between 12 and 24 hours post infection and its levels seemed to be related to the nature of the cytopathic effect. The effect of inhibitors (cycloheximide and actinomycin D) was inconclusive because their concentrations were again too low.

3.5.12. Protein Kinase

Randall *et al.* (1972) reported on the presence of an "endogenous" protein kinase in equine abortion herpesvirions. Demonstration of optimal activity of this enzyme requires purification of the virus with respect to a phosphohydrolase enzyme by centrifugation of virus suspended in a Tris-EDTA buffer in a tartrate gradient. The kinase activity requires Mg^{2-} and is enhanced by the addition of exogenous acceptors (protamine or arginine-rich histone) but not by cyclic AMP.

In an *in vitro* test, when purified preparations of frozen, thawed, then sonicated virus were used as substrates, all 17 proteins detected in their polyacrylamide gels served as acceptors of ^{32}P. At least two-thirds of the incorporated ^{32}P was rendered acid soluble by snake venom diesterase, suggesting that the incorporated phosphate may be involved in a phosphodiester bond linking proteins to small, acid-soluble molecules. Parenthetically, the profile of ^{32}P-labeled proteins *in vitro* seems to be vastly different from that seen *in vivo* (Sect. 2), suggesting many more proteins are phosphorylated *in vitro* than *in vivo*. The source of genetic information for this enzyme is not known. Rubenstein *et al.* (1972) reported the presence of protein kinase in enveloped HSV particles and, like Randall *et al*, (1972), they noted that it was not stimulated by cyclic AMP—an indication that a regulatory subunit enhancing the activity of the enzyme found in uninfected cells was absent or inoperative. However, several other conclusions regarding this enzyme seem far-fetched. Rubenstein *et al.* found that capsids de-enveloped with a detergent did not possess protein kinase activity and concluded that the enzyme is in the envelope of the virus. Although this is likely, since the kinase is normally associated with cellular membranes, the experiment itself does not offer compelling evidence on this point since the NP-40 extract containing the solubilized envelope proteins was not tested for the presence of protein kinase. This test is important as an indication that the detergents did not inactivate the enzyme. The authors also suggest that protein kinase is essential for infectivity and offer in support of this postulate the observation that heat treatment which reduced protein kinase activity also reduced infectivity. Quite obviously heat treatment denatures more than just protein kinase.

3.5.13. Proteases

In the course of concentration of highly purified HSV-1 polypeptides following their solubilization in SDS it was observed (Spear, Gibson, and Roizman, unpublished data) that several proteins appeared to be degraded. The degradation was made less severe or did not occur in the absence of SDS. The data suggested that the herpesvirions may contain a protease which is either activated or becomes accessible to the substrate in the presence of SDS. The source of this enzyme is not clear; it resembles superficially the enzyme reported to be present in other viruses (Holland *et al.*, 1972) but could be a host contaminant.

3.6. Herpesvirus DNA Synthesis

3.6.1. Introduction

Most of the information on herpesvirus DNA synthesis comes from studies on pseudorabies-infected rabbit kidney cells (Kaplan and Ben-Porat, 1963, 1964, 1966*a,b*; Kaplan *et al.*, 1967; Kaplan, 1964; Ben-Porat and Kaplan, 1963) and on HSV-1-infected HEp-2 cells (Roizman and Roane, 1964; Frenkel and Roizman, 1972*a*; Cohen *et al.*, 1971). Both HSV and pseudorabies virus contain DNA with high G+C content and these are readily separable from host DNA containing a significantly lower G+C content. With the exception of studies on EAV DNA (Lawrence, 1971*a,b*; O'Callaghan *et al.*, 1968*b*), little has been reported on the synthesis of viral DNAs more closely approximating host DNA with respect to base composition. The pseudorabies–rabbit kidney cell system is particularly advantageous because only viral DNA is synthesized in rabbit kidney cells arrested prior to infection by contact (density) inhibition; host DNA synthesis along with cell division remain inhibited throughout the infectious cycle (Kaplan and Ben-Porat, 1960). The studies on the synthesis of HSV DNA were done in HEp-2 cells (Roizman and Roane, 1964; Frenkel and Roizman 1972*a*). Alas, optimal virus yields are obtained from young rapidly growing cells (Roizman and Spear, 1968), in which host DNA synthesis persists for several hours. For accurate determination of viral DNA it is, therefore, necessary to centrifuge the extract containing DNA in CsCl solution (Roizman and Roane, 1964). However, by infecting synchronized KB cells at different times during the mitotic cycle, Cohen *et al.* (1971) have shown that host DNA synthesis is completely precluded if the cells are infected in G2; and, therefore, only viral DNA is made in those cells. O'Callaghan *et al.* (1968*b*) studied the synthesis of EAV virus DNA in L-M cells. They separated newly synthesized viral and cellular DNA on methylated albumin-kieselguhr columns.

3.6.2. The Site of Synthesis of Viral DNA

Viral DNA synthesis takes place in the nucleus. For many years this conclusion was based on histochemical evidence (Newton and Stoker, 1958; Munk and Sauer, 1964; Chopra *et al.*, 1970). Biochemical evidence was obtained in HEp-2 cells infected with HSV.

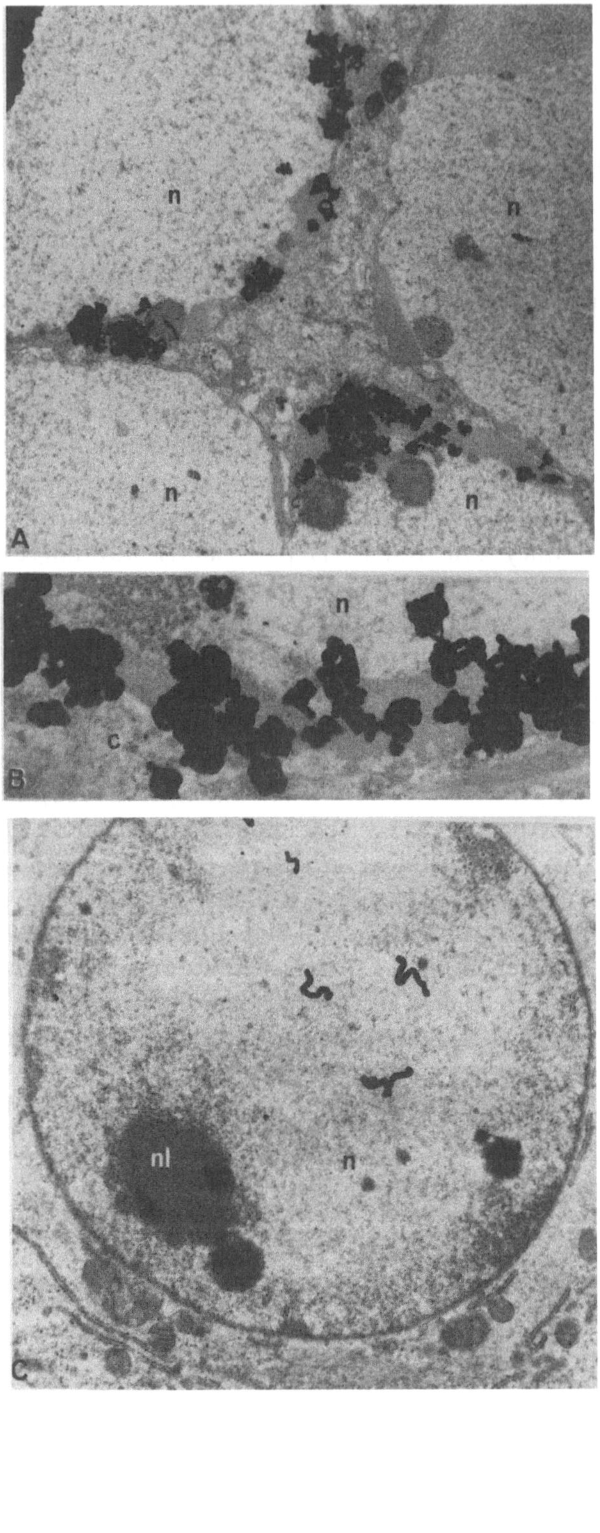
n
n
n
n
A
n
c
B
nl
n
C

Thus, in cells fractionated immediately after a 5-minute pulse-labeling with ^{3}H-thymidine, labeled DNA was found in the nuclear fraction (Roizman, 1969). Autoradiographic data (Fig. 30) indicate that DNA synthesis is associated with an electron-translucent matrix within the nucleus and not with any of the numerous electron-opaque bodies present in the infected cell nucleus (Roizman, 1969). Analyses of the total DNA made in the infected cell during the interval of labeling indicated that more than 95% of the label was incorporated into viral DNA.

3.6.3. Pattern of Synthesis of Viral DNA

The most reliable method for the estimation of the pattern of synthesis of viral DNA used to date involved determination of the specific activity of viral DNA in cells pulse-labeled with radioactive thymidine at intervals after infection. The pattern of DNA synthesis in HSV-infected cells is shown in Fig. 31. Similar patterns have been published for EAV-infected cells (O'Callaghan *et al.*, 1968*b*) and pseudorabies-infected cells (Kaplan and Ben-Porat, 1963). In each instance, a few hours after infection host DNA synthesis is almost completely replaced by viral DNA synthesis. The bulk of viral DNA is synthesized between 4 and 7 hours after infection; thereafter all DNA synthesis declines slowly and irreversibly. On the basis of data shown in Fig. 31 it has been calculated that 6–9 hours elapse between the synthesis of DNA and formation of infectious virus (Roizman, 1969). These determinations suffer from the assumption that the size of the nucleotide pools remain constant throughout the infectious cycle, and their usefulness is limited to measurements of the relative amounts of host and viral DNA synthesized at discrete intervals. By other techniques involving

Fig. 30. Autoradiographic studies of DNA synthesis in HSV-1-infected cells. Autoradiography of thin sections of HEp-2 cells infected with HSV-1 virus. (A) Portions of 4 nuclei of 18-hour-infected cells labeled with ^{3}H-thymidine methyl prior to infection. The grains identify the location of cell chromosomes. They indicate further that host DNA is not degraded to small diffusable fragments. (B) Same as (A) except the magnification is greater. Grains indicating the presence of host DNA are localized over marginated chromatin. (C) A 4-hour-infected cell pulse-labeled for 15 minutes with ^{3}H-thymidine methyl prior to fixation. Grains indicating the location of viral DNA are generally seen in the central, electron-translucent part of the nucleus and are apparently not associated with any distinctive structures. Abbreviations: n, nucleus; nl, nucleolus; c, cytoplasm (Roizman, 1969).

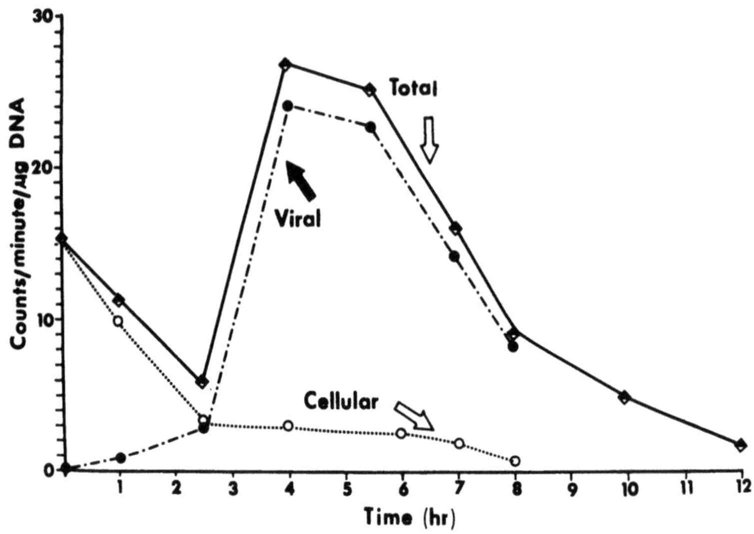

Fig. 31. The pattern of incorporation of ^{3}H-thymidine into DNA of HEp-2 cells infected with HSV-1 (mP). The cells were pulse-labeled for 15 minutes at different times after infection. The DNA extracted after the pulse was centrifuged to equilibrium in CsCl density gradients (Roizman, 1969).

both direct and indirect measurements it was shown that the minimal interval between the synthesis of DNA and its incorporation into mature virions is 2 hours (Olshevsky *et al.*, 1967; Roizman *et al.*, 1963).

3.6.4. Requirements for Viral DNA Synthesis

Two series of experiments have been reported showing that protein synthesis after infection is required to initiate the synthesis of viral DNA. In the first series, the exposure of HSV-infected HEp-2 cells to puromycin at any time between 0 and 3–4 hours after infection blocked the synthesis of viral DNA (Roizman and Roane, 1964). However, cells exposed to puromycin between 4 and 6 hours after infection, i.e., after the onset of viral DNA synthesis, continued to incorporate radioactive thymidine into viral DNA, albeit at a reduced rate. In view of the difference in size, base composition, and general structure of viral and cellular chromosomes, the observation that puromycin treatment immediately after infection blocks the onset of synthesis of viral DNA seems entirely reasonable. It indicates that viral DNA synthesis requires the participation of new enzymes made after infection. However, there is no simple, straightforward explanation for the observation that in cells treated with puromycin after the onset of viral DNA synthesis, the rate of incorporation of thymidine into viral DNA is reduced. The greatest reduction in rate of incorporation of thymidine into viral DNA occurs within a short time after addition of puromycin and could reflect either a sudden change in the pool size of thymidine or a real change in the rate of synthesis of viral DNA. The first hypothetical explanation is probably trivial but it must be considered, particularly in view of the possibility raised by Newton *et al.* (1962) that the size of the thymidine pool in the infected cells may change during the reproductive cycle. The second hypothesis, if true, would indicate that puromycin interferes with the availability of a necessary rate-limiting constituent such as an enzyme or possibly the primer itself; it is perhaps pertinent to note that viral DNA extracted from puromycin-treated cells behaves on isopycnic centrifugation in CsCl as if it were highly fragmented (Roizman and Roane, 1964). The objection to the second hypothesis is that the transfer of proteins from the cytoplasm into the nucleus is rather slow (Spear and Roizman, 1968), and, hence, the absence of this putative protein should not have become manifest immediately after addition of puromycin.

Similar results were obtained by Kaplan *et al.* (1967) in rabbit kidney cells infected with pseudorabies virus. During the first hours after exposure to puromycin the rate of incorporation of thymidine into viral DNA dropped to one half that observed in untreated cells. Thereafter the rate decreased, but not very appreciably. The authors concluded that changes in the rate of thymidine incorporation reflect changes in DNA synthesis and that pseudorabies virus DNA synthesis requires concomitant synthesis of one or more proteins. They speculate that the function of the short-lived protein which must be synthesized continuously is to "prime" or initiate the synthesis of DNA. The conclusion is based on the observation that extracts from cells exposed to puromycin appear to contain all of the enzymes necessary to maintain DNA synthesis in a cell-free system consisting of cell extract, added deoxynucleotide triphosphates, and heat-denatured primer.

To sum up, it seems clear that viral DNA synthesis requires *de novo* protein synthesis. Once initiated, viral DNA synthesis continues in the absence of concomitant protein synthesis, but at a reduced rate. The reduction in rate is not readily interpretable for lack of information concerning (1) thymidine pool size before and after addition of puromycin, (2) integrity of replicating DNA in puromycin-treated cells, and (3) secondary effects of puromycin on the intranuclear environment of the cell. There is, incidentally, no information concerning the stability *in situ* of the enzymes involved in the synthesis of viral DNA.

A second possible requirement emerged in recent studies by Lawrence (1971*a,b*). In these studies, the onset of synthesis of EAV DNA in human (KB) cells varied as much as 5 hours, depending on the stage of the cell cycle. Thus synchronized KB cells infected in G1 or G2 replicated viral DNA at a time corresponding to the cellular S phase determined independently in mock-infected cells. Lawrence concluded that the initiation of EAV DNA synthesis was dependent upon some cellular functions which were related to the S phase of the KB cell. It is noteworthy that the S-phase cell did not fulfill all the requirements since viral DNA synthesis did not immediately follow infection of cells in S phase. The dependence on host cell functions does not appear to be universal for all herpesviruses or in all kinds of cells. Thus, Cohen *et al.* (1971) demonstrated that HSV-1 DNA synthesis proceedes independently of the mitotic cycle in the same KB cell line, whereas O'Callaghan *et al.* (1972) showed that EAV "adapted" to hamsters does not require host DNA synthesis for its own multiplication. It is conceivable that host DNA synthesis is required only in "heterologous" hosts for reasons that at present remain obscure.

3.6.5. Characteristics of the Replication of Viral DNA

Kaplan *et al.* (1967) summarized studies reported between 1963 and 1967 showing that pseudorabies virus DNA replicates in a semiconservative fashion and that less than one-half of the DNA not incorporated into virions and presumed available to function as template for replication is actually replicating.

Recent studies (Frenkel and Roizman, 1972*a*) on sedimentation of the alkali-denatured HSV-1 DNA showed that nascent DNA is always found as small fragments which become elongated only after prolonged incubation of cells. In these studies, viral DNA extracted from nuclei of cells labeled with ^{3}H-thymidine for intervals of 3–120 minutes from 4.5 to 6.5 hours post-infection was centrifuged in alkaline sucrose density gradients. The DNA labeled for 3 minutes was roughly 0.5×10^6 to 10×10^6 daltons in molecular weight (Fig. 32). As the duration of the labeling period increased, the DNA became larger. However, despite the fact that at 2 hours the nuclear DNA was still largely in

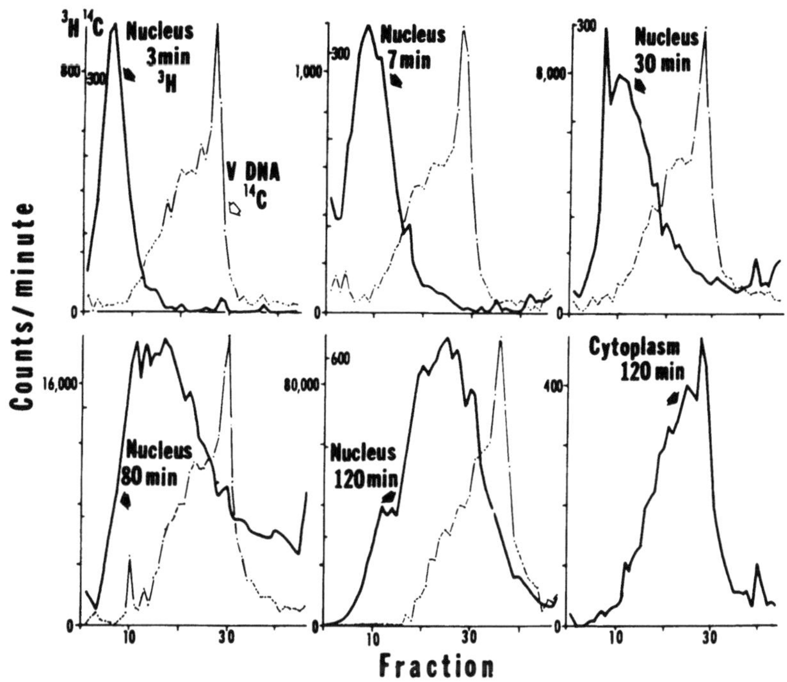

Fig. 32. Profile of labeled DNA in nuclear and cytoplasmic lysates labeled with ^{3}H-thymidine (10 μCi/ml) for intervals ranging from 3 to 120 minutes between 4.5 and 6.5 hours post-infection with HSV-1. The nuclear and cytoplasmic lysates were centrifuged in 5–20% (wt/wt) alkaline sucrose density gradients. The lower right panel shows the profile of the labeled DNA in the cytoplasm of cells labeled for 120 minutes. To facilitate comparison of the various profiles, ^{14}C-labeled HSV-1 DNA extracted from virions contained in cytoplasm of 24-hour-infected cells was cocentrifuged with all nuclear lysates. (——) ^{3}H-labeled DNA, (- · -) ^{14}C-labeled DNA. Sedimentation from left to right. Data from Frenkel and Roizman (1972*a*).

fragments smaller than those generated by alkaline denaturation of DNA extracted from virions, the DNA labeled during the same interval and extracted from the virions in the cytoplasm could not be differentiated from the DNA contained in virions collected at 24 hours post-infection. The data suggest that viral replication begins at numerous initiation sites along each strand and that the elongation beyond the size of the replication unit involves repair (excision of ribonucleotides serving as primers?) or ligation, or both, and that only "mature" DNA is incorporated into the virion. At the time of writing of this discussion, it is still not clear whether the fragmentation by alkaline denaturation of DNA extracted from virions is due to residual ribonucleotides, nicks or gaps, or to some other reason (Sect. 2.3).

3.7. Biosynthesis of Other Constituents of the Virion

3.7.1. Polyamines

As pointed out in Sect. 2.5, HSV produced in HEp-2 cells contains both spermidine and spermine. The accumulated data on the metabolism of spermine and spermidine in infected cells emerged from attempts to determine whether the stringent requirement for arginine cited earlier (Sect. 2.5) is related to the utilization of polyamines for virus assembly. Most of the experiments deal with utilization of ornithine as a precursor for polyamines in infected and uninfected cells and may be summarized as follows:

(1) Gibson and Roizman (1971) noted that ornithine serves as an efficient precursor for viral polyamines only if added to the cells prior to infection. Subsequently, they reported 2 series of experiments designed to examine the possibility that the decrease in ornithine utilization is a consequence of modifications of the normal cellular metabolism by the virus (Gibson and Roizman, 1973). In the first, ^{3}H ornithine was added to uninfected cells and to 1-hour-infected cells, and its conversion to spermine and spermidine was then followed in parallel. As indicated in Fig. 33, the specific activities of the infected-

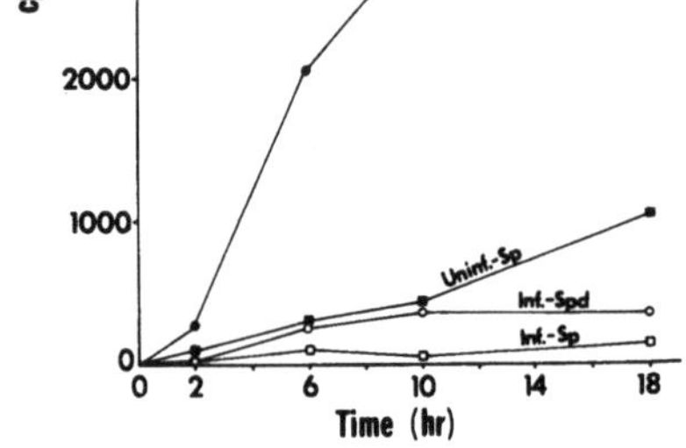

Fig. 33. Changes in polyamine-specific activities in infected cells labeled after infection, and uninfected cells labeled during the same interval. Four cultures of HEp-2 cells were infected with HSV-1. One hour after infection, fresh medium containing 5 μCi/ml of 3-^{3}H-D,L-ornithine (2.38 Ci/mmole, New England Nuclear Corp., Boston, Mass.) was added. At the same time, four uninfected cultures were similarly overlayed with ^{3}H-ornithine-containing medium. At 2, 6, 10, and 18 hours after infection, one culture each of the infected and uninfected cells was collected, as described in the legend to Table 3 and extracted with perchloric acid. Polyamine analyses were done as described in the footnotes to Table 3. The amount of radioisotope incorporated into each polyamine was determined by scraping its fluorescent spot from the thin-layer plate, drying it at 70°C for 1 hour, suspending it in a toluene-base scintillation fluid, and then measuring the radioactivity in a Packard Tri-Carb scintillation spectrometer. Spd, spermidine; Sp, spermine. From Gibson and Roizman (1973).

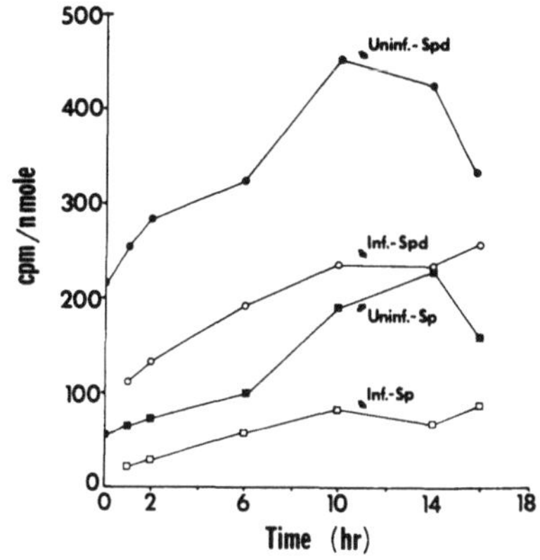

Fig. 34. Changes in the specific activity of polyamines in infected cells labeled before infection and in uninfected cells labeled during the same time interval. During the 18-hour period immediately prior to infection, HEp-2 cells were incubated in medium containing ^{3}H-ornithine (5 μCi/ml, as described in Fig. 33). The labeled medium was then replaced with unlabeled medium, and half of the cultures were infected with HSV-1. At the times indicated, one culture each of infected and uninfected cells was collected, extracted with perchloric acid, and analyzed for its polyamine content and radioactivity as described in Fig. 33. Spd, spermidine; Sp, spermine. Data from Gibson and Roizman (1973).

cell polyamines increased only slightly as compared with those of uninfected cells. At the end of the 18-hour labeling period, the specific activities of spermine and spermidine in the infected cells were only 13 and 6% of those in uninfected cells, respectively. It is unlikely that the decrease in the utilization of ornithine for polyamine synthesis resulted from an alteration of ornithine uptake by the infected cells since nearly equal amounts of the radioisotope became incorporated into cold, trichloroacetic acid (7.5%)-precipitable material in both infected and uninfected cells. A plausible interpretation of these data is based on the observations that ornithine decarboxylase has an extremely rapid rate of turnover (Snyder *et al.*, 1970) and that HSV shuts off host protein synthesis after infection. These observations predict that ornithine decarboxylase would disappear very rapidly from cells after infection and that, therefore, the biosynthesis of putrescine from ornithine would cease, as is apparently the case. The objective of the second experiment was to determine whether polyamine metabolism subsequent to the formation of putrescine is affected by HSV-1 infection. The cells in this instance were labeled with ^{3}H-ornithine for 18 hours to develop an intracellular pool of labeled putrescine. Half were then infected with HSV-1, and the others left uninfected. As shown in Fig. 34, the specific activities of spermine and spermidine increased in both the infected and uninfected cultures. It is not known why the specific activities of the infected-cell polyamines were lower than those of the uninfected cultures. However, one explanation consistent with the idea that ornithine decarboxylase becomes rapidly depleted is that in the infected cell less of the intracellular ornithine becomes "fixed" into the polyamine

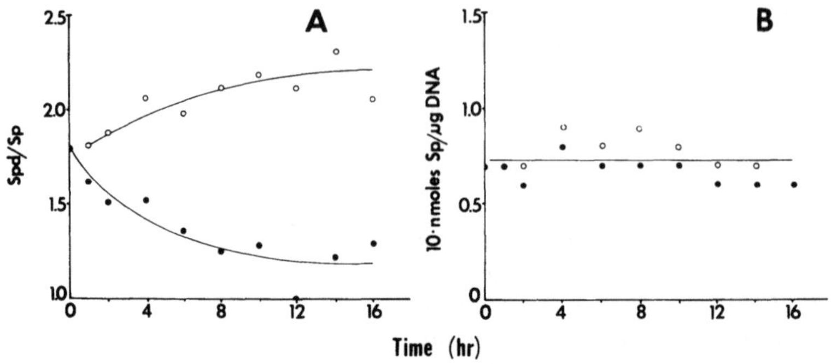

Fig. 35. The ratios of spermidine to spermine and of spermine to DNA in infected and uninfected cells. (A) Spermidine to spermine ratios calculated from the data obtained from the experiment described in Fig. 34. (B) The amount of spermine relative to DNA calculated for both infected and uninfected cells. The procedures used to quantitate the amounts of polyamine and DNA were the same as described in the footnotes to Table 3. Infected cell ratios (○——○), uninfected cell ratios (●——●). From Gibson and Roizman (1973).

pathway by conversion to putrescine. A second possibility is that the process of infection may specifically alter either the ornithine or putrescine pool.

Several interesting observations emerged from analyses of the change in the ratio of spermidine to spermine in infected and uninfected cells. Figure 35A shows that this ratio increased slightly in the infected cultures but decreased in the uninfected ones. These changes in the ratios are significant. Thus, as indicated in Fig. 35B, the amount of spermine relative to DNA remained essentially constant in both infected and uninfected cells, even though the amount of DNA increased about 40% in the uninfected cells and nearly doubled in the infected cells. Moreover, while the amount of spermidine remained constant in uninfected cells, it more than doubled in infected cells. Therefore, the decreasing ratio of spermidine to spermine in the uninfected cells appears to be due to an increasing amount of spermine in the presence of a constant amount of spermidine, whereas the increase in the ratio of spermidine and spermine observed in the infected cells appears to be due to simultaneous increases of both spermine and spermidine, but with spermidine accumulating at a slightly faster rate.

(2) The foregoing studies involving the utilization of ornithine as a precursor of polyamines in infected cells argue strongly against the hypothesis that arginine serves as a precursor for viral polyamines and is required for that purpose in infected cells. However, an interesting

observation worth noting is that the polyamine content of infected cells deprived of arginine becomes similar to that of uninfected cells (Gibson and Roizman, 1973). Specifically, as indicated in Table 11, the ratio of spermidine to spermine in uninfected cells changed from a value of 0.85 with arginine present to 1.35 in its absence. This second ratio, as shown in the table, is close to that of the polyamines in infected cells maintained either in the presence of absence of arginine.

Just why the effects of arginine deprivation on the polyamine content of uninfected cells mimic the effects of infection is not clear. It is perhaps noteworthy that Russell (1971) and Gfeller and Russell (1970) showed that in Xenopus cells a relationship may exist between the structure of the nucleolus and polyamine metabolism. As was previously reported (Schwartz and Roizman, 1969*a*), in HSV-infected cells the nucleolus disaggregates during infection. If an association does indeed exist between the nucleolus and polyamine metabolism, then it is perhaps not surprising that conditions such as arginine starvation (Granick and Granick, 1971) and infection with HSV, both of which disrupt nucleolar integrity, also appear to cause changes in the cellular polyamine composition.

TABLE 11

The Effect of Arginine Starvation on the Polyamine Composition of Infected and Uninfected Cells[a]

	Ratio Spd/Sp[b]	
Cells	Arginine present	Arginine absent
Uninfected	0.85	1.35
Infected	1.56	1.42

[a] Four cultures of HEp-2 cells were used. One was left uninfected and incubated in the presence of fresh, complete medium; a second was infected and then incubated in fresh, complete medium; a third was left uninfected and incubated with fresh medium devoid of the amino acid arginine; and a fourth was infected and also incubated in arginine-deficient medium. After 20 hours, the cells were scraped from the culture bottles, pelleted by centrifugation at 1500 × g for 10 minutes at 5°C, and extracted with perchloric acid. Data from Gibson and Roizman (1973).

[b] Polyamine analyses were done as described in the footnote to Table 3.

3.7.2. Lipids

Studies on the effect of viral infection on lipid metabolism are scarce and the results reported to date are inconsistent. Thus, selective increase in sphingolipid synthesis (Ben-Porat and Kaplan, 1971), generalized decrease in phospholipid synthesis (Kieff and Roizman, unpublished studies), or no change at all (Asher *et al.*, 1969) have been observed. No adequate studies of neutral lipid or glycolipids are available. Preliminary studies (Kieff and Roizman, unpublished data) suggested significant changes in the extent of glycosylation of glycolipids following infection.

3.8. The Morphogenesis and Egress of the Virion from Infected Cells

3.8.1. Electron Microscopic Studies

3.8.1a. Assembly of the Core and Capsid

The core and capsid of herpesviruses are assembled in the nucleus. There have been numerous studies of the morphogenesis of the capsid and the core (Zee and Talens, 1972; Abraham and Tegtmeyer, 1970 Dillard *et al.*, 1972; Leetsma *et al.*, 1969; Mannweiler and Palacios, 1969; Shipkey *et al.*, 1967; Cook and Stevens, 1970; Fong and Hsiung, 1972; Fong *et al.*, 1973; Iwasaki *et al.*, 1973; Trung and Lardemer, 1972; Kanich and Craighead, 1972; Patrizi *et al.*, 1965; Lunger *et al.*, 1965; Zambernard and Vatter, 1966; McCracken and Clarke, 1971; Maré and Graham, 1973; Strandberg and Carmichael, 1965). It is convenient to discuss the morphogenesis of the capsid and core from the point of view of major structures seen in the nucleus. These are: (1) capsids varying in amount and shape of the core, and in the sharpness of the outline of the capsomeric layer; (2) "arcs" of partially assembled capsids; (3) filaments (DNA?) and material associated with them; (4) small and large granules; and (5) tubular structures.

(1) Eight kinds of capsids have been described, differing chiefly in the core structure (Fig. 36) (Kazama and Schornstein, 1973; Kanich and Craighead, 1972; Kristensson *et al.*, 1970; Lunger *et al.*, 1965; Leetsma *et al.*, 1969; Chopra *et al.*, 1970; Cook and Stevens, 1970; Dillard *et al.*, 1972; Falke *et al.*, 1972; McCracken and Clarke, 1971; Mannweiler and Palacios, 1969; Miyamoto, 1971; Miyamoto *et al.*, 1971; Nayak, 1971; Nazerian and Burmester, 1968; Nazerian *et al.*, 1971; Nii *et al.*, 1968*a,b*; Nii, 1971*a,b*; Schwartz and Roizman,

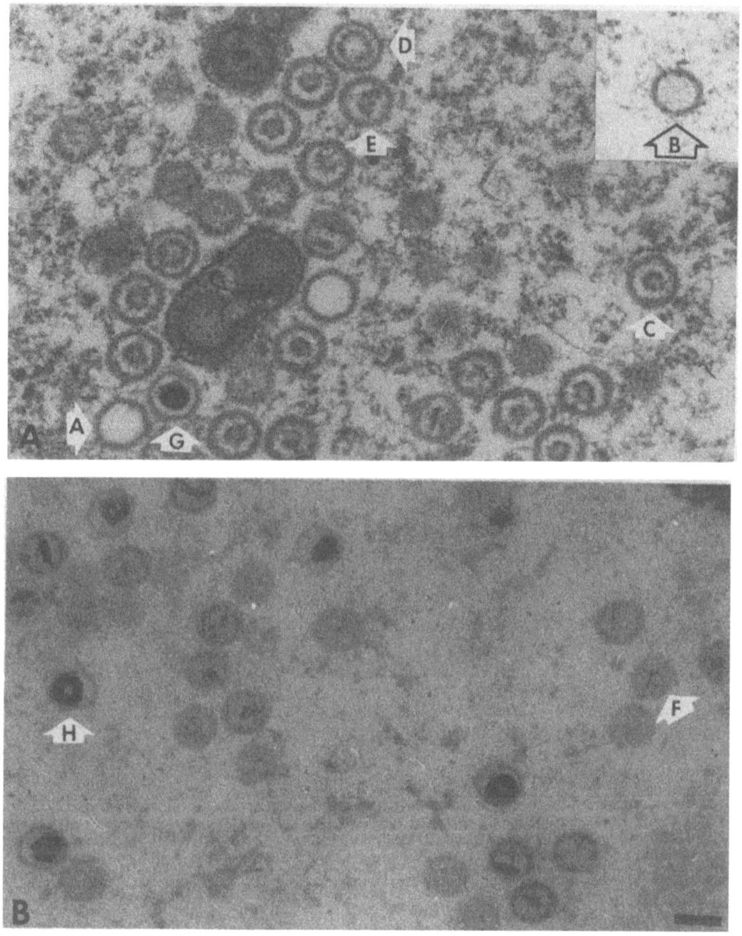

Fig. 36. Variability in the structure of the capsid and core during morphogenesis. Thin sections show HSV-1 capsids in nuclei of infected HEp-2. The letters inside arrows refer to descriptions of the various types of capsids in the text.

1969*a*; Shipkey *et al.*, 1967; Wolf and Darlington, 1971; Becker *et al.*, 1965). Five types are prominent early in infection. These are (A) empty capsids, i.e., with electron-translucent centers, (B) capsids with slightly dense, finely granular centers, (C) capsids with small (35–45 nm in diameter) cores, usually centrally located but sometimes lying against the capsid, (D) capsids with ringlike cores 45–55 nm in diameter; (E) capsids with internal granules described as forming pin-wheel structures, crescents, or outlining electron-translucent crosses. These have been lumped together because they could be descriptions of the same core, either seen from different angles of sectioning or following modifications introduced during fixation. Late in infection, in addition to the five types described above, three more types of capsids become prominent. Characteristically all contain in the core material (DNA?) that is much denser than the capsid. These are (F) capsids with dense fibrils or tails frequently extending from inside the capsid to outside (through?) the capsid, (G) capsids with small very dense cores 40–60 nm in diameter, and (H) capsids with toroidal cores resembling those of virions. (Capsid types A–H are shown in Fig. 36.) There is some variation of size of capsids even within the same nucleus. The variation has been correlated to core structure, empty capsids generally being larger than those containing cores of any sort (Becker *et al.*, 1965; Swanson *et al.*, 1966).

All these capsids were seen randomly scattered throughout the nucleus, in small clumps, and in large groups or arranged in crystalline arrays (Fig. 37). They occurred in the vicinity of the clumped chromatin, free in the nuclear sap, and in association with clumps of granular material. Within a particular group of capsids there was often heterogeneity of core morphology and of capsid morphology (some fluffy and some crisply defined) (Fig. 36B). All forms were seen enveloped later in infection (Fig. 38).

(2) Arcs formed by partially completed capsids have been described by several authors (Kazama and Schornstein, 1972, 1973; Miyamoto, 1971). These were rarely seen; when seen, they were usually among loose groups of completed capsids or near collections of granular material. Based on their rarity, it has been suggested that capsids form very rapidly. Nearly complete arcs frequently showed material extended out from the incomplete side of the capsid to filamentous structures nearby. This material extending from the arcs could not be identified by immunologic reactivity as a capsid component but it could not be differentiated immunologically from granules forming "crescents" inside capsid type E described above (Miyamoto *et al.*, 1971).

(3) *Granules*. Two types of granules appeared separately or together in aggregates or collections, but they are not totally interdispersed (Fig. 39). One type of particle, occurring either at the side or center of the aggregates, was small and indistinct. The other type of granule was larger and accumulated often in rows at the periphery of the aggregate, and had a center to center spacing of 4.8 nm. Partially complete capsids were most often seen in the immediate vicinity of or adjacent to aggregations of large granules (Miyamoto, 1971).

The data on the interrelationships of these structures are limited. Miyamoto and Morgan (1971) showed that one type of granular ma-

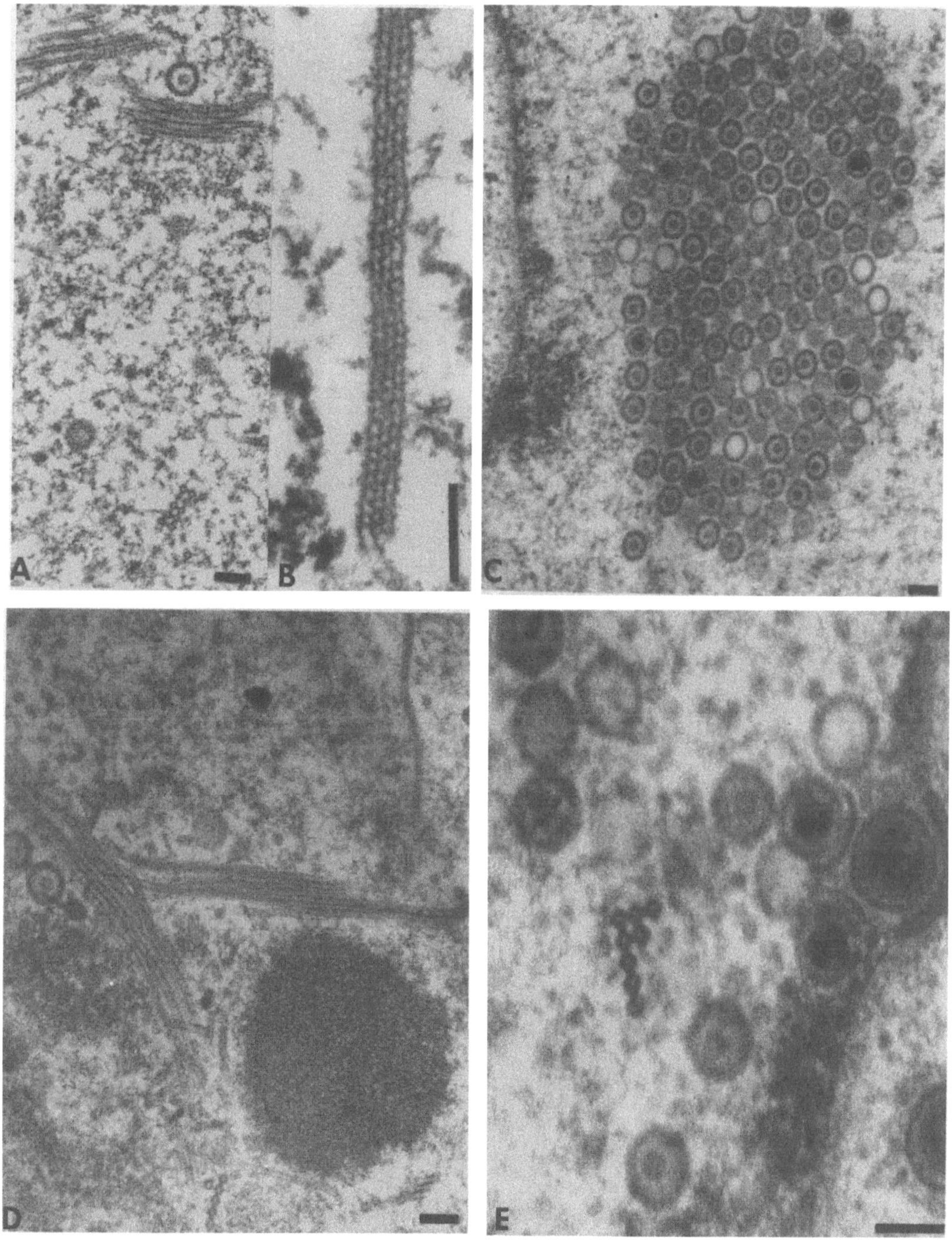

Fig. 37. (A) Thin sections of nuclei of HSV-2-infected HEp-2 cells. (B) Higher magnification of tubules showing capsidlike substructure. (C) Crystal of capsids varying in core morphology (D). Tubules with nearby capsids. (E) Crystal of unknown material which superficially appears more electron-opaque than the capsid material present in the same section. A and B (Aurelian and Strandberg, 1974), C, D, and E (Schwartz and Roizman, unpublished data).

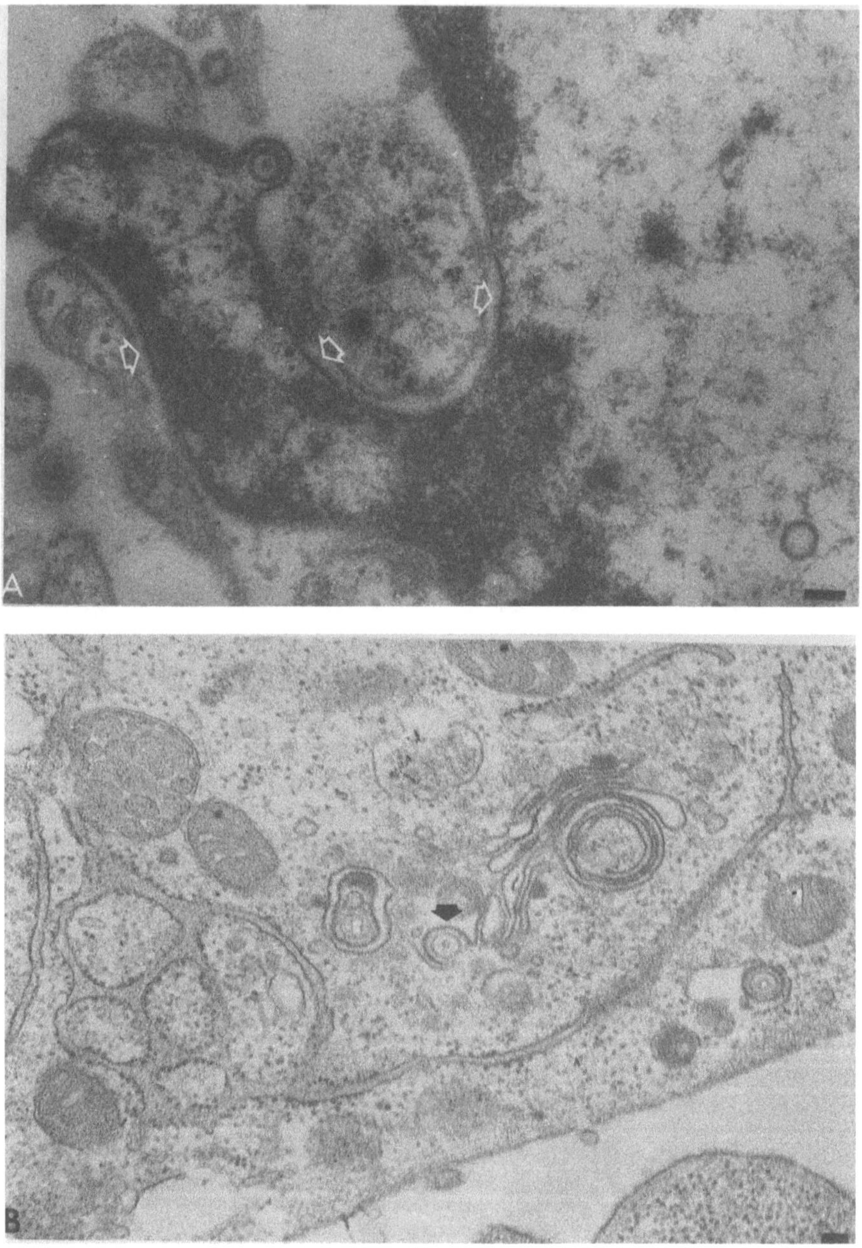

Fig. 38. Electron micrograph of thin section showing capsid in the process of envelopment at one of the altered maculae, i.e., stretches of increased density and altered curvature on the inner lamella of the nuclear membrane (arrows). Note spike projections into the perinuclear space at the site of envelopment.

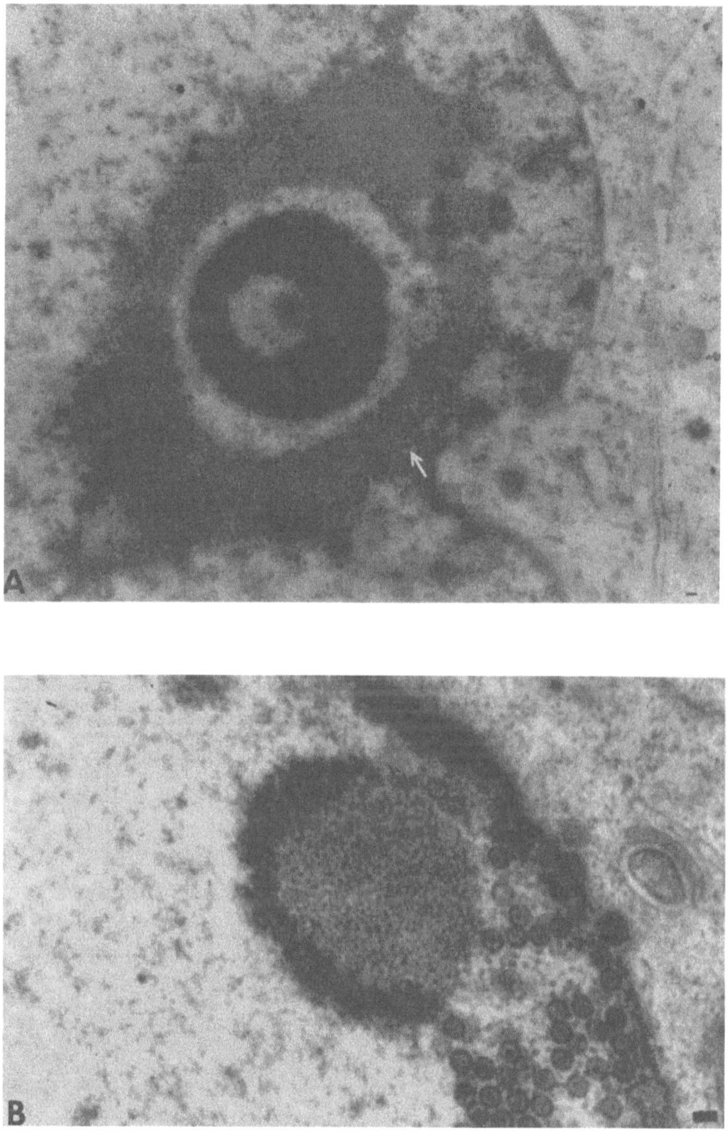

Fig. 39. (A) Electron micrograph of distorted nucleolus. Note large dense bodies (arrow) not commonly present in the nucleolus of uninfected cells (20,000×). (B) Electron micrograph of thin section showing capsid budding into smooth membrane vesicle in the Golgi region (arrow) and enveloped capsid in membrane-bound vesicle.

terial shared immunologic specificity with capsids, whereas another type of granules shared specificity with the material associated with filaments and with the low-density (protein?) core material. The two sets of antigenic material aggregated separately and were first seen together only in arcs formed by nascent capsids. The flow of events leading to capsid assembly has not been unequivocally established. The immunological reactivity of the nascent capsid could lead one to believe that the protein components of the core (i.e., the low- and medium-density core components) are incorporated into the capsid during initial assembly. The mode of entry of the DNA, i.e., the high-density component of the core, is not known. The finding of high-density filaments extending from capsids has been insufficient to differentiate between the hypothesis that DNA is stuffed into the completed capsid and the alternative that DNA is incorporated into the capsid during the initial assembly.

(4) Nuclei of cells infected with some herpesviruses contain tubular structures and crystals which could be transverse sections of tubular aggregates or aggregates of tubular subunits (Lunger *et al.*, 1965; Abraham and Tegtmeyer, 1970; Murphy *et al.*, 1967; Nazerian, 1971; Nii and Ono, 1971; Skinner and Mizell, 1972; Zambernard and Vatter, 1966; Parker *et al.*, 1973; Schwartz and Roizman, 1969*a*) (Fig. 37A,C). Two sizes have been reported. The smaller are much more common; the diameter of the tubules and the units arranged in crystalline arrays is 30–35 nm. No surface structure was discerned in negatively stained preparations of isolated tubules of this size (Stackpole and Mizell, 1968).

The small tubules occur independently of large ones. The large tubules have been seen only in thin sections containing small tubules. The diameter of the tubules and of the units arranged in crystalline arrays is 55–65 nm. In negatively stained preparations, the large tubular elements have the appearance of capsomers (Stackpole and Mizell, 1968). The crystals of both large and small tubular material appear to occur frequently in the vicinity of granular material of similar density (Fig. 37A, B, D).

3.8.1b. Envelopment and Egress

Although there is some disagreement concerning the site at which the capsid acquires its envelope, it is generally accepted that the envelope is acquired by the capsid budding through, or becoming wrapped in (Cook and Stevens, 1970; Leetsma *et al.*, 1969; Gershon *et al.*, 1973), a cellular membrane which has become suitably altered during the course of infection. Budding has been reported most consistently at the inner nuclear membrane (Fig. 38A) (Cook and Stevens, 1970; Dillard *et al.*, 1972; Jasty and Chang, 1972*b*; McCracken and Clarke, 1971; Mannweiler and Palacios, 1969; Nazerian and Witter, 1970; Nii *et al.*, 1968*a*; Patrizi *et al.*, 1965; Morgan *et al.*, 1959; Schwartz and Roizman, 1969*b*; Siegert and Falke, 1966; Shipkey *et al.*, 1967; Zambernard and Vatter, 1966; Zee and Talens, 1972; Abraham and Tegtmeyer, 1970; Kanich and Craighead 1972; Nazerian and Burmester, 1968; Epstein, 1962*b*; Roizman and Spear, 1973; Nazerian and Chen, 1973; Darlington and Moss, 1968, 1969; Wolf and Darlington, 1971). Other sites of budding are the Golgi apparatus, (Fig. 38B), the endoplasmic reticulum, and into vacuoles in the cytoplasm (Fong and Hsuing, 1972; Fong *et al.*, 1973; Gershon *et al.*, 1973; Iwasaki *et al.*, 1973; Kazama and Schorstein, 1973; Miyamoto *et al.*, 1971; Nayak, 1971) (Fig. 39). At the nuclear envelope the capsid buds through patches or macula wherein the inner lamella has been altered (Fig. 38). In the case of HSV, these patches can be seen as regions of increased density and altered curvature (Fig. 38A) (Dillard *et al.*, 1972). On the nuclear side there is often dense, amorphous material present, and on the cisternal side spikes morphologically resembling those on the surface of the virion can be seen. After budding through the inner lamella, the now enveloped virion is contained in the perinuclear space. Capsids budding through the inner nuclear membrane pick up the dense material that is present on the nuclear side of this lamella as well as some of the membrane itself. Although this material, by virtue of its location would be defined as the tegument, it is not known whether this is the only material present in the tegument. This is surely not the only site at which the tegument is acquired since unenveloped capsids covered with a layer of material can be seen in the nucleus (Nazerian, 1971; Nayak, 1971; Nazerian and Burmester, 1968; Heine *et al.*, 1971) and in the cytoplasm (Fig. 1) (Cook and Stevens, 1970; Craighead *et al.*, 1972; Fong and Hsiung, 1972; Kazama and Schornstein, 1973). It is not known whether the material seen around capsids not undergoing envelopment was the same as the material present in the tegument or whether the material seen on the cytoplasmic capsids was identical to, in addition to, or a replacement of the material seen around capsids in the nucleus. There is variability among herpesviruses concerning the site of acquisition of the tegument. Tegument-like material was seen in HSV-1-infected cells along the inner nuclear membrane as well as along the cytoplasmic membranes; this material could not be differentiated from the tegument of virions. In the guinea pig herpesvirus this material was not seen along the inner nuclear membrane or on capsids accumulated in the nucleus, and only those capsids that were exposed to the cytoplasmic sap and were enveloped at cytoplasmic membranes had a tegument (Fong *et al.*, 1973).

In general, electron micrographs of capsids in apposition to or partially surrounded by altered membranes have been interpreted as showing capsids undergoing envelopment. This interpretation, particularly of structures found at the nuclear membrane, seems reasonable. The question arises, however, whether the same interpretation is valid for similar structures seen at cytoplasmic membranes, since particles in the process of being deenveloped could give rise to similar images. The chief objections to envelopment at the cytoplasmic membranes are twofold. *A priori*, one would predict that cytoplasmic membranes would differ structurally not only among themselves but also from nuclear membranes, and it could be expected that the structural requirements for envelopment would be somewhat rigid. The second objection stems from lack of information as to how unenveloped capsids arrive in the cytoplasm to become enveloped. These objections are not compelling since some herpesviruses do have a wide host range, implying that the requirement for a particular chemical composition of the membrane enveloping the capsid may not be rigid, or that the virus is able to alter it to suit its needs. Further, it is conceivable, as suggested by Stackpole (1969), that the capsid undergoes sequential envelopment and deenvelopment at each membrane barrier in its passage from the nucleus to extracellular space.

The foregoing discussion should not be interpreted to mean that the site of envelopment is completely random. The impression gained from both examination of the literature and of infected cells in this laboratory (Schwartz and Roizman, 1969*a,b*), is that the inner lamella of the nuclear membrane is the predominant site of envelopment and that the images interpretable as envelopment at the cytoplasmic membrane are seen more frequently with some herpesviruses than with others.

Available data do not allow us to describe unequivocably the pathway of the virion from the site of envelopment to the extracellular space. There have been several pathways proposed, all with supporting data. One is that throughout the virion's journey, from the time it enters the perinuclear space, it is encased in some component of the cytoplasmic vesicular system (Fong and Hsiung, 1972; Gershon *et al.*, 1973; Hill *et al.*, 1972; Jasty and Chang, 1972*a*; Kanich and Craighead, 1972; Nii *et al.*, 1968*a*; Schwartz and Roizman, 1969*b*; Shipkey *et al.*, 1967). It has been suggested that it remains in the endoplasmic reticulum which connects with the perinuclear space until it is released into the extracellular space via a direct connection of the endoplasmic reticulum (Schwartz and Roizman, 1969*b*; Strandberg and Aurelian, 1969; Jasty and Chang, 1972*a*; Abraham and Tegtmeyer, 1970). This is a continuous route. The other variation of this scheme is that the route is discontinuous, i.e., the virions are thought to travel in vacuoles and be secreted in a manner similar to other exported cellular products (Morgan *et al.*, 1959). The virion would be nonetheless always within some membranous vesicle, protected from the cytoplasmic sap which initially served release the core from the capsids in the inoculum (Roizman, 1969). In support of these release mechanisms are many pictures of virions in the endoplasmic reticulum and in vacuoles in the cytoplasm, and pictures of virions in the cytoplasmic sap which appear to be disaggregating.

The alternative proposed mechanism of release is based on the assumption that the cytoplasmic sap is not necessarily an unsafe place for virions or capsids, and suggests that capsids may acquire their tegument there (Craighead *et al.*, 1972; Fong *et al.*, 1973; Iwasaki *et al.*, 1973; Kazama and Schornstein, 1973; Leetsma *et al.*, 1969; Lunger *et al.*, 1965; McCracken and Clarke, 1971; Mannweiler and Palacios, 1969; Nazerian and Burmester, 1968; Wolf and Darlington, 1971; Zee and Talens, 1972). This hypothesis proposes that capsids exit by budding into and out of whatever membranes they encounter and that the virion envelope comes from the last membrane encountered. This requires that all membranes become identically altered at sites which serve for envelopment or that there be heterogeneity among virion envelopes. For almost every herpesvirus studied, virus particles have been seen inside and outside all membranous structures (excluding mitochondria), although the proportion of particles in each region varies a great deal from one infected-cell system to another. A conservative evaluation of the available data is that, while no unique mode of egress from infected cells prevails, a particular mode of release may be more prevalent in some infected-cell systems than in others.

3.8.2. Biochemical Aspects of Viral Morphogenesis

A striking feature of viral morphogenesis is its inefficiency. Approximately 5- to 10-fold more viral DNA is produced than is utilized. The utilization of viral proteins is not known, but a similar order of inefficiency would not be too surprising.

Nothing is known concerning capsid assembly. Some data are available on envelopment. These may be summarized as follows:

(1) As pointed out in Sect. 2, unenveloped capsids containing viral DNA differed from deenveloped capsids in two respects (Gibson and Roizman, 1972, 1974). First, neither the deenveloped capsids nor the enveloped ones contain protein 22a present in unenveloped capsids. Based on staining properties of polypeptides in polyacrylamide gels, efficiency of incorporation of *in vivo*-labeled polypeptides into virions as a function of the time of addition of labeled amino acids into the medium, and *in vivo* phosphorylation, it has been deduced that polypeptide 22a is cleaved and that at least one product of this cleavage, polypeptide 22, is incorporated into the virion. Unenveloped capsids also lack polypeptides 1–3, which are quantitatively recovered in the deenveloped capsids. Topologically, proteins 1–3 and 22a are in the tegument. These data suggest that the cleavage of polypeptide 22a and the addition of at least polypeptides 1–3 and possibly other tegument polypeptides occurs during envelopment.

(2) In cells infected with HSV-1, viral glycoproteins have been found associated with all of the membranes fractionated to date, i.e., smooth endoplasmic reticulum, rough endoplasmic reticulum, and plasma membranes (Spear *et al*., 1970; Heine *et al*., 1972; Roizman and Heine, 1972). In cells infected with "wild" strains of HSV-1, the glycoproteins in the membranes appear to be at least qualitatively, identical to those found in purified virions. The available evidence on the composition of the viral polypeptides in the cytoplasmic membranes does not preclude the possibility that herpesvirions could be enveloped at any membrane of the infected cell although, as pointed out in the preceding section, the bulk of the envelopment takes place at the inner lamella of the nuclear membrane.

4. ALTERATIONS OF CELLULAR FUNCTION AND STRUCTURE DURING PRODUCTIVE INFECTION

4.1. General Considerations

The emphasis of the preceding section was on the virus, its structure, and replication. In this section the emphasis shifts from the virus to the infected host. All available evidence indicates that the consequence of productive infection with herpesvirus is cell death. The causes of cell death are obscure. The infected cell cannot divide; host DNA and protein synthesis are drastically reduced or cease altogether, and host RNA synthesis is greatly altered. This is accompanied by gross alterations in cell morphology and in the structure of cellular organelles. Information concerning structural and functional modifications of the infected cell have comc from direct visualization of the infected cell in the electron and light microscopes and biochemical studies on synchronously and parasynchronously infected cell populations. Many of the electron microscopic examinations were done very late in infection, when all biochemical events ceased or were no longer of any significance. Further, the biochemical and electron microscopic measurements are not comparable since microscopic examinations involve very few cells, and by and large the photomicrographs tend to depict isolated instances of the more extreme rather than average situation.

4.2. Alterations in Host Macromolecular Metabolism

4.2.1. General Description

Inhibition of host DNA synthesis has been observed in cells infected with pseudorabies virus (Kaplan and Ben-Porat, 1960, 1963), HSV-1 (Aurelian and Roizman, 1965; Roizman, 1969; Roizman and Roane, 1964), EAV (O'Callaghan *et al.*, 1968*b*), and EBV (Gergely *et al.*, 1971*a,b*; Nonoyama and Pagano, 1972*b*). A typical example of the decline of incorporation of thymidine into host DNA is shown in Fig. 31. The significance of the trace amounts of DNA synthesized late in infection is not clear; it could represent either asynchronously infected cells or a species of host DNA which escapes inhibition.

Host protein synthesis also declines rapidly after infection. This may be deduced from analyses of autoradiograms of infected-cell lysates shown in the preceding section. Cessation of host protein synthesis is accompanied by a decrease in the number of polyribosomes and a concurrent drop in the amino acid incorporation rate soon after infection (Sydiskis and Roizman, 1966; Roizman *et al.*, 1965). Cessation of host protein synthesis is accompanied by a cessation of glycosylation of host proteins (Spear *et al.*, 1970; Heine *et al.*, 1972).

The effects of viral infection on host RNA metabolism have been studied by numerous authors (Aurelian and Roizman, 1965; Flanagan, 1967, Hay *et al.*, 1966; Kaplan and Ben-Porat, 1960; Rakusanova *et al.*, 1971, 1972; Wagner and Roizman, 1969*a*). The emerging picture is complex and poorly understood. Very briefly, there is an overall decrease in host RNA synthesis, but the decrease is not uniform for all types of RNA. The decrease in the rate of synthesis is least for RNA greater than 28 S and most pronounced for 4 S RNA.

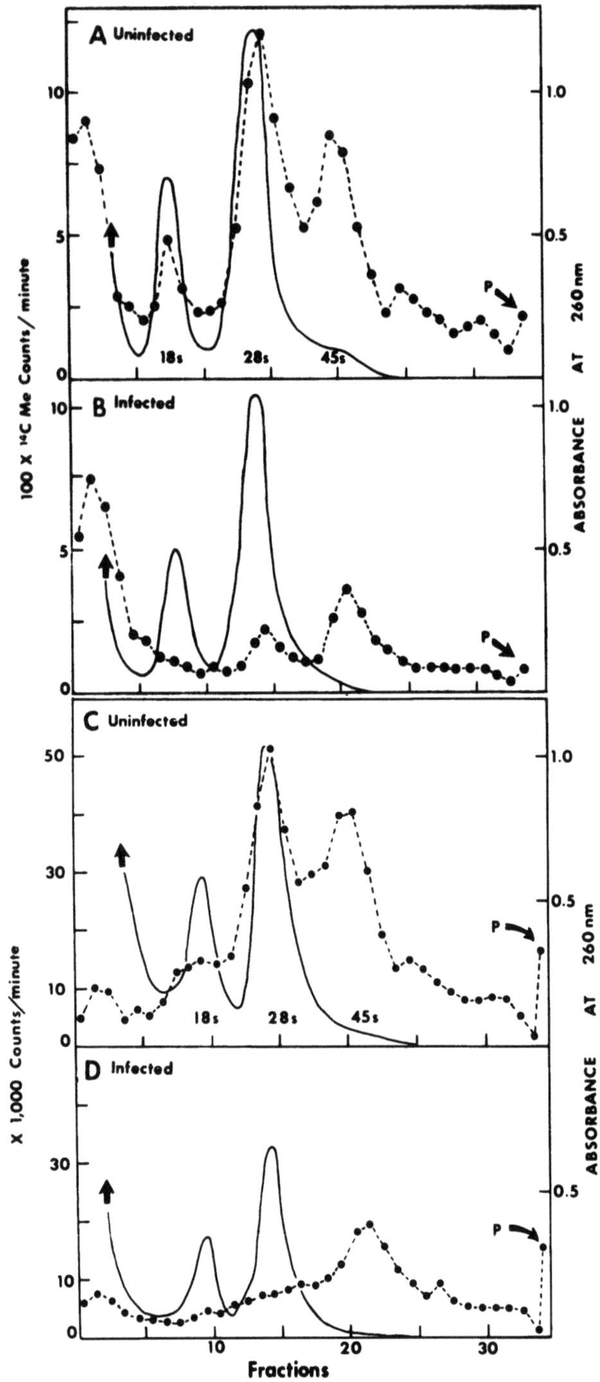

Fig. 40. Comparison of the size distribution of nuclear RNA synthesized in uninfected and HSV-1-infected HEp-2 cells. Here "P" marks material found as a pellet at the bottom of the tube. (——), optical density; (- · -), cpm. (A, B) Cells labeled with ^{3}H-uridine, (C, D) cells labeled with ^{14}C-methylmethionine. (A, B) Duplicate cultures of 3×10^7 HEp-2 cells were mock-infected or infected with 100 PFU/cell. After 4.5 hours, the cultures were incubated for 60 minutes in medium containing 5 μCi of ^{3}H-uridine/ml and then harvested. The nuclei of the cells were isolated and extracted as described by Wagner and Roizman (1969*a*). The extracted nuclear RNA was precipitated with ethyl alcohol and centrifuged in an 0.5% SDS 15–30% (wt/wt) sucrose density gradient for 15 hours at 25°C and at 21,000 rpm in a Spinco SW25.3 rotor. The 1-ml fractions were collected through a Gilford recording spectrophotometer. The RNA was precipitated with trichloroacetic acid and assayed for radioactivity. (C, D) Duplicate cultures of 5×10^7 cells were either mock-infected or infected with a multiplicity of 50 PFU/cell. The cells were incubated for 4 hours in Eagle's minimal essential medium containing 50% the normal amount of methionine and then for 60 minutes in methionine-free Eagle's minimal essential medium containing 0.8 μCi of ^{14}C-methylmethionine/ml and 2×10^{-4} unlabeled adenosine and guanosine. The isolation of nuclei and the extraction and fractionation of nuclear RNA were as described by Wagner and Roizman (1969*a*).

The appearance of new ribosomal RNA is nearly completely inhibited, but the site of action is not at the level of synthesis of 45 S precursor nor its methylation, but rather at the level of subsequent processing (Fig. 40) (Wagner and Roizman, 1969*a*; Kaplan, 1973). The nonribosomal RNA which continues to be made does not appear to be functional in the sense of being capable of directing host protein synthesis in the infected cell (Roizman *et al.*, 1970*a*; Rakusanova *et al.*, 1972). The report by Roizman *et al.* (1970*a*) that host RNA is transported from the nucleus to the cytoplasm without delay, as compared to the bulk of viral RNA which lingers in the nucleus some 10–20 minutes after it is made, indicates an alteration in processing of host RNA. Just why host RNA synthesis is not inhibited as completely as host DNA or host protein synthesis remains a mystery.

4.2.2. The Role of Viral Gene Expression in the Inhibition of Host Macromolecular Metabolism

The substantive observation is that the infected cell discriminates between host and viral macromolecular synthesis. The fundamental question raised here is the mechanism by which this discrimination takes place.

(1) The first and obvious question is whether the inhibition of host macromolecular synthesis requires gene expression following infection. In spite of a brief report by Newton (1969) summarized in the proceedings of a meeting, there is no evidence that UV-irradiated virus inhibits host functions (Sydiskis and Roizman, 1967). The data are not conclusive since rather high dosages of nonionizing radiation are required to substantially reduce the infectivity of virus preparations (Duff and Rapp, 1971), and, therefore, UV light irradiation might inactivate a sensitive component required for this function. There is no simple way to get around this objection. One expectation of the hypothesis that the inhibitor is a structural component of the virus is that the rate of inhibition would be multiplicity-dependent. In fact, this is the case for RNA synthesis (Wagner and Roizman, 1969*a*). However, in contrast with adenovirus-infected cells, in which high multiplicity of infection results in cessation of both host and viral macromolecular synthesis (Levine and Ginsberg, 1967, 1968), increasing the multiplicity of herpesvirus infection does not have an adverse effect on virus yield and by extension, on viral macromolecular synthesis.

(2) There is in fact ample evidence that viral gene expression augments the rate of inhibition of host functions. Ben-Porat and Kaplan (1965) demonstrated that puromycin (20 μg/ml) decreased the

rate of inhibition of host DNA synthesis in rabbit kidney cells infected with pseudorabies virus. Similar data are also available for the inhibition of host protein synthesis. Thus, actinomycin D, *p*-fluorophenylanine, and 6-azauridine prevented the inhibition of host protein synthesis in dog kidney cells abortively infected with HSV-1 (MP) at relatively high (1000 PFU/cell) multiplicities of infection, although the drugs were ineffective in HEp-2 cells infected at low multiplicities, presumably because the gene product may be more effective in human (HEp-2) cells. In another type of experiment, Ben-Porat *et al.* (1971) showed that in pseudorabies-infected rabbit kidney cells treated with cycloheximide during the first 5 hours following infection, host polyribosomes function immediately after cycloheximide is removed but disappear entirely 15 minutes after removal of the drug. The authors concluded that the inhibition of host protein synthesis is mediated by a protein synthesized early after infection.

(3) A question arises as to whether the inhibition of host DNA, RNA, and protein synthesis is mediated by a single viral gene product or by multiple, independently acting viral gene products. Kaplan (1973) found most appealing the hypothesis that the inhibition of host DNA synthesis is due to inhibition of synthesis of host proteins required for replication of host DNA, largely because the inhibition of host cell DNA synthesis is a relatively slow process. It is more difficult to postulate that the inhibition of host RNA synthesis is also the result of cessation of host protein synthesis since that does not explain the diversity of observed effects. Moreover, we cannot at the moment disregard the possibility that the margination of chromatin, the reported chromosome breakages, disaggregation of nucleoli, etc., dealt with in the next section are contributory causes to the overall inhibition of host macromolecular metabolism and not the consequence of inhibition of host protein synthesis. For heuristic reasons, if none other, it seems more productive to regard the inhibition of synthesis of these macromolecules as being interrelated, although obviously the answer is not known.

(4) The principal fact that emerges from these studies is that a viral gene product discriminates between one or more self (viral) and nonself (host) macromolecular syntheses. This point should be pursued further. We have already cited the various considerations which lead us to speculate that the viral DNA is transcribed by a host polymerase and that, therefore, the initial transcription and translation need not be discriminatory. However, host macromolecular synthesis is reduced once the initial viral mRNA is transcribed. The data would suggest that the initial viral mRNA has the same cognitive signals for the

initiation of protein synthesis as does host mRNA or, at least, that the host translational machinery cannot discriminate between host and this viral mRNA. But what happens to this mRNA after translation of host mRNA is terminated? If the cognitive signals are identical, this mRNA should disappear from polyribosomes at the same rate as host RNA. If the host does not discriminate between viral and host mRNA carrying different cognitive signals, one might predict that host mRNA would continue to function long after host mRNA ceases to be translated. These two possibilities, albeit simple in substance, are not easily differentiated, even though there are classes of viral proteins (α polypeptides, see Sect. 3.4.7) which show similar declines in the rate of synthesis in comparison to that of the host, and, more important, these syntheses persist as do those of the host following removal of cycloheximide added at the time of infection of HEp-2 cells with HSV-1 (F) (Honess and Roizman, unpublished observation). The temptation to conclude that early viral mRNA carries the same cognitive signals as does host RNA and that late viral RNA carries different cognitive signals which allow discrimination to take place must be tempered by the recognition that other explanations may fit the facts equally well. The concept that the virus in the course of evolution develops a macromolecule capable of discriminating between self and nonself in the context defined earlier in the text is in itself rather interesting, for in order to function it must interact with a host rather than a viral structure. This hypothesis would predict that inhibition of host function would not be equally effective in all species, particularly in those which in nature the virus does not infect. Both herpes simplex and pseudorabies viruses have a rather wide host range, and in only one instance, i.e., the abortive infection of dog kidney cells with HSV-1 (MP), was there evidence that inhibition of host macromolecular synthesis is host dependent. In those cells, the inhibition was very much less effective and required infection at very high multiplicities to reduce host macromolecular synthesis to a level comparable to that observed in human cells infected at relatively low multiplicities (Aurelian and Roizman, 1965).

(5) The hypothesis that host macromolecular syntheses of DNA, RNA, and protein are each inhibited independently has not been adequately tested. The hypothesis predicts that in each instance a putative inhibitor must be able to discriminate between viral and nonviral macromolecules. Only one hypothesis, namely that herpesviruses specify histone-like proteins which bind to the host chromatin, has been exploited, and indeed such polypeptides have been found, but the data are not conclusive. In BHK 21 cells which were infected with pseu-

dorabies virus, two viral acid-soluble polypeptides were reported to be associated with isolated chromatin (Chandler and Stevely, 1973). However, evidence that they play a role in the alteration of the structure and function of host DNA is lacking.

Not to be overlooked in this discussion of the properties of the putative viral inhibitors of host macromolecular metabolism is the fact that herpesviruses do contain genetic information for these functions. We wonder why viruses have acquired, retained, and expressed the capacity to inhibit the host. The only available data pertinent to this problem would seem to indicate that inhibition of host macromolecular synthesis is a prerequisite for virus multiplication. The conclusion is based on the observation that the HSV-1 (MP) strain of herpes simplex virus multiplies and effectively inhibits cells of human derivation but does not produce infectious progeny and does not effectively inhibit macromolecular syntheses in dog kidney cells, largely because, as indicated elsewhere in this section, some viral products specified in dog kidney cells malfunction. The significant finding is that in dog kidney cells, viral DNA and proteins are synthesized only in cells infected at a multiplicity sufficiently high to inhibit host functions. At low multiplicities of infection the cell makes interferon only (Aurelian and Roizman, 1965). The data would seem to indicate that (1) host response to infection and inhibition of host macromolecular synthesis are competing processes initiated on infection, and that (2) at low multiplicities of infection the cells attain the upper hand only because the amount of effective inhibitor specified by the virus is insufficient to inhibit the host in time to prevent it from inhibiting the virus (Aurelian and Roizman, 1965; Sydiskis and Roizman, 1967).

4.3. Alteration in the Structure and Function of Cellular Membranes

4.3.1. General Description

The first inkling that herpesviruses alter the structure of the plasma membrane (Roizman, 1962*a,b*) was based on the numerous reports of isolation of mutants or variants of HSV, pseudorabies, and herpes B virus strains which differed from the "wild" or parental strains with respect to their effects on cells (Tokumaru, 1957; Gray *et al.*, 1958; Hoggan and Roizman, 1959*a*; Hinze and Walker, 1961; Falke, 1961; Nii and Kamahora, 1961; Kohlhage and Siegert, 1962; Schneweiss, 1962; Kohlhage, 1964; Wheeler, 1964; Kohlhage and Schieferstein, 1965; Ejercito *et al.*, 1968; Schiek, 1967; Schneweis *et* *al.*, 1972; Schiek and Schneweis, 1968). Whereas the parental strains caused the cells to round up and clump, the variants caused cells to fuse. Ejercito *et al.* (1968) classified HSV strains into four groups, i.e., (1) strains causing rounding of cells but no adhesion or fusion, (2) strains causing loose aggregation of rounded cells, (3) strains causing tight adhesion of cells, and (4) strains causing polykaryocytosis. The viruses comprising each group may differ in fine detail with respect to their effects on cells. Thus, polykaryocytes induced by various strains of HSV differ in size and morphology (Roizman and Aurelian, 1965; Kohlhage, 1964; Wheeler, 1964). The term "social behavior of infected cells" was introduced (Ejercito *et al.*, 1968; Roizman, 1969) to describe the different interactions of cells among themselves. The alteration in the social behavior of infected cells is readily demonstrable in monolayer cultures of cells infected at very low multiplicities; in this instance the interaction of infected cells is readily apparent and easily differentiated from the surrounding lawn of uninfected cells (Fig. 41). So striking is the difference between the appearance of the foci of infected cells that the technique was used not only for differentiation of strains differing in the effects on the social behavior of cells but also as a precise and reproducible plaque assay for measuring the infectivity of virus preparations (Hoggan and Roizman, 1959*a*; Roizman and Roane, 1961*a*, 1963).

The reasoning that viruses alter the structure of the plasma membrane (Roizman, 1962*a,b*) was based on the necessary conclusion that the shape, adhesiveness, and social behavior of cells must reflect the structure of the plasma membrane and the observation that the social behavior of infected cells is genetically determined by the virus. All subsequent studies were directed toward elucidation of the nature of the structural function and immunologic properties of the infected-cell membranes.

4.3.2. Leakage of Macromolecules from Infected Cells

The evidence that macromolecules leak from infected cells emerged from a series of studies by Kaplan and co-workers (Kamiya *et al.*, 1964, 1965; Zemla, *et al.*, 1967), whose objectives were to elucidate the regulation of viral macromolecules. The first of the papers reported on an *in vitro* system measuring collectively the enzymes involved in the incorporation of deoxynucleotides into DNA. The authors showed that the limiting enzyme increased in activity during the first 6 hours, leveled off between 6 and 10 hours, and subsequently decreased.

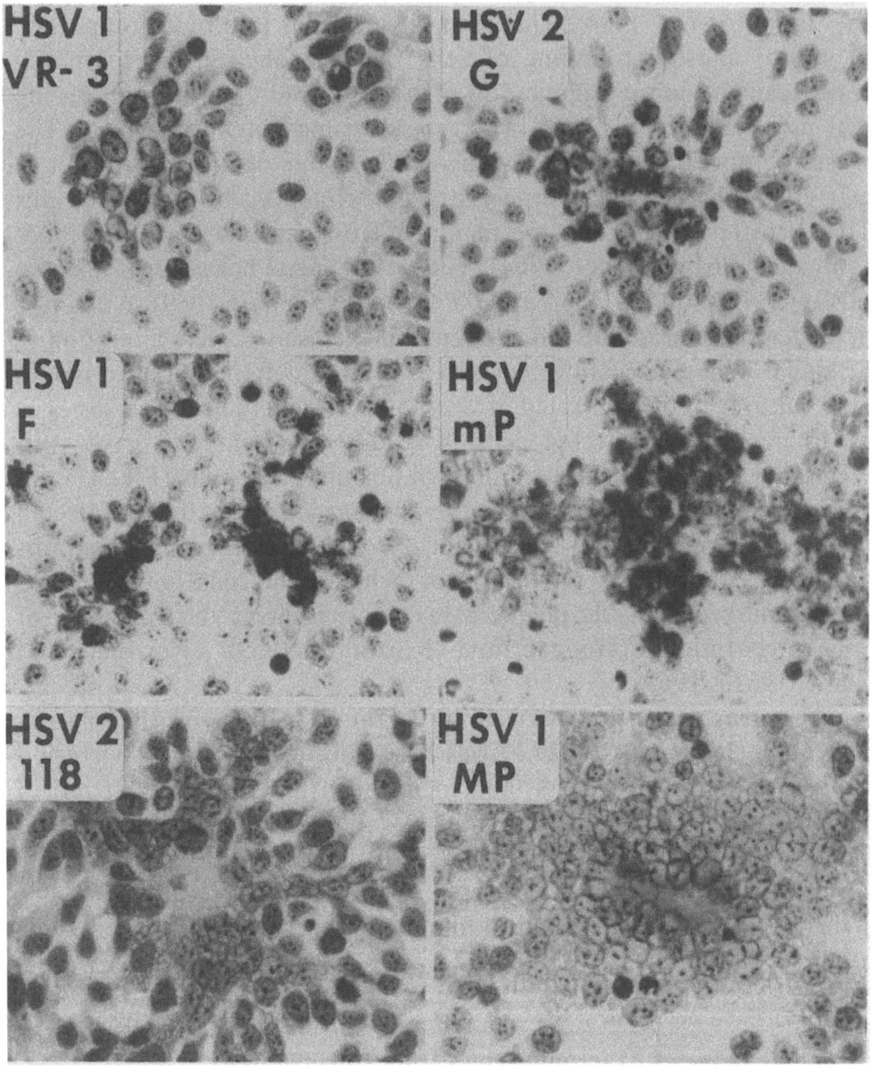

Fig. 41. The social behavior of HEp-2 cells infected with HSV-1 and HSV-2. VR-3 virus, rounding but little or no clumping of cells; G virus, loose clumps of rounded cells; mP virus, tight large clumps of rounded cells; 118 virus, small polykaryocytes which tend to fragment; MP, large polykaryocytes. HEp-2 cells were infected at a multiplicity of 1PFU/1000 cells and stained with Giemsa 24 hours after infection. Photomicrographs were taken and printed at the same magnification. The size of clumps, polykaryocytes, etc., is representative except for that produced by MP virus, which is the smallest found in that culture.

However, no leveling off of decrease in enzyme activity was observed in extracts of infected cells grown in a medium containing BUdR. This led to the conclusion that (1) the enzymes are regulated, (2) substitution of BUdR for thymidine interferes with the regulation, and (3) regulation is dependent not on the presence of viral DNA *per se* but upon the presence of newly synthesized complement of viral DNA. In subsequent studies designed to account for the leveling and decrease in the enzyme activity, Kamiya *et al.* (1965) and Zemla *et al.* (1967) found significant leakage of proteins from infected cells beginning 5–8 hours post-infection. On the basis of experiments with inhibitors, they also concluded that the leakage is somehow due to one or more proteins synthesized approximately 4 hours post-infection. Although the evidence for the involvement of proteins in the leakage is inadequate (Roizman, 1969), leakage of RNA was also manifestly higher in HEp-2 cells infected with HSV-1 than in uninfected cells (Wagner and Roizman, 1969*a*). It is conceivable that the change in transmembrane potential observed in HSV-infected cells reported recently is related to the leakiness of infected-cell membranes (Fritz and Nahmias, 1972).

4.3.3. Changes in the Surface Structure of Infected Cells

As noted in the preceding section, Wilbanks and Campbell (1972) differentiated between uninfected and HSV-infected cells by the number of microvilli. Changes in the reactivity of the cells with concanavalin A following infection was also reported (Tevethia *et al.*, 1972). In this instance the cells became agglutinable by concanavalin A by 2 hours post-infection.

4.3.4. Incorporation of Viral Proteins into Plasma Membrane

The presence of viral proteins in cellular membranes was reported in several publications (Spear *et al.*, 1970; Keller *et al.*, 1970; Heine *et al.*, 1972; Heine and Roizman, 1973). The pertinent data may be summarized as follows:

(1) Purified fractionated cytoplasmic membranes (Spear *et al.*, 1970) and plasma membranes (Heine *et al.*, 1972) extracted from HSV-1 (F)-infected cells were found to cosediment with proteins made after infection and absent from uninfected cells. In polyacrylamide gels, these proteins migrated with the major noncapsid virion proteins. Moreover, the glycosylation profile of the viral proteins associated

with the plasma membrane was similar to that of the major noncapsid virion proteins. The studies of the plasma membranes of infected and uninfected cells did not reveal major changes in the composition of the host membrane proteins remaining associated with the plasma membrane after infection.

(2) That the virus-specific proteins cosedimenting with the plasma membranes throughout purification do not constitute either adventitious contaminants adhering to the plasma membranes nor fragments of viral envelopes stripped off virions during the preparation of the infected-cell plasma membranes can be demonstrated as follows: Mixtures of infected and uninfected plasma membranes are readily separated by isopycnic centrifugation in sucrose gradients following reaction with antiviral antibody. The function of the antiviral antibody is to bind to viral proteins in the membranes, thereby augmenting the total protein mass relative to that of lipids. Consequently, membranes binding antiviral antibody band at a higher density (Fig. 42) than those

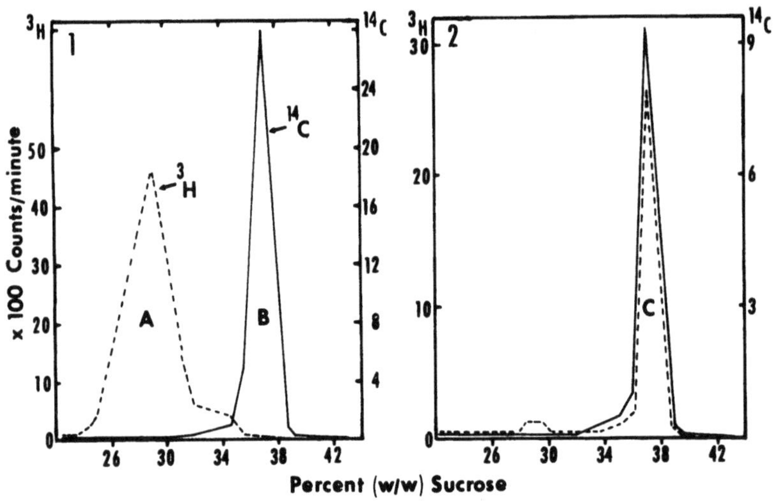

Fig. 42. Distribution of plasma membranes labeled with radioactive amino acids and reacted with anti-HSV-1 serum after centrifugation in sucrose density gradients. The serum was prepared by immunization of rabbits with virus produced in rabbit kidney cells and was the kind gift of Douglas Watson. The radioactivity was determined by drying 25 μl of each fraction onto glass fiber disks (Whatman) and placing the disks into vials containing 3 ml of a toluene-based scintillation fluid for counting in a scintillation counter. Panel 1, artificial mixture of plasma membrane vesicles from uninfected HEp-2 cells labeled with ^{3}H-amino acid and from HSV-1-infected HEp-2 cells labeled with ^{14}C-amino acids. Panel 2, plasma membrane vesicles from HSV-1 infected HEp-2 cells labeled with ^{3}H-amino acids before infection and with ^{14}C-amino acids after infection. The plasma membrane vesicles were prepared as described by Heine *et al.* (1972). Data from Heine and Roizman (1973).

lacking these antigens (Roizman and Spear, 1971). In a series of experiments, Heine and Roizman (1973) have shown that whereas artificial mixtures of labeled, infected- and uninfected-cell membranes are readily separable by isopycnic centrifugation following reaction with antiviral antibody, the host polypeptides labeled before infection band with the infected-cell membrane and not with the host proteins contained in the membranes extracted from uninfected cells (Figs. 42 and 43). Moreover, microcavitation employed for preparation of plasma membranes does not inactivate infectivity. The data indicated that the viral proteins are contained on the same membrane fragments as host proteins and that the binding or incorporation of viral proteins to membranes is sufficiently tenacious to withstand considerable hydrodynamic stress augmented by the presence of antibody bound to the polypeptides. The fact that not all viral glycoproteins appear in the plasma membrane, as discussed in a subsequent section, suggests that the association of viral polypeptides is not only tenacious but also specific.

4.3.5. Alteration in Immunologic Specificity of the Plasma Membrane

The studies on immunologic specificity of cells infected with herpesviruses were prompted by the observation that herpesviruses alter the social behavior of cells (Roizman, 1962*b*) as discussed in the preceding section. The evidence that infected cells acquire a new immunologic specificity was obtained with the aid of a test based on the observation that viruses fail to multiply in somatic cells injured by antibody and complement (Roizman and Roane, 1961*b*). In practice, cells were infected with HSV-1, suspended, washed, and incubated at 37°C with appropriate amounts of antibody and complement. After 1 hour the cells were diluted in an appropriate medium and seeded on monolayer cultures of HEp-2 cells. The survivors produced plaques whereas injured cells did not. Both antibody and complement were required for immune injury; antibody alone or complement alone was ineffective. The sole function of infected cells was to provide a measurement of the fraction of the test population that remains viable after exposure to antibody and complement. The sensitivity of the test stemmed from the fact that very few cells were needed since nearly every infected cell produces a plaque. The assay was initially standardized with 2-hour-infected cells and antibody made against uninfected cells (Roizman and Roane, 1961*b*). Pertinent here are the findings that the immunologic specificity of 2-hour-infected cells could not be differentiated from that of uninfected cells and that the differentiation between "viable" and

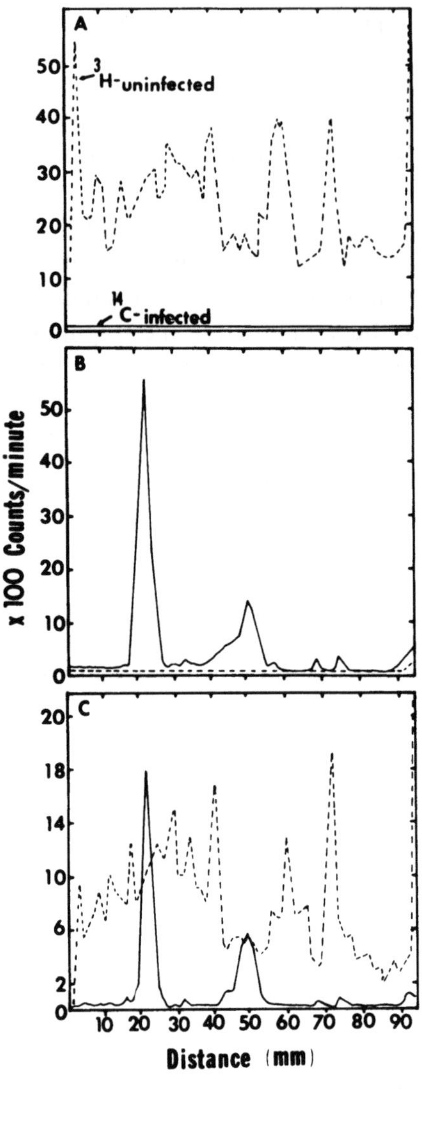

Fig. 43. Polyacrylamide gel electrophorograms of the plasma membrane preparations banded in sucrose density gradients shown in Fig. 42. The designations A, B, and C refer to the bands A and B of Fig. 42, panel 1, and the band C in Fig. 42, panel 2. The acrylamide gels were cross-linked with methylene-(bis)acrylamide. Data from Heine and Roizman (1973).

"killed" cells was not affected by the type of monolayer culture used for enumeration of infective centers.

The alteration of immunologic specificity after infection was demonstrated in tests employing 20- to 24-hour-infected cells and rabbit sera prepared against HSV-1-infected cells (Roane and Roizman, 1964). The tests showed that complement and unabsorbed anti-infected cell serum precluded the formation of plaques by 2-, 24-, and 48-hour-infected cells. However, following absorption with uninfected cells, the serum and complement precluded the formation of plaques by 24- and

48-hour-infected cells only; the absorbed serum was not effective against 2-hour-infected cells, leading to the conclusion that 24- and 48-hour-infected cells contained on their surface one or more antigens absent in uninfected cells. The conclusion that the membranes of infected cells become altered with respect to structure and immunologic specificity was corroborated in a study by Watkins (1964) showing that HeLa cells infected with the HFEM strain of herpes simplex acquire "stickiness" for sheep erythrocytes sensitized with rabbit anti-sheep erythrocyte serum. The adhesion of sensitized erythrocytes to the infected cells could be abolished by exposing the infected HeLa cells to antiviral serum. Normal rabbit serum and rabbit anti-sheep erythrocyte serum failed to prevent the adhesion of the sensitized erythrocytes to infected HeLa cells. Virus-specific antigens have since been demonstrated by a variety of techniques, not only in cells infected with HSV (Ito and Barron, 1972; Smith *et al.*, 1972*a,b*; Nahmias *et al.*, 1971; Wildy, 1973; Espmark, 1965; Brier *et al.*, 1971), but also with EBV (Klein *et al.*, 1968), MDV (Chen and Purchase, 1970; Ahmed and Schidlovaky, 1972; Nazerian, 1973), and herpes zoster virus (Ito and Barron, 1973; Nii *et al.*, 1968*b*).

The weight of the evidence favors the hypothesis that the new antigen is a structural component of the viral envelope. The evidence consists of the following findings: (1) Absorption of serum with partially "purified" HSV-1 virions removed both neutralizing and cytolytic antibody (Roane and Roizman, 1964). However, the significance of this observation is limited by the fact that the most purified-virus preparations available at that time were not free of host antigens. Similar evidence was also obtained for the antigens appearing on the surface of EBV-infected lymphocytes and the antigens on the surface of EB virions (Pearson *et al.*, 1970). (2) Assays of hyperimmune sera prepared against a variety of antigens extracted from infected permissive and nonpermissive cells showed excellent correlation between neutralizing and cytolytic titers (Roizman and Spring, 1967). (3) Dog kidney cells abortively infected with HSV-1 (mP) virus produced naked nucleocapsids only; envelopment did not take place (Spring *et al.*, 1968; Spring and Roizman, 1967). Rabbit hyperimmune sera produced against extracts of abortively infected DK cells lacked both neutralizing and cytolytic antibody (Roizman and Spring, 1967). (4) In accordance with the hypothesis that the plasma membranes of infected cells contain virus-specific antigens, purified plasma membranes from HSV-1-infected cells were found to contain herpesvirus glycoproteins (Heine *et al.*, 1972; Heine and Roizman, 1973). The infected-cell membranes readily react with antiviral antibody as evidenced by the studies described earlier in the text, showing that the infected-cell membrane–antibody complex has a higher buoyant density than the unreacted membrane or those reacted with a nonimmune serum (Roizman and Spear, 1971). Evidence that the antibody reactive with isolated membranes is capable of neutralizing virus emerged from the observation that addition of purified infected-cell membrane to a mixture of virus and antibody competed with the infectious virus for a neutralizing antibody (Fig. 44) (Roizman *et al.*, 1973*a*). Lastly, direct evidence that it was the glycoproteins and not some other constituent of the membranes which reacted with the antibody was furnished by the observation that glycoproteins were removed from solution during passage through immunoabsorbent columns prepared from immune sera (Savage *et al.*, 1972).

4.3.6. Viral Membrane Proteins and the Social Behavior of Infected Cells

Since the studies of infected-cell membranes were prompted by the hypothesis that structural alterations in the plasma membrane underlie the observed changes in the social behavior of cells following infection, it is of interest to review the data bearing on this hypothesis. Initial comparisons of smooth membranes from cells infected with HSV-1 (mP), HSV-2 (MP), and HSV-2 (G) revealed differences in the viral

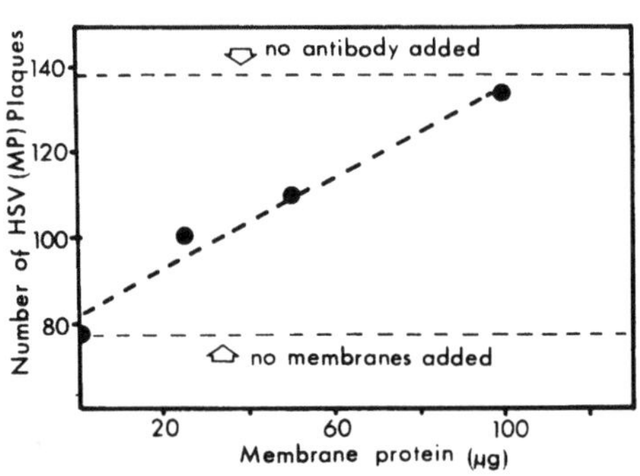

Fig. 44. Evidence for the identity of surface antigens on virions and on the smooth membranes of infected cells. Smooth membranes from HEp-2 cells infected with HSV-1 (mP) were purified according to the procedure of Spear *et al.* (1970). The test system consisted of 50 μl of challenge virus [HSV-1 (MP)] containing 500 PFU, 5 μl of anti HSV-1 (mP) serum (No. 39), and 200 μl of variable amounts of purified membranes as shown. In the absence of membranes, 5 μl of serum reduced challenge virus plaque count by 50%. The membrane antigen competed with the virus for neutralizing antibody, thereby sparing the infectivity. The plaques formed by the challenge virus were polykaryocytes, i.e., readily differentiated from the plaques produced by HSV-1 (mP) virus which might have contaminated the smooth membrane preparation. In this instance the smooth membranes were free of demonstrable virus. Data from Roizman *et al.* (1973*a*).

polypeptides bound to those membranes. In particular, the smooth membranes of cells infected with HSV-1 (MP), which causes cells to fuse, differed from the membranes of cells infected with HSV-1 (mP), which causes cell to clump, with respect to the number and electrophoretic mobility of viral proteins (Keller *et al.*, 1970). Because the resolution of the individual proteins in polyacrylamide gels was poor, it was not possible to identify which glycoproteins were absent. Pertinent also is the observation that in cells doubly infected with HSV-1 (mP) and HSV-1 (MP), both viruses multiply (Roizman, 1963*a*), but the social behavior of cells, as well as the composition of the proteins in the cellular membranes, was that of HSV-1 (mP)-infected cells (Roizman, 1962*b*, 1971*b*).

More recent analyses employing polyacrylamide gels cross-linked with DATD showed that HSV-1 (mP) virions contain glycoprotein 8 whereas HSV-1 (MP) virions lack this polypeptide. Moreover (Fig. 45), the plasma membranes from cells infected with HSV-1 (MP) differed from those of cells infected with HSV-1 (mP) in their total absence of polypeptide 8 and also in their very much reduced levels of polypeptides 7 and 8.5 when compared to glycoproteins 17–18. The data would seem to indicate that HSV strains cause different kinds of social behavior among infected cells and specify different membrane glycoproteins, however, not enough strains have been tested to determine unequivocably whether a change in membrane glycoprotein composition invariably corresponds to a change in the social behavior of infected cells. It would also seem that the presence of glycoprotein 8 inhibits fusion, but just what predisposes fusion of cells is not entirely clear since the membranes of cells infected with HSV-1 (mP) and HSV-1 (MP) differ in more than just the qualitative absence of VP8 from HSV-1 (MP)-infected cells. (Heine, Honess, and Roizman, unpublished data). Unquestionable resolution of this problem will require suitable temperature-sensitive mutants and, possibly, *in vitro* work with isolated viral membrane proteins.

4.4. The Structural Changes

4.4.1. Alterations in the Structure of the Nucleus

In addition to partially assembled, "mature" and "immature" capsids, and the various granules described in Sect. 3.8.2, which are related to virion morphogenesis, the infected-cell nucleus exhibits gross changes in the structure of the nucleolus, chromatin, and the nuclear

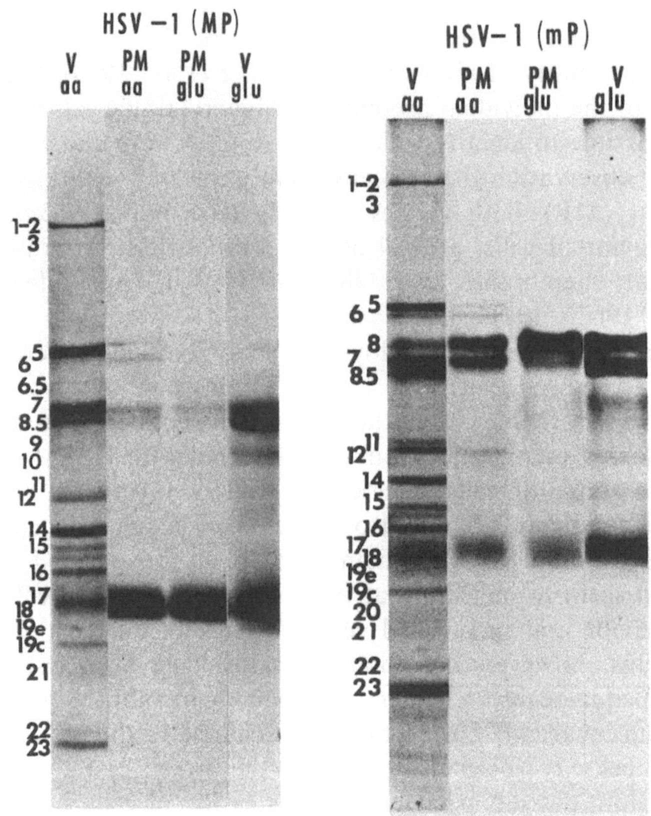

Fig. 45. Comparisons of the viral polypeptides in the purified HSV-1 (mP) and HSV-1 (MP) virions and purified plasma membranes obtained from infected cells. The infected HEp-2 were labeled with either ^{14}C-leucine, -isoleucine, and -valine; ^{14}C-glucosamine; or ^{14}C-fucose. Virions and plasma membranes were purified according to Spear and Roizman (1972) and Heine *et al.* (1972), respectively. The solubilized polypeptides were subjected to electrophoresis in gels cross-linked with DATD as described by Heine *et al.* (1974). Although each set of solubilized polypeptides was subjected to electrophoresis on the same slab gel, because of the differences in the intensities of the autoradiographic image, the autoradiographic image of each sample of electrophoretically separated polypeptides was printed at different exposure time. Note the absence of polypeptides in preparations of HSV-1 (MP) material and the large amounts of polypeptides 17–18 relative to polypeptides 7 and 8.5 in plasma membranes derived from HSV-1 (MP)-infected cells as compared with those derived from cells infected with HSV-1 (mP). Data from Heine and Roizman, in preparation.

membrane. During infection the nucleolus becomes enlarged and is often found displaced toward the nuclear membrane (Schwartz and Roizman, 1969*a*). The nucleolar material appears to be altered as well. The amount of granular material, in comparison to the amorphous component of the nucleolus, increases (Sirtori and Bosisio-Bestetti, 1967; Cook and Stevens, 1970). It is not clear whether this is a conversion of amorphous material to granules or whether the granular material is additional. Large, dense granules have been frequently seen along the periphery of the nucleolus (Fig. 39). The presence of viral material in the nucleolus has been demonstrated with ferritin-labeled antiviral antibodies (Miyamoto *et al.*, 1971). Late in infection the nucleolus appears disaggregated or fragmented (Nii, 1971*a,b*; Schwartz and Roizman, 1969*a*). Although the changes in the structure of the nucleolus appear to correlate with the inhibition of processing of ribosomal precursor RNA (Sect. 4.2), the biochemical basis for the changes in the nucleolar structure is not known; it is also not known whether the inhibition of processing of ribosomal RNA and the alterations which ultimately lead to the disaggregation of the nucleolus are causally related.

It has been reported by Moretti *et al.* (1972) and Zitelli *et al.* (1970) that ribosomes contained in the nuclei of cells from chick chorioallantoic membranes infected with HSV-2 formed crystals in the presence of rifampicin or subsequent to hypothermia, whereas the ribosomes of uninfected cell nuclei do not. Latticelike structures which resemble the crystallized ribosomes can sometimes be seen in HEp-2 cells infected with HSV-2 (Fig. 37D) (Zitelli *et al.*, 1970; Schwartz and Roizman, 1969*b*). It is not known whether the behavior of ribosomes in infected cells reflects the alterations in the processing of ribosomal RNA, the altered specificity in directing protein synthesis exhibited by the functioning cytoplasmic ribosomes after infection, or simply the physiological state of the nucleus after infection.

A diagnostic feature of the infected cell is the displacement and condensation of the chromatin at the nuclear membrane. The disaplacement occurs quite early in infection and is, in fact, one of the very early signs that the cell is infected. Nothing is known of the mechanism of displacement of the chromatin although it has been suggested that it is not an invariant feature of cells infected with all herpesviruses (Smith and deHarven, 1973). Correlated with the displacement of chromatin are two phenomena. The first, erroneously labeled as amitotic nuclear division (Scott *et al.*, 1953; Kaplan and Ben-Porat, 1959; Reissig and Kaplan, 1960; Nii and Kamahora, 1963), is very probably an extreme consequence of the distortion in the shape of the nucleus resulting in its fragmentation. The fragments, frequently unequal in size, remain attached to each other. Little is known of the second phenomenon, the chromosome breakages, also described by numerous authors (Hampar and Ellison, 1961, 1963; Tanzer *et al.*, 1964; Stich *et al.*, 1964; Benyesh-Melnick *et al.*, 1964; Boiron *et al.*, 1966; Nichols, 1966; Mikharlova, 1967; Waubke *et al.*, 1968; Sablina and Bocharov, 1968; Aya *et al.*, 1967; Mazzone and Yerganian, 1963; O'Neill and Miles, 1969; Rapp and Hsu, 1965; O'Neill and Rapp, 1971*a,b*; Huang and Minowada, 1972). The mechanisms underlying the aberrations induced by herpesviruses cannot be differentiated with respect to their site of occurrence and their nature from those occurring spontaneously or those induced by a variety of mutagenic agents (Stich *et al.*, 1964; Huang, 1967). A noteworthy observation by O'Neill and Rapp (1971*a*) is that the amount of chromosome breakages in cells infected with HSV-2 and treated with cytosine arabinoside was greater than in either uninfected cells treated with the drug or untreated infected cells. The second conclusion is that some product made in the cell early after infection is responsible for the breakage of the chromosomes (O'Neill and Rapp, 1971*b*) since it was reduced in cells pretreated with interferon but not in those treated with cytosine arabinoside.

Parenthetically, many of the experiments designed to demonstrate chromosome damage are perplexing and difficult to understand. In typical experiments, chromosome analyses are done on cells arrested in mitosis a few hours after infection at multiplicities of 2–5 PFU/cell. The problem arises from the consideration of reports by Stoker (1959), Stoker and Newton (1959), and by Vantis and Wildy (1962) that infection can both prevent and abort mitosis. If we take these observations at their face value, and we think they are valid, chromosome studies are done on an unknown, but presumably small, fraction of cells which undergo mitosis either because the virus failed to inhibit mitosis and in that particular cell the infection aborted, or the cell did not become infected to begin with. The question arises, therefore, whether the infected cell undergoing mitosis is representative of the culture, i.e., whether the mitotic cell was truly infected and whether the mitotic infected and uninfected cells belong to the same or to entirely different populations. It could be, for example, that in the infected culture a small fraction of cells characterized by broken chromosomes is resistant to infection and goes into mitosis, whereas the bulk of the cells with intact chromosomes are infected and do not go into mitosis. Whatever the case, it would obviously be expected that interferon-treated cells would resemble the uninfected cell population rather than the infected one. Obviously, comparisons of the chromosomes of infected and uninfected cells under these conditions would not be very meaningful. Clearly it remains to be shown that the conclusions derived from the chromosome studies are valid.

A characteristic of productively infected cells late in infection is the presence of long stretches of nuclear membranes folded upon themselves (Fig. 46). Such stretches have been described by numerous authors (Nii *et al.*, 1968*b*; Leetsma *et al.*, 1969; Schwartz and Roizman, 1969*a*; Wolf and Darlington, 1971; Morgan *et al.*, 1959; Cook and Stevens, 1970; McCracken and Clarke, 1971; Shipkey *et al.*, 1967) as reduplicated nuclear membranes. The staining properties of these membranes superficially resemble those of nuclear membranes containing the "macula" at which the capsids become enveloped. The resemblance derives from the fact that both the macula and the folded membranes contain stretches of electron-opaque material. It remains to be shown that the folded membranes contain viral glycoproteins. Moreover, it is not at all clear whether these membranes are synthesized *de novo* after infection or whether they represent modified stretches of the host nuclear membrane. In any event, it is probable that these refolded stretches represent an abberation of some cellular process necessary for the envelopment of the virus or that the multifolded stretches of membranes play no functional role in the maturation of the virus.

4.4.2. Alterations in Cytoplasmic Structures

Other than the virions and capsids which abound late in infection, the cytoplasmic landscape contains no unique features characteristic of herpesvirus-infected cells. The most striking features are the increase in the size of polyribosomes and in the amount of membraneous structures (Fig. 47A,B). The cytoplasm of infected cells often appears filled with polyribosomes. The polyribosomes appear to be much longer than those contained in uninfected cells; indeed, spirals containing 15–23 ribosomes have been seen (Fig. 47A) and this finding agrees well with the increased sedimentation rate of infected-cell polyribosomes reported by Sydiskis and Roizman (1966, 1968). Both free and bound polyribosomes have been seen in infected cells. At present there is no information as to whether these polymerize the same or different classes of polypeptides.

The cytoplasmic sap of infected cells appears to be more granular

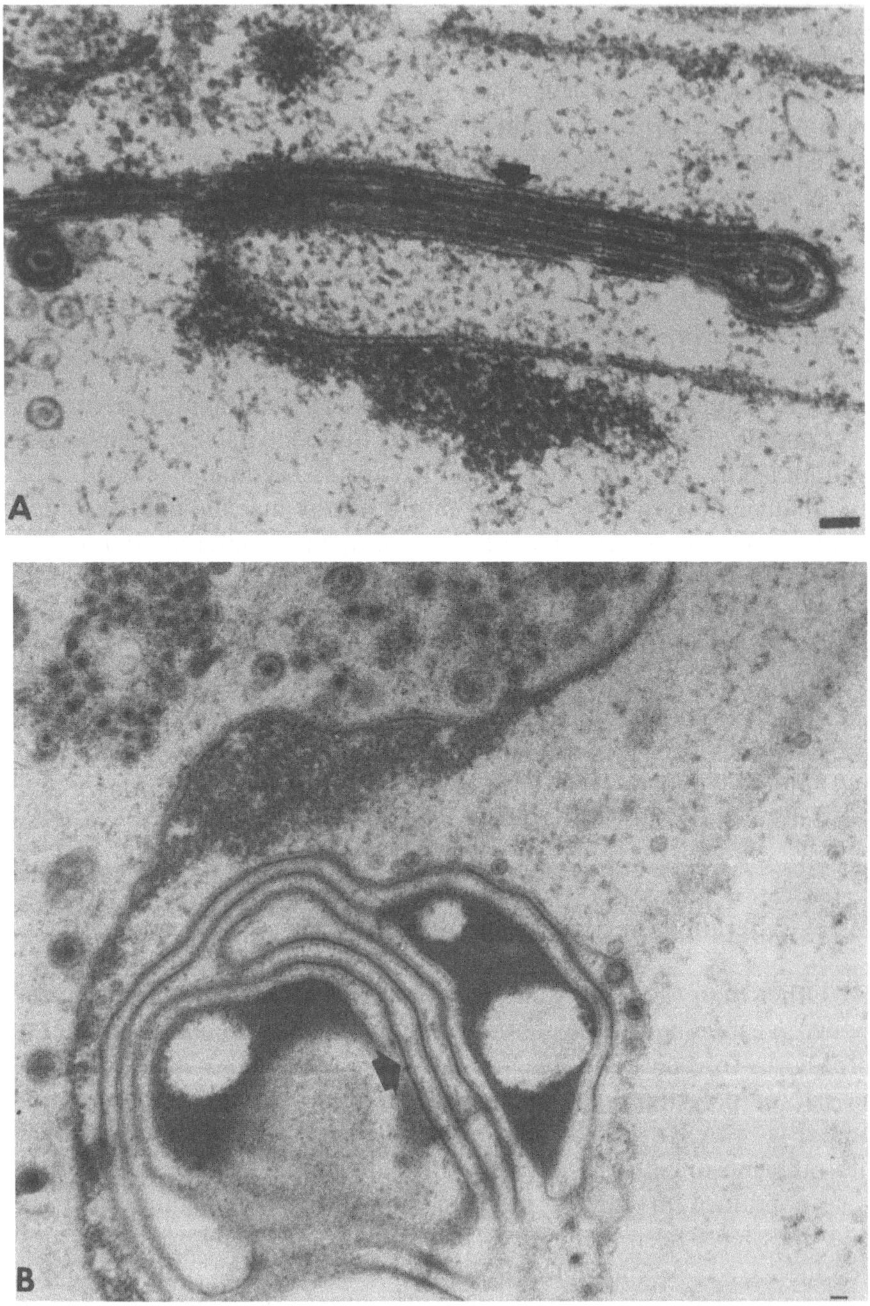

Fig. 46. Folding and modification of the nuclear membranes in HSV-1-infected HEp-2 cells. (A) Thin section showing cut perpendicular to the plane of the nuclear membrane. (B) Thin section showing cut nearly tangentially to the nuclear membrane. Note dense material between layers of membranes (arrows). The nature of this high density material is unknown.

than that of uninfected cells. In addition, aggregates of granular material are prevalent throughout the cytoplasm but especially near the nuclear pores. Microfibrils, often arranged in large bundles, are also prominent (Fig. 47A) (Schidlovsky *et al.*, 1969; Abraham and Tegtmeyer, 1970; Cook and Stevens, 1970; Stackpole and Mizell, 1968).

Although there are many reports of swollen or otherwise damaged mitochondria in infected cells, these all refer to changes which take place late in infection, after the cytoplasmic membrane becomes so fragile that the changes may simply reflect fixation artefacts or the general demise of the cell (McCracken and Clarke, 1970; Rabin *et al.*, 1968*a,b*). Early in infection the mitochondria seem little changed. Kazama and Schornstein (1973) reported that mitochondria were much larger in size but fewer in number in the productively infected cell as compared with the nonproductively infected cell. It is unlikely, but it has been suggested, that the changes in productively infected cells reflect the possibility that there is no further need for energy production since all the products necessary for assembly of virions are already available. There have been no systematic studies to determine the change in number or activity of mitochondria as a result of herpesvirus infection.

Electron-opaque, "dense," unidentified bodies have been seen occasionally in cells infected with nearly all herpesviruses, but they are particularly prominent in cells infected with cytomegaloviruses (Kanich and Craighead, 1972) and MDV (Nazerian *et al.*, 1971; Nii *et al.*, 1973). The electron-opaque bodies appear to be membrane-bound, 200–500 nm in diameter, and to contain very dense, slightly granular material. The nature of this material contained in the electron-opaque bodies is not known; based on its reactivity with antiviral antibody it has been suggested that it is virus-specific (Kanich and Craighead, 1972). Iwasaki *et al.* (1973) suggested that these bodies are lysosomes. The suggestion was based on their density and the fact that they were first seen in the vicinity of the Golgi region. However, the available data suggests that this is unlikely. Their morphogenesis as described by Iwasaki *et al.*, (1973) and by Nii *et al.* (1973) differs considerably from that generally accepted for lysosomes and secretory vacuoles (Jamieson and Palade, 1967*a,b*). Lysosomes are thought to form by concentration in the Golgi region of material contained within the cell's reticular system. The dense bodies, in contrast, are formed by budding of granular material free in the cytoplasmic sap through the membrane. Another point of contrast between the dense body and lysosomes and other secretory vacuoles is their tendency to fuse with the plasma membrane, spilling their contents into the extracellular space. Dense-

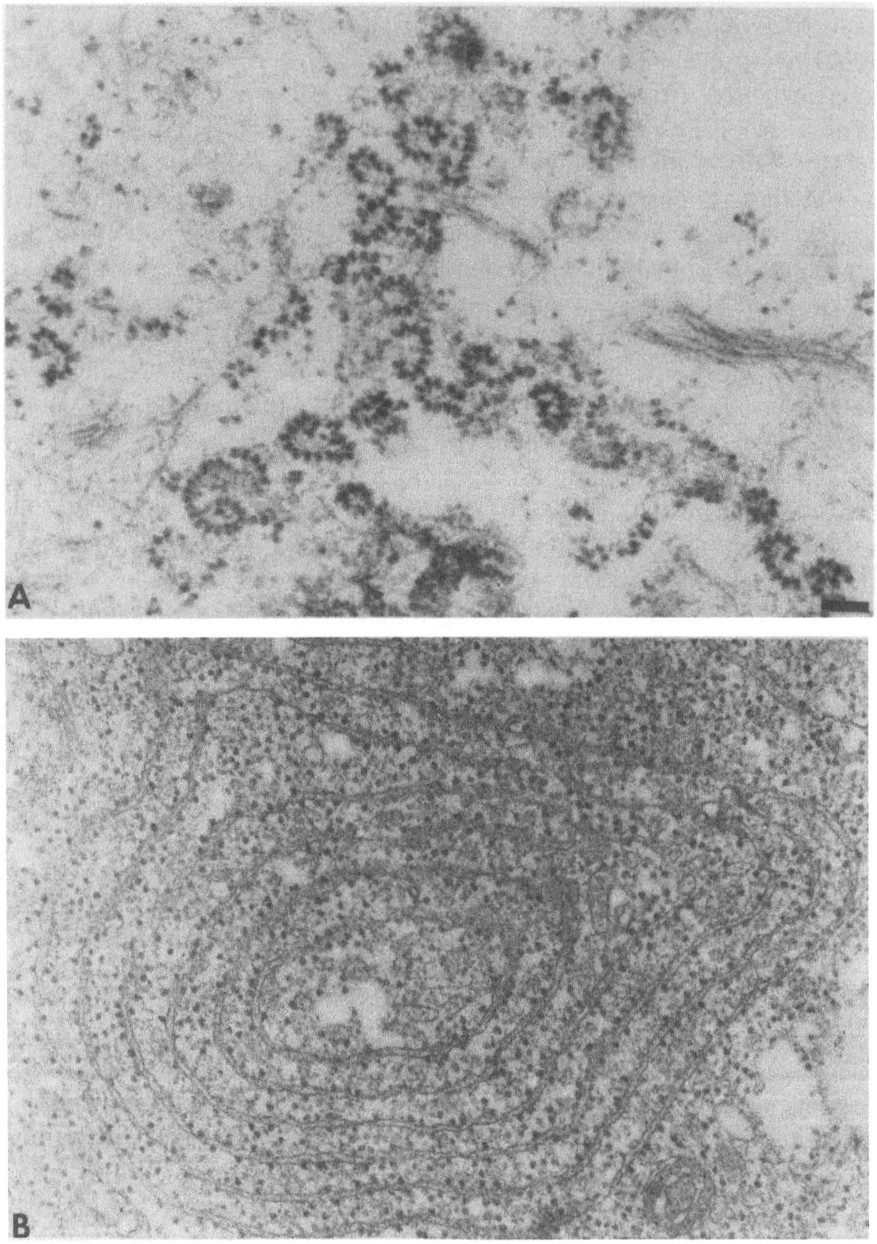

Fig. 47. (A) Electron micrograph of thin section of HSV-1 (F1)-infected HEp-2 cell showing polyribosomes and bundles of fibrils. Note polyribosomes with more than 15 ribosomes. (B) Electron micrograph of thin section showing extensive rough endoplasmic reticulum.

body vacuoles, however, are frequently seen undisrupted in the extracellular space and, indeed, even in the inocula used in these studies of Iwasaki *et al.* (1973). The presence of the dense bodies only late in infection and then mostly in cells which contain many virions suggests that they are viral products and not a response of the cell's defense system. Craighead *et al.* (1972) showed that among the dense bodies in infected cells some contained acid phosphatase, a marker for lysosomes, while others in the same cell did not contain the enzyme. As might be expected, this study showed that, in addition to these dense bodies, infected cells do contain lysosomes, which are often seen at the periphery of the Golgi region. Viral particles contained within lysosomal structures often seemed to be in the process of being disaggregated.

The increase in the amount of cytoplasmic membraneous structures, particularly endoplasmic reticulum, in the infected cells has been described by numerous authors as "extensive" and occurring "often." There is seldom much smooth endoplasmic reticulum, and in fact most of the surface of the cytoplasmic membranes seems to be studded with ribosomes (Fig. 47B). There are, however, smooth, membrane-bound vacuoles present in infected cells. The endoplasmic reticulum occasionally contains a very dense, finely granular material. The origin, nature, and function of this material are not known. The Golgi apparatus remains prominent throughout infection (Lunger *et al.*, 1965, 1966; Zambernard and Vatter, 1966; Iwasaki *et al.*, 1973; Kazama and Schornstein, 1973).

Plasma membranes from infected cells exhibit three kinds of alterations visible under the electron microscope. These are the appearance of patches of altered membranes, corresponding in all likelihood to regions of the membranes containing viral glycoproteins described in the preceding section; a redistribution of receptors reacting with concanavalin A; and an increase in the length and frequency of the microvilli. Appearance of altered patches or plaques along the plasma membrane was shown very clearly by Nii *et al.* (1968*a*). Following reaction of cells with ferritin-labeled antibody to HSV, the label was seen only along patches of the plasma membrane of infected cells which were denser and more curved than adjacent, unreacted areas. Filamentous material was often seen on the inside of the cell surface along these patches. These patches of altered membranes are very reminiscent of the macula seen on the nuclear membranes. The pattern of binding of concanavalin A to infected cells has been visualized using the diaminobenzidine reaction. Horseradish peroxidase, which binds to remaining sites of cell-bound concanavalin A, reacts with diaminobenzidine in the presence of peroxide to form an electron-dense precipitate. The precipitate is seen in patches along the plasma membrane and on virions. Wilbanks and Campbell (1972) observed with the aid of the scanning electron microscope that in cervical epithelium infected with HSV-2 the microvilli were longer, more numerous, and more irregularly spaced.

4.5. Conclusions

Of all the aspects of herpesvirus replication in eukaryotic cells, the alterations in the host structure and metabolism are the least well understood and show the least progress in the past several years.

(i) We must sadly acknowledge that we know virtually nothing about the mechanisms by which the virus shuts off the synthesis of host macromolecules. The problem is very tantalizing; one would predict that at least some macromolecules in the infected cells descriminate between host and viral informational molecules and that the latter must carry appropriate cognitive signals. As already emphasized earlier in the text, this is an extremely interesting area, not because herpesviruses shut off host macromolecular synthesis in productively infected cells, but because this phenomenon is both a useful probe and a convenient point of entry into the eukaryotic cell to study the mechanisms by which it regulates and differentiates its macromolecular synthesis.

(ii) Of necessity, the infected cell must become modified in order for herpesvirus multiplication to take place. The modifications, as exemplified by the alterations in the infected cell membranes, involve complex interactions between viral and cellular macromolecules. The impression that the cell is a malleable menstrum readily parasitized by the invading virus is correct only to a certain point. It is heuristically profitable to consider as more correct the view that productive infection with viruses is complex and that herpesvirus infection is a form of extreme, irreversible differentiation resulting in the synthesis of one product, the virus, and in cell death. If we accept the requirement for intricate and complex interaction between viral and host macromolecules, and if we view virus multiplication as a form of cellular differentiation which could not be expected to occur in just any cell, the question arises as to why infection is productive as often as it is. In point of fact, this chapter has dealt with productive infections.

We have no basis for comparing productive infections in culture with those in multicellular organisms to determine which are more restrictive. However, it is clear from the limited knowledge we do have that failure to multiply, because the cell is partially restrictive or entirely nonpermissive, is common and constitutes a plausible explanation for latent infection and the neoplastic growths caused by herpesviruses. These necessarily result from lack of appropriate interactions between viral and host macromolecules which would permit productive infection to ensue. Failure of such interactions to take place must necessarily account for nonpermissiveness, i.e., for the inability of some agents like frog herpesvirus 1 to grow in culture and for the narrow host range of many other. Transient nonpermissiveness resulting from failure of such interactions to take place because of the absence or inappropriate structure of host components may account for latent herpesvirus infections (Roizman, 1966, 1971*a*) and for the changes in the permissiveness of the cell mediated by temperature in the case of the frog herpesvirus 1 (Breidenbach *et al.*, 1971) and the fungal herpes-like agent (Kazama and Schornstein, 1973). In the same vein, the development in the course of evolution of numerous herpesviruses infecting different topologic areas or different organs of the same host must also be viewed as a selective process designed to ensure the best possible fit between the interacting macromolecules of the infecting virus and of the target cell. Because of their enormous potential significance in our understanding of the regulation of cell functions and conversion of cells from the normal to the malignant state, it is likely that the interactions of host and viral macromolecules at numerous levels, i.e., transcription, translation, enzyme aggregations, membrane modification, etc., will receive considerable attention in the years to come.

ACKNOWLEDGMENT

The authors are particularly indebted to Mrs. Patricia Wiedner for the preparation of illustrations and to Drs. R. W. Honess, M. Kozak, S. Silverstein, and S. Wadsworth for their thoughtful advice. The studies conducted at the University of Chicago and discussed in this review were aided by grants from the National Cancer Institute, United States Public Health Service (CA 08494), the American Cancer Society (VC 1031), and the Whitehall Foundation.

5. REFERENCES

Abodeely, R. A., Palmer, E., Lawson, L. A., and Randall, C. C., 1970, The proteins of enveloped and de-enveloped equine abortion (herpes) virus and the separated envelope, *Virology* **44,** 146–152.

Abraham, A., and Tegtmeyer, P., 1970, Morphologic changes in productive and abortive infection by feline herpesvirus, *J. Virol.* **5,** 617–623.

Ahmed, M., and Schidlovsky, G., 1972, Detection of virus associated antigens on membrane of cells productively infected with Marek's disease virus, *Cancer Res.* **32,** 187–192.

Ames, B. N., and Dubin, D. T., 1960, The role of polyamines in the neutralization of bacteriophage deoxyribonucleic acid, *J. Biol. Chem.* **235,** 769–775.

Ames, B. N., Dubin, D. T., and Rosenthal, S., 1958, Presence of polyamines in certain bacterial viruses, *Science* (*Wash., D.C.*) **127,** 814–816.

Anker, H. S., 1970, A solubilizable acrylamide gel for electrophoresis, *FEBS* (*Fed. Eur Biochem. Soc.*) *Lett.* **7,** 293.

Asher, Y., Heller, M., and Becker, Y., 1969, Incorporation of lipids into herpes simplex virus particles, *J. Gen. Virol.* **4,** 65–76.

Aurelian, L., and Roizman, B., 1965, Abortive infection of canine cells by herpes simplex virus. II. The alternative suppression of synthesis of interferon and viral constituents, *J. Mol. Biol.* **11,** 539–548.

Aurelian, L., and Strandberg, J. D., 1974, Biologic and immunologic comparison of two HSV-2 variants, and an isolate from cervical tumor cells, *Archiv. Ges. Virusforsch.*, in press.

Aurelian, L., Royston, I., and Davis, H. J., 1970, Antibody to genital herpes simplex virus: Association with cervical atypia and carcinoma *in situ*, *J. Natl. Cancer Inst.* **45,** 455–464.

Aya, T., Makino, S., and Yamada, M., 1967, Chromosome aberrations induced in cultured human leucocytes by herpes simplex virus infection, *Proc. Jap. Acad.* **43,** 239–244.

Bachenheimer, S. L., and Roizman, B., 1972, Ribonucleic acid synthesis in cells infected with herpes simplex virus. VI. Polyadenylic acid sequences in viral messenger ribonucleic acid, *J. Virol.* **10,** 875–879.

Bachenheimer, S. L., Kieff, E. D., Lee, L., and Roizman, B., 1972, Comparative studies on DNAs of Marek's disease and herpes simplex virus, *in* "Oncogenesis and Herpesviruses" (P. M. Biggs, G. de The, and L. N. Payne, eds.), pp. 74–81, International Agency for Research on Cancer, Lyon.

Becker, Y., and Olshevsky, U., 1972, Localization of structural viral peptides in the herpes simplex virion, *in* "Oncogenesis and Herpesviruses" (P. M. Biggs, G. de The, and L. N. Payne, eds.), pp. 420–423, International Agency for Research on Cancer, Lyon.

Becker, P., Melnick, J. L., and Mayor, H. D., 1965, A morphologic comparison between the developmental stages of herpes zoster and human cytomegalovirus, *Exp. Mol. Pathol.* **4,** 11–23.

Becker, Y., Dym, H., and Sarov, I., 1968, Herpes simplex virus DNA, *Virology* **36,** 184–192.

Bell, D., Wilkie, N. M., and Subak-Sharpe, J. H., 1971, Studies on arginyl transfer ribonucleic acid in herpesvirus-infected baby hamster kidney cells, *J. Gen. Virol.* **13,** 463–475.

Benda, R., 1966, Effect of heparin on B virus multiplication *in vitro*, *Acta Virol. (Prague)* **10**, 376.
Ben-Porat, T., and Kaplan, A. S., 1962, The chemical composition of herpes simplex and pseudorabies viruses, *Virology* **16**, 261–266.
Ben-Porat, T., and Kaplan, A. S., 1963, The synthesis and fate of pseudorabies virus DNA in infected mammalian cells in the stationary phase of growth, *Virology* **20**, 310–317.
Ben-Porat, T., and Kaplan, A. S., 1965, Mechanism of inhibition of cellular DNA synthesis by pseudorabies virus, *Virology* **25**, 22–29.
Ben-Porat, T., and Kaplan, A. S., 1971, Phospholipid metabolism of herpesvirus-infected and uninfected rabbit kidney cells, *Virology* **45**, 252–264.
Ben-Porat, T., Shimono, H., and Kaplan, A. S., 1969, Synthesis of proteins in cells infected with herpesvirus. II. Flow of structural viral proteins from cytoplasm to nucleus, *Virology* **37**, 56–61.
Ben-Porat, T., Rakusanova, T., and Kaplan, A. S., 1971, Early functions of the genome of herpesvirus. II. Inhibition of the formation of cell-specific polysomes, *Virology* **46**, 890–899.
Benyesh-Melnick, M., Stich, H. F., Rapp, F., and Hsu, T. C., 1964, Viruses and mammalian chromosomes. III. Effect of herpes zoster virus on human embryonal lung cultures, *Proc. Soc. Exp. Biol. Med.* **117**, 546–549.
Bernhard, W., 1969, A new staining procedure for electron microscopical cytology, *J. Ultrastruct. Res.* **27**, 250–265.
Beswick, T. S. L., 1962, The origin and the use of the word herpes, *Med. Hist.* **6**, 214–232.
Boiron, M., Tanzer, J., Thomas, M., and Hampe, A., 1966, Early diffuse chromosome alterations in monkey kidney cells infected *in vitro* with herpes simplex virus, *Nature (Lond.)* **209**, 737–738.
Breidenbach, G. P., Skinner, M. S., Wallace, J. H., and Mizell, M., 1971, *In vitro* induction of a herpes-type virus in "summer-phase" Lucke tumor explants, *J. Virol.* **7**, 679–682.
Bretcher, M. S., 1971, Major human erythrocyte glycoprotein spans the cell membrane, *Nature (Lond.)* **231**, 229–232.
Brier, A. M., Wohlenberg, C., and Rosenthal, J., 1971, Inhibition or enhancement of immunological injury of virus-infected cells, *Proc. Natl. Acad. Sci. USA* **68**, 3073–3077.
Britten, R. J., and Smith, J., 1970, A bovine genome, Carnegie Inst., Washington Yearb. **68**, 378–386.
Bronson, D. L., Graham, B. J., Ludwig, H., Benyesh-Melnick, M., and Biswal, N., 1972, Studies on the relatedness of herpes viruses through DNA–RNA hybridization, *Biochim. Biophys. Acta* **259**, 24–34.
Buchan, A., and Watson, D. H., 1969, The immunological specificity of thymidine kinases in cells infected by viruses of the herpes group, *J. Gen. Virol.* **4**, 461–463.
Buchan, A., Watson, D. H., Dubbs, D. R., and Kit, S., 1970*a*, Serological study of a mutant of herpesvirus unable to stimulate thymidine kinase, *J. Virol.* **5**, 817–818.
Buchan, A., Luff, S., and Wallis, C., 1970*b*, Failure to demonstrate the interaction of subunits of thymidine kinase in cells simultaneously infected with herpes virus and a kinaseless mutant, *J. Gen. Virol.* **9**, 239–242.
Cassai, E., and Bachenheimer, S., 1973, Effect of isotopic label on buoyant density determination of viral DNA in the preparative ultracentrifuge, *J. Virol.* **11**, 610–613.

Chandler, J. K., and Stevely, W. S., 1973, Virus-induced proteins in pseudorabies-infected cells. I. Acid-extractable proteins of the nucleus, *J. Virol.* **11,** 815–822.

Chen, J. H., and Purchase, H. G., 1970, Surface antigens on chick kidney cells infected with the herpes virus of Marek's disease, *Virology* **40,** 410–412.

Chopra, H. C., Shibley, G. A., and Walling, M. J., 1970, Electron microscopic cytochemistry of herpes simplex virus using enzyme extraction and autoradiography, *J. Microscop.* **9,** 167–176.

Cohen, G. H., 1972, Ribonucleotide reductase activity of synchronized KB cells infected with herpes simplex virus, *J. Virol.* **9,** 408–418.

Cohen, G. H., Vaughn, R. K., and Lawrence, W. C., 1971, Deoxyribonucleic acid synthesis in synchronized mammalian KB cells infected with herpes simplex virus, *J. Virol.* **6,** 783–791.

Collard, W., Thornton, H., Mizell, M., and Green, M., 1973, Virus-free adenocarcinoma of the frog (summer phase tumor) transcribes Lucké tumor herpesvirus-specific RNA, *Science (Wash., D.C.)* **181,** 447–449.

Cook, M. L., and Stevens, J. G., 1970, Replication of varicella-zoster virus in cell cultures an ultrastructural study, *J. Ultrastruct. Res.* **32,** 334–350.

Courtney, R. J., McCombs, R. M., and Benyesh-Melnick, M., 1971, Antigens specified by herpesviruses. II. Effect of arginine deprivation on the synthesis of cytoplasmic and nuclear proteins, *Virology* **43,** 356–365.

Craighead, J. E., Kanich, R. E., and Almeida, J. D., 1972, Nonviral microbodies with viral antigenicity produced in cytomegalovirus-infected cells, *J. Virol.*, **10,** 766–775.

Crawford, L. V., and Lee, A. J., 1964, The nucleic acid of human cytomegalovirus, *Virology* **23,** 105–107.

Dales, S., and Silverberg, H., 1969, Viropexis of herpes simplex virus by HeLa cells, *Virology* **37,** 475–480.

Darlington, R. W., and Moss, L H., III, 1968, Herpesvirus envelopment, *J. Virol.* **2,** 48–55.

Darlington, R. W., and Moss, L. H., III, 1969, The envelope of herpesvirus, *Progr. Med. Virol.* **11,** 16–45.

Davis, B. J., 1964, Disc electrophoresis. II. Method and application to serum proteins, *Ann. N. Y. Acad. Sci.* **121,** 404–427.

Dillard, S. H., Cheatham, W. J., and Moss, L. H., 1972, Electron microscopy of zosteriform herpes simplex infection in the mouse, *Lab. Invest.* **26,** 391–402.

Dimmock, N. J., and Watson, D. H., 1969, Proteins specified by influenza virus in infected cells: Analysis by polyacrylamide gel electrophoresis of antigens not present in the virus particle, *J. Gen. Virol.* **5,** 499–509.

Dion, A. S., and Herbst, E. J., 1967, The localization of spermidine in salivary gland cells of ***Drosophila melanogaster*** and its effect on ^{3}H-uridine incorporation, *Proc. Natl. Acad. Sci. USA* **58,** 2367–2371.

Dreesman, G. R., Suriano, J. R., Swartz, S. K., and McCombs, R. M., 1972, Characterization of the herpes virion. I. Purification and amino acid composition of nucleocapsids, *Virology* **50,** 528–534.

Dubbs, D. R., and Kit, S., 1964, Mutant strains of herpes simplex deficient in thymidine kinase-inducing activity, *Virology* **22,** 493–502.

Dubbs, D. R., and Kit, S., 1965, The effect of temperature on induction of deoxythymidine kinase activity by herpes simplex mutants, *Virology* **25,** 256–270.

Duff, R., and Rapp, F., 1971, Properties of hamster embryo fibroblasts transformed *in vitro* after exposure to ultraviolet-irradiated herpes simplex virus type 2, *J. Virol.* **8,** 469–477.

Dundaroff, S., and Falke, D., 1972, Thymidine-, uridine- and choline-kinase in rabbit kidney cells infected with herpesvirus hominis, Type I and II, *Arch. Ges. Virusforsch.* **38,** 56–66.

Ejercito, P. M., Kieff, E. D., and Roizman, B., 1968, Characterization of herpes simplex virus strains differing in their effect on social behavior of infected cells, *J. Gen. Virol.* **3,** 357–364.

Epstein, M. A., 1962*a*, Observations on the fine structure of mature herpes simplex virus and on the composition of its nucleoid, *J. Exp. Med.* **115,** 1–11.

Epstein, M. A., 1962*b*, Observations on the mode of release of herpes virus from infected HeLa cells, *J. Cell Biol.* **12,** 589–597.

Espmark, J. A., 1965, Rapid serological typing of herpes simplex virus and titration of herpes simplex antibody by the use of mixed hemadsorption—A mixed antiglobulin reaction applied to virus-infected tissue cultures, *Arch. Ges. Virusforsch.* **17,** 89–97.

Falke, D., 1961, Isolation of two variants with different cytopathic properties from a strain of herpes B virus, *Virology* **14,** 492–495.

Falke, D., Siegert, R., and Vogell, W., 1959, Elektronenmikroskopische Befunde zur Frage der Doppelmembranbildung des Herpes-simplex-virus, *Arch. Ges. Virusforsch.* **9,** 484–496.

Falke, D., Heicke, B., and Bassler, R., 1972, The effect of arabinofuranosyl-cytosine upon the synthesis of herpesvirus hominis, *Arch. Ges. Virusforsch.* **39,** 48–62.

Farley, C. A., Banfield, W. G., Kasnic, G., Jr., and Foster, W. S., 1972, Oyster herpes-type virus, *Science (Wash., D.C.)* **178,** 759–760.

Farnham, A. E., and Newton, A. A., 1959, The effect of some environmental factors on herpes virus grown in HeLa cells, *Virology* **7,** 449–461.

Fisher, T. N., and Fisher, E., Jr., 1973, Marked phenylalanyl-tRNA activity of herpesvirus type 2 infected preparations, *Proc. Soc. Exp. Biol. Med.* **143,** 208–211.

Flanagan, J. F., 1966, Hydrolytic enzymes in KB cells infected with poliovirus and herpes simplex virus, *J. Bacteriol.* **91,** 789–797.

Flanagan, J. F., 1967, Virus-specified ribonucleic acid synthesis in KB cells infected with herpes simplex virus, *J. Virol.* **1,** 583–590.

Fleissner, E., 1971, Chromatographic separation and antigenic analysis of proteins on the oncornaviruses, *J. Virol.* **8,** 778–785.

Fong, C. K. Y., and Hsiung, G. D., 1972, Development of an equine herpesvirus in two cell culture systems: Light and electron microscopy, *J. Infect. Immun.* **6,** 865–876.

Fong, C. K. Y., Tenser, R. B., Hsiung, G. D., and Gross, P. A., 1973, Ultrastructural studies of the envelopment and release of guinea pig herpes-like virus in cultured cells, *Virology* **52,** 468–477.

Frearson, P. M., Kit, S., and Dubbs, D. R., 1965, Deoxythymidylate synthetase and deoxythymidine kinase activities of virus-infected animal cells, *Cancer Res.* **25,** 737–744.

Freifelder, D., 1970, Molecular weights of coliphages and coliphage DNA. IV. Molecular weights of DNA from bacteriophages T4, T5, T7 and general problem of determination of M, *J. Mol. Biol.* **54,** 569–577.

Frenkel, N., and Roizman, B., 1971, Herpes simplex virus: Studies of the genome size and redundancy by renaturation kinetics, *J. Virol.* **8,** 591–593.

Frenkel, N., and Roizman, B., 1972*a*, Separation of the herpesvirus deoxyribonucleic acid on sedimentation in alkaline gradients, *J. Virol.* **10,** 565–572.

Frenkel, N., and Roizman, B., 1972*b*, Ribonucleic acid synthesis in cells infected with herpes simplex virus: Control of transcription and of RNA abundance, *Proc. Natl. Acad. Sci. USA* **69,** 2654–2658.

Frenkel, N., Roizman, B., Cassai, E., and Nahmias, A., 1972, A herpes simplex 2 DNA fragment and its transcription in human cervical cancer tissue, *Proc. Natl. Acad. Sci. USA* **69,** 3784–3789.

Frenkel, N., Silverstein, S., Cassai, E., and Roizman, B., 1973, RNA synthesis in cells infected with herpes simplex virus. VII. Control of transcription and of transcript abundancies of unique and common sequences of herpes simplex 1 and 2, *J. Virol.* **11,** 886–892.

Fritz, M. E., and Nahmias, A. J., 1972, Reversed polarity in transmembrane potentials of cells infected with herpesviruses, *Proc. Soc. Exp. Biol. Med.* **139,** 1159–1161.

Fujiwara, S., and Kaplan, A. S., 1967, Site of protein synthesis in cells infected with pseudorabies virus, *Virology* **32,** 60–68.

Furlong, D., Swift, H., and Roizman, B., 1972, Arrangement of herpesvirus deoxyribonucleic acid in the core, *J. Virol.* **10,** 1071–1074.

Gergely, L., Klein, G., and Einberg, I., 1971*a*, Host cell macromolecular synthesis in cells containing EBV-induced early antigens studied by combined immunofluorescence and autoradiography, *Virology* **45,** 22–29.

Gergely, L., Klein, G., and Einberg, I., 1971*b*, The action of DNA antagonists on Epstein-Barr virus (EBV) associated early antigen (EA) in Burkitt lymphoma lines, *Intl. J. Cancer* **7,** 293–302.

Gershon, A., Casio, L., and Branell, P. A., 1973, Observations on the growth of varicella zoster virus in human diploid cells, *J. Gen. Virol.* **18,** 21–31.

Gfeller, E., and Russell, D. H., 1970, Distribution of ^{3}H-polyamines in a xenopus laevis liver cell line, *Anat. Rec.* **166,** 306.

Gibson, W., and Roizman, B., 1971, Compartmentalization of spermine and spermidine in herpes simplex virion, *Proc. Natl. Acad. Sci. USA* **68,** 2818–2821.

Gibson, W., and Roizman, B., 1972, Proteins specified by herpes simplex virus. VIII. Characterization and composition of multiple capsid forms of subtypes 1 and 2, *J. Virol.* **10,** 1044–1052.

Gibson, W., and Roizman, B., 1973, The structural and metabolic involvement of polyamines with herpes simplex virus, *in* "Polyamines in Normal and Neoplastic Growth" (D. H. Russell, ed.), pp. 123–135, Raven Press, New York.

Gibson, W., and Roizman, B., 1974, Proteins specified by herpes simplex virus. X. Staining and radiolabeling properties of B-capsid and virion proteins in polyacrylamide gels, *J. Virol.* **13,** 155–165.

Gillespie, D., and Spiegelman, S., 1965, A quantitative assay for DNA–RNA hybrids with DNA immobilized on a membrane, *J. Mol. Biol.* **12,** 829–842.

Goodheart, C. R., 1970, Herpesviruses and cancer, *J. Am. Med. Assoc.* **211,** 91–96.

Goodheart, C., and Plummer, G., 1974, The densities of herpes viral DNAs, *in* "Progress in Medical Virology" (J. L. Melnick, ed.), Vol. 19, S. Karger, Basel.

Goodheart, C. R., Plummer, G., and Waner, J. L., 1968, Density difference of DNA of human herpes simplex viruses Types I and II, *Virology* **35,** 473–475.

Gordin, M., Olshevsky, U., Rosenkranz, H. S., and Becker, Y., 1973, Studies on herpes simplex virus DNA: Denaturation properties, *Virology* **55,** 280–284.

Graham, B. J., Ludwig, H., Bronson, D. L., Benyesh-Melnick, M., and Biswal, N., 1972, Physicochemical properties of the DNA of herpes viruses, *Biochim. Biophys. Acta* **259,** 13–23.

Granick, S., and Granick, D., 1971, Nucleolar necklaces in chick embryo myoblasts formed by lack of arginine, *J. Cell Biol.* **51,** 636–642.

Gravell, M., 1971, Viruses and renal carcinoma of *Rana pipiens.* X. Comparison of herpes-type viruses associated with Lucke tumor-bearing frogs, *Virology* **43,** 730–733.

Gray, A., Tokumaru, T., and Scott, T. F. McN., 1958, Different cytopathogenic effects observed in HeLa cells infected with herpes simplex virus, *Arch. Ges. Virusforsch.* **8,** 60–76.

Grüter, W., 1924, Das Herpesvirus, seine ätiologische und klinische Bedeutung, *Munch. Med. Wschr.* **71,** 1058–1060.

Hadhazy, G., Lehel, P., and Gergely, L., 1966, Studies on the influence of endogenous regulatory factors on the growth of herpes simplex virus, *Acta Microbiol. Acad. Sci. Hung.* **13,** 145–150.

Hamada, C., Kamiya, T., and Kaplan, A. S., 1966, Serological analysis of some enzymes present in psuedorabies virus-infected and non-infected cells, *Virology* **28,** 271–281.

Hampar, B., and Ellison, S. A., 1961, Chromosomal aberrations induced by an animal virus, *Nature* (*London.*) **192,** 145–147.

Hampar, B., and Ellison, S. A., 1963, Cellular alterations in the MCH line of Chinese hamsters cells following infection with herpes simplex virus, *Proc. Natl. Acad. Sci. USA* **49,** 474–480.

Hay, J., Icoteles, G., Mikeir, H. M., and Subak-Sharpe, H., 1966, Herpesvirus-specified ribonucleic acids, *Nature* (*Lond.*) **210,** 387–390.

Hay, J., Subak-Sharpe, H., and Shepherd, W. M., 1967, New transfer ribonucleic acid in BHK21/C13 cells infected with herpes virus, *Biochem. J.* **103,** 69.

Hay, J., Perera, P. A. J., Morrison, J. M., Gentry, G. A., and Subak-Sharpe, J. H., 1971, Herpes virus-specified proteins, *in* "Ciba Symposium on Strategy of the Viral Genome" (G. E. W. Wolstenholme and M. O'Connor, eds.), Churchill Livingstone, London.

Heine, J. W., and Roizman, B., 1973, Proteins specified by herpes simplex virus. IX. Contiguity of host and viral proteins in the plasma membrane of infected cells, *J. Virol.* **11,** 810–813.

Heine, U., Ablashi, V., and Armstrong, G. R., 1971, Morphological studies on herpesvirus saimiri in subhuman and human cell cultures, *Cancer Res.* **31,** 1019–1029.

Heine, J. W., Spear, P. G., and Roizman, B., 1972, The proteins specified by herpe simplex virus. VI. Viral proteins in the plasma membrane. *J. Virol.* **9,** 431–439.

Heine, J. W., Honess, R. W., Cassai, E., and Roizman, B., 1974, The proteins specified by herpes simplex virus. XII. The virion polypeptides of type 1 strains, *J. Virol.* September issue.

Hill, T. J., Field, H. J., and Roome, A. P. C., 1972, Intra-axonal location of herpes simplex virus particles, *J. Gen. Virol.* **15,** 253–255.

Hinze, H. C., and Walker, D. L., 1961, Variation of herpes simplex virus in permanently infected tissue cultures, *J. Bacteriol.* **82,** 498–504.

Hochberg, E., and Becker, Y., 1968, Adsorption penetration and uncoating of herpes simplex virus, *J. Gen. Virol.* **2,** 231–241.

Hoggan, M. D., and Roizman, B., 1959*a*, The effect of the temperature of incubation on the formation and release of herpes simplex virus in infected FL cells, *Virology* **8,** 508–524.

Hoggan, M. D., and Roizman, B., 1959*b*, The isolation and properties of a variant of herpes simplex producing multinucleated giant cells in monolayer cultures in the presence of antibody, *Am. J. Hyg.* **70,** 208–219.

Holland, J. J., Doyle, M., Perrault, J., Kinsbury, D. T., and Etchison, J., 1972, Proteinase activity in purified animal viruses, *Biochem. Biophys. Res. Commun.* **46,** 634–639.

Holmes, I. H., and Watson, D. H., 1961, An electron microscope study of the attachment and penetration of herpes virus in BHK 21 cells, *Virology* **21,** 112–123.

Honess, R. W., and Roizman, B., 1973, Proteins specified by herpes simplex virus. XI. Identification and relative molar rates of synthesis of structural and non-structural herpesvirus polypeptides in the infected cell, *J. Virol.* **12,** 1347–1365.

Honess, R. W., and Roizman, B., 1974, Regulation of herpesvirus macromolecular synthesis. I. Cascade regulation of the synthesis of three groups of viral proteins, *J. Virol.* **14,** 8–19.

Honess, R. W., and Watson, D. H., 1974, Herpes simplex virus-specific polypeptides studied by polyacrylamide gel electrophoresis of immune precipitates, *J. Gen. Virol.* **22,** 171–183.

Huang, C. C., 1967, Induction of a high incidence of damage to X chromosomes of Rattus (Mastomys) natalensis by base analogues, viruses and carcinogens, *Chromosoma* **23,** 162–179.

Huang, C. C., and Minowada, J., 1972, Differential effects of infection with herpes simplex virus on the chromosomes of human hematopoietic cell, *Cancer Res.* **32,** 1218–1225.

Huang, H. L., Szabocsik, J. M., Randall, C. C., and Gentry, G., 1971, Equine abortion (herpes) virus-specific RNA, *Virology* **45,** 381–389.

Hummeler, K., Tomassian, N., and Zajac, B., 1969, Early events in herpes simplex virus infection: A radioautographic study, *J. Virol.* **4,** 67–74.

Inglis, V. B. M., 1968, Requirement of arginine for replication of herpesvirus, *J. Gen. Virol.* **3,** 9–18.

Ito, M., and Barron, A. L., 1972, Surface antigen produced by herpes simplex virus (HSV), *J. Immunol.* **108,** 711–718.

Ito, M., and Barron, A. L., 1973, Surface antigens produced by herpesviruses: Varicella-zoster virus, *Infect. Immun.* **8,** 48–52.

Iwasaki, Y., Furukawa, T., Plotkin, S., and Koprowski, H., 1973, Ultrastructural study on the sequence of human cytomegalovirus infection in human diploid cells, *Arch. Ges. Virusforsch.* **40,** 311–324.

Jacobson, M. F., Asco, J., and Baltimore, D., 1970, Further evidence on the formation of poliovirus proteins, *J. Mol. Biol.* **49,** 657–669.

Jamieson, J. D., and Palade, G. E., 1967*a*, Intracellular transport of secretory proteins in the pancreatic exocrine cell. I. Role of the peripheral elements of the Golgi complex, *J. Cell Biol.* **34,** 577–596.

Jamieson, J. D., and Palade, G. E., 1967*b*, Intracellular transport of secretory proteins in the pancreatic exocrine cell. II. Transport to condensing vacuoles and zymogen granules, *J. Cell Biol.* **34,** 597–615.

Jasty, V., and Chang, P. W., 1972*a*, Release of infectious bovine rhinotracheitis virus from productively infected bovine kidney cells: An electron microscopic study, *J. Ultrastruct. Res.* **38,** 433–443.

Jasty, V., and Chang, P. W., J. M., 1972*b*, Effects of hydroxyurea on replication of infectious bovine rhinotracheitis virus in bovine kidney cells: An electron microscopic study, *Am. J. Vet. Res.* **33,** 1945–1953.

Jehn, U., Lindahl, T., and Klein, G., 1972, Fate of virus DNA in the abortive infection of human lymphoid cell lines by Epstein-Barr virus, *J. Gen. Virol.* **16,** 409–412.

Kamiya, T., Ben-Porat, T., and Kaplan, A. S., 1964, The role of progeny viral DNA in the regulation of enzyme and DNA synthesis, *Biochem. Biophys. Res. Commun.* **16**, 410–415.

Kamiya, T., Ben-Porat, T., and Kaplan, A. S., 1965, Control of certain aspects of the infective process by progeny viral DNA, *Virology* **26**, 577–589.

Kanich, R. E., and Craighead, J. E., 1972, Human cytomegalovirus infection of cultured fibroblasts. II. Viral replicative sequence of a sild and an adopted strain, *Lab. Invest.* **27**, 273–282.

Kaplan, A. S., 1964, Studies on the replicating pool of viral DNA in cells infected with pseudorabies virus, *Virology* **24**, 19–25.

Kaplan, A. S., 1969, Herpes simplex and pseudorabies viruses, *in* "Virology Monographs," Springer-Verlag, New York.

Kaplan, A. S., 1973, A brief review of the biochemistry of herpesvirus–host cell interaction, *Cancer Res.* **33**, 1393–1398.

Kaplan, A. S., and Ben-Porat, T., 1959, The effect of pseudorabies virus on the nucleic acid metabolism and on the nuclei of rabbit kidney cells, *Virology* **8**, 352–366.

Kaplan, A. S., and Ben-Porat, T., 1960, The incorporation of C^{14}-labeled nucleosides into rabbit kidney cells infected with pseudorabies virus, *Virology* **11**, 12–27.

Kaplan, A. S., and Ben-Porat, T., 1963, The pattern of viral and cellular DNA synthesis in pseudorabies virus-infected cells in the logarithmic phase of growth, *Virology* **19**, 205–214.

Kaplan, A. S., and Ben-Porat, T., 1964, Mode of replication of pseudorabies virus DNA, *Virology* **23**, 90–95.

Kaplan, A. S., and Ben-Porat, T., 1966*a*, The replication of the double-stranded DNA of an animal virus during intracellular multiplication, *Symp. Intl. Congr. Microbiol., Moscow*, 463–482.

Kaplan, A. S., and Ben-Porat, T., 1966*b*, Mode of antiviral action of 5-iodouracil deoxyriboside, *J. Mol. Biol.* **19**, 320–332.

Kaplan, A. S., and Vatter, A. E., 1959, A comparison of herpes simplex and pseudorabies viruses, *Virology* **7**, 394–407.

Kaplan, A. S., Ben-Porat, T., and Coto, C., 1967, Studies on the control of the infective process in cells infected with psuedorabies virus, *in* "Molecular Biology of Viruses" (J. Colter, ed.), pp. 527–545, Academic Press, New York.

Kaplan, A. S., Shimono, H., and Ben-Porat, T., 1970, Synthesis of proteins in cells infected with herpesvirus. III. Relative amino-acid content of various proteins formed after infection, *Virology* **40**, 90–101.

Kazama, F. Y., and Schornstein, K. L., 1972, Herpes-type virus particles associated with a fungus, *Science (Wash., D.C.)* **177**, 696–697.

Kazama, F. Y., and Schornstein, K. L., 1973, Ultrastructure of a fungus herpes-type virus, *Virology* **52**, 478–482.

Keir, H. M., 1965, DNA polymerases from mammalian cells, *Progr. Nucleic Acid Res. Mol. Biol.* **4**, 81–128.

Keir, H. M., 1968, Virus-induced enzymes in mammalian cells infected with DNA viruses, *in* "Molecular Biology of Viruses," Vol. 18, pp. 67–99, Cambridge University Press, Cambridge.

Keir, H. M., and Gold, E., 1963, Deoxyribonucleic acid nucleotidyltransferase and deoxyribonuclease from cultured cells infected with herpes simplex virus, *Biochim. Biophys. Acta* **72**, 263–276.

Keir, H. M., Subak-Sharpe, H., Shedden, W. I. H., Watson, D. H., and Wildy, P.,

1966*a*, Immunological evidence for a specific DNA polymerase produced after infection by herpes simplex virus, *Virology* **30,** 154–157.

Keir, H. M., Hay, J., Morrison, J. M., and Subak-Sharpe, H., 1966*b*, Altered properties of deoxyribonucleic acid nucleotidyltransferase after infection of mammalian cells with herpes simplex virus, *Nature* (*Lond.*) **210,** 369–371.

Keller, J. M., Spear, P. G., and Roizman, B., 1970, The proteins specified by herpes simplex virus. III. Viruses differing in their effects on the social behavior of infected cells specify different membrane glycoproteins, *Proc. Natl. Acad. Sci. USA* **65,** 865–871.

Kieff, E. D., Bachenheimer, S. L., and Roizman, B., 1971, Size, composition and structure of the DNA of subtypes 1 and 2 herpes simplex virus, *J. Virol.* **8,** 125–132.

Kieff, E. D., Hoyer, B., Bachenheimer, S. L., and Roizman, B., 1972, Genetic relatedness of type 1 and type 2 herpes simplex viruses, *J. Virol.* **9,** 738–745.

Kit, S., and Dubbs, D. R., 1963*a*, Acquisition of thymidine kinase activity by herpes simplex infected mouse fibroblast cells, *Biochem. Biophys. Res. Commun.* **11,** 55–59.

Kit, S., and Dubbs, D. R., 1963*b*, Non-functional thymidine kinase cistron in bromodeoxyuridine resistant strains of herpes simplex virus, *Biochem. Biophys. Res. Commun.* **13,** 500–504.

Klein, G., Pearson, G., Nadkarni, J. S., Nadkarni, J. J., Klein, E., Henle, G., Henle, W., and Clifford, P., 1968, Relation between Epstein-Barr viral and cell membrane immunofluorescence of Burkitt tumor cells. I. Dependence of cell membrane immunofluorescence on presence of EB virus, *J. Exp. Med.* **128,** 1011–1020.

Klemperer, H. G., Haynes, G. R., Shedden, W. I. H., and Watson, D. H.: A virus-specific thymidine kinase in BHK21 cells infected with herpes simplex virus, *Virology* **31,** 120–128.

Kohlhage, H., 1964, Differentiation of plaque variants of the herpes simplex virus by gradient centrifugation and column chromatography, *Zbl. Bakt.* **191,** 252–256 (*Arch. Ges. Virusforsch.* **14,** 358–365).

Kohlhage, H., and Schieferstein, G., 1965, Untersuchungen über die genetische Stabilität des Plaquebildes beim Herpes-simplex-virus in Zellkulturen (Investigations on the genetic stability of the plaque picture of herpes simplex virus in cell cultures), *Arch. Ges. Virusforsch.* **15,** 640–650.

Kohlhage, H., and Siegert, R., 1962, Zwei genetisch determinierte Varianten eines Herpes-simplex-Stammes, *Arch. Ges. Virusforsch.* **12,** 273–286.

Kozak, M., and Roizman, B., 1974, Regulation of herpesvirus macromolecular synthesis: Nuclear retention of non-translated viral RNA sequences, *Proc. Natl. Acad. Sci. USA,* in press.

Kristensson, K., Hansson, H.-A., and Sourander, P., 1970, Observations on cultures of rabbit retina infected with vaccinia and herpes simplex virus, *J. Gen. Virol.* **6,** 41–49.

Laemmli, U. K., 1970, Cleavage of structural proteins during the assembly of the head of bacteriophage T4, *Nature* (*Lond.*) **227,** 680–684.

Lampert, F., Bahr, G. F., and Rabson, A. S., 1969, Herpes simplex virus: Dry mass, *Science* (*Wash., D.C.*) **166,** 1163–1165.

Lando, D., and Ryhiner, M.-L., 1969, Pouvoir infectieux du DNA d'Herpesvirus hominis en culture cellulaire, *C.R. Hebd. Seances Acad. Sci. Ser. D Sci. Nat.* **269,** 527–530.

Laver, W. G., 1970, Isolation of an arginine-rich protein from particles of adenovirus type 2, *Virology* **41,** 488–500.

Lawrence, W. C., 1971*a*, Nucleic acid and protein synthesis in KB cells infected with equine abortion virus (equine herpesvirus type 1), *Am. J. Vet. Res.* **32,** 41–44.
Lawrence, W. C., 1971*b*, Evidence for a relationship between equine abortion (herpes) virus deoxyribonucleic acid synthesis and the S phase of the KB cell mitotic cycle, *J. Virol.* **7,** 736–748.
Lee, L., Kieff, E. D., Bachenheimer, S. L., Roizman, B., Spear, P. G., Burmester, B. R., and Nazerian, K., 1971, The size and composition of Marek's disease virus DNA, *J. Virol.* **7,** 289–294.
Lee, L. F., Armstrong, R. L., and Nazerian, K., 1972, Comparative studies of six avian herpesviruses, *Avian Dis.* **16,** 799–805.
Leestma, J. E., Bornstein, M. B., Sheppard, R. D., and Feldman, L. A., 1969, Ultrastructural aspects of herpes simplex virus infection in organized cultures of mammalian nervous tissue, *Lab. Invest.* **20,** 70–78.
Levine, A. J., and Ginsberg, H. S., 1967, Biochemical studies on the mechanism by which the fiber antigen inhibits multiplication of type 5 adenovirus, *J. Virol.* **1,** 747–757.
Levine, A. J., and Ginsberg, H. S., 1968, Role of adenovirus structural proteins in the cessation of host cell biosynthetic functions, *J. Virol.* **2,** 430–439.
Liquori, A. M., Constantino, L., Crescenzi, V., Elia, V., Giglio, E., Puliti, R., Savino, D. S., and Vitagliano, V., 1967, Complexes between DNA and polyamines: A molecular model, *J. Mol. Biol.* **24,** 113–122.
Ludwig, H., 1972, Untersuchungen am genetischen Material von Herpesviren. I. Biophysikalischchemische Charakterisierung von Herpesvirus-Desoxyribonucleinsäuren, *Med. Microbiol. Immunol.* **157,** 186–211.
Ludwig, H., Biswal, N., Bryans, J. T., and McCombs, R. M., 1971*a*, Some properties of the DNA from a new equine herpesvirus, *Virology* **45,** 534–537.
Ludwig, H., Biswal, N., and Benyesh-Melnick, M., 1971*b*, Characterization of DNA isolated from metaphase chromosomes of cells containing Epstein-Barr virus, *Biochim. Biophys. Acta* **232,** 261–270.
Ludwig, H., Haines, H. G., Biswal, N., and Benyesh-Melnick, M., 1972*a*, The characterization of varicella-zoster virus DNA, *J. Gen. Virol.* **14,** 111–114.
Ludwig, H. O., Biswal, N., and Benyesh-Melnick, M., 1972*b*, Studies on the relatedness of herpesviruses through DNA–DNA hybridization, *Virology* **49,** 95–101.
Lunger, P. D., Darlington, R. W., and Granoff, A., 1965, Cell–virus relationships in the Lucké renal adenocarcinoma: An ultrastructure study, *Ann. N.Y. Acad. Sci.* **126,** 289–314.
Lunger, P. D., 1966, A new intranuclear inclusion body in the frog renal adenocarcinoma, *J. Morphology* **118,** 581–588.
McAuslan, B. R., Herde, P., Pett, D., and Ross, J., 1965, Nucleases of virus-infected animal cells, *Biochem. Biophys. Res. Commun.* **20,** 586–591.
McCombs, R., Brunschwig, J. P., Mirkovic, R. and Benyesh-Melnick, M., 1971, Electron microscopic characterization of a herpes-like virus isolated from tree shrews, *Virology* **45,** 816–820.
McCracken, R. M., and Clarke, J. K., 1971, A thin section study of the morphogenesis of Aujeszky's disease virus in synchronously infected cell cultures, *Arch. Ges. Virusforsch.* **34,** 189–201.
Mannini-Palenzona, A., Costanzo, F., and LaPlaca, M., 1971, Impairment of herpesvirus growth in chick embryo fibroblast cultures by α-amanitin, *Arch. Ges. Virusforsch.* **34,** 381–384.

Mannweiler, K., and Palacios, O., 1969, Züchtung and Vermehrung von Herpes-simplex-virus in Zellkulturen vom Nervensystem, *Acta Neuropathol.* **12,** 276–299.

Maré, C. J., and Graham, D. L., 1973, Falcon herpesvirus, the etiologic agent of inclusion body disease of falcons, *Infect. Immun.* **8,** 118–126.

Mark, G. E., and Kaplan, A. S., 1971, Synthesis of proteins in cells infected with herpesviruses. VII. Lack of migration of structural viral proteins to the nucleus of arginine-deprived cells, *Virology* **45,** 53–60.

Mark, G. E., and Kaplan, A. S., 1972, Synthesis of proteins in cells infected with herpesvirus. VIII. Absence of virus-induced alteration of nuclear membrane in arginine-deprived cells, *Virology* **49,** 102–111.

Martin, W. B., Hay, D., Crawford, L. V., Le Bouvier, G. L., and Crawford, E. M., 1966, Characteristics of bovine mammilitis virus, *J. Gen. Microbiol.* **45,** 325–332.

Mazzone, H. M., and Yerganian, G., 1963, Gross and chromosomal cytology of virus-infected Chinese hamster cells, *Exp. Cell Res.* **30,** 591–592.

Melli, M., Whitfield, C., Rao, K. V., Richardson, M., and Bishop, J. O., 1971, DNA–RNA hybridization in vast DNA excess, *Nat. New Biol.* **231,** 8–12.

Melnick, J. L., 1973, Classification and nomenclature of viruses, *in* "Ultrastructure of Animal Viruses and Bacteriophages: An Atlas" (A. J. Dalton and F. Haguenau, eds.), pp. 7–20, Academic Press, New York.

Mikharlova, G. R., 1967, Action of viruses on the karyotype of man and animals, *Genetika* **7,** 129–137.

Miyamoto, K., 1971, Mechanism of intranuclear crystal formation of herpes simplex virus as revealed by the negative staining of thin sections, *J. Virol.* **8,** 534–550.

Miyamoto, K., and Morgan, C., 1971, Structure and development of viruses as observed in the electron microscope. XI. Entry and uncoating of herpes simplex virus, *J. Virol.* **8,** 910–918.

Miyamoto, K., Morgan, C., Hsu, K. C., and Hampar, B., 1971, Differentiation by immunoferritin of herpes simplex virion antigens with the use of rabbit 7S and 19S antibodies from early (7-day) and late (7-week) immune sera, *J. Natl. Cancer Inst.* **46,** 629–646.

Moretti, G. F., Zitelli, A., and Baroni, A., 1972, Ribosome crystallization without hypothermia in chorioallantoic membranes infected with type 2 herpesvirus hominis and treated with rifampicin, *J. Submicroscop. Cytol.* **4,** 215–219.

Morgan, C., Ellison, S. A., Rose, H. M., and Moore, D. H., 1954, Structure and development of viruses as observed in the electron microscope. I. Herpes simplex virus, *J. Exp. Med.* **100,** 195–202.

Morgan, C., Rose, H. M., Holden, M., and Jones, E. P., 1959, Electron microscopic observations on the development of herpes simplex virus. *J. Exp. Med.* **110,** 643–656.

Morgan, C., Rose, H. M., and Mednis, B., 1968, Electron microscopy of herpes simplex virus. I. Entry, *J. Virol.* **2,** 507–516.

Morris, V. L., Wagner, E. K., and Roizman, B., 1970, RNA synthesis in cells infected with herpes simplex-virus. III. Absence of virus-specified arginyl- and seryl-tRNA in infected HEp-2 cells, *J. Mol. Biol.* **52,** 247–263.

Morrison, J. M., and Keir, H. M., 1966, Heat-sensitive deoxyribonuclease activity in cells infected with herpes simplex virus, *Biochem. J.* **98,** 37C–39C.

Morrison, J. M., and Keir, H. M., 1967, Characterization of the deoxyribonuclease activity induced by infection with herpes virus, *Biochem. J.* **103,** 70–71.

Morrison, J. M., and Keir, H. M., 1968*a*, Further studies on deoxyribonucleic acid exonuclease induced by herpes virus, *Biochem. J.* **110,** 39P.

Morrison, J. M., and Keir, H. M., 1968*b*, A new DNA-exonuclease in cells infected with herpes virus partial purification and properties of enzyme, *J. Gen. Virol.* **3**, 337.

Mosmann, T. R., and Hudson, J. B., 1973, Some properties of the genome of murine cytomegalovirus (MCV), *Virology* **54**, 135–149.

Munk, K., and Sauer, G., 1964, Relationship between cell DNA metabolism and nucleocytoplasmic alterations in herpesvirus-infected cells, *Virology* **22**, 153–154.

Munyon, W., and Kit, S., 1965, Inhibition of thymidine kinase formation in LM(TK$^-$) cells simultaneously infected with vaccinia and a vaccinia mutant, *Virology* **26**, 374–377.

Munyon, W., Buchsbaum, R., Paoletti, E., Mann, J., Kraiselburd, E., and Davis, D., 1972, Electrophoresis of thymidine kinase activity synthesized by cells transformed by herpes simplex virus, *Virology* **49**, 683–689.

Murphy, F. A., Harrison, A. K., and Whitfield, S. G., 1967, Intranuclear formation of filaments in herpesvirus hominis infection of mice, *Arch. Ges. Virusforsch.* **21**, 463–468.

Nahmias, A. J., and Dowdle, W. R., 1968, Antigenic and biologic differences in herpesvirus hominis, *Progr. Med. Virol.* **10**, 110–159.

Nahmias, A. J., and Kibrick, S., 1964, Inhibitory effect of heparin on herpes simplex virus, *J. Bacteriol.* **87**, 1060–1066.

Nahmias, A., Kibrick, S., and Bernfeld, P., 1964, The effect of synthetic and biological polyanions on herpes simplex virus, *Proc. Soc. Exp. Biol. Med.* **115**, 993–996.

Nahmias, A. J., del Buono, I., Scheweiss, K., Gordon, D., and Thies, D., 1971, Type specific surface antigens of cells infected with herpes simplex virus type 1 and 2, *Proc. Soc. Exp. Biol. Med.* **138**, 21–27.

Nayak, D. P., 1971, Isolation and characterization of a herpesvirus from leukemic guinea pigs, *J. Virol.* **8**, 579–588.

Nazerian, K., 1971, Further studies on the replication of Marek's disease virus in the chicken and in cell culture, *J. Natl. Cancer Inst.* **47**, 207–217.

Nazerian, K., 1973, Studies of intracellular and membrane antigens induced by Marek's disease virus, *J. Gen. Virol.* **21**, 193–195.

Nazerian, K., and Burmester, B., 1968, Electron microscopy of a herpesvirus associated with the agent of Marek's disease in cell culture, *Cancer Res.* **28**, 2454–2462.

Nazerian, K. and Chen, J. H., 1973, Immunoferritin studies of Marek's disease virus directed intracellular and membrane antigens, *Arch. Ges. Virusforsch.* **41**, 59–65.

Nazerian, K., and Witter, R. L., 1970, Cell-free transmission and *in vivo* replication of Marek's disease virus, *J. Virol.* **5**, 388–397.

Nazerian, K., Lee, F. C., Witter, R. L., and Burmester, B. R., 1971, Ultrastructural studies of a herpesvirus of turkeys antigenically related to Marek's disease virus, *Virology* **43**, 442–452.

Newton, A. A., 1964, Synthesis of DNA in cells infected by virulent DNA viruses, *in* "Acidi Nucleici e Lora Funzione Biologica," p. 109, Istituto Lombardo, Accademia di Scienze e Lettere, Convegno Antonio Baselli., Milano.

Newton, A. A., 1969, Report of A. A. Newton's Paper by Cohen and Joklik, *in* "International Virology," Vol. I (J. L. Melnick, ed.), pp. 65 and 253, Karger, Basel.

Newton, A. A., and Stoker, M. G. P., 1958, Changes in nucleic acid content of HeLa cells infected with herpesvirus, *Virology* **5**, 549–560.

Newton, A., Dendy, P. P., Smith, C. L., and Wildy, P., 1962, A pool size problem associated with the of tritiated thymidine, *Nature* (*Lond.*) **194**, 886–887.

Nichols, W. W., 1966, One role of viruses in the etiology of chromosomal abnormalities, *Am. J. Human Genet.* **18,** 81–92.

Nii, S., 1971*a*, Electron microscopic observations on FL cells infected with herpes simplex virus. I. Viral forms, *Biken. J.* **14,** 177–190.

Nii, S., 1971*b*, Electron microscopic observations on FL cells infected with herpes simplex virus, II. Envelopment, *Biken. J.* **14,** 325–348.

Nii, S., and Kamahora, J., 1961, Cytopathic changes induced by herpes simplex virus, *Biken. J.* **4,** 255–270.

Nii, S., and Kamahora, J., High frequency appearance of amitotic nuclear divisions in PS cells induced by herpes simplex virus, *Biken. J.* **6,** 33–36.

Nii, S., Morgan, C., and Rose, H. M., 1968*a*, Electron microscopy of herpes simplex virus. II. Sequence of development, *J. Virol.* **2,** 517–536.

Nii, S., Morgan, C., Rose, H. M., and Hsu, K. C., 1968*b*, Electron microscopy of herpes simplex virus. IV. Studies with ferritin-conjugated antibodies, *J. Virol.* **2,** 1172–1184.

Nii, S., Rosenkranz, H. S., Morgan, C., and Rose, H. M. 1968*c*, Electron microscopy of herpes simplex virus. III. Effect of hydroxyurea, *J. Virol.* **2,** 1163–1171.

Nii, S. and Ono, N., 1971, Viral crystalline arrays in FL cells infected with herpes simplex virus, *Biken J.* **14,** 51–63.

Nii, S., Katsume, I., and Ono, K., 1973, Dense bodies in duck embryo cells infected with turkey herpes, *Biken. J.* **16,** 111–116.

Nohara, M., and Kaplan A. S., 1963, Induction of a new enzyme in rabbit kidney cells by pseudorabies virus, *Biochem. Biophys. Res. Commun.*, **12,** 189–193.

Nonoyama, M., and Pagano, J. S., 1971, Detection of Epstein-Barr viral genome in nonproductive cells, *Nat. New Biol.* **233,** 103–106.

Nonoyama, M., and Pagano, J. S., 1972*a*, Separation of Epstein-Barr virus DNA from large chromosomal DNA in non-virus-producing cells, *Nat. New Biol.* **238,** 169–171.

Nonoyama, M., and Pagano, J. S., 1972*b*, Replication of viral deoxyribonucleic acid and breakdown of cellular doxyribonucleic acid in Epstein-Barr virus infection, *J. Virol.* **9,** 714–716.

O'Callaghan, D. J., Hyde, J. M., Gentry, G. A., and Randall, C. C., 1968*a*, Kinetics of viral deoxyribonucleic acid, protein, and infectious particle production and alterations in host macromolecular syntheses in equine abortion (herpes) virus-infected cells, *J. Virol.* **2,** 793–804.

O'Callaghan, D. J., Cheevers, W. P., Gentry, G. A., and Randall, C. C., 1968*b*, Kinetics of cellular and viral DNA synthesis in equine abortion (herpes) virus infection of L-M cells, *Virology*, **36,** 104–114.

O'Callaghan, R., Randall, C. C., and Gentry, G. A., 1972, Herpesvirus replication *in vivo*, *Virology* **49,** 784–793.

Ogino, T., and Rapp, F., 1971, Differences in thermal stability of deoxythimidine kinase activity in extracts from cells infected with herpes simplex virus type 1 or 2, *Virology* **46,** 953–955.

Olshevsky, U., and Becker, Y., 1970, Herpes simplex virus structural proteins, *Virology* **40,** 948–960.

Olshevsky, U., Levitt, J., and Becker, Y., 1967, Studies on the synthesis of herpes simplex virions, *Virology* **33,** 323–334.

O'Neill, F. J., and Miles, C. P., 1969, Chromosome changes in human cells induced by herpes simplex types 1 and 2, *Nature* (*Lond.*) **223,** 851–853.

O'Neill, F. J., and Rapp, F., 1971*a*, Synergistic effect of herpes simplex virus and cytosine arabinoside on human chromosomes, *J. Virol.* **7,** 692–695.

O'Neill, F. J., and Rapp, F., 1971*b*, Early events required for induction of chromosome abnormalities in human cells by herpes simplex virus, *Virology* **44,** 544–553.

Parker, J. C., Vernon, M. L., and Cross, S. S., 1973, Classification of mouse thymic virus as a herpesvirus, *Infect. Immun.* **7,** 305–308.

Patrizi, G., Middlekamp, J. N., Herwig, J. C., and Thornton, H. K., 1965, Human cytomegalovirus: Electron microscopy of a primary viral isolate, *J. Lab. Clin. Med.* **65,** 825–838.

Pearson, G., Deney, F., Klein, G., Henle, G., and Henle, W., 1970, Correlation between antibodies to Epstein-Barr virus (EBV)-induced membrane antigens and neutralization of EBV infectivity, *J. Natl. Cancer Inst.* **45,** 989–995.

Plummer, G., Goodheart, C. R., Henson, D., and Bowling, C. P., 1969, A comparative study of the DNA density and behavior in tissue cultures of fourteen different herpesviruses, *Virology* **39,** 134–137.

Plummer, G., Waner, J. L., Phuangrab, A., and Goodheart, C. R., 1970, Type 1 and type 2 herpes simplex viruses: Serological and biological differences, *J. Virol.* **5,** 51–59.

Plummer, G., Goodheart, C. R., and Studdert, M. J., 1973, Equine herpesviruses: Antigenic relationships and DNA densities, *Infect. Immun.* **8,** 621–627.

Prage, L., and Petterson, U., 1971, Structural proteins of adenoviruses. VI. Purification and properties of an arginine-rich core protein from adenovirus type 2 and type 3, *Virology* **45,** 364–373.

Prusoff, W. H., Bakhle, Y. S., and Sekely, S., 1965, Cellular and antiviral effects of halogenated deoxyribonucleosides, *Ann. N.Y. Acad. Sci.* **130,** 135–150.

Rabin, E. R., Jenson, A. B., and Melnick, J., 1968*a*, Neural spread of herpes simplex virus in mice—An electron microscopic study, *Am. J. Pathol.* **52,** A6.

Rabin, E. R., Jenson, A. B., Phillips, C. A., and Melnick, J. L., 1968*b*, Herpes simplex virus hepatitis in mice: An electron microscopic study, *Exp. Mol. Pathol.* **8,** 34–48.

Rakusanova, T., Ben-Porat, T., Himeno, M., and Kaplan, A. S., 1970, Early functions of the genome of herpesvirus. I. Characterization of the RNA synthesized in cycloheximide-treated, infected cells, *Virology* **46,** 877–889.

Rakusanova, T., Ben-Porat, T., and Kaplan, A. S., 1972, Effect of herpesvirus infection on the synthesis of cell-specific RNA, *Virology* **49,** 537–548.

Randall, C. C. Rogers, H. W., Downer, D. N., and Gentry, G. A., 1972, Protein kinase activity in equine herpesvirus, *J. Virol.* **9,** 216–222.

Rapp, F., and Hsu, T. C., 1965, Viruses and mammalian chromosomes. IV. Replication of herpes simplex virus in diploid hamster cells, *Virology* **25,** 401–411.

Reichard, P., Canellakis, Z. N., and Cannellakis, E. S., 1960, Regulatory mechanisms in the synthesis of deoxyribonucleic acid *in vitro*, *Biochim Biophys. Acta* **41,** 558–559.

Reichard, P., Canellakis, Z. N., and Canellakis, E. S., 1961, Studies on a possible regulatory mechanism for the biosynthesis of deoxyribonucleic acid, *J. Biol. Chem.* **236,** 2514–2519.

Reissig, M., and Kaplan, A. S., 1960, The induction of amitotic nuclear division by pseudorabies virus multiplying in single rabbit kidney cells, *Virology* **11,** 1–11.

Roane, P. R., Jr., and Roizman, B., 1964, Studies of the determinant antigens of viable cells. II. Demonstration of altered antigenic reactivity of HEp-2 cells infected with herpes simplex virus, *Virology* **22,** 1–8.

Roizman, B., 1962*a*, Polykaryocytosis induced by viruses, *Proc. Natl. Acad. Sci. USA* **48,** 228–234.
Roizman, B., 1962*b*, Polykaryocytosis, *Cold Spring Harbor Symp. Quant. Biol.* **27,** 327–342.
Roizman, B., 1963*a*, The programming of herpesvirus multiplication in doubly-infected and in puromycin-treated cells, *Proc. Natl. Acad. Sci. USA* **49,** 165–171.
Roizman, B., 1963*b*, The programming of herpesvirus multiplication in mammalian cells, *in* "Viruses, Nucleic Acids and Cancer (Proceedings of the 17th Annual Symposium, M. D. Anderson Hospital and Tumor Institute)," pp. 205–223, Williams & Wilkins, Baltimore.
Roizman, B., 1965*a*, Abortive infection of canine cells by herpes simplex virus. III. The interference of conditional lethal virus with an extended host range mutant, *Virology* **27,** 113–117.
Roizman, B., 1965*b*, Extracellular *p*H and herpes simplex virus multiplication, *Proc. Soc. Exp. Biol. Med.* **119,** 1021–1023.
Roizman, B., 1966, An inquiry into the mechanisms of recurrent herpes infections of man, *in* "Perspectives in Virology," Vol. IV (M. Pollard, ed.), pp. 283–304, Harper and Row, Hoeber Medical Division, New York.
Roizman, B., 1969, The herpesviruses—A biochemical definition of the group, *in* "Current Topics in Microbiology and Immunology," Vol. 49, pp. 1–79, Springer Verlag, Heidelberg.
Roizman, B., 1971*a*, Herpesvirus, man and cancer—Or the persistence of the viruses of love, *in* "Of Microbes and Life" (J. Monod and E. Borek, eds.), pp. 189–214, Columbia University Press, New York.
Roizman, B., 1971*b*, Herpesviruses, membranes and the social behavior of infected cells, *in* "Proceedings of the 3rd International Symposium on Applied and Medical Virology (Fort Lauderdale, Fla.)," pp. 37–72, Warren Green, St. Louis, Mo.
Roizman, B., 1972, Biochemical features of herpesvirus-infected cells particularly as they relate to their potential oncogenicity, in "Oncogenesis and Herpesviruses" (P. M. Biggs, G. de The, and L. N. Payne, eds.), pp. 1–20, International Agency for Research on Cancer, Lyon.
Roizman, B., and Aurelian, L., 1965, Abortive infection of canine cells by herpes simplex virus. I. Characterization of viral progeny from cooperative infection with mutants differing in ability to multiply in canine cells, *J. Mol. Biol.* **11,** 528–538.
Roizman, B., and Frenkel, N., 1973, Herpes simplex virus DNA: its transcription and state in productive infection and in human cervical cancer tissue, *Cancer Res.* **33,** 1402–1416.
Roizman, B., and Heine, J. W., 1972, Modification of human cell membranes by herpes viruses, *in* "Proceedings of the First California Membrane Conference" (C. F. Fox, ed.), pp. 203–237, Academic Press, New York.
Roizman, B., and Roane, P. R., Jr., 1961*a*, A physical difference between two strains of herpes simplex virus apparent on sedimentation in cesium chloride, *Virology* **15,** 75–79.
Roizman, B., and Roane, P. R., Jr., 1961*b*, Studies of the determinant antigens of viable cells. I. A method and its application in tissue culture studies, for enumeration of killed cells, based on the failure of virus multiplication following injury by cytotoxic antibody and complement, *J. Immunol.* **87,** 714–727.
Roizman, B., and Roane, P. R., Jr., 1963, Demonstration of a surface difference between virions of two strains of herpes simplex virus, *Virology* **19,** 198–204.

Roizman, B., and Roane, P. R., Jr., 1964, The multiplication of herpes simplex virus. II. The relation between protein synthesis and the duplication of viral DNA in infected HEp-2 cells, *Virology* **22,** 262–269.
Roizman, B., and Spear, P. G., 1968, Preparation of herpes simplex virus of high titer, *J. Virol.* **2,** 83–84.
Roizman, B., and Spear, P. G., 1971, Herpesvirus antigens on cell membranes detected by centrifugation of membrane-antibody complexes, *Science* (*Wash., D.C.*) **171,** 298–300.
Roizman, B., and Spear, P. G., 1973, Herpesviruses, *in* "Ultrastructure of Animal Viruses and Bacteriophage: An Atlas" (A. J. Dalton and F. Haguenau eds.), pp. 83–107, Academic Press, New York.
Roizman, B., and Spring, S. B., 1967, Alteration in immunologic specificity of cells infected with cytolytic viruses, *in* "Proceedings of the Conference on Cross-Reacting Antigens and Neoantigens" (J. J. Trentin, ed.), pp. 85–96, Williams & Wilkins, Baltimore.
Roizman, B., Aurelian, L., and Roane, P. R., Jr., 1963, The multiplication of herpes simplex virus. I. The programming of viral DNA duplication in HEp-2 cells, *Virlogy* **21,** 482–498.
Roizman, B., Borman, G. S., and Kamali-Rousta, M., 1965, Macromolecular synthesis in cells infected with herpes simplex virus, *Nature* (*Lond.*) **206,** 1374–1375.
Roizman, B., Spring, S. B., and Roane, P. R., Jr., 1967, Cellular compartmentalization of herpesvirus antigens during viral replication, *J. Virol.* **1,** 181–192.
Roizman, B., Bachenheimer, S. L., Wagner, E. K., and Savage, T., 1970*a*, Synthesis and transport of RNA in herpesvirus infected mammalian cells, *Cold Spring Harbor Symp. Quant. Biol.* **35,** 753–771.
Roizman, B., Keller, J. M., Spear, P. G., Terni, M., Nahmias, A. J., and Dowdle, W., 1970*b*, Variability of structural glycoproteins and classification of hepes simplex virus, *Nature* (*Lond.*) **227,** 1253–1254.
Roizman, B., Spear, P. G., and Kieff, E. D., 1973*a*, Herpes simplex viruses I and II: A biochemical definition, *in* "Perspectives in Virology," Vol. VIII (M. Pollard, ed.), pp. 129–169, Academic Press, New York.
Roizman, B., Bartha, A., Biggs, P. M., Carmichael, L. E., Granoff, A., Hampar, B., Kaplan, A. S., Melendez, L. V., Munk, K., Nahmias, A., Plummer, G., Rajcani, J., Rapp, F., Terni, M., de The, G., Watson, D. H., and Wildy, P., 1973*b*, Provisional labels for herpesviruses, *J. Gen. Virol.* **20,** 417–419.
Roizman, B., Kozak, M., Honess, R. W., and Hayward, G., 1974, Regulation of herpesvirus macromolecular synthesis: Evidence for multilevel regulation of herpes simplex 1 RNA and protein synthesis, *in:* "Tumor Viruses," Cold Spring Harbor Symposium on Quantitative Biology, Vol. 39.
Rosenkranz, H. S., Rose, H. M., Morgan, C., and Hsu, K. C., 1968, The effect of hydroxyurea on virus development. II. Vaccinia, *Virology* **28,** 510–519.
Ross, L. J. N., Watson, D. H., and Wildy, P., 1968, Development and localization of virus-specific antigens during the multiplication of herpes simplex virus in BHK21 cells, *J. Gen. Virol.* **2,** 115–122.
Rubenstein, D. S., Gravell, M., and Darlington, R., 1972, Protein kinase in enveloped herpes simplex virions, *Virology* **50,** 287–290.
Russell, D. H., 1971, Putrescine and spermidine biosynthesis in the development of normal and anucleate mutants of xenopus laevis, *Proc. Natl. Acad. Sci. USA* **68,** 523–527.

Russell, W. C., 1962, Herpesvirus nucleic acid, *Virology* **16,** 355–357.

Russell, W. C., and Crawford, L. V., 1964, Properties of the nucleic acids from some herpes group viruses, *Virology* **22,** 288–292.

Russell, W. C., Gold E., Keir, H. M., Omura, H., Watson, D. H., and Wildy, P., 1964, The growth of herpes simplex virus and its nucleic acid, *Virology* **22,** 103–110.

Sablina, O. V., and Bocharov, E. F., 1968, Chromosomal aberrations in the HeLa and HEp-2 tissue culture cells induced by herpes simplex virus, *Genetics* **4,** 129–134.

Sauer, G., and Munk, K., 1966, Interference of actinomycin D with the replication of the DNA of herpesvirus. II. Relationship between yield of virus and time of actinomycin treatment, *Biochim. Biophys. Acta* **119,** 341–346.

Sauer, G., Orth, H. D., and Munk, K., 1966, Interference of actinomycin D with the replication of the herpesvirus DNA. I. Difference in behaviour of cellular and viral nucleic acid synthesis following treatment with actinomycin D, *Biochim. Biophys. Acta* **119,** 331–340.

Savage, T., Roizman, B., and Heine, J. W., 1972, The proteins specified by herpes simplex virus. VII. Immunologic specificity of the glycoproteins of subtypes I and II, *J. Gen. Virol.* **17,** 31–48.

Schaffer, P. A., Courtney, R. J., McCombs, R. M., and Benyesh-Melnick, M., 1971, A temperature-sensitive mutant of herpes simplex virus defective in glycoprotein synthesis, *Virology*, **46,** 356–368.

Schaffer, P. A., Aron, G. M., Biswal, N., and Benyesh-Melnick, M., 1973, Temperature-sensitive mutants of herpes simplex virus type 1: Isolation, complementation and partial characterization, *Virology* **52,** 57–71.

Schidlovsky, G., Slattery, S., Leech, J., Mason, R., and Ahmed, M., 1969, Herpes-type virus in Marek's disease tumors and cell culture, *Proc. Am. Assoc. Cancer Res.* **10,** 76.

Schiek, W., 1967, Die Massendichte des Herpes simplex-Virus im Caesiumchlorid-Wasser-Gradienten. Beziehungen zur Plaquemorphologie auf HeLa-Zellen und zum serologischen Typ, *Z. I mun.* **132,** 207–217.

Schiek, W., and Schneweis, K. E., 1968, Beitrag zur Massendichte von Plaque Varianten des Herpesvirus hominis im Caesiumchlorid-Wasser-Gradienten (Density of plaque variants in type herpes simplex virus under cesium chloride-water gradients), *Arch. Ges. Virusforsch.* **23,** 280–283.

Schneweis, K. E., 1962, Der cytopatische Effekt Herpes simplex Virus, *Zbl. Bakt.* (*Orig.*) **186,** 467–485, 1962.

Schnewis, K. E., Sommerhauser, H., and Huber, D., 1967, Biological and immunologic comparison of two plague variants of herpes simplex virus type 1, *Arch. Ges. Virusforsch.* **38,** 338–346.

Schulte-Holthausen, H., and Zur Hausen, H., 1970, Purification of the Epstein-Barr virus and some properties of its DNA, *Virology* **40,** 776–779.

Schwartz, J., and Roizman, B., 1969*a*, Similarities and differences in the development of laboratory strains and freshly isolated strains of herpes simplex virus in HEp-2 cells: Electron microscopy, *J. Virol.* **4,** 879–889.

Schwartz, J., and Roizman, B., 1969*b*, Concerning the egress of herpes simplex virus from infected cells: Electron microscope observations, *Virology* **38,** 42–49.

Scott, T. F. McN., Burgoon, C. F., Coriell, L. L., and Blank, M., 1953, The growth curve of the virus of herpes simplex in rabbit corneal cells grown in tissue culture with parallel observations on the development of the intranuclear inclusion body, *J. Immunol.* **71,** 385–396.

Seiler, N., and Wiechmann, M., 1967, Determination of amines in 10^{-10} molar measurement, *Experimentia* **21,** 203–204.

Shapiro, A. L., Viñuela, E., and Maizel, J. V., Jr., 1967, Molecular weight estimation of polypeptide chains by electrophoresis in SDS-polyacrylamide gels, *Biochem. Biophys. Res. Commun.* **23,** 815–820.

Shedden, W. I. H., Subak-Sharpe, H., Watson, D. H., and Wildy, P., 1966, Immunological evidence for a specific DNA polymerase produced after infection by herpes simplex virus, *Virology* **30,** 154–157.

Sheldrick, P., Laithier, M., Lando, D., and Ryhiner, M. L.,1973, Infectious DNA from herpes simplex virus: Infectivity of double-stranded and single-stranded molecules, *Proc. Natl. Acad. Sci. USA* **70,** 3621–3625.

Sheldrick, P., and Berthelot, N., 1974, Inverted repetitions in the chromosome of herpes simplex virus, *in:* "Tumor Viruses," Cold Spring Harbor Symposium on Quantitative Biology, Vol. 39.

Shipkey, F. H., Erlandson, R. A., Bailey, R. B., Babcock, V. I., and Southam, C. M., 1967, Virus biographies. II. Growth of herpes simplex virus in tissue culture, *Exp. Mol. Pathol.* **6,** 39–67.

Siegert, R. S., and Falke, D., 1966, Electron microscopic investigation of the development of herpes virus hominis in culture cells, *Arch. Ges. Virusforsch.* **19,** 230–249.

Silverstein, S., Bachenheimer, S. L., Frenkel, N., and Roizman, B. 1973, The relationship between post-transcriptional adenylation of herpesvirus RNA and mRNA abundance (paper No. 8 in the series on RNA synthesis in cells infected with herpes simplex virus), *Proc. Natl. Acad. Sci. USA* **70,** 2101–2104.

Siminoff, P., and Menefee, M. G., 1966, Normal and 5-bromo-deoxyuridine-inhibited development of herpes simplex virus. An electron microscope study, *Exp. Cell Res.* **44,** 241–255.

Sirtori, C., and Bosisio-Bestetti, M., 1967, Nucleolar changes in KB tumor cells infected with herpes simplex virus, *Cancer Res.* **27,** 367–376.

Skinner, M. S., and Mizell, M., 1972, The effect of different temperatures on herpes virus induction and replication in the Lucké tumor explants, *Lab. Invest.* **26,** 671–681.

Smith, J. D., and de Harven, E., 1973, Herpes simplex virus and human cytomegalovirus replication in WI-38 cells. I. Sequence of viral replication, *J. Virol.* **12,** 919–930.

Smith, J. W., Adam, E., Melnick, J. L., and Rawls, W. E., 1972*a*, Use of the ^{51}Cr release test to demonstrate patterns of antibody response in humans to herpesvirus types 1 and 2, *J. Immunol.* **109,** 554–564.

Smith, J. W., Lowry, S. P., Melnick, J. L., and Rawls, W. E., 1972*b*, Antibodies to surface antigens of herpesvirus type 1 and type 2 infected cells among women with cervical cancer and control women, *Infect. Immun.* **5,** 305–310.

Smith, K. O., and Dukes, C. D., 1963, Some biological aspects of herpesvirus–cell interactions in the presence of 5-iodo-2-desoxyuridine (IDU). Demonstration of a cytotoxic effect by herpesvirus, *J. Immunol.* **91,** 582–590.

Smith, K. O., 1964, Effects of 5-iodo-2-desoxyuridine (IDU) on herpesvirus synthesis and survival in infected cells, *J. Immunol.* **92,** 550–554.

Snyder, S. H., Krenz, D. S., Medina, V. J., and Russell, D. H., 1970, Polyamine synthesis and turnover in rapidly growing tissues, *Ann. N.Y. Acad. Sci.* **171,** 749–771.

Soehner, R. L., Gentry, C. A., and Randall, C. C., 1965, Some physicochemical characteristics of equine abortion virus nucleic acid, *Virology* **26,** 394–405.

Southern, E. M., 1971, Effects of sequence divergence on the reassociation properties of repetitive DNAs, *Nat. New Biol.* **232,** 82–83.

Spear, P. G., and Roizman, B., 1968, The proteins specified by herpes simplex virus. I. Time of synthesis, transfer into nuclei, and proteins made in productively infected cells, *Virology* **36,** 545–555.

Spear, P. G., and Roizman, B., 1970, The proteins specified by herpes simplex virus. IV. The site of glycosylation and accumulation of viral membrane proteins, *Proc. Natl. Acad. Sci. USA* **66,** 730–737.

Spear, P. G., and Roizman, B., 1972, Proteins specified by herpes simplex virus. V. Purification and structural proteins of the herpesviron, *J. Virol.* **9,** 143–159.

Spear, P. G., Keller, J. M., and Roizman, B., 1970, The proteins specified by herpes simplex virus. II. Viral glycoproteins associated with cellular membranes, *J. Virol.* **5,** 123–131.

Spring, S. B., and Roizman, B., 1967, Herpes simplex virus products in productive and abortive infection. I. Stabilization with formaldehyde and preliminary analysis by isopycnic centrifugation in CsCl, *J. Virol.* **1,** 294–301.

Spring, S. B., and Roizman, B., 1968, Herpes simplex virus products in productive and abortive infection. III. Differentiation of infectious virus derived from nucleus and cytoplasm with respect to stability and size, *J. Virol.* **2,** 979–985.

Spring, S. B., Roizman, B., and Schwartz, J., 1968, Herpes simplex virus products in productive and abortive infection. II. Electron microscopic and immunological evidence for failure of virus envelopment as a cause of abortive infection, *J. Virol.* **2,** 384–392.

Stackpole, C. W., 1969, Herpes-type virus of the frog renal adenocarcinoma. I. Virus development in tumor transplants maintained at low temperature, *J. Virol.* **4,** 75–93.

Stackpole, C. W., and Mizell, M., 1968, Electron microscopic observations on herpes-type virus-related structures in the frog renal adenocarcinoma, *Virology* **36,** 63–72.

Stein, S., Todd, P., and Mahoney, J., 1970, The arginine requirement for nucleocapsid maturation in herpes simplex virus development, *Can. J. Microbiol.* **16,** 851–854.

Stevens, J. G., and Jackson, N. L., 1967, Studies on a temperature sensitive step essential to herpesvirus DNA replication, *Virology* **32,** 654–661.

Stitch, H. F., Hsu, T. C., and Rapp, F., 1964, Viruses and mammalian chromsomes. I. Localization of chromosome aberrations after infection with herpes simplex virus, *Virology* **22,** 439–445.

Stoker, M. G. P., 1959, Growth studies with herpes virus, *in* "Ninth Symposium of the Society of General Microbiology," pp. 142–170, Cambridge University Press, Cambridge.

Stoker, M. G. P., and Newton, A. A., 1959, Mitotic inhibition in HeLa cell and caused by herpesvirus, *Ann. N.Y. Acad. Sci.* **81,** 129–132.

Strandberg, J. D., and Aurelian, L., 1969, Replication of canine herpesvirus. II. Viral development and release in infected dog kidney cells, *J. Virol.* **4,** 480–489.

Strandberg, J. D., and Carmichael, L. E., 1965, Electron microscopy of a canine herpesvirus, *J. Bacteriol.* **90,** 1790–1792.

Subak-Sharpe, H., 1966, Virus-induced changes in translation mechanisms, *in* "The Molecular Biology of Viruses" (L. V. Crawford and M. G. P. Stoker, eds.), pp. 47–66, Cambridge University Press, Cambridge.

Subak-Sharpe, H., and Hay, J., 1965, An animal virus with DNA of high guanine + cytosine content which codes for sRNA, *J. Mol. Biol.* **12,** 924–928.

Subak-Sharpe, H., Shepherd, W. M., and Hay, J., 1966, Studies on sRNA coded by herpes virus, *Cold Spring Harbor Symp. Quant. Biol.* **31,** 583–594.

Summers, D. F., and Maizel, J. V., Jr., 1968, Evidence for large precursor proteins in poliovirus syntheis, *Proc. Natl. Acad. Sci. USA* **59,** 966–971.
Summers, D. F., Maizel, J. V., Jr., and Darnell, J. E., Jr., 1965, Virus-specific proteins in poliovirus-infected HeLa cells, *Proc. Natl. Acad. Sci. USA* **54,** 505–513.
Summers, D. F., Shaw, E. N., Stewart, M. L., and Maizel, J. V., Jr., 1972, Inhibition of cleavage of large poliovirus-specific precursor proteins in infected HeLa cells by inhibotors of proteolytic enzymes, *J. Virol.* **10,** 880–884.
Sutton, W. D., and McCallum, M., 1971, Mismatching and the reassociation rate of mouse satellite DNA, *Nat. New Biol.* **232,** 83–85.
Swanson, J. L., Craighead, J. E., and Reynolds, E. A., 1966, Electron microscopic observations on herpesvirus hominis (herpes simplex virus) encephalitis in man, *Lab. Invest.* **15,** 1966–1981.
Sydiskis, R. J., and Roizman, B., 1966, Polysomes and protein synthesis in cells infected with a DNA virus, *Science* (*Wash., D. C.*) **153,** 76–78.
Sydiskis, R. J., and Roizman, B., 1967, The disaggregation of host polyribosomes in productive and abortive infection with herpes simplex virus, *Virology* **32,** 678–686.
Sydiskis, R. J., and Roizman, B., 1968, The sedimentation profiles of cytoplasmic polyribosomes in mammalian cells productively and abortively infected with herpes simplex virus, *Virology* **34,** 562–565.
Takemoto, K. K., and Fabisch, P., 1964, Inhibition of herpesvirus by natural and synthetic acid polysaccharides, *Proc. Soc. Exp. Biol. Med.* **116,** 140–144.
Tankersley, R. W., Jr., 1964, Amino acid requirements of herpes simplex virus in human cells, *J. Bacteriol.* **87,** 609–613.
Tanzer, J., Thomas, M., Stoitchkov, Y., Boiron, M., and Bernard, J., 1964, Alterations chromosomiques observées dans des cellules de rein de Singe infectées *in vitro* par le virus de l'herpes, *Ann. Inst. Pasteur* **107,** 366–373.
Tevethia, S. S., Lowry, S., Rawls, W. E., Melnick, J. L., and McMillan, V., 1972, Detection of early cell surface changes in herpes simplex virus infected cells by agglutination with concanavalin A., *J. Gen. Virol.* **5,** 93–97.
Thouless, M. E., 1972, Serological properties of thymidine kinase produced in cells infected with type 1 or type 2 herpesvirus, *J. Gen. Virol.* **17,** 307–315.
Thouless, M. E., and Skinner, G. R. B., 1971, Differences in the properties of thymidine kinase produced in cells infected with type 1 and type 2 herpesvirus, *J. Gen. Virol.* **12,** 195–197.
Tokumaru, T., 1957, Pseudorabies virus in tissue culture: Differentiation of two distinct strains of virus by cytopathogenic pattern induced. *Proc. Soc. Exp. Biol. Med.* **96,** 55–60.
Toplin, I., and Schidlovsky, G., 1966, Partial purification and electron microscopy of virus in the EB-3 cell line derived from a Burkitt lymphoma, *Science* (*Wash., D.C.*) **152,** 1084–1085.
Trung, P. H., and Lardemer, F., 1972, Ultrastructure du virus cytomegalique en culture de tissu sur fibroblastes humains: Etude cytochimique par digestion enzymatique, *J. Microscop.* **14,** 271–278.
Tytell, A. A., and Neuman, R. F., 1963, A medium free of agar, serum and peptone for plaque assay of herpes simplex virus (28362), *Proc. Soc. Exp. Biol. Med.* **113,** 343–346.
Underwood, P. A., 1972, Herpes simplex virus infection of HEp-2 and L-929 cells, 3. Envelopment of virus particles and particle infectivity ratios, *Microbios* **5,** 231–235.
Vaheri, A., and Cantell, K., 1963, The effect of heparin on herpes simplex virus, *Virology* **21,** 661–662.

Vaheri, A., and Penttinen, K., 1962, Effect of polyphloroglucinol phosphate, an acid polymer, on herpes simplex virus, *Ann. Med. Exp. Biol. Fenniae* (*Helsinki*) **40,** 334–341.

Vantis, J. T., and Wildy, P., 1962, Interaction of herpes virus and HeLa cells: Comparison of cell killing and infective center formation, *Virology* **17,** 225–232.

Wagner, E. K., 1972, Evidence for transcriptional control of the herpes simplex genome in infected human cells, *Virology* **47,** 502–506.

Wagner, E. K., and Roizman, B., 1969*a*, RNA synthesis in cells infected with herpes simplex virus. I. The patterns of RNA synthesis in productively infected cells, *J. Virol.* **4,** 36–46.

Wagner, E. K., and Roizman, B., 1969*b*, RNA synthesis in cells infected with herpes simplex virus. II. Evidence that a class of viral mRNA is derived from a high molecular weight precursor synthesized in the nucleus, *Proc. Natl. Acad. Sci. USA* **64,** 626–633.

Wagner, E. K., Roizman, B., Savage, T., Spear, P. G., Mizell, M., Durr, F. E., and Sypowicz, D., 1970, Characterization of the DNA of herpesviruses associated with Lucké adenocarcinoma of the frog and Burkitt lymphoma of man, *Virology* **42,** 257–261.

Wagner, E. K., Swanstrom, R. I., and Stafford, M. G., 1972, Transcription of the herpes simplex virus genome in human cells, *J. Virol.* **10,** 675–682.

Watkins, J. F., 1964, Adsorption of sensitized sheep erythrocytes of HeLa cells infected with herpes simplex virus, *Nature* (*Lond.*) **202,** 1364–1365.

Watson, D. H., 1968, The structure of animal viruses in relation to their biological functions, *Symp. Soc. Gen. Microbiol.* **18,** 207–229.

Watson, D. H., and Wildy, P., 1963, Some serological properties of herpes virus particles studied with electron micrscope, *Virology* **21,** 100–111.

Watson, D. H., and Wildy, P., 1969, The preparation of "monoprecipitin" antisera of herpes virus-specific antigens, *J. Gen. Virol.* **4,** 163–168.

Watson, D. H., Wildy, P., and Russell, W. C., 1964, Quantitative electron microscope studies on the growth of herpes virus using the techniques of negative staining and ultramicrotomy, *Virology*, **24,** 523–538.

Watson, D. H., Sheden, W. I. H., Elliot, A., Tetsuka, T., Wildy, P., Bourgaux-Ramoisy, D., and Gold, E., 1966, Virus-specific antigens in mammalian cells infected with herpes simplex virus, *Immunology* **11,** 399–408.

Waubke, R., Zur Hausen, H., and Henle, W., 1968, Chromosomal and autoradiographic studies of cells infected with herpes simplex virus, *J. Virol.* **2,** 1047–1054.

Weber, K., and Osborn, M., 1969, The reliability of molecular weight determinations by dodecyl sulfate-polyacrylamide gel electrophoresis, *J. Biol. Chem.* **244,** 4406–4412.

Weinberg, A., and Becker, Y., 1969, Studies on EB virus of Burkitt's lymphoblasts, *Virology* **39,** 312–321.

Wetmur, J. G., and Davidson, N., 1968, Kinetics of renaturation of DNA, *J. Mol. Biol.* **31,** 349–370.

Wheeler, C. E., 1964, Biological comparisom of a syncytial and a small giant cell-forming strain of herpes simplex, *J. Immunol.* **93,** 749–756.

Wilbanks, G. D., and Campbell, J. A., 1972, Effect of herpesvirus hominis type 2 on human cervical epithelium: Scanning electron microscopic observations, *Am. J. Obstet. Gynec.* **112,** 924–929.

Wildy, P., 1973, Antigens of herpes simplex virus of oral and genital origin, *Cancer Res.* **33,** 1465–1468.

Wildy, P., Russell, W. C., and Horne, R. W., 1960, The morphology of herpes virus, *Virology* **12,** 204–224.

Witter, R. L., Nazerian, K., and Solomon, J. J., 1972, Studies on the *in vivo* replication of turkey herpesvirus, *J. Natl. Cancer Inst.* **49,** 1121–1130.

Wolf, K., and Darlington, R. W., 1971, Channel catfish virus: A new herpesvirus of ictalurid fish, *J. Virol.* **8,** 525–533.

Zambernard, J., and Vatter, A. E., 1966, The fine structural cytochemistry of virus particles found in renal tumors of leopard frogs. I. An enzymatic study of the viral nucleoid, *Virology* **28,** 318–324.

Zee, Y. C., and Talens, L., 1972, Electron microscopic studies on the development of infectious bovine rhinotracheitis virus in bovine kidney cells, *J. Gen. Virol.* **17,** 333–336.

Zemla, J., Coto, C., and Kaplan, A. S., 1967, Correlation between loss of enzymatic activity and of protein from cells infected with pseudorabies virus, *Virology* **31,** 736–738.

Zitelli, A., Baroni, A., and Moretti, G. F., 1970, Intranuclear ribosome crystals in chorioallantoic membrane infected with herpesvirus hominis, type 2, *Sept. Cong. Int. Microscop. Electron, Grenoble,* 75–76.

Zur Hausen, H., and Schulte-Holthausen, H., 1970, Presence of EB virus nucleic acid homology in a "virus-free" line of Burkitt tumor cells, *Nature (Lond.)* **227,** 245–248.

Zur Hausen, H., Schulte-Holthausen, H., Klein, G., Henle, W., Henle, G., Clifford, P., and Santesson, L., 1970, EBV DNA in biopsies of Burkitt tumours and anaplastic carcinomas of the nasopharynx, *Nature (Lond.)* **228,** 1056–1058.

CHAPTER 5

Reproduction of Poxviruses

Bernard Moss

Laboratory of Biology of Viruses
National Institute of Allergy and Infectious Diseases
National Institutes of Health
Bethesda, Maryland 20014

1. INTRODUCTION

The poxviruses are distinguished principally by their unusual morphology, large DNA genome, and cytoplasmic site of replication. Smallpox, which is caused by variola virus, was recognized many centuries ago as a distinct disease entity because of its striking symptoms and high mortality. Elimination of the disease from most parts of the world stemmed in large measure from the classic report of Jenner in 1798 on smallpox vaccination. Approximately 80 years have passed since poxvirions were observed by microscopy, and it is about 35 years since vaccinia was obtained in a purified state for chemical characterization. Studies with poxviruses have been greatly facilitated by their cytoplasmic site of replication, which is unusual for a DNA virus, and the ability of some members of the group to rapidly terminate host macromolecular synthesis. Biosynthetic events are temporally regulated, and an ordered sequence of development begins with the apparent *de novo* formation of viral membranes. Further development proceeds with morphopoietic processes that in their complexity resemble cell or organelle differentiation. The resulting virions contain, in addition to their large DNA genomes, many proteins, some of which have enzymatic functions, lipids, and small amounts of carbohydrate. Several reviews describing the state of knowledge regarding poxvirus replication as of 1966–1968 have been published (Joklik, 1966, 1968; Fenner, 1968; Woodson, 1968; McAuslan, 1969*a*). This chapter, although intended as a comprehensive review, emphasizes biochemical aspects of poxvirus reproduction under active investigation during the past five years.

2. SUBGROUPS

The vertebrate poxviruses, many of which are listed in Table 1, show a similarity in size, morphology, chemical composition, including DNA content, and mode of development in the cytoplasm of host cells. Members of the defined groups and some of the ungrouped mammalian poxviruses have been shown to share a common group "NP" antigen (Takahashi *et al.*, 1959; Woodroofe and Fenner, 1962) and to have the capacity to rescue other heat-inactivated poxviruses by a process of nongenetic reactivation (Hanafusa *et al.*, 1959; Fenner and Woodroofe, 1960). Within groups, the poxviruses are closely related, both morphologically and antigenically, and generally have quite similar biological properties (Fenner and Burnet, 1957; Downie and Dumbell, 1956; Andrewes and Pereira, 1972). Recombination, which

TABLE 1

Poxviruses

Variola-vaccinia subgroup	*Sheeppox subgroup*
Variola major (smallpox)	Sheeppox
Variola minor (alastrim)	Goatpox
Vaccinia	Lumpy skin disease
Cowpox	
Rabbitpox	*Avian subgroup*
Ectromelia (mousepox)	Fowlpox
Monkeypox	Canarypox
	Pigeonpox
Myxoma-fibroma subgroup	Turkeypox
Rabbit myxoma	
Rabbit fibroma	*Ungrouped mammalian poxviruses*
Squirrel fibroma	Swinepox
Hare fibroma	Molluscum contagiosum
	Yaba virus
Orf subgroup	Tanapox
Orf (contagious pustular dermatitis)	
Bovine papular stomatitis	*Ungrouped insect poxviruses*
Milkers nodules	Melolontha poxvirus
	Amsacta poxvirus

has only been shown within one group, may also be valuable for subclassification. More recently, Bellett and Fenner (1968) have used nucleic acid-hybridization techniques to compare base-sequence homologies. By this criterion individual members of the variola-vaccinia subgroup are 30–100% related, and the degree of relatedness correlates well with previous immunological studies. Vaccinia and rabbitpox were virtually indistinguishable by these hydridization techniques. Similar studies have not yet been reported either within or between the other poxvirus groups, and the further utility of this approach for subclassification is not known. As expected, none of the tested poxviral DNAs showed base-sequence homology with frog virus 3, a morphologically distinct, cytoplasmic DNA virus. In a recent review, Bergoin and Dales (1971) have stressed the similarity of morphological and preliminary biochemical properties of certain insect viruses to those of vertebrate poxviruses, and have suggested that they be grouped together. Further studies of these viruses may indicate the usefulness and propriety of such a classification.

Although characteristics of various poxviruses will be discussed in subsequent sections, attention has been directed to the variola-vaccinia subgroup. This emphasis stems from the fact that much of what is known about poxvirus reproduction has come from studies with vaccinia and rabbitpox. Comparative descriptions of individual vertebrate (Andrewes and Pereira, 1972) and invertebrate (Bergoin and Dales, 1971) poxviruses have been recently published.

3. VIRIONS

3.1. Morphology

The virions of all poxviruses appear as large oval or brick-shaped bodies. Vaccinia measures approximately 270 × 218 nm in cross section (Westwood *et al.*, 1964). Other members of the variola-vaccinia, myxoma-fibroma, and avian subgroups are quite similar in size and appearance. Orf appears more narrow (252 × 158 nm; Nagington *et al.*, 1964) and sheeppox smaller (194 × 115 nm; Abdussalam, 1957) than vaccinia. Electron microscopy of either negatively stained or thin-sectioned specimens reveals a complex structure consisting essentially of a biconcave core, two lateral bodies which fit into the concavities, and an envelope (Peters, 1956, Epstein, 1958; Dales and Siminovich, 1961). The three major structural elements of vaccinia virus are clearly illustrated in Fig. 1A, B. Easterbrook (1966) further demonstrated that

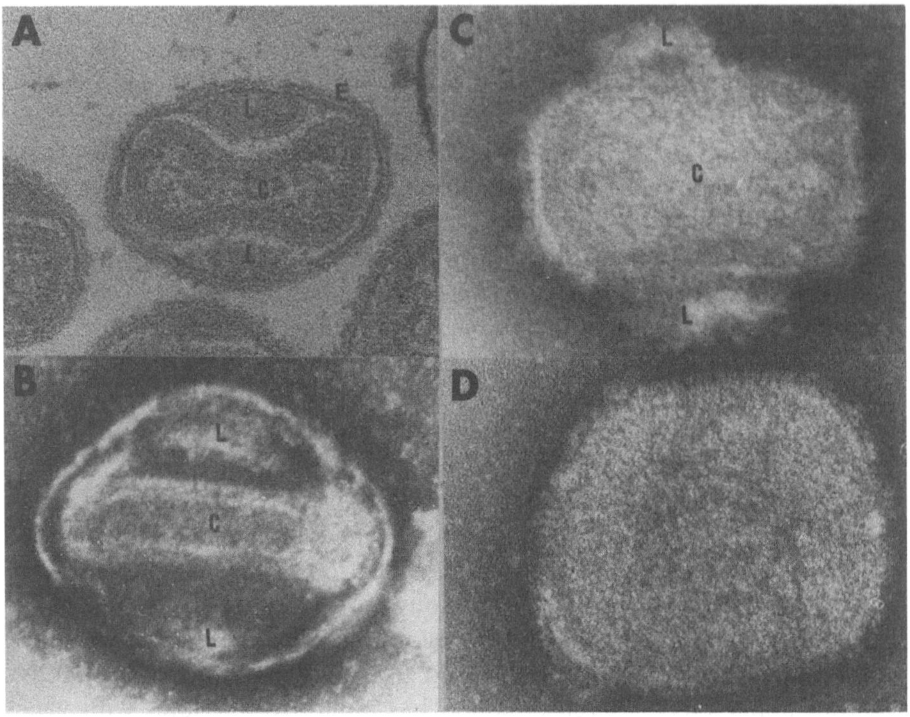

Fig. 1. Morphology of vaccinia virus as observed in (A) thin section or negatively stained preparation of (B) whole particle, (C) particle stripped of its envelope, (D) isolated core. E, envelope; L, lateral body; C, core (×165,600). From Pogo and Dales (1969).

the envelope could be removed by treatment with a reducing agent and nonionic detergent and that the lateral bodies could then be digested with trypsin, liberating a viral core which no longer has a biconcave appearance. Similar results obtained by Pogo and Dales (1969) are shown in Fig. 1C, D.

The external surface of virions may appear beaded when negatively stained (Noyes, 1962*a,b*). Westwood and co-workers (1964) suggested that the surface projections are composed of loops of threadlike structures which are themselves double helices formed from two coiled strands. The technique of freeze-etching, which eliminates some possible artifacts resulting from chemical fixation and dehydration, reveals parallel ridges composed of small globular units 5 nm in diameter superimposed on a smoother surface (Medzon and Bauer, 1970). The ridges appear to be randomly oriented, but they do not cross each other. A comparison of the surface structure of vaccinia virus as seen by negative staining and freeze-etching is shown in Fig. 2.

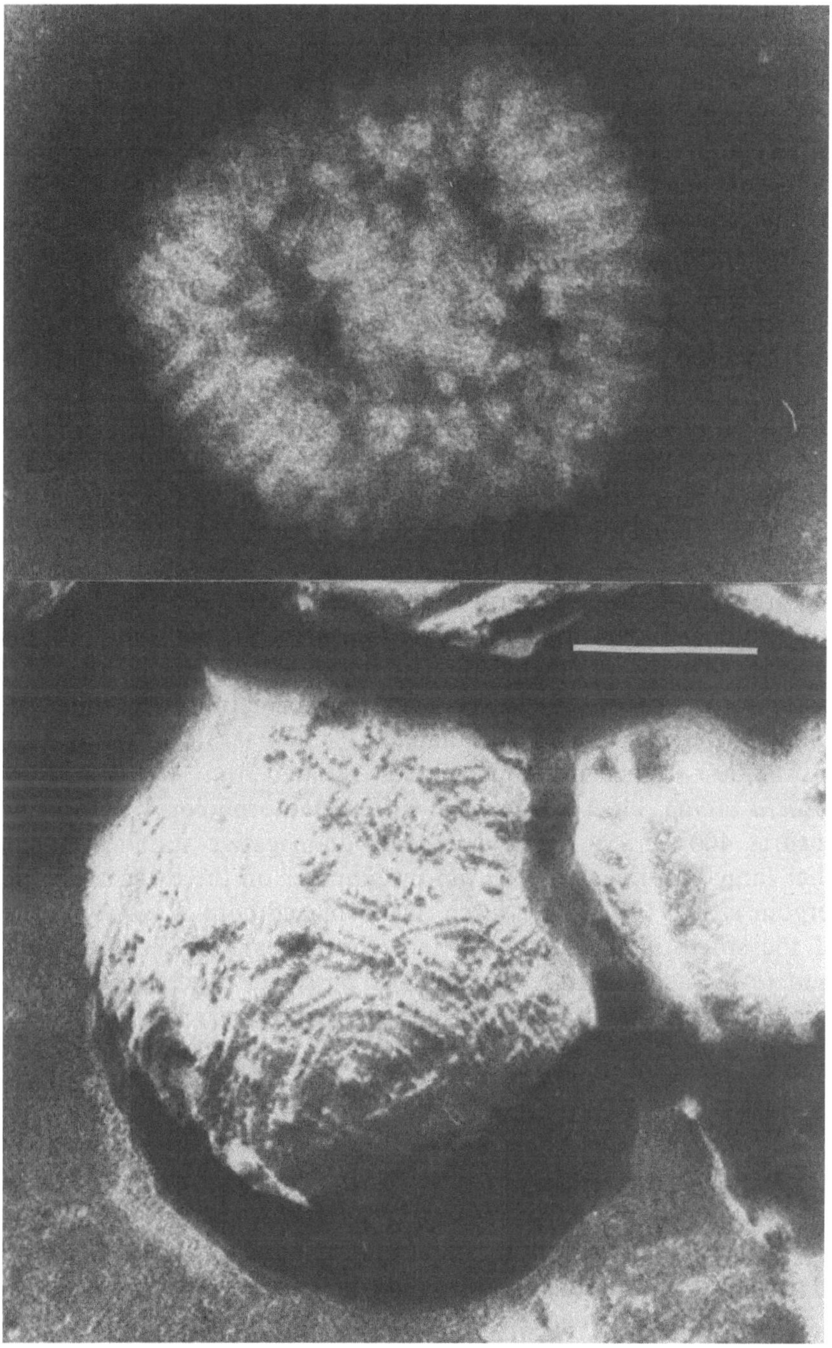

Fig. 2. Surface structure of vaccinia virus revealed by negative staining (top) and freeze-etching (bottom). The bar indicates 100 nm. From Medzon and Bauer (1970).

In constrast to vaccinia, orf virus exhibits what appears to be a single tubular thread in the form of a left-handed spiral coiled around the external surface (Nagington *et al.*, 1964). The latter structure has not yet been studied by freeze-etching.

The viral genome is located within the core. Phosphotungstate staining of intact and carefully degraded vaccinia virus particles indicate that the core wall is composed of cylindrical pegs which can be detached from an inner smooth layer (Dales, 1963; Westwood *et al.*, 1964; Easterbrook, 1966). Freeze-etched preparations of particles in which the cleavage plane passses below the envelope also reveal a closely packed array of units on the external surface of the core (Medzon and Bauer, 1970). On the basis of morphological data obtained by organic extraction and enzymatic digestion, Mitchiner (1969) suggested that the core membrane is composed of protein covered by lipid. The inner portion of the core or nucleoid contains cylindrical elements which may take an S-shaped form (Peters and Müller, 1963). Under specified conditions of fixation, a more complex flowerlike structure can be observed in fowlpox cores (Hyde and Peters, 1971). After prolonged trypsin or ultrasonic treatment the cores may rupture to release a mass of tangled threads which is presumably the DNA (Easterbrook, 1966). Relatively little information regarding the packing of the nucleic acid is available.

The best-studied insect poxviruses, isolated from *Melolontha* and *Amsacta* larvae, have a somewhat different morphology. The former measures 400 × 250 nm and appears on negative staining to have rather large spherical units 22 nm in diameter on the external surface (Bergoin *et al.*, 1971). The eccentrically placed core is kidney-shaped and the single lateral body fits into the concavity. The virus from *Amsacta* larvae is slightly smaller and possesses a beaded surface, a dense core, and material of intermediate density surrounding the core in place of discrete lateral bodies (Fig. 3; Granados and Roberts, 1970).

3.2. Physical Properties

The rapid sedimentation rate of approximately 5×10^3 S allowed early workers to obtain relatively pure preparations of vaccinia virus by differential centrifugation (Craigie, 1932; Smadel *et al.*, 1939). In more recent years, purification procedures have been refined principally by use of zonal sedimentation in sucrose density gradients (Joklik, 1962*a*,*b*; Zwartouw *et al.*, 1962) and equilibrium centrifugation

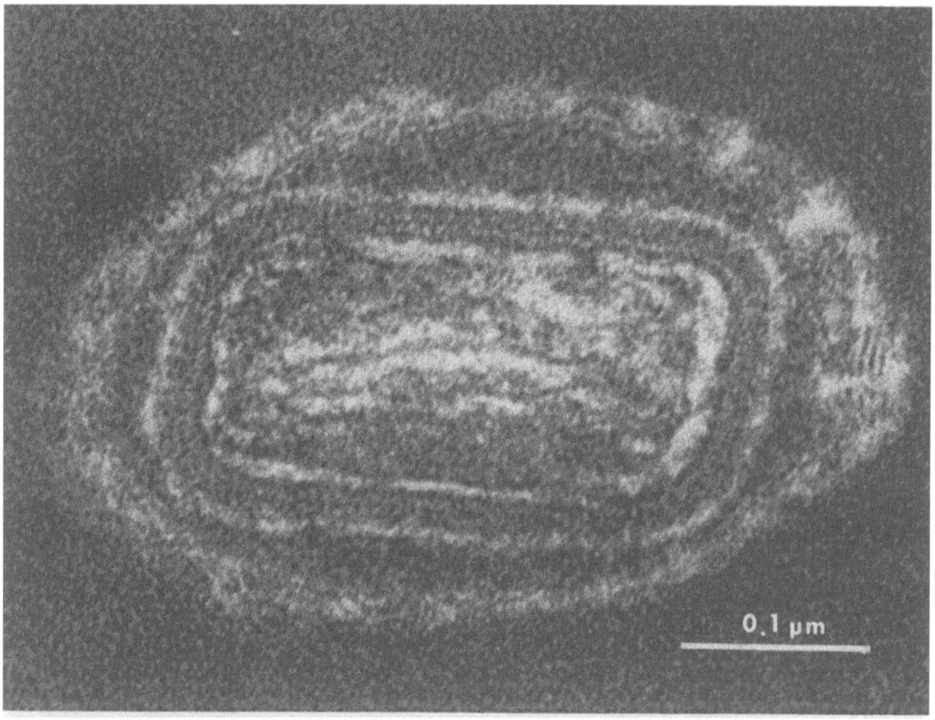

Fig. 3. Longitudinal section of an insect virus originally isolated from larvae of *Amsacta moorei*. The rectangular core, which is composed of two layers, is clearly differentiated from the viral envelope. The virion is surrounded by an electron-dense protein matrix. From Granados and Roberts (1970).

in cesium chloride (Planterose *et al.*, 1962) or potassium tartrate (Pfau and McCrea, 1963) gradients. Aggregation in high salt severely limits the amount of virus that may be purified by the latter equilibrium procedures; however, using sucrose gradient sedimentation, quite large quantities may be easily obtained.

The density of vaccinia virions responds to osmotic influences and has been estimated to be 1.16 g/cm^3 in dilute buffers, 1.25 g/cm^3 in 53% sucrose (Smadel *et al.*, 1938), and 1.20 g/cm^3 in potassium tartrate (Sharp and McGuire, 1970). Using data obtained by centrifugation, the average diameter of vaccinia virus was calculated to be 236 nm (Pickles and Smadel, 1938), and the dry weight was calculated to be 5.34×10^{-15} g (Smadel *et al.*, 1939). There have been some reports of physical heterogeneity as determined by sedimentation of unpurified vaccinia virus released from L cells (Sharp and McGuire, 1970; McGuire and Sharp, 1972; McGuire *et al.*, 1973).

A diameter of 350–380 nm was obtained by small- and wide-angle light-scattering studies of vaccinia virus (Fiel *et al.*, 1970). A particle weight of 4.5×10^{-15} g was obtained in the same study, assuming the refractive index increment to be that of nucleoprotein.

The shape of electrophoretic mobility *vs.* pH curves in molar sucrose suggests that the surface of poxvirions contains amphoteric protein or lipoprotein (Douglas *et al.*, 1969). All of the tested poxviruses have a negative charge at physiological pH values. The isoelectric point is 2.3 for rabbitpox, 3.4 for variola, alastrim and monkey-pox, and 4 for cowpox and vaccinia.

3.3. Chemical Composition

The pioneering work of members of the Rockefeller Institute on the chemical composition of vaccinia virus was summarized by Smadel and Hoagland (1942). Their conclusions regarding the makeup of the major components were supported and refined by more recent studies of Zwartouw (1964), which are presented in Table 2. The principal component is protein, which accounts for about 90% of the dry weight.

TABLE 2

Chemical Composition of Vaccinia Virions[a]

Components	Percentage of total dry weight
Principal components	
Nitrogen	14.7
Phosphorus	0.49
Sulfur	0.76
DNA	3.2
Cholesterol	1.2
Phospholipid	2.1
Neutral fat	1.7
Trace substances	
Carbohydrate	0.2
Copper	0.02
Riboflavin	5.0×10^{-4}
Biotin	1.3×10^{-5}
RNA	0.1

[a] From Zwartouw (1964).

DNA was reported to comprise 2.1–7.3% of vaccinia and closely related viruses (Hoagland *et al.*, 1940; Wyatt and Cohen, 1953; Joklik 1962*a,b*; Allison and Burke, 1962; Zwartouw, 1964). Zwartouw (1964) considered that this large range is probably due to indirect methods of particle-weight determination and protein and DNA standarization by various workers, and that 3.2% was the most accurate valve. Assuming a particle weight of about 5×10^{-15} g, approximately 5% DNA would correspond to the size of the isolated DNA (Sect. 3.4).

Lipid constitutes about 5% of vaccinia virus and is present as cholesterol, phospholipid, and neutral fat (Table 2). The phospholipid is present predominantly as lecithin (Hoagland *et al.*, 1940; Dales and Mosbach, 1968). The latter workers also measured the ratios of palmitic, stearic, and oleic acids in vaccinia virus and concluded that it differed from that of whole cells or microsomes. Lipid was reported to comprise about one-third of the weight of fowlpox, and about half of it was squalene (White *et al.*, 1968). Squalene is present in only trace amounts in normal epithelial tissues in which fowlpox is grown.

Trace amounts of other substances have been detected in purified vaccinia virus. The non-DNA carbohydrate appears to be a true constituent since one of the proteins contains covalently bound glucosamine (Sect. 3.5.2). Although copper, riboflavin, and biotin may only be tenaciously adsorbed impurities, the possibility that they are cofactors of core-associated enzymes (Sect. 3.5.3) has not been excluded.

RNA, if present in purified poxvirus preparations, comprises less than 0.1–0.2% of the dry weight (Table 2). Similar conclusions were reached using radioactive tracers (Joklik, 1962*b*; Planterose *et al.*, 1962). Although there is no evidence that RNA of any significance is present, even 0.1% would amount to more RNA than present in the genome of a picornavirus.

3.4. Size and Structure of DNA

The DNA of poxviruses appears to be intimately associated with the core proteins, but it can be released by treatment with various combinations of reducing agents, detergents, proteolytic enzymes, and phenol. The DNA has both the physical properties and base composition expected for a double-stranded molecule (Joklik, 1962*a,c*; Pfau and McCrea, 1962*a,b*). Approximately 5–10% was reported to be single-stranded; however, it was not associated with the major infectious band of vaccinia virus on potassium tartrate gradients, and its viral identity and significance are unknown (Pfau and McCrea, 1963).

The mole % of G+C is 36–37 for members of the vaccinia-variola subgroups (Joklik, 1962*a,b*) and 35 for fowlpox (Randall *et al.*, 1962; Szybalski *et al.*, 1963). The lowest value, 32.5, was obtained for Yaba virus DNA (Yohn and Gallagher, 1969), and the highest, 40.4, for rabbit fibroma DNA (Jaquemont *et al.*, 1972). Since the latter viruses are both tumor-producing, there is no correlation of G+C content with this property. Subak-Sharpe and co-workers (1966) reported that the doublet base frequencies of vaccinia DNA as determined by nearest-neighbor analysis were significantly different from those of mammalian DNA and speculated on possible evolutionary implications of this. No unusual bases have been reported in any poxvirus DNA.

The technical problems associated with the isolation of the very large DNA molecules without subjecting them to significant shear forces have been solved. The molecular weight of fowlpox DNA was determined to be approximately 200×10^6 daltons by sedimentation and viscometry (Gafford and Randall, 1967, 1970); the DNA appeared as a linear duplex with a contour length of 100 μm (Hyde *et al.*, 1967). These values indicate that fowlpox DNA exists as a single molecule and is the largest viral nucleic acid yet known. Sedimentation studies with vaccinia virus DNA indicated that it has a molecular weight of 150×10^6 (Sarov and Becker, 1967). The molecule is 75–80 μm in length (McCrea and Lipman, 1967; Easterbrook, 1967; Becker and Sarov, 1968). Sedimentation studies with DNA from rabbit fibroma virus indicated a molecular weight of about 153×10^6, and a molecule as long as 80.3 μm was detected (Jaquemont *et al.*, 1972).

Studies with both fowlpox (Szybalski *et al.*, 1963; Gafford and Randall, 1970) and vaccinia (Jungwirth and Dawid, 1967; Berns and Silverman, 1970) indicated that the DNA molecules are capable of rapid renaturation. Gafford and Randall (1970) concluded that this property did not result from covalent cross-linking of fowlpox DNA. Nevertheless, Berns and Silverman (1970) provided strong evidence for covalent cross-linking of vaccinia DNA. More recently P. Geshelin and K. Berns (personal communication) concluded that the cross-links are located at or very near the termini of the DNA molecule. This unusual feature, not previously described as occurring naturally in any DNA, would mean that the two strands are joined to form one continuous molecule. Although the native DNA molecule appeared as a linear duplex on electron microscopy, Geshelin and Berns found that after denaturation it opened to form a large, single-stranded circle. An electron micrograph of such a circle produced by formamide denaturation of vaccinia DNA is shown in Fig. 4. More details regarding these findings should soon be available.

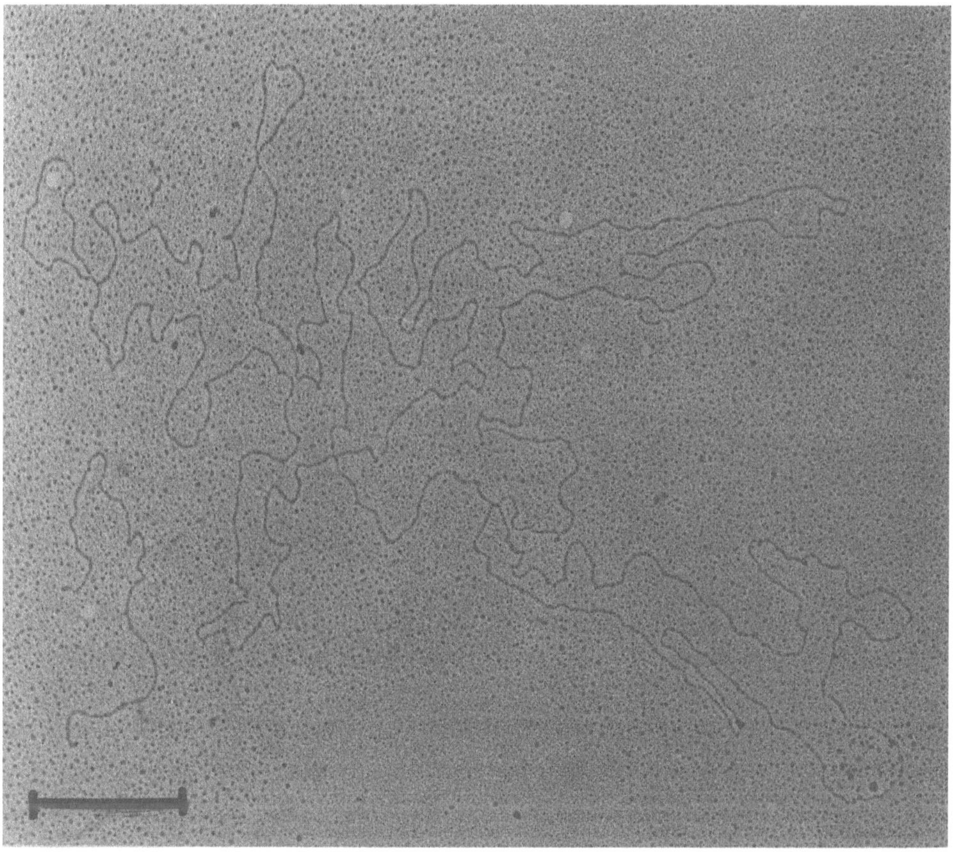

Fig. 4. Electron micrograph of a denatured vaccinia virus DNA molecule prepared by the formamide-protein film technique. No free ends are visible. The bar represents 1 μm. From unpublished data of P. Geshelin and K. Berns.

3.5. Proteins

3.5.1. Polypeptide Composition

A large amount of protein relative to DNA is present within poxviruses. In fact, the ratio of DNA to protein is quite similar to that of free-living organisms (Roberts *et al.*, 1955). Elucidation of the functions of these viral proteins represents an interesting area of current research. For a long while, poxvirus particles remained essentially insoluble masses, and polypeptides could be released only by harsh methods. The "nucleoprotein antigen" described by Smadel *et al.* (1942) contained DNA and about half of the viral protein. It was obtained by incubation of the virus suspension in 0.04 N NaOH at 56°C.

This preparation contained many proteins among which was the group antigen (Woodroofe and Fenner, 1962). The so-called LS antigen was purified by isoelectric precipitation from cytoplasmic extracts (Craigie and Wishart, 1936) and is now believed to be a complex of antigens, some of which may be located on the surface of virions (Marquardt *et al.*, 1965; Cohen and Wilcox, 1968). Somewhat milder alkali treatment or partial digestion with trypsin resulted in an extract which contained 20% of the protein and produced eight precipitin lines by immunodiffusion against antiserum (Zwartouw *et al.*, 1965). Marquardt *et al.*, (1965) suceeded in releasing eight immunoprecipitinogens by means of an "X press," a device in which frozen material is passed through a narrow orifice under high pressure. In a similar manner, five antigens were obtained from Yaba virus (Olsen and Yohn, 1970). As an alternative approach, vaccinia virus was degraded more completely with sodium dodecyl sulfate (SDS) and, after dialysis, antiserum was prepared against the proteins (Wilcox and Cohen, 1967; Cohen and Wilcox, 1968). The antiserum formed several precipitin lines with soluble proteins from infected cells. Generally, immunological studies have thus far been limited by an inability to solubilize and/or resolve more than a relatively small number of proteins from poxvirions. Readers with greater interest in this approach may refer to a recent review by Wilcox and Cohen (1969). More antigens are detected by examining soluble proteins from cytoplasmic extracts with antiserum prepared against infectious virus (Westwood *et al.*, 1965; Rodriguez-Burgos *et al.*, 1966). However, this antiserum may be directed against non-structural as well as structural proteins.

Significant progress in defining the polypeptide components of virions was made following the introduction of SDS-polyacrylamide gel electrophoresis (Summers *et al.*, 1965). Up to 30 bands can be resolved from dissociated vaccinia virus particles (Holowczak and Jolkik, 1967*a*; Moss and Salzman, 1968; Sarov and Joklik, 1972*a*). A photograph of a stained gel together with the approximate molecular weights and relative amounts of each resolved polypeptide is presented in Fig. 5. Of the 30 identified bands, 3 accounted for 35%, 8 for an additional 50%, and 19 for the remaining 15% of the total protein. The sum of their molecular weights, which ranged from 10,000 to 200,000 daltons, equals nearly 2×10^6 daltons, or about 20% of the coding capacity of the vaccinia genome. Approximately 8–10 proteins were separated by SDS-hydroxyapatite chromatography of whole vaccinia virions (Moss and Rosenblum, 1972). However, in contrast to SDS-gel electrophoresis, the proteins did not all elute in order of molecular weight and thus utilization of the two techniques in succession provided

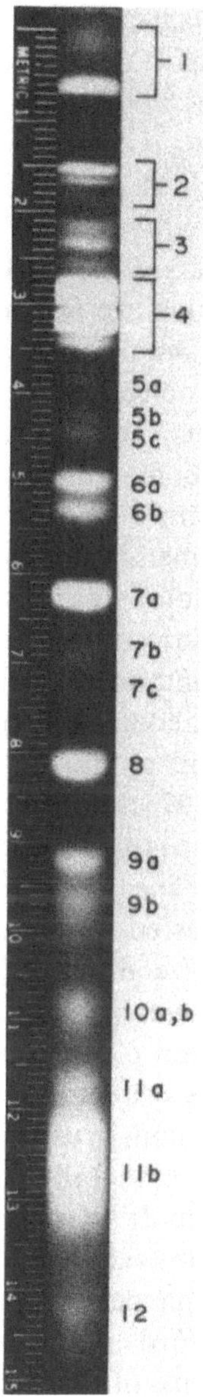

	Polypeptide	MW	Percent
	1a	200,000–	~0.5
	1b	250,000	
	1c	152,500	2.4
	1d	145,000	
	2a	94,000	2.2
	2b	90,000	
	2c	84,000	0.3
	3a	78,500	2.0
	3b	73,500	
	3c	70,000	0.7
	4a	63,000	13.8
	4b	58,500	11.1
	4c	56,000	2.4
	4d	54,000	
	5a	51,000	
	5b	47,500	3.4
	5c	46,000	
	5P	46,750	< 0.5
	6a	41,000	9.1
	6b	39,000	
	7P	31,500	< 0.1
	7a	31,000	6.5
	7b	27,000	1.0
	7c	26,000	
	8	23,000	7.0
	9a	18,500	7.3
	9b	17,000	
	9P	17,000	< 0.1
	10a	16,000	10.5
	10b	14,500	
	11a	~11,800	1.5
	11b	~11,000	11.4
	12	~8,000	6.6
Total	30	$\sim 2 \times 10^6$	99.0

Fig. 5. Coomassie brilliant-blue-stained vaccinia virus proteins separated by electrophoresis in polyacrylamide gels. Molecular weights were determined from the electrophoretic mobilities, and the percentage values were determined from autoradiograms of gels containing polypeptides labeled with ^{14}C-amino acids. From Sarov and Joklik (1972*a*).

additional resolving power (Moss *et al.*, 1973*a*). The polypeptide composition of fowlpox virus appears to be of a similar order of complexity, and resemblance with that of vaccinia virus has been noted (Obijeski *et al.*, 1973). Comparable analyses of the proteins of other poxviruses have not yet been published.

It is appropriate to ask whether all of the resolved proteins have unique amino acid sequences and are coded by the viral genome. The first question is particularly relevant in the case of poxviruses since, as will be discussed in succeeding sections, some proteins are formed by cleavage of higher-molecular-weight precursors and others are modified by glycosylation or phosphorylation. Incompletely processed viral proteins, perhaps present in immature virions, could produce polypeptide heterogeneity. Probably tryptic peptide mapping of the proteins is the simplest approach to this question. Only small amounts of purified proteins are required, particularly if ^{35}S-methionine is used as a label and tryptic peptides are analyzed by radioautography on thin-layer cellulose sheets. In this manner, two major components of the polypeptide 4 complex and three major components of the polypeptide 6 complex, which together comprise about 40% of the virion protein mass, were shown to contain unique amino acid sequences (Moss *et al.*, 1973*a*; Moss and Rosenblum, 1973).

It is evident from numerous studies that the virion polypeptides are labeled and, hence, synthesized after infection. Since host protein synthesis is rapidly turned off (Sect. 5.2), it is suggested that the host proteins are coded by the viral genome. More definitive answers can be obtained by studying the induction of viral proteins in enucleated cells, by isolation of specific mutants and analysis of altered proteins, or by synthesis of viral proteins in a DNA-directed cell-free system.

The possibility that some minor components are non-virus-induced proteins which are tenaciously adsorbed to or accidentally sealed within the developing virions is difficult to substantiate. It has been suggested that some surface components can be removed by tryptic digestion without loss of and with some apparent enhancement of vaccinia virus infectivity (Zwartouw *et al.*, 1965; Gifford and Klapper, 1969; Moss *et al.*, 1973*a*). At this time, the possibility of determining the essential polypeptides by the *in vitro* reconstruction of infectious virus from isolated components seems remote.

3.5.2. Modified Proteins

A glycoprotein with a molecular weight of 38,000 daltons was found to be a component of vaccinia virions by labeling infected HeLa cells with radioactive glucosamine (Holowczak, 1970; Garon and Moss, 1971). This protein migrates with other polypeptides of the polypeptide 6 complex (Fig. 6) but can be purified by SDS-hydroxyapatite chromatography (Moss *et al.*, 1973*a*). The virion glycoprotein was not labeled with other radioactive sugars, suggesting a very simple carbohydrate composition, and only a single major glycopeptide was resolved after pronase digestion (Garon and Moss, 1971). Two very closely spaced glycoprotein bands were resolved by electrophoresis of proteins from virus grown in L cells, but they have not been further characterized (Sarov and Joklik, 1972*a*).

Analysis by SDS-polyacrylamide gel electrophoresis of vaccinia virions, purified from infected cells labeled with $^{32}P_i$, revealed the presence of several labeled bands (Sarov and Joklik, 1972*a*; Rosemond and Moss, 1973). This is illustrated in Fig. 6. The major component has a molecular weight of about 11,000 daltons and contains phosphoserine residues (Rosemond and Moss, 1973). The specificity of phosphorylation was indicated by the demonstration of a single major tryptic phosphopeptide on two-dimensional electrophoresis and

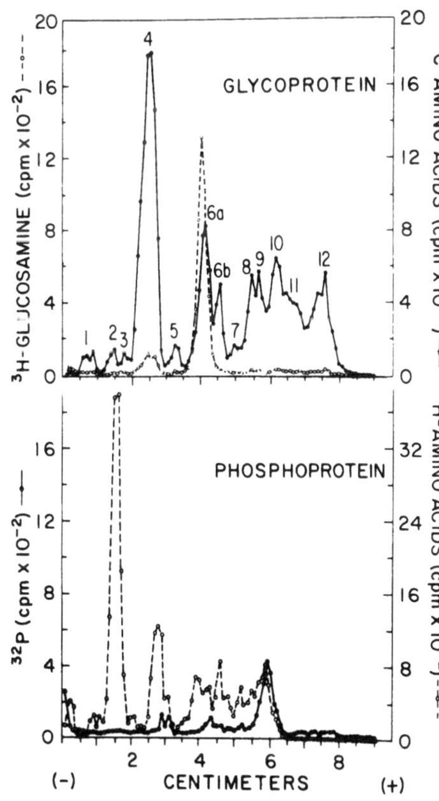

Fig. 6. Glycoprotein and phosphoprotein components of vaccinia virions. Vaccinia virions were purified from infected cells after labeling with ^{3}H-glucosamine or $^{32}P_i$. The extracted proteins were separated by polyacrylamide gel electrophoresis after mixing with ^{3}H- or ^{14}C-amino acid-labeled viral proteins. From Moss *et al.* (1973*a*) and Rosemond and Moss (1973).

chromatography (Rosemond and Moss, 1973). The discovery of a phosphoprotein has special significance because of the presence of protein kinase within the virion (Sect. 3.5.3).

3.5.3. Enzymes

3.5.3a. Identification

The large size and complex structure of poxviruses led some early workers to question whether they might possess enzymes or even independent metabolic activity. Macfarlane and Salaman (1938) reported the presence of alkaline phosphatase and several other enzymatic activities associated with their preparations of vaccinia virus. However, Hoagland and co-workers (1942) noted the ready adsorption of enzymes to purified preparations of virus and decided that it was not possible at that time to distinguish between adsorbed and integral components. Interest in virion enzymes subsided (1) as more purified preparations of poxviruses were found to contain no detectable alkaline phosphatase (Joklik, 1962*a*), (2) as a consequence of the classic Hershey-Chase experiment, which demonstrated the injection of principally the DNA component of T-even bacteriophage, and (3) following the demonstration of infectious nucleic acids from smaller viruses. This attitude was changed radically in 1967 when a DNA-dependent RNA polymerase was discovered in virions of rabbitpox (Kates and McAuslan, 1967*c*) and vaccinia virus (Munyon *et al.*, 1967). These results have had a profound influence on the whole field of animal virology. Vaccinia virus was subsequently shown to contain phosphohydrolase activities for all four ribonucleoside triphosphates (Munyon *et al.*, 1968; Gold and Dales, 1968): acid and alkaline *p*H deoxyribonuclease (DNase) (Pogo and Dales, 1969), polyadenylate [poly(A)] polymerase (Kates and Beeson, 1970*b*), and protein kinase (Paoletti and Moss, 1972*a*; Downer *et al.*, 1973) activities. The enzymes all appear to reside within the virion since they are stimulated by detergents and reducing agents which disrupt the viral envelope and they remain associated with the viral core. Until the enzymes could be extracted from the core, it was not possible to determine whether a separate enzyme was responsible for each activity. For example, one or two enzymes might be responsible for RNA and poly(A) polymerase activities, one to four for nucleoside triphosphate phosphohydrolase activities, and one or two for acid and alkaline *p*H DNase activities. Furthermore, an apparent single activity such as hydrolysis of ATP might result from the combined action of two enzymes, e.g., a protein kinase and a phosphatase. Clarification of some of these problems has come from the recent discovery that release of the viral DNA and solubilization of several active enzymes occurs following treatment of cores with sodium deoxycholate and reducing agents (Paoletti and Moss, 1972*b*). Thus far, two nucleoside triphosphate phosphohydrolases (Paoletti *et al.*, 1974), a DNase (Rosemond-Hornbeak *et al.*, 1974), a poly(A) polymerase (Moss *et al.*, 1973*b*), and a protein kinase (Kleiman and Moss, 1973, and unpublished) have been separated (Fig. 7).

3.5.3b. DNA-Dependent RNA Polymerase

Thus far, this enzyme has been described in members of the variola-vaccinia subgroup (Kates and McAuslan, 1967*c*; Munyon *et al.*, 1967) and Yaba virus (Schwartz and Dales, 1971), but it is presumed to be present in all poxviruses. A DNA-dependent RNA polymerase has not been described in any other group of viruses. The RNA polymerase is associated with viral DNA within the core and does not respond to added DNA templates. All attempts to extract the enzyme have thus far been unsuccessful and RNA polymerase activity is lost upon disruption of the core with deoxycholate (Moss *et al.*, 1973*b*). Synthesis of RNA is dependent on all four ribonucleoside triphosphates [however, see poly(A) polymerase activity, Sect. 3.5.3c] and divalent cations. RNA synthesis proceeds linearly for many hours, producing considerably more RNA than DNA template. Nevertheless, DNA–RNA hybridization experiments indicate that only 14% of the genome is transcribed (Kates and Beeson, 1970*a*). These RNA sequences correspond to the very early RNA made in infected cells (Sect. 4.3). The *in vitro*-synthesized RNA is 8–10 S and thus is similar in size to early RNA made *in vivo*. Apparently both ATP and CTP are capable of chain initiation (Kates and Beeson, 1970*a*), although initiation of an RNA chain with a pyrimidine is quite unusual. Each particle may have about 40 RNA-chain growing points, and an ATP-dependent extrusion mechanism seems to be involved in the release of the RNA from the core (Kates and Beeson, 1970*a*). Since conditions optimal for RNA polymerase activity are also suitable for poly(A) polymerase, nucleoside triphosphate phosphohydrolase, and protein kinase activities, it is not known whether the latter enzymes are also involved in RNA synthesis. The possibility that the RNA polymerase is activated by the protein kinase or other yet unknown core enzyme may explain the difficulties in isolating it.

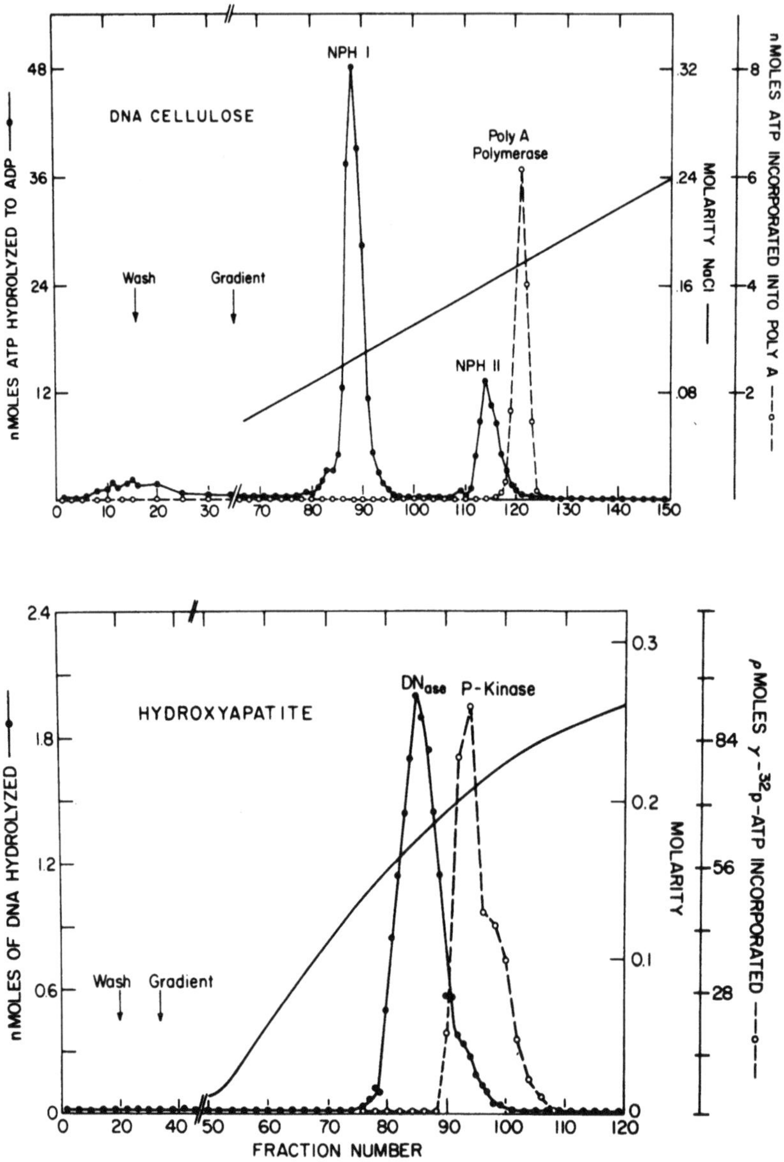

Fig. 7. Purification of enzymes from vaccinia virions. Vaccinia cores were disrupted with sodium deoxycholate. The DNA was removed by DEAE-cellulose chromatography and the soluble enzyme mixture was chromatographed on an affinity column made of denatured DNA bound to cellulose. Nucleoside triphosphate phosphohydrolase I (NPH I), nucleoside triphosphate phosphohydrolase II (NPH II) and the poly(A) polymerase eluted separately. The DNase and protein kinase which were not adsorbed to DNA-cellulose were separated by hydroxyapatite chromatography. From Paoletti, Rosemond, Kleiman, Rosenblum, and Moss (unpublished data).

3.5.3c. Poly(A) Polymerase

Kates and Beeson (1970*b*) found that poly(A) was formed when vaccinia cores were incubated with ATP and divalent cations. When all four ribonucleoside triphosphates were present, 100–200 residues of adenylic acid were attached to the synthesized RNA chain. Similar poly(A) tracts are found on most eukaryotic and viral mRNA species. The localization of the poly(A) on the 3′ terminus of the vaccinia RNA was supported by end-group analysis, specific enzymatic digestion, and kinetic data (Kates and Beeson, 1970*b*; Sheldon *et al.*, 1972*b*). Kates and Beeson (1970*b*) suggested that poly(A) is formed by a transcriptional mechanism, possibly by the RNA polymerase itself. Evidence for such a mechanism came from the finding that poly(A) synthesis is inhibited by agents which bind by intercalation to double-stranded nucleic acids. The possibility that vaccinia DNA contains some poly(dA-dT) tracts that might serve as a template is consistent with hybridization data and with results obtained by transcription of vaccinia DNA with *Escherichia coli* RNA polymerase (Kates and Beeson, 1970*b*). However, the low level of hybridization (1% of the DNA) would preclude the presence of a long (dA-dT) tract in each cistron. The nearest-neighbor frequencies previously published by Subak-Sharpe *et al.* (1966) also do not indicate more ApA and TpT doublets in vaccinia DNA than in DNA of other viruses or animal cells.

Brown *et al.* (1973) have suggested that the poly(A) polymerase is distinct from the RNA polymerase because of differences in divalent ion requirements, *p*H optima, and heat stabilities of the core-associated enzymes. The latter workers also reported that oligo A stimulates the poly(A) polymerizing activity of cores.

The purified poly(A) polymerase (Fig. 7) has no detectable RNA-polymerizing ability and cannot utilize UTP, CTP, or GTP as substrates (Moss *et al.*, 1973*b*). SDS-polyacrylamide gel electrophoresis reveals 2 polypeptides of 35,000 and 50,000 daltons, corresponding to minor bands of the 5 and 7 complex, respectively. The poly(A) polymerizing activity is stimulated by addition of poly(dA-dT) or poly(dT), but this is prevented by intercalating agents such as ethidium bromide and proflavine, suggesting a template mechanism. These data are consistent with the previous idea that poly(dA-dT) tracts could serve to initiate poly(A) synthesis in viral cores. However, the purified poly(A) polymerase is also stimulated by poly(C) (Moss *et al.*, 1973*b*), suggesting a terminal addition mechanism. Recent experiments (B. Moss and E. N. Rosenblum, in preparation) demonstrate attachment of the poly(A) to poly(C) and to the poly(dT) strand of poly(dA-dT), indicating terminal addition and making template-directed synthesis unlikely. The vaccinia poly(A) polymerase appears to be unique in its ability to catalyze the attachment of adenylate residues to both polyribonucleotide and polydeoxyribonucleotide primers. It is likely, however, that only the former has biological significance.

3.5.3d. Nucleic-Acid-Dependent Nucleoside Triphosphate Phosphohydrolase

An ATP hydrolytic activity has been described in vaccinia virus (Munyon *et al.*, 1968; Gold and Dales, 1968), in Yaba virus (Schwartz and Dales, 1971), and in an insect poxvirus (Pogo *et al.*, 1971). Since UTP, GTP, and CTP are also hydrolyzed, the enzymatic activity is properly called a nucleoside triphosphate phosphohydrolase. After solubilization, the enzyme was found to be dependent upon addition of nucleic acid to the reaction mixture (Paoletti and Moss, 1972*b*). Two such nucleic-acid-dependent enzymes were separated during purification (Fig. 7; Paoletti *et al.*, 1974; Paoletti and Moss, 1974). Although both are monomeric enzymes with molecular weights of about 68,000 daltons, phosphohydrolase I hydrolyzed only ATP or dATP while phosphohydrolase II hydrolyzed all tested ribo- or deoxyribonucleoside triphosphates. Furthermore, phosphohydrolase I was stimulated only by DNA while DNA or RNA stimulated phosphohydrolase II. Studies with a variety of DNAs and synthetic polynucleotides suggested that a partially single-stranded, partially double-stranded structure was required for stimulation of phosphohydrolase I; a single-stranded structure was required for phosphohydrolase II. Both enzymes produce nucleoside diphosphates and P_i in stoichiometric amounts, and it is not yet known with what function ATP hydrolysis is coupled. Similar enzymes have not been described in eukaryotic cells or other viruses, and the vaccinia enzyme is clearly different from the DNA-dependent ATPases of bacteria which are also ATP-dependent DNases. One possible function is the ATP-dependent extrusion of RNA from poxvirus cores (Kates and Beeson, 1970*a*). Another possibility is that the enzymes are involved in assembly and maturation of virus particles.

3.5.3e. Deoxyribonucleases

Deoxyribonuclease activities have been detected in vaccinia virus (Pogo and Dales, 1969), in Yaba virus (Schwarz and Dales, 1971), and in an insect poxvirus (Pogo *et al.*, 1971). The activity is associated with the vaccinia virus core, specifically hydrolyzes single-stranded DNA, and displays a biphasic *p*H optimum. Some evidence indicating that the neutral (*p*H 7.8) activity is endonucleolytic and that the acid (*p*H 5) activity is exonucleolytic has been presented (Pogo and Dales, 1969; Aubertin and McAuslan, 1972). Only one DNase has thus far been purified from vaccinia virus (Rosemond-Hornbeak *et al.*, 1974; Rosemond-Hornbeak and Moss, 1974). It has a molecular weight of 100,000, may be composed of two similar or identical subunits, has a *p*H optimum of 4.4, is specific for single-stranded DNA, and has both endonucleolytic and exonucleolytic activities at *p*H 4.4, as demonstrated by direct analysis of the digestion products of a variety of DNAs. In many respects the enzyme is similar to the S_1 nuclease of *Aspergillus oryzae* (Sutton, 1971). Possible functions of the enzymes will be discussed in subsequent sections.

3.5.3f. Protein Kinase

Incubation of vaccinia virus cores with γ^{32}P-ATP and divalent cations led to the phosphorylation of serine and threonine residues of specific endogenous viral proteins (Paoletti and Moss, 1972*a*; Downer *et al.*, 1973). The solubilized enzyme retains its substrate specificity but requires protamine as an activator (Kleiman and Moss, 1973, and unpublished results). Presumably the function of the enzyme is to phosphorylate the virion phosphoprotein (Sect. 3.5.2), but the biological role of this is unknown. Attempts to demonstrate protein kinase activity in fowlpox virions have been unsuccessful (Downer *et al.*, 1973); it is not known whether the latter virus contains a phosphoprotein.

3.5.4. Location

It is not yet possible to correlate the detailed morphology of poxviruses as observed by electron microscopy with individual polypeptide components. The approaches used so far to locate proteins within the virion have been limited to extraction, digestion, and labeling of the surface components. As already noted (Sect. 3.1; Fig. 1), incubation of vaccinia virions with a reducing agent and nonionic detergent results in the degradation of the envelope and release of the core, which may still have attached lateral bodies. All of the virion enzymes described in the preceding section are associated with cores prepared in this manner. The polypeptide components of the two fractions have been analyzed by immunological methods (Cohen and Wilcox, 1968) and by SDS-polyacrylamide gel electrophoresis (Sarov and Joklik, 1972*a*; Moss *et al.*, 1973*a*). Stained polyacrylamide gels shown in Fig. 8 indicate that at least six major polypeptides are released by this treatment. Not all of the proteins released by nonionic detergent and reducing agents are located on the surface, as indicated by labeling with ^{125}I or fluoroscein isothiocyanate (Sarov and Joklik, 1972*a*; Katz and Margalith, 1973) or by digestion of virions with chymotrypsin (Sarov and Joklik, 1972*a*) and trypsin (Moss *et al.*, 1973*a*). The glycoprotein may fall into this category since it does not react with some of the surface-labeling reagents.

The ability to elicit neutralizing antibody has also been used to determine internal or external location of vaccinia antigens (Marquardt *et al.*, 1965; Cohen and Wilcox, 1968). A more powerful immu-

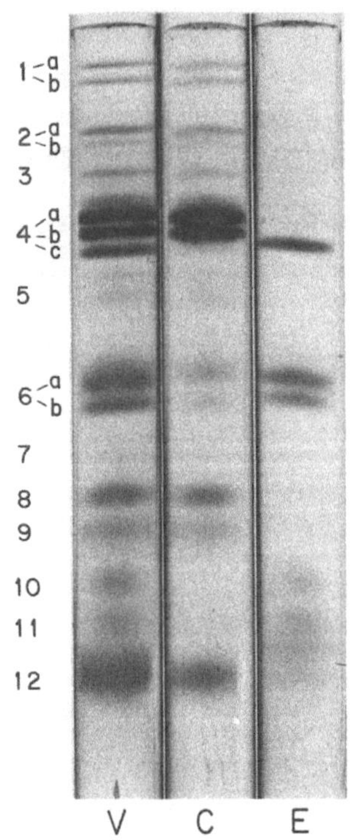

Fig. 8. Polypeptide composition of vaccinia virus cores. Whole virus (V), cores (C), and envelope (E) fractions were obtained by degradation with a nonionic detergent and reducing agent. The polypeptides were resolved by electrophoresis on SDS-polyacrylamide gels and stained with Coomassie brilliant blue. From Moss *et al.* (1973*a*).

nological method not yet attempted would be to prepare conjugated antibodies to individual polypeptides and then use the resolving power of the electron microscope to locate their precise sites of attachment to whole virions, partially degraded particles, and thin sections of virions.

4. GROWTH CYCLE

4.1. Single-Step Infection

The replication of many poxviruses can be studied in primary or continuous lines of mammalian and avian cells. Single-step growth curves may be obtained under conditions of simultaneous infection. Adsorption is characteristically followed by a latent period in which infectivity diminishes and then by a period of exponential growth. The precise timing varies not only with the particular virus but also with the passage level, multiplicity, and host cell. Under suitable conditions, a rise in infectious vaccinia virus is detectable in 4–6 hours and a maximum yield of about 10,000 particles per cell is obtained by 12–15 hours. Studies with cells synchronized by excess thymidine indicate that vaccinia virus could replicate in cells during any period of the growth cycle except perhaps from late prophase through telophase (Groyon and Kniazeff, 1967; Mantani *et al.*, 1968). In general, the rate of replication decreases in the order variola-vaccinia > myxoma-fibroma > avian poxviruses. Some viruses, such as *Molluscum contagiosum* and the insect poxviruses have not been grown *in vitro*.

The association of infectivity with poxvirions was made more than 40 years ago (Woodruff and Goodpasture, 1929; Eagles and Ledingham, 1932; Elford and Andrewes, 1932; Parker and Rivers, 1936). An average ratio of particles to infectious units of almost 4:1 was obtained by titration on rabbit skin (Smadel *et al.*, 1939). Values commonly reported by plaque titration of sucrose gradient-purified vaccinia virus on cell monolayers are somewhat higher. Sharp and coworkers (1964) have reported changes in plaquing efficiency associated with viral adaptation to cell lines. Statistical studies indicate that infection is a single-hit process (Parker, 1938). In view of the RNA polymerase function of intact virions, reports made several years ago that poxviral DNA or subviral particles (Abel, 1963; Abel and Trautner, 1964; Babbar *et al.*, 1966; Randall *et al.*, 1966; Takehara and Schwerdt, 1967) are infectious for eukaryotic and even prokaryotic cells must be regarded with skepticism until widely confirmed.

4.2. Adsorption, Penetration, and Uncoating

Detailed studies of adsorption, penetration, and uncoating have been carried out mainly with vaccinia virus and suspended HeLa or L cells. The first step, adsorption to the cell membrane, is a rapid process and more than half the inoculum may be attached within 15 minutes (Joklik, 1964*a*). Electron microscopic studies indicate that attachment occurs without specific orientation of the virion with respect to its long and short axes (Dales and Siminovich, 1961; Dales, 1965*a*). Next, the entire particle appears to be engulfed in vesicles formed at the cell surface by the process of phagocytosis. This is an active process which may be inhibited by NaF, in which case the virus particles remain outside the cell and susceptible to neutralization by antiserum. Some insect poxviruses enter by fusion of the viral envelope with the microvillus membrane, in addition to phagocytosis (Granados, 1973). Within 20 minutes, the outer coat of some vaccinia virions is disrupted, the vacuolar membrane disintegrates, and the virus core lacking lateral bodies passes into the cytoplasmic matrix (Dales, 1963). The latter process must be rapid since relatively few particles are seen enclosed within vacuoles. Heat-denatured, nonionic-detergent-treated, and antibody-coated viruses are all engulfed by phagocytic vacuoles, but the majority are digested and eventually disintegrate (Dales and Kajioka, 1964; Easterbrook, 1966). Parallel studies with radioactively labeled virus indicate that breakdown of the phospholipid and dissociation of about 50% of the protein occurs without an appreciable lag, and probably during transit from the vacuoles (Joklik, 1964*a*). This process, which is not prevented by inhibitors of RNA or protein synthesis, is referred to as the first stage of uncoating.

The second stage of uncoating involves the release of the viral genome. Some electron microscopic images suggest that this might occur by rupture of the core (Dales, 1963, 1965*b*). Joklik (1964*a,b,c*) defined the second stage of uncoating operationally as changes which lead to the susceptibility of the DNA to DNase. A lag period, varying from 0.5 to 2 hours depending on the multiplicity of infection, occurs before significant uncoating is detected. The time needed for 50% uncoating ranges from 1.25 to 2.5 hours; however, uncoating is never complete and varies from 50% to 70%. Inhibitors of RNA and protein synthesis prevent second-stage uncoating, as does UV irradiation of virus, suggesting that a specific uncoating protein is induced (Joklik, 1964*b*; Dales, 1965*a*). Joklik (1966) proposed that poxvirions contain an UV-sensitive inducer protein which derepresses a portion of the host genome which codes for the uncoating protein. Although this was a reasonable and imaginative suggestion at the time it was made, an alternative hypothesis now appears more likely. Poxvirus cores have been found to contain a polymerase capable of synthesizing viral RNA prior to uncoating (Sect. 4.3). Moreover, uncoating, as judged by progression of the infection to the stage of viral DNA synthesis, occurs in enucleated cells which lack the host genome (Prescott *et al.*, 1971, 1972). In addition, the ultraviolet action spectrum of virions indicates that the prime target is nucleic acid (Sime and Bedson, 1973). Thus, if there is a specific uncoating protein, it is likely to be a viral gene product. Some reports of uncoating activity in crude lysates of infected cells has been reported (Abel, 1963; Abel and Trautner, 1964); however, they are apparently difficult to reproduce (McAuslan, 1967).

The failure of heat-inactivated poxviruses to be uncoated and the phenomenon of nongenetic reactivation whereby an inactivated virus may be rescued by another untreated poxvirus were previously explained by effects on the inducer protein (Joklik, 1966); inactivation of the core RNA polymerase or other virion enzymes is now considered the more likely possibility. Munyon and co-workers (1970) have correlated the loss of infectivity by heating with loss of RNA polymerase and nucleoside triphosphate phosphohydrolase activities. Furthermore, they showed that all were protected by addition of nucleoside triphosphates, which presumably could bind to the active sites of virion enzymes.

More recent work on uncoating has focused on the isolation and characterization of intermediates isolated by sucrose gradient centrifugation (Dahl and Kates, 1970*a*; Holowczak, 1972; Sarov and Joklik, 1972*b*). Some of these intermediates appear similar but not identical in polypeptide composition to those derived *in vitro* by successive application of reducing agents, nonionic detergents, and trypsin.

4.3. Transcription

The question of how poxviruses initiate synthesis of macromolecules posed a problem until quite recently. The logical alternatives were (1) utilization of a cellular RNA polymerase, (2) the presence of mRNA in the virion which could code for a viral RNA polymerase, or (3) the presence of an RNA polymerase in the virion.

The finding that protein synthesis is not a prerequisite for viral RNA synthesis ruled out the second possibility (Munyon and Kit, 1966; Kates and McAuslan, 1967*b*; Woodson, 1967). A definitive answer to the problem came with the demonstration of DNA-dependent RNA polymerase within the viral core (Sect. 3.5.3), thus confirming the third possibility.

The study of transcription *in vivo* is greatly facilitated by the cytoplasmic location of the site of vaccinia replication (Salzman *et al.*, 1964; Becker and Joklik, 1964). Becker and Joklik (1964) took advantage of the fact that appreciable amounts of newly synthesized cellular RNA larger than 4 S are not transported from the nucleus to the cytoplasm during a pulse of 20 minutes or less with radioactive uridine. By carefully rupturing the cells, the cytoplasmic contents can be obtained free of nuclear contamination, and the viral RNA can then be separated from 4 S RNA by sedimentation in sucrose gradients (Becker and Joklik, 1964). Somewhat better separations are obtained (Fig. 9) by gel electrophoresis (Moss and Filler, 1970). The cytoplasmic RNA can be further identified by base composition (Salzman *et al.*, 1964) and by specific hybridization with vaccinia DNA (Becker and Joklik, 1964).

The initial burst of RNA synthesis detected immediately after infection is followed by a second rise in RNA synthesis at about two

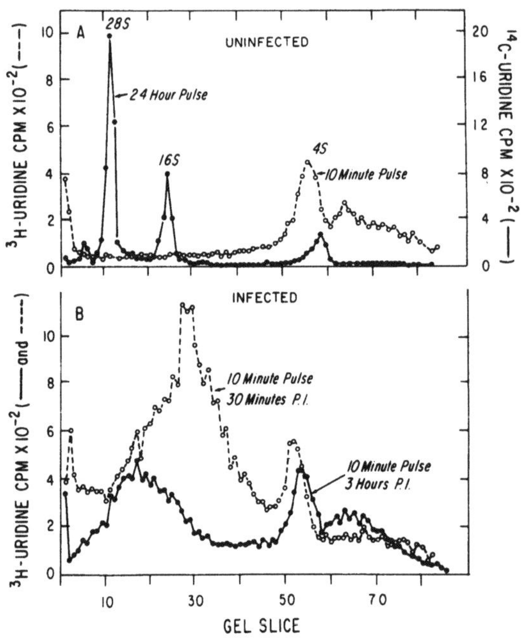

Fig. 9. Polyacrylamide gel electrophoresis of vaccinia virus RNA. (A) Uninfected HeLa cells were labeled for 24 hours with ^{14}C-uridine or 10 minutes with ^{3}H-uridine, and the cytoplasmic fractions were dissociated with sodium dodecyl sulfate and analyzed by electrophoresis in composite 2.5% polyacrylamide–0.5% agarose gels. (B) Similar analysis of cytoplasmic fraction from cells pulse-labeled at 30 minutes or 3 hours after infection. From Moss and Filler (1970).

hours later. The relative amounts of RNA made at the two times depends on the multiplicity of infection and the cell type (Oda and Joklik, 1967; Woodson, 1967). The early and late species of RNA differ both in median size and RNA sequence (Oda and Joklik, 1967). The latter studies indicated that of the viral DNA sequences transcribed at late times about half are also transcribed early. It is not known whether the entire genome is transcribed, whether nontranscribable spacer regions exist, or whether there is strand specificity. Some insight into the regulating factors was obtained by use of metabolic inhibitors. When DNA synthesis was inhibited, the late sequences were not made, suggesting that they were transcribed from progeny DNA (Oda and Joklik, 1967). Late sequences were also not made at the restrictive temperature after infection with a temperature-sensitive mutant blocked in DNA synthesis (Stevenin *et al.*, 1970). The formation of viral DNA may not be sufficient for late RNA synthesis since under certain conditions of arginine deprivation DNA was made but late RNA was not (Obert *et al.*, 1970). When inhibitors of protein synthesis were added at the start of infection only a portion of the early sequences were made, but these were synthesized in a sustained, uncontrolled fashion (Kates and McAuslan, 1967*b*). The latter RNA species were presumably made by the viral cores which do not undergo second-stage uncoating in the absence of protein synthesis. Uncoating does occur in the presence of inhibitors of DNA synthesis, and under these conditions the formation of early RNA gradually declines (Woodson, 1967). A model consistent with present information is that (1) RNA sequences corresponding to 14% of the genome are made by the core, (2) the rate of early RNA synthesis decreases as the DNA is released from the core but additional early sequences are transcribed, and (3) the entire genome is transcribed from progeny DNA. Alternatively, only specific late sequences may be made from progeny DNA and the continuing synthesis of some early RNA may occur by uncoated viral genomes or released parental DNA.

Kates and co-workers (Dahl and Kates, 1970*b*; Polisky and Kates, 1972) have isolated rapidly sedimenting structures containing newly synthesized DNA and protein which synthesized both early and late RNA sequences *in vitro*. The "transcriptional complexes" were not stimulated by addition of DNA and were inhibited by deoxyribonuclease treatment or sonication. Whether a single polymerase is responsible for synthesis of early and late RNA species is not known. Mature or maturing progeny vaccinia virus particles isolated at 5 hours or later after infection contain a "masked" RNA polymerase which can be activated *in vitro* by detergent treatment and is probably not responsible for RNA synthesis *in vivo* until the start of a new round of infection (Kates *et al.*, 1968; Pitkanen *et al.*, 1968; Dahl and Kates, 1970*b*). Another activity which catalyzes the polymerization of ATP and UTP in the presence of a copolymer template has been described in a soluble form at early times after infection, but its relationship to RNA polymerase is unknown (Pitkanen *et al.*, 1968). Increased RNA polymerase activity in fibroma virus-infected cells has also been described (Chang and Hodes, 1968).

Both early and late classes of vaccinia RNA, like most eukaryotic messenger RNA species, contain poly(A) tracts of 100–200 nucleotides at the 3′ terminus (Sheldon *et al.*, 1972*a,b*). The polyadenylate residues attached to early mRNA are probably formed by the virion poly(A) polymerase (Sect. 3.5.3). It is not known whether the same enzyme or an additional one is responsible for adding adenylate residues to late mRNA.

A small percentage of the RNA made by cores *in vitro* and in infected cells at early and late times contains complementary sequences (Colby and Duesberg, 1969; Duesberg and Colby, 1969; Colby *et al.*, 1971). Some of the complementary RNA exists in a heterogeneous, partially double-stranded form. It was suggested that synthesis might occur by convergent or divergent transcription along opposite strands of the DNA. The significance of double-stranded RNA in poxvirus replication is unknown. The possibility that it is responsible for the natural induction of interferon has been disputed (Bakay and Burke, 1972).

Although T-even bacteriophages induce new transfer RNA species and specifically modify old ones, a similar process has not been definitively demonstrated for an animal virus. It has been known for a long time that synthesis of 4 S RNA continues during vaccinia infection (Salzman *et al.*, 1964; Becker and Joklik, 1964). The base composition and hybridization properties of the radioactively labeled 4 S RNA are consistent with it being a mixture of transfer RNAs from uninfected cells and viral RNA species. An intriguing question is whether the viral RNA of this size contains unique sequences and functions as transfer RNA or whether it is heterogeneous and consists of degraded messenger. Klagsbrun (1971) demonstrated that uridine-labeled 4 S RNA from vaccinia-infected HeLa cells eluted from BD-cellulose in the same position as tRNA. He also reported that the tRNA synthesized during infection is very highly methylated. It is not known whether this reflects differences in synthesis, transport, or methylation of RNA. Although Kit *et al.*, (1970) found no alteration in tRNA methylase activity in CV-1 cells infected with vaccinia virus, this was only measured with yeast and *E. coli* tRNA. Clarkson and Runner (1971) reported changes in the reversed phase chromatography elution profiles of certain aminoacylated tRNAs after vaccinia infection. Quantitative changes in aspartyl-tRNA and a qualitative change in phenylalanine tRNA were detected. Hybridization experiments with vaccinia DNA were not performed either by Klagsbrun (1971) or Clarkson and Runner (1971).

The stability of vaccinia or rabbitpox mRNA has been measured by addition of actinomycin D to stop new RNA synthesis. The half life of existing RNA was then measured either functionally from the continued rate of synthesis of viral proteins (McAuslan, 1963*b*; Shatkin *et al.*, 1965; Sebring and Salzman, 1967), or directly by DNA–RNA hybridization (Oda and Joklik, 1967). These studies have indicated that the early mRNA species, including species responsible for synthesis of thymidine kinase and several antigens, are quite stable. In HeLa cells early mRNA has a half life of several hours and late RNA less than 30 min. In L cells early and late mRNA appear equally stable with a half life of 2–3 hours. The possibility that actinomycin D itself affects the stability of the RNA has not been evaluated.

4.4. DNA Replication

Studies concerning the replication of vaccinia DNA have been concerned largely with temporal and spatial matters and relatively little is known regarding the molecular mechanisms involved. The cytoplasmic location of poxvirus DNA replication has been determined by autoradiography with the light microscope (Cairns, 1960; Kato *et al.*, 1960) and electron microscope (Harford *et al.*, 1966; Scherrer, 1968), as well as by cytochemical methods (Kato *et al.*, 1960; Loh and Riggs, 1961). These studies have shown that replication occurs within discrete foci. Although nuclear involvement has been suggested (Walen, 1971), vaccinia DNA is synthesized in enucleated cells, thus ruling out this possibility (Prescott *et al.*, 1971, 1972). Cairns (1960) presented evidence that each infecting particle initiated a separate center of DNA synthesis, that DNA synthesis occurred similtaneously in all centers of each multiply infected cell, and that the process could be synchronized among all cells by increasing the multiplicity of infection. Salzman (1960) added 5-fluorodeoxyuridine at varying times after infection to inhibit thymidylate synthetase and block DNA synthesis and then measured the effect on the final yield of virus. He reasoned that since any virus formed must contain DNA synthesized prior to addition of inhibitor, a time course of DNA replication could be derived. The results indicated that the amount of viral DNA made in 6.5 hours was sufficient for a full yield of virus (Fig. 10A). Using biochemical techniques, Joklik and Becker (1964) determined more directly that cytoplasmic and hence viral DNA synthesis began at about 1.5 hours, reached a maximum rate at 2–2.5 hours, and then decreased sharply (Fig. 10B). More than 90% of the viral DNA was made by 4.5 hours, the time at which formation of virions began. During the subsequent 10 or more hours of virus assembly, viral DNA appeared to be randomly selected from the preformed pool. With all other animal and bacterial viruses studied, DNA synthesis and particle formation are simultaneous events. The basis for the abrupt cessation of poxviral DNA

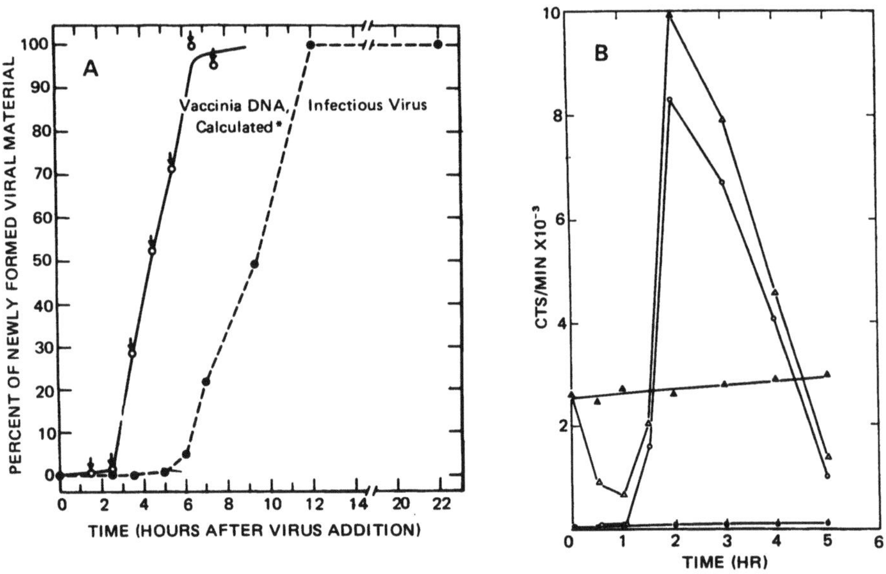

Fig. 10. (A) The rate of formation of vaccinia DNA as determined by inhibition of virus growth with 5-fluorodeoxyuridine. The filled circles indicate the time course of infectious virus formation in an uninhibited control. Each point represented by an unfilled circle indicates the final yield of virus in the inhibited culture when 5-fluorodeoxyuridine was added at the time indicated by the arrow compared to the final yield in the control. From Salzman (1960). (B) The rate of formation of vaccinia DNA as determined by incorporation of ^{14}C-thymidine. Infected and uninfected cells were labeled with ^{14}C-thymidine at the indicated times for 10-minute intervals and then disrupted with a homogenizer. The trichloroacetic acid-precipitable radioactive DNA present in the total cell extract or separated cytoplasmic fraction was determined. (—▲—▲—) Control, total; (—●—●—) control, cytoplasm; (—Δ—Δ—) infected, total; (—O—O—) infected, cytoplasm. From Joklik and Becker (1964).

synthesis is unknown. Possibly it is related to the switch-off of early protein synthesis or the accumulation of late proteins. Experimentally, viral DNA synthesis can be interrupted by addition of an inhibitor of protein synthesis such as puromycin. However, if early proteins are allowed to accumulate in the presence of 5-fluorodeoxyuridine, DNA synthesis may then proceed for a limited time in the presence of puromycin (Kates and McAuslan, 1967*a*). Assuming no important changes in the specific activity of the DNA precursors, Joklik and Becker (1964) estimated that the rate of vaccinia viral DNA synthesis is several times that of the host cell population and that the total amount of viral DNA made is about half that of the host cell. The precise timing of DNA synthesis depends on the virus strain, cell type, and multiplicity. As expected from the longer growth cycles, a greater delay and a more prolonged period of DNA synthesis occurs in cells infected with rabbit fibroma (Ewton and Hodes, 1967), fowlpox (Gafford *et al.*, 1969), and Yaba (Yohn *et al.*, 1970) viruses.

The structure of replicating poxviral DNA has not yet been investigated in spite of the fact that the cytoplasmic site of replication would be a definite advantage in such studies. Joklik and Becker (1964) observed that the parental replicating DNA is associated with large aggregates, possibly identical to the inclusions seen by electron microscopy. The rapid sedimentation results from the association of the DNA with protein since treatment with either pronase or detergent releases DNA molecules which sediment at the same rate as virion DNA (Dahl and Kates, 1970*a*; Polisky and Kates, 1972). No evidence was obtained in these studies for long, concatenated DNA molecules, as has been shown for replicating T4; only vaccinia DNA labeled for 3 hours was examined and further studies with pulse-labeled DNA are needed. The evidence provided by Geshelen and Berns (Sect. 3.4) that the two strands of vaccinia DNA are linked at their termini raises several interesting problems. An important first question that needs to be answered is whether poxviral DNA replication is semiconservative or conservative. Semiconservative replication would require the breaking and sealing of DNA strands.

An increase in the level of several enzymatic activities which may be concerned with DNA replication has been reported to follow vaccinia virus infection. Hanafusa (1961) reported increased thymidine incorporation into DNA and enhancement of deoxyribonuclease activity in cell-free extracts of L cells infected with vaccinia virus. This was soon shown to result from induction of thymidine kinase (Magee, 1962; Kit *et al.*, 1962; McAuslan and Joklik, 1962), DNA polymerase (Green and Piña, 1962), and several deoxyribonucleases (Jungwirth and

Joklik, 1965; McAuslan, 1965; McAuslan *et al.*, 1965). Induction of thymidine kinase and DNA polymerase has also been shown to follow rabbit fibroma infection (Barbanti-Brodano *et al.*, 1968; Chang and Hodes, 1967). Substantial evidence indicates that thymidine kinase is coded by viral DNA. This includes a requirement for protein synthesis for its induction and continued increase in activity (Jungwirth and Joklik, 1965; Kit *et al.*, 1963*b*), differences between the enzymatic and immunological properties of the host and viral enzymes (McAuslan, 1963*b*; Kit and Dubbs, 1965), and, most importantly, both the induction of enzymatic activity in thymidine kinase-negative (TK^-) cells and the isolation of TK^- vaccinia mutants (Kit *et al.*, 1963*c*; Dubbs and Kit, 1964). Under laboratory conditions, thymidine kinase induction is gratuitious, i.e., TK^- vaccinia can grow in TK-negative cells. This, of course, does not exclude the posssibility that induction of the enzyme plays a significant role under certain conditions of growth. Indeed, Shatkin and Salzman (1963) noted that levels of thymidine that were suboptimal for cellular DNA synthesis supported a high rate of viral DNA synthesis. The partially purified enzyme is feedback-inhibited by its distal product, deoxythymidine triphosphate, suggesting that it may be a complex enzyme of the allosteric type (McAuslan, 1967). The presence of subunits was also suggested as a possible explanation for the observation that thymidine kinase activity is lower in cells mixedly infected with TK^+ and TK^-, as compared with TK^+ alone (Munyon and Kit, 1965). According to this model, hybrid enzymes composed of active and inactive subunits would have low activity.

A DNA polymerase has been partially purified from vaccinia-infected cells and appears to differ from enzymes obtained from uninfected cells with regard to chromatographic properties, molecular weight, response to DNA template concentrations, heat stability, and immunological properties (Magee and Miller, 1967; Berns *et al.*, 1969; Citarella *et al.*, 1972). The isolated enzyme was unable to carry out net synthesis of DNA but appeared limited to the repair of gaps in activated DNA primer (Citarella *et al.*, 1972). Although DNA polymerase can bind adventitiously to virions, it is not an integral component (Tan and McAuslan, 1970). Chang and Hodes (1967) have described DNA polymerase activities utilizing native and denatured DNA induced by rabbit fibroma virus.

Polynucleotide ligase, an enzyme capable of sealing single-stranded breaks in duplex DNA, is thought to be required for the repair and replication of prokaryotic and eukaryotic DNA. In addition, T-even phases induce a new ligase with a different cofactor requirement. Sambrook and Shatkin (1969) observed more than a tenfold increase in polynucleotide ligase activity in the cytoplasm of vaccinia-infected cells. This was measured in an *in vitro* assay by the conversion of simian virus 40 component II DNA to component I. Although the ATP cofactor requirement was the same as for the polynucleotide ligase of uninfected cells, induction of the enzyme did not occur in the presence of puromycin, suggesting that it was newly synthesized. Induction occurred in the presence of cytosine arabinoside, an inhibitor of DNA synthesis, indicating that it should be classed as an early enzyme. The activity also increased in the presence of rifampicin, an inhibitor of poxviral assembly (Sect. 6.2). It is possible that this enzyme is involved in the synthesis, repair, or cross-linking of vaccinia DNA.

A variety of DNase activities capable of hydrolyzing native DNA at alkaline *p*H, denatured DNA at neutral *p*H, and denatured DNA at low *p*H have been described in extracts of cells infected with cowpox, rabbitpox, or vaccinia virus (McAuslan, 1965, McAuslan *et al.*, 1965; Jungwirth and Joklik, 1965; Eron and McAuslan, 1966; Jungwirth *et al.*, 1969). Only the DNase active at low *p*H values has been purified (McAuslan and Kates, 1967). It is distinct from a low-*p*H-optimum host DNase and was reported to be an exonuclease. However, the enzyme is very similar to one purified from vaccinia virions (Sect. 3.5.3) which has both exonucleolytic and endonucleolytic activities. The neutral DNase has been described as an endonuclease and is also present in the viral core but has not been purified (Sect. 3.5.3). It is not known whether any of the DNases are involved in the processing or replication of poxviral DNA. Since the alkaline DNase is an early enzyme and is not a virion component, it is a candidate for such a function. Pogo and Dales (1973) have suggested a role in inhibition of host DNA synthesis for the neutral virion-associated DNase.

It is likely that many additional enzymes necessary for DNA replication will be found. Synthesis of vaccinia DNA requires continuing protein synthesis, and requirements for undefined protein factors have been based on inhibitor studies (Kates and McAuslan 1967*a*; Bedson and Cruikshank, 1968). Studies with a temperature-sensitive mutant also indicated that at the nonpermissive temperature DNA synthesis was inhibited despite the induction of a DNA polymerase that was not temperature-sensitive (Braunwald *et al.*, 1969, 1970).

Recombination has been demonstrated between vaccinia and rabbitpox viruses (Fenner, 1959), between white pock (*u*) mutants of rabbitpox virus (Gemmel and Fenner, 1960; Sambrook *et al.*, 1965), between temperature-sensitive mutants of rabbitpox (Padgett and Tomkins, 1968), but not between poxviruses belonging to different subgroups (Woodroofe and Fenner, 1960). Scoring of pairwise crosses as positive or negative on the basis of appearance of wild-type pocks allowed the linear arrangement of *u* mutants. The temperature-sensitive mutants of rabbitpox could not be arranged in a map based on two-factor crosses, and the use of three-factor crosses was suggested. More complete discussions of conditional lethal mutants of animal viruses have been written by Fenner (1969) and Ghendon (1972).

4.5. Regulation of Protein Synthesis

The initiation of early and late viral protein synthesis appears to be under transcriptional control. Thus, only early proteins are made when synthesis of progeny DNA and late RNA are prevented. The burst of early RNA, which occurs almost immediately after vaccinia infection, ensures the prompt initiation of viral protein synthesis. The high rate of early polypeptide synthesis was demonstrated by quantitative immunoprecipitin studies (Shatkin 1963*a*; Salzman and Sebring, 1967). Three to five early antigens, which include some structural proteins, have been identified by immunodiffusion (Appleyard and Westwood, 1964; Cohen and Wilcox, 1966, 1968). Temporal studies are facilitated by pulse-labeling infected cells with radioactive amino acids followed by immunodiffusion analysis (Salzman and Sebring, 1967; Moss and Salzman, 1968). Radioautographs of this type of experiment are shown in Fig. 11. Early proteins can also be detected by labeling with amino acids and by radioautographic analysis of polyacrylamide gels (Moss and Salzman, 1968; Katz and Moss, 1970*b*; Esteban and Metz, 1973*a*). The latter approach is possible because of the rapid inhibition of host protein synthesis (Sect. 5.2). Similarly, by labeling with radioactive sugars, several early virus-induced glycoproteins were detected (Moss *et al.*, 1971*a*). The latter are distinct from a virion glycoprotein which is made at late times (Moss *et al.*, 1973*a*).

The induced enzymes made at early times include thymidine kinase, DNA polymerase, alkaline deoxyribonuclease, and polynucleotide ligase (Sect. 4.4). It is significant that these enzymes are not virion components, and are all involved in the synthesis or modification of DNA. Experiments interpreted as indicating that thymidine kinase is coded by sequences transcribed by intact cores and that DNA polymerase is coded by sequences transcribed from uncoated parental DNA have been reported (Kates and McAuslan, 1967*b*). In contrast,

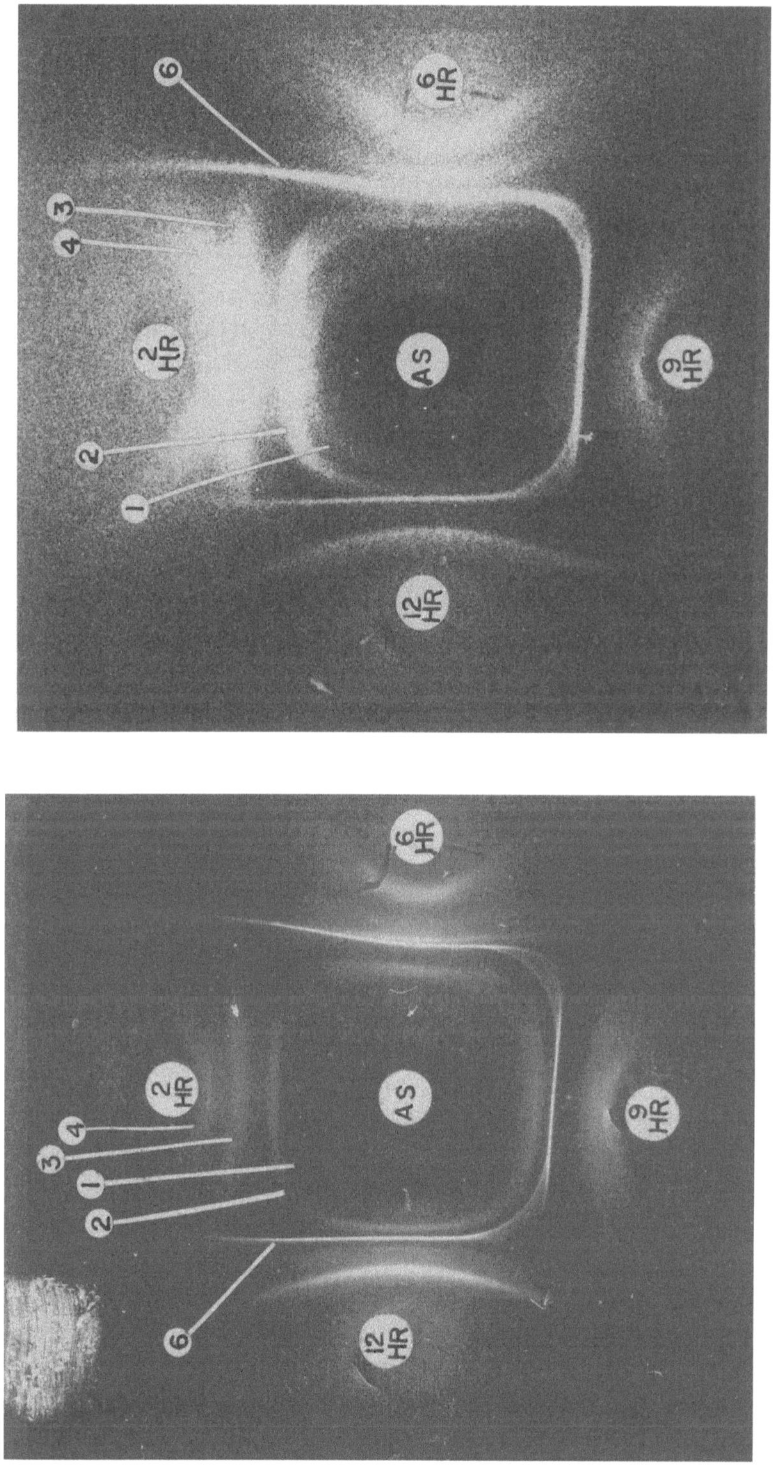

Fig. 11. Temporal synthesis of vaccinia virus-induced antigens. Vaccinia virus-infected cells were exposed to ^{14}C-phenylalanine for 1-hour periods at the indicated times and then disrupted by sonic vibrations. The soluble proteins were placed in the outer wells; antiserum prepared from infected rabbits was placed in the center well of an agar-coated slide. (Left) Stained immunoprecipitin lines. (Right) Radioautograph of immunoprecipitin lines. From Salzman and Sebring (1967).

RNA polymerase (Kates *et al.*, 1968; Pitkanin *et al.*, 1968), acid deoxyribonuclease (McAuslan and Kates, 1967), and nucleoside triphosphate phosphohydrolase (Pogo and Dales, 1969), which are all virion components, appear to be late proteins. It should be noted, however, that the latter studies only indicate the times at which each activity was detected with an *in vitro* assay and not necessarily the time at which constituent polypeptides were formed.

Pulse-chase experiments indicate that most of the structural protein is made at late times after infection (Salzman and Sebring, 1967). Four of the virion proteins, at least three of which are components of the viral core, were labeled during the first 1–2 hours after infection and were subsequently incorporated into virions (Holowczak and Joklik, 1967*b*; Katz and Moss, 1970*b*; Moss *et al.*, 1973*a*). The remaining structural proteins were labeled starting at 3 hours post-infection. Results obtained by SDS-polyacrylamide gel electrophoresis followed by SDS-hydroxyapatite chromatography indicating that polypeptides 1a, 1b, 2a, and a component of the 6 complex designated HE are early structural proteins are shown in Fig. 12 (Moss *et al.*, 1973*a*). SDS-hydroxyapatite chromatography was used to separate the polypeptides of the 6 complex into four components.

The mechanisms involved in the cessation of early protein synthesis are not well understood. Synthesis of early enzymes (McAuslan, 1963*a*,*b*; Jungwirth and Joklik, 1965; McAuslan and Kates, 1967) and some early proteins (Salzman and Sebring, 1967; Moss and Salzman, 1968; Esteban and Metz, 1973*a*) appears to be almost completely terminated several hours after infection, while synthesis of early structural proteins declines gradually (Holowczak and Joklik, 1967*b*; Moss *et al.*, 1973*a*). The termination of early protein synthesis occurs despite the intrinsic stability of early mRNA and the continued synthesis of early RNA sequences (Sect. 4.3). Simple competition by late mRNA is not thought to account for the relatively small amount of early sequences in late polyribosomes (Oda and Joklik, 1967). The switch-off of early enzyme and antigen synthesis fails to occur when DNA synthesis is prevented by inhibitors or when UV-irradiated poxvirus is used for infection (McAuslan, 1963*a*,*b*; McAuslan and Kates, 1967; Jungwirth and Joklik 1965; Salzman and Sebring, 1967). In a series of elegant experiments McAuslan (1963*b*, 1969*a*) presented evidence for translational control of thymidine kinase. Essentially he found that repression of thymidine kinase formation is not established in the presence of inhibitors of RNA or protein synthesis (Fig. 13), and he therefore postulated a repressor protein. Inhibitor studies also suggest that proteins made at 30–60 minutes after infection might be

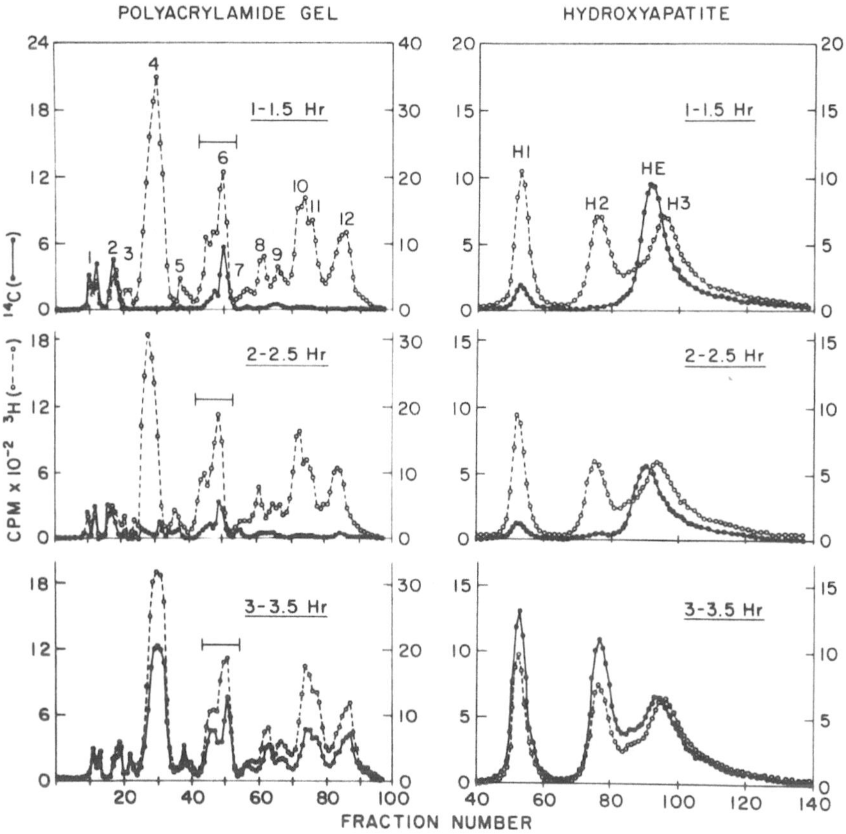

Fig. 12. Identification of early and late vaccinia virus structural proteins. Progeny virus was purified from cells labeled with ^{14}C-phenylalanine during specific 30-minute periods of the growth cycle and from cells labeled for 7 hours with ^{3}H-phenylalanine. The viral proteins were first separated by SDS-polyacrylamide gel electrophoresis and then the polypeptide 6 complex (indicated by brackets) was further analyzed by SDS-hydroxyapatite chromatography. From Moss *et al.* (1973*a*).

necessary for subsequent viral protein synthesis (Dales, 1965*b*, Moss and Filler, 1970). Direct evidence for translational control will undoubtedly require cell-free protein synthesizing systems. Cytoplasmic extracts from infected cells prepared at early and late times incorporated amino acids into the appropriate viral proteins (Katz and Moss, 1969; Moss and Katz, 1969); however, attempts have not yet been made to use purified ctyoplasmic RNA as an *in vitro* messenger, as has been done in phage systems. Vaccinia RNA made *in vitro* from cores stimulated amino acid incorporation by a cell-free system from uninfected cells (Beaud *et al.*, 1972; Fournier *et al.*, 1973). Discrete polypeptide bands, some of which correspond, to virus-induced pro-

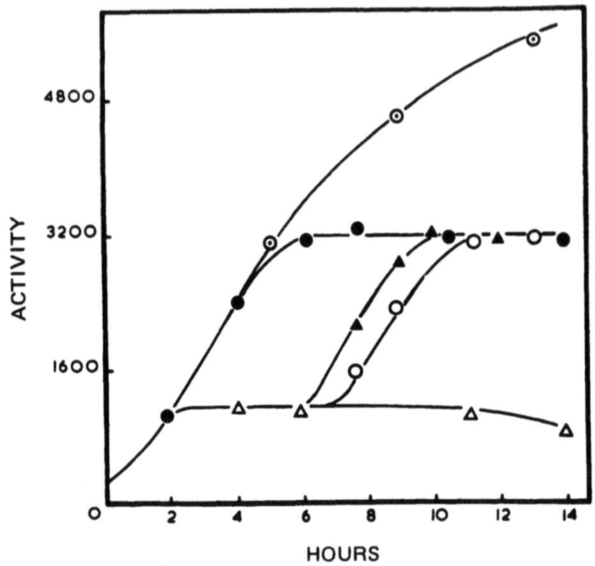

Fig. 13. Effects of inhibitors of RNA or protein synthesis on the repression of thymidine kinase formation in cowpox-infected cells. (●) Thymidine kinase activity in untreated cowpox-infected cells; (○) activity after addition of actinomycin D at 2 hours; (△) activity after addition of puromycin at 2 hours; (▲) activity after removal of puromycin inhibition at 5.5 hours; (○) activity after addition of actinomycin 30 minutes before removal of puromycin. From McAuslan (1963*b*).

teins have been detected by radioautography. Presumably, the core RNA should code for thymidine kinase, but assays for this enzyme were apparently not made.

4.6. Post-Translational Modification

Several types of post-translational modification of vaccinia viral proteins, including cleavage, glycosylation, and phosphorylation, have been reported. Formation of viral polypeptides from higher-molecular-weight precursors was demonstrated first for poliovirus and other small RNA viruses (Jacobson *et al.*, 1970). At least three major late proteins, comprising more than a third of the protein mass of vaccinia virions, are formed by cleavage of higher-molecular-weight precursors (Katz and Moss, 1970*a,b*; Pennington, 1973; Moss and Rosenblum, 1973). Evidence for such a process came from electrophoretic analysis of cytoplasmic polypeptides obtained after a short pulse with a radioactive amino acid and then after a chase. Radioautographs in Fig. 14a show the formation of polypeptides 4a, 4b, and 10 from precursors. The half life of the precursors is about 1–2 hours. Several lines of evidence suggested that cleavage and viral morphogenesis are related. Rifampicin, a drug shown to interrupt morphogenesis (Sects. 4.7, 6.2), did not prevent the formation of the precursors, but it blocked the cleavages needed to form core polypeptides 4a and 4b (Fig. 14b). Rifampicin did not prevent the formation of polypeptide 10, which is not a component of the core. Use of the drug provided a unique way to accumulate sufficient amounts of the core precursors for structural analysis. Tryptic peptide maps have demonstrated that P4a (105,000 daltons) is the precursor of 4a (68,000 daltons) and that P4b (74,000 daltons) is the

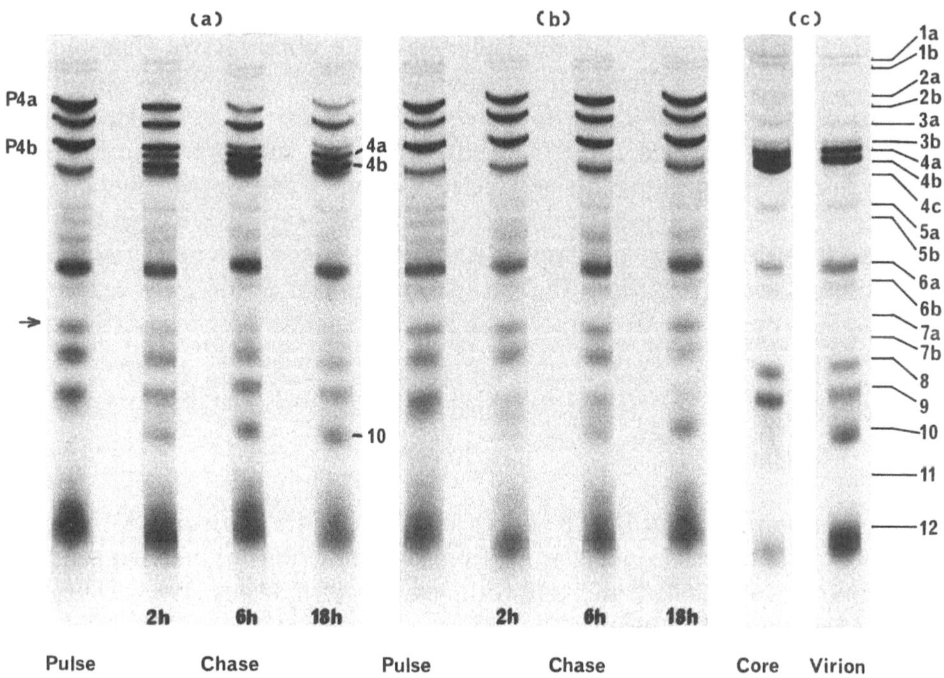

Fig. 14. Formation of vaccinia virus polypeptides from higher-molecular-weight precursors. HeLa cells were infected in the presence (b) or absence (a) of rifampicin and incubated for 20 minutes with ^{35}S-methionine at 6 hours after infection. Cytoplasmic proteins obtained at the conclusion of the labeling period and at intervals thereafter were analyzed by polyacrylamide gel electrophoresis and radioautography. Radioautographs of core and total virion proteins are also shown (c). The gradual disappearance of P4a, P4b, and the polypeptide indicated by an arrow and the appearance of 4a, 4b, and 10 occur during the chase period in the absence of the drug rifampicin. Rifampicin prevents the cleavage of P4a and P4b and the formation of 4a and 4b. From Moss and Rosenblum (1973).

precursor of 4b (62,000 daltons). These results are shown in Fig. 15. (Moss and Rosenblum, 1973). All peptides from the products are also obtained by digestion of the precursors. In addition, P4a contains many more peptides than 4a, as expected from the large differences in their molecular weights. Nothing is yet known concerning the proteolytic enzymes or the fates of the smaller fragments produced by cleavage. Maturational cleavages occur during assembly of at least 3 DNA bacteriophages, T4 (Laemmli, 1970), lambda (Murialdo and Siminovitch, 1972), and T5 (Zweig and Cummings, 1973), and have recently been demonstrated for adenovirus (Anderson *et al.*, 1973).

The virion and nonvirion glycoproteins made after vaccinia virus infection differ in sugar composition and may be glycosylated at different intracellular sites. The finding that the virion glycoprotein can be labeled with glucosamine but not fucose, galactose, or mannose (Garon and Moss, 1971) is consistent with the observations that the viral envelope is not formed in association with cell membranes which contain glycosyltransferases for the latter sugars. In contrast, the nonvirion glycoprotein, which bands with cell membranes during isopycnic centrifugation, can be labeled with all 4 sugars (Moss *et al.*, 1972). The effects of poxvirus infection on the level and location of glycosyltransferases are not known.

Speculations regarding the phosphorylation of virion polypeptide 12 by the core-associated protein kinase were made previously (Sect. 3.5.3).

4.7. Morphogenesis

Information regarding poxvirus morphogenesis has come largely from electron microscopic examination of thin sections of infected cells (Morgan *et al.*, 1954; Dales and Siminovich, 1961; Dales, 1963; Dales and Mosbach 1968). Within 2–3 hours after a synchronous infection, granular or fibrous electron-dense regions of the cytoplasm, referred to as factories or viroplasm, are evident. These areas, which are the sites of DNA synthesis, do not contain the usual cytoplasmic organelles and are clearly demarcated from the surrounding cytoplasm. Within 3 hours, arcs and circles, composed of membranes with distinct coating on the convex surface, appear. Serial sectioning (Morgan *et al.*, 1955), high-voltage electron microscopy of thick sections (Grimley, 1971*a,b*), and freeze-etching (Easterbrook and Rozee, 1971; Easterbrook, 1972) indicate that in three dimensions the arcs and circles would appear as cupules and spheres. These viral membranes are unique for their lack

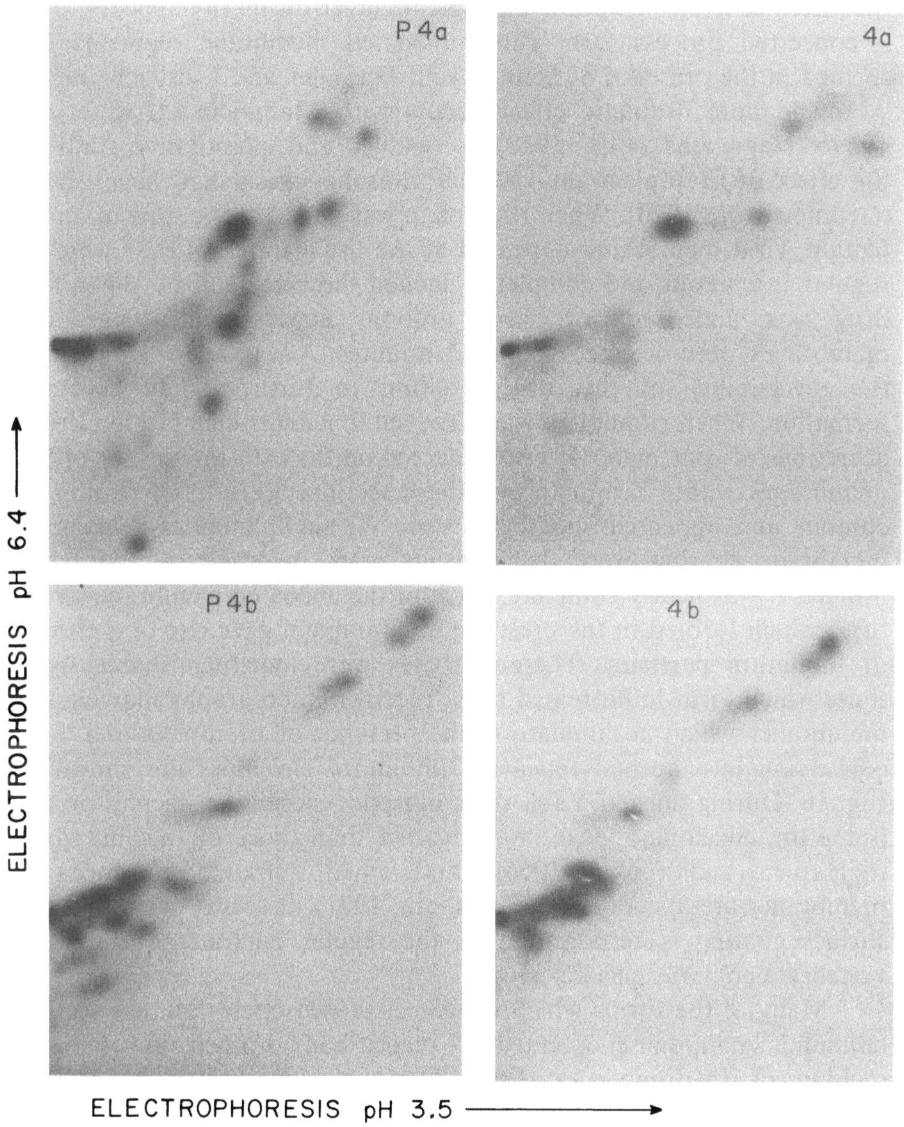

Fig. 15. Tryptic peptide analysis of two vaccinia virus core proteins and their precursors. Polypeptides P4a and P4b, labeled with ^{35}S-methionine, were purified by polyacrylamide gel electrophoresis from the cytoplasm of cells infected with vaccinia virus in the presence of rifampicin. Polypeptides 4a and 4b were purified from ^{35}S-methionine-labeled virions. All proteins were then oxidized with performic acid, digested with trypsin, and analyzed by two-dimensional thin-layer electrophoresis. Radioautographs are shown. From Moss and Rosenblum (1973).

of obvious continuity with cellular membranes. Although formation of the vaccinia viral membrane and spicule layer ordinarily appears to be a concerted process, very short uncoated membrane segments are formed in the presence of actinomycin D (Dales and Mosbach, 1968). A much more dramatic effect occurs with rifampicin (Moss *et al.*, 1969*b*; Nagayama *et al.*, 1970). A detailed and quantitative study of the effect of rifampicin on vaccinia morphogenesis has been made (Grimley *et al.*, 1970). When rifampicin was added at the start of an infection, viral membranes appeared at the usual time but they were irregular in contour and completely lacked the spicule coat. When this drug was added during virus growth, newly formed uncoated membranes were detected within 15 minutes. Advantage was taken of the reversibility of this drug's action to further study envelope formation. When rifampicin was removed in a controlled fashion, focal accretions of coat material were detected on the exterior surface of the membranes within 1 minute and these sections were more regular in contour and appeared slightly curved. Within 2 minutes regions of membrane coating were detected on most membranes, and by 5 minutes it was nearly complete. Each of the uncoated membrane structures which formed in the presence of rifampicin gave rise to a cluster of immature particles. These particles were indistinguishable from those which form in untreated cells. Electron micrographs showing the membranes which accumulate in the presence of rifampicin and their conversion into normal-appearing immature envelopes are shown in Fig. 16. During normal Yaba virus morphogenesis, the spicules on the immature envelope appear more distinct than those of vaccinia virus (deHarven and Yohn, 1966), and small micelles of uncoated membranes are also present (Tsuruhara, 1971). It seems likely that the surface coating is responsible for the regular contour of immature vertebrate poxvirus envelopes.

Many of the forms which appear as closed circles on thin section contain a small, dense, eccentrically placed body or nucleoid composed of filaments (Morgan *et al.*, 1954). One such particle is visible in Fig. 16A. Thick, 1-μm sections, examined by high-voltage electron microscopy, indicate that the nucleoids are curved, probably cupular, structures (Grimley, 1971*a,b*). Much larger crystalloid bodies with a fine structure similar or identical to that of nucleoids are sometimes found in the viroplasm during vaccinia (Morgan *et al.*, 1954), Yaba (deHarven and Yohn, 1966), and rabbit fibroma (Scherrer, 1968) virus infections. They are particularly abundant when vaccinia virus morphogenesis is interrupted by rifampicin (Nagayama *et al.*, 1970; Grimley *et al.*, 1970). In 1-μm sections the crystalloids appear as

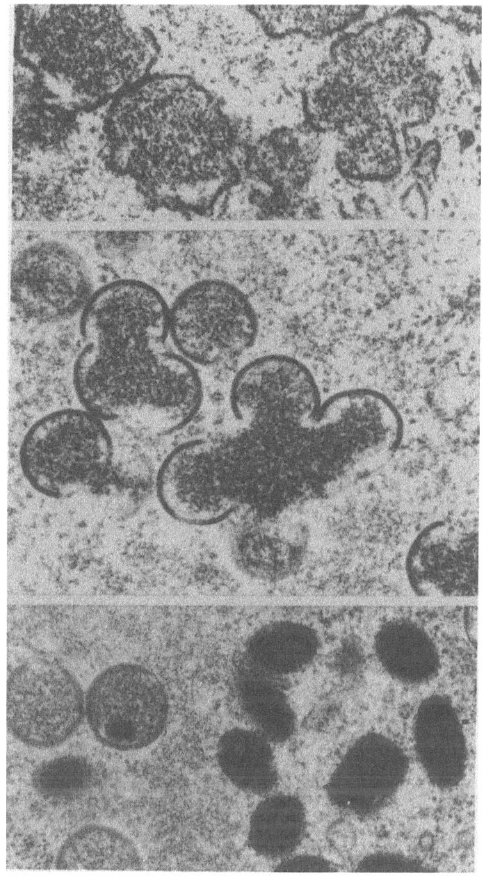

Fig. 16A. Electron micrographs of immature vaccinia virus particles. Thin sections were prepared from cells at 8 hours after infection in the presence of rifampicin (top) and at 10 minutes (middle) and 4 hours (bottom) after removal of the drug. From Moss *et al.* (1969*b*).

laminated sheets which are irregularly bent or folded (Grimley, 1971*b*). Evidence that the nucleoids contain DNA comes from labeling of the morphologically identical but larger crystalloids with ^{3}H-thymidine (Scherrer, 1968; Nagayama *et al.*, 1970; Grimley *et al.*, 1970). As expected, immature particles containing nucleoids do not form when inhibitors of DNA synthesis are used (Rosenkranz *et al.*, 1966; Pogo and Dales, 1971). Envelopes lacking nucleoids eventually develop in the presence of such inhibitors, but even their formation is greatly delayed under these conditions.

Further maturation appears to occur within the immature viral envelope although it is difficult to reconstruct in detail all the necessary transitional forms (Dales and Mosbach, 1968). The synchronization of virus development obtained by removal of rifampicin is particularly useful for such studies (Grimley *et al.*, 1970). Images suggesting that the biconcave core forms by modeling of the nucleoid can be seen by

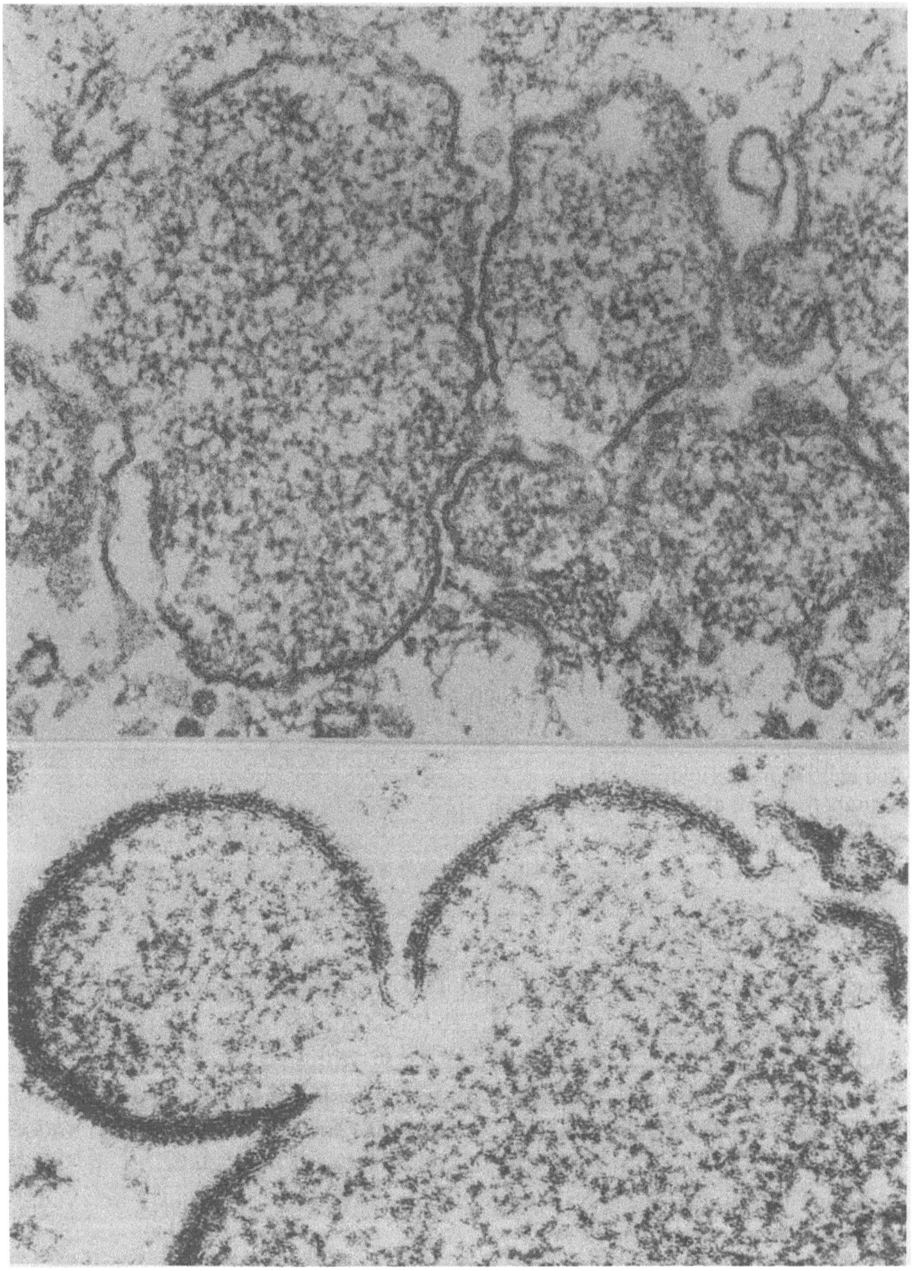

Fig. 16B. Electron micrographs of immature vaccinia virus particles. Higher magnification than in Fig. 16A of viral envelopes formed in presence of rifampicin (top) and 10 minutes after rifampicin was removed (bottom). Note the residual loops of uncoated membrane in the lower photograph. From Moss *et al.* (1969*b*) and Grimley *et al.* (1970).

high-voltage electron microscopy of thick sections (Grimley, 1971*b*). The maturing virus particles appear to migrate away from the viroplasm. During development of many poxviruses, including cowpox, ectromelia, and fowlpox but not vaccinia, they become occluded in dense, proteinaceous masses distinct from the viroplasm and called A-type occlusions (Kato *et al.*, 1963; Ichihashi and Matsumoto, 1966). Integration of virus particles into these structures is a separate genetic property which is independent of their formation (Ichihashi and Matsumoto, 1969). Electron microscopy studies indicate that the insect poxviruses also form by a process of envelope formation followed by internal differentiation (Bergoin *et al.*, 1969; Granados and Roberts, 1970; Fig. 17).

Nucleoprotein complexes are the first structures formed that are identifiable by biochemical methods. Polisky and Kates (1972) isolated nucleoprotein complexes which contain two principal polypeptides from vaccinia virus-infected cells. Sarov and Joklik (1973) identified and characterized three additional particles, one of which lacked DNA and contained relatively few polypeptides. None of these structures formed

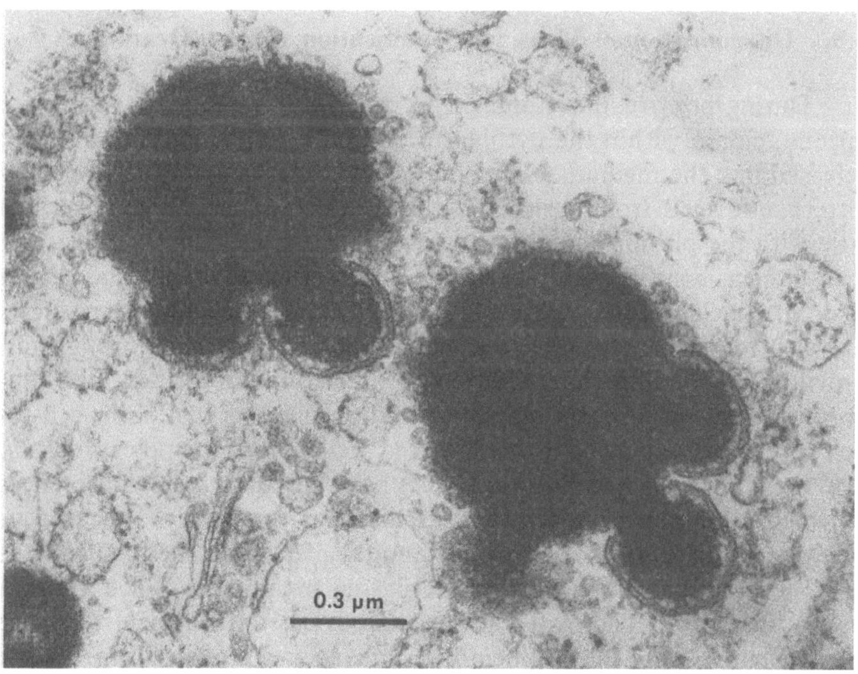

Fig. 17. Electron micrograph of developing insect poxvirus originally isolated from larvae of *Amsacta moorei*. From Granados and Roberts (1970).

in the presence of inhibitors of DNA synthesis or when assembly was interrupted by rifampicin. Identification of these structures as true intermediates has not yet been established by kinetic experiments. Katz and Moss (1970*b*) noted that certain polypeptides changed from a soluble to a particulate form within minutes after the removal of rifampicin. Synchronization of assembly obtained in this manner should be useful for the isolation of intermediate structures.

The formation of several of the major structural polypeptides by cleavage of higher-molecular-weight precursors was described in a preceding section (Sect. 3.6). What role this process plays in assembly is not known. One possibility is that "polypeptide tails" on the structural proteins either enhance their solubility or prevent their self-assembly until correctly positioned at the site of virion formation where cleavage then occurs. The peculiar mode of formation of poxviruses within a membrane barrier may explain the presence of certain proteins within the mature virions. Enzymes such as the protein kinase, nucleoside triphosphate phosphohydrolases, and deoxyribonucleases might function during maturation and be retained within the finished particle.

4.8. Dissemination of Virus and Modification of Host Membranes

During poxvirus development in cell cultures most of the mature virions remain within the cytoplasm and only very small amounts are released into the medium. Some mature particles become enveloped by cisternae formed from vesicles of the Golgi complex. These may then fuse with the plasma membrane, creating a channel for the release of virus which is still enveloped by the inner cisternal membrane (Dales and Siminovich, 1961; Ichihashi *et al.*, 1971). Unenveloped particles are also found near the plasma membrane and may exit through microvilli. The finding that addition of antiserum soon after virus adsorption does not prevent plaque formation suggests that some cell-to-cell spread may also occur (Nishmi and Keller, 1962). Appleyard *et al.* (1971) and Turner and Squires (1971) have noted antigenic differences between rabbitpox or vaccinia virus that was "naturally" released or released by disruption of cells. The antigen present on the former virus may be derived from the modified plasma membrane. Since antisera produced against inactivated intracellular virus did not neutralize the extracellular virus, the implications for vaccine production are quite important.

Cell-surface alterations induced by poxviruses have been demonstrated by a variety of methods. Immune hemadsorption, fluorescent-antibody, and ^{125}I-labeled-antibody techniques reveal the presence of viral antigens (Ueda *et al.*, 1969; Miyamoto and Kato, 1971; Hayashi *et al.*, 1972). In addition, vaccinia-infected cells become agglutinatable with the plant lectin concanavalin A (Zarling and Tevethia, 1971; Bandlow *et al.*, 1973). Moss *et al.* (1971*a*) have suggested that the nonvirion glycoproteins induced after vaccinia infection may be associated with cell membranes. All of the changes mentioned in this paragraph are early functions and thus may represent common membrane changes analyzed in different ways. It is not known whether these changes are involved in virus dissemination; if so it is not clear why they should occur so early in infection.

5. EFFECTS ON HOST CELL METABOLISM

5.1. Cytopathic Effects

Poxviruses produce extensive cytopathic effects which usually lead to cell death. Rounding of cells is detected within a few hours and is an early viral function (Bernkopf *et al.*, 1959; Brown *et al.*, 1959; Hanafusa, 1960; Appleyard *et al.*, 1962) which requires viral protein synthesis (Bablanian, 1968, 1970). Ball and Medzon (1973) have demonstrated a shift in the distribution of cells in Ficoll density gradients within 2 hours after infection. It may be useful to try to correlate these early cytopathic effects with the early plasma membrane changes described in Sect. 4.8. Cell fusion is a later cytopathic effect which occurs after infection with certain poxviruses (Appleyard *et al.*, 1962; Kaku and Kamahora, 1964).

5.2. Macromolecular Synthesis

Biochemical studies indicate an inhibition of host protein, RNA and DNA synthesis after infection. The effect on cell protein synthesis can be measured by the effect on host enzyme synthesis (Kit *et al.*, 1964; Dubbs and Kit, 1964) by determining the amount of newly synthesized protein not precipitated by virus-specific antiserum (Salzman and Sebring, 1967) and most sensitively by direct examination of newly synthesized polypeptides by gel electrophoresis (Moss and Salzman, 1968; Moss, 1968; Esteban and Metz, 1973*a*). Inhibition of host protein synthesis can be established in the presence of actinomycin D (Shatkin, 1963*b*; Moss, 1968), cycloheximide (Moss 1968), or cordycepin (Esteban and Metz, 1973*a*), suggesting that *de novo* viral RNA or protein synthesis is not needed. The rate of inhibition is dependent on the multiplicity of infection, may be essentially complete in 30 minutes, and is associated with a decrease in size and breakdown of host polyribosomes (Moss, 1968). It is possible that a component of the virus particle is directly responsible for the effect on host protein synthesis, but proof is lacking.

Inhibition of nuclear RNA synthesis occurs within several hours, but transport of nuclear RNA into the cytoplasm is blocked in 2–3 hours (Salzman *et al.*, 1964; Becker and Joklik, 1964). Jefferts and Holowczak (1971) reported that processing of precursor ribosomal RNA and maturation of nucleoprotein particles is slowed within 2 hours. This may result from inhibition of ribosomal protein synthesis which occurred earlier.

Reduction in nuclear DNA synthesis has been reported to follow infection with many poxviruses (Kit *et al.*, 1963*a*; Kato *et al.*, 1964; Jungwirth and Launer, 1968). The latter workers took advantage of the more rapid renaturation of vaccinia DNA to separate it from cell DNA for their measurements. Inhibition of host DNA synthesis occurs within a few hours after vaccinia infection, even when viral DNA synthesis is prevented or viral infectivity is reduced by ultraviolet irradiation. Although nuclear DNA is not converted to acid-soluble products and transferred to the cytoplasm in large amounts (Sheek and Magee, 1961), some viral re-utilization of the nuclear DNA may occur under certain conditions (Walen, 1971; Oki *et al.*, 1971). Oki and co-workers (1971) showed that this was a result of intranuclear irradiation caused by ^{3}H in the prelabeled cell. Re-utilization is not detected when cells are labeled with ^{14}C and does not occur under any circumstances in some cell lines.

5.3. Cell Proliferation

Under certain conditions many poxviruses appear to stimulate cell proliferation. This is particularly evident in the case of rabbit fibroma (Shope, 1932*a,b*) and Yaba (Bearcroft and Jamieson, 1958; Andrewes *et al.*, 1959) viruses, which produce benign tumors in rabbits and monkeys, respectively. Active virus replication occurs in the cells composing the tumors which spontaneously regress by necrosis and without evidence of malignant transformation. The cellular foci have been considered to form by aggregation rather than proliferation (Israeli and Sachs, 1964; Verna, 1965; Israeli, 1966). Mitotic cells are not evident in such foci and nuclear DNA synthesis is suppressed (Kato *et al.*, 1965, 1966; Scherrer, 1968; Takehara, 1970). Nevertheless, cell proliferation and stimulation of host DNA synthesis have been described in cells persistently infected with fibroma virus (Hinze and Walker, 1964, 1971) and as a transient phenomenon after fowlpox infection (Cheevers *et al.*, 1968; Gafford *et al.*, 1972). Koziorowska *et al.* (1971) reported the occurrence of morphological changes indicative of cell transformation in a high percentage of mouse embryo cell cultures persistently infected with vaccinia virus. The mechanisms of these effects are not known.

6. INTERFERENCE WITH POXVIRUS REPLICATION

6.1. Antiviral Substances

6.1.1. Thiosemicarbazones

Growth of poxviruses may be inhibited by many substances which are general inhibitors of macromolecular synthesis. These drugs, while not selective antiviral substances, nevertheless have proven extremely valuable in studying poxvirus replication, as indicated in previous sections. In contrast isatin β-thiosemicarbazone (IBT) and its derivatives are specific inhibitors of poxvirus growth (Thompson *et al.*, 1953*a,b*; Sheffield *et al.*, 1960; Bauer, 1961) and have found clinical use in the prophylaxis of smallpox and treatment of complications of vaccination. Bauer and Sadler (1960) have carried out extensive studies on the relationship of thiosemicarbazone structure and antipoxviral activity. Evidence that the action of IBT is indeed specific comes from the isolation of both resistant (Appleyard and Way, 1966; Ghendon and Chernos, 1972; Katz *et al.*, 1973*b*) and dependent (Katz *et al.*, 1973*b*) vaccinia mutants. DNA and some viral antigens form, indicating that the drug effects a late step in virus growth (Bach and Magee, 1962; Easterbrook, 1962; Magee and Bach, 1965; Appleyard *et al.*, 1965). Mature virus particles were not detected by electron microscopy, although immature forms may (Easterbrook, 1962) or may not (Harford *et al.*, 1972) be present. Appleyard and co-workers (1965) reported that IBT prevented or greatly reduced the synthesis of at least five late viral antigens. This was consistent with the finding that early and late viral messenger RNA species formed but that the functional half life of the latter was shortened by IBT (Woodson and Joklik, 1965). In contrast, Katz and co-workers (1973*a*) utilizing SDS-polyacrylamide gel electrophoresis determined that both early and late vaccinia proteins were made and processed in the presence of IBT. Further work with this drug will be required to elucidate its precise mode of action. Evidence has been obtained in some systems for an action of IBT which is dependent on its ability to chelate metals [review, Levinson (1973)].

6.1.2. Rifampicin

Rifamycins are broad-spectrum antibiotics which inhibit the replication of bacteria by binding to DNA-dependent RNA polymerase and blocking transcription [review, Wehrli and Staehelin (1971)]. Rifampicin, a member of this class of antibiotics, does not inhibit mammalian RNA polymerases, but at 50–100 μg/ml it prevents poxvirus growth (Heller *et al.*, 1969; Subak-Sharpe *et al.*, 1969). As in the case of IBT, the isolation of rifampicin-resistant mutants (Subak-Sharpe *et al.*, 1969) provided convincing evidence for the virus-specific nature of this effect. The antipoxviral action of rifampicin is still incompletely understood. Subak-Sharpe and co-workers (1969) proposed originally that the drug inhibits poxvirus RNA synthesis. In agreement with this, Pogo (1971) reported a 40% reduction in uridine incorporation in cells infected with wild-type but not mutant vaccinia virus. Although the RNA polymerase activity of intact vaccinia virus cores is insensitive to rifampicin (Moss *et al.*, 1969*a*; McAuslan, 1969*b*), Pogo (1971) found that the *in vitro* RNA polymerase activity of a particulate cytoplasmic fraction isolated 6–8 hours after infection was sensitive to rifampicin. In contrast, some other investigators found that, under conditions in which rifampicin blocks poxvirus growth, viral DNA and early and late species of RNA (Moss *et al.*, 1969*a,b*; McAuslan, 1969*b*; Ben-Ishai *et al.*, 1969) and all early and late proteins detectable by gel electrophoresis (Moss *et al.*, 1969*a,b*, 1971*c*) are made. Formation of virions (Moss, 1969*a*) and some virion-associated enzymes (McAuslan, 1969*b*; Nagayama *et al.*, 1970) are not detected. Clearly, if there is an effect on transcription in this system the result is quite different than in bacteria and only certain RNA species are not formed. No DNA–RNA hybridization experiments to test this possibility have yet been reported and the RNA polymerase has not yet been solubilized. The nucleoside triphosphate phosphohydrolase and the poly(A) polymerase which have been purified are not inhibited by rifampicin (Paoletti and Moss, 1974; E. N. Rosenblum and Moss, unpublished).

Some of the most striking *in vivo* effects of rifampicin are on virus morphogenesis (Sect. 4.6) and processing of structural proteins (Sect. 4.5). Initially Moss and co-workers (Moss *et al.*, 1969*a,b*, 1971*a,b*; Katz *et al.*, 1970; Grimley *et al.*, 1970) and subsequently others (Tan and McAuslan, 1970, Pennington *et al.*, 1970) have considered that the primary effect of rifampicin on poxviruses is related to assembly or morphogenesis, and that other effects are secondary. This idea is consistent with the demonstrated synthesis of viral DNA, RNA, and proteins, the specific and total block at a unique stage in viral envelope formation, and the rapid reversal of the effect on viral envelope formation even in the presence of inhibitors of RNA and protein synthesis. Direct proof of this hypothesis, however, is lacking. A more detailed discussion of the effects of rifamycins on poxvirus replication may be found in a recent review (Moss, 1973).

Rifampicin does not appear to be a therapeutically useful antiviral agent (Moshkowitz *et al.*, 1971). However, in contrast to the effect on bacteria, relatively small changes in one of the side chains of rifampicin have a marked effect on antipoxviral activity (Subak-Sharpe *et al.*, 1969; Zakay-Rones and Becker, 1970; Pennington and Follett, 1971; Grimley and Moss, 1971; Moss *et al.*, 1972), and it is possible that analogues better suited for clinical use will be found. Despite some earlier claims to the contrary by Thiry and Lancini and co-workers (Thiry and Lancini 1970; Lancini *et al.*, 1971), the macrocyclic ring structure of rifampicin appears to be essential for activity (Follett and Pennington, 1971; Pennington and Follett, 1971; Grimley and Moss, 1971).

Although rifampicin does not inhibit the core-associated RNA polymerase of vaccinia virions, Szilagyi and Pennington (1971) found that certain other derivatives were inhibitory. Since the enzyme activity of rifampicin-resistant mutants were inhibited to the same extent as that of wild-type virus, these effects on core RNA synthesis were considered to be distinct from the specific antiviral effect of rifampicin in infected cells. Many of the rifamycin derivatives studied by Szilagyi and Pennington (1970) are particularly toxic to tissue culture cells and may inhibit cellular polymerases as well.

6.1.3. Interferons

Interferons are species-specific proteins with broad-spectrum antiviral properties, and no attempt will be made here to review the vast literature concerning them. In interferon- or poly(I):poly(C)-treated cells, uncoating and synthesis of both early and late poxvirus proteins are inhibited (Ghosh and Gifford, 1965; Ohno, 1967; Bodo and Jungwirth, 1967; Levine *et al.*, 1967; Magee *et al.*, 1968; Bodo *et al.*, 1972; Hiller *et al.*, 1973). Studies by Joklik and Merigan (1966), Jungwirth *et al.* (1972), Metz and Esteban (1972), and Esteban and Metz (1973*b*) indicate that early vaccinia messenger RNA is formed and that the block must occur at the level of translation. These results appear to conflict with a report that early vaccinia RNA synthesis is inhibited in interferon-treated cells (Bialy and Colby, 1972). A poxvirus-specific cytopathic effect occurs in interferon-treated L cells which makes interpretation of results in this system somewhat difficult (Horak *et al.*, 1971).

6.1.4. Other Substances

Streptovaricins are macrolide antibiotics structurally related to rifamycins. Quintrell and McAuslan (1970) reported that a mixture of streptovaricins inhibited cowpox replication. In contrast to rifampicin, the streptovaricins are effective only when added prior to DNA synthesis and appear to act by inhibition of early RNA synthesis.

Acetone added to the cell culture medium reduces the yield of infectious rabbitpox virus in L-cell monolayer cultures by 90–97% (Ghendon and Samilova, 1968; Chernos *et al.*, 1972). The effect appears to be at a late state in reproduction, and virus particles with a slighly altered buoyant density and a 10- to 25-fold lower infectivity are produced. Mutants resistant to acetone can be isolated (Ghendon and Chernos, 1972).

6.2. Dual Viral Infections

Several investigators have examined the effects of dual infections with the hope of understanding biochemical mechanisms involved in suppression. Dales and Silverberg (1968) concluded on the basis of competitive suppression of progeny formation that mengovirus is more aggressive than vaccinia, and vaccinia more than reovirus. Under the conditions used by Freda and Buck (1971), vaccinia and mengovirus interfered with each other, but the effect on vaccinia was the more complete. Interference by mengovirus appeared to be a late event since early vaccinia DNA and proteins were formed. In contrast, vaccinia interfered with the early production of mengovirus RNA. Giorno and Kates (1971) reported that early vaccinia RNA is not translated in

HeLa cells preinfected 18 hours earlier with adenovirus. Vaccinia DNA synthesis is also inhibited in KB cells simultaneously infected with frog virus 3 at 37°C, a nonpermissive temperature for the amphibian virus (Aubertin *et al.*, 1970). Vilagines and McAuslan (1970) concluded that frog virus 3 inhibits the replication of vaccinia DNA and transcription from uncoated genomes.

Facilitation of growth of certain viruses in fibroma- and Yaba virus-infected cells has also been reported (Padgett and Walker, 1970; Tsuchiya and Tagaya, 1972*a,b*).

7. SUMMARY

An outline of the steps currently believed to occur during the reproduction of vaccinia and some other poxviruses is presented in Fig. 18 and outlined below.

1. Virions are engulfed by the host cell, the envelope layers are degraded, and the resulting cores are released into the cytoplasmic matrix.
2. Enzymes prepackaged within the cores are activated. Early RNA species corresponding to about 14% of the genome and containing terminal poly(A) sequences are synthesized and extruded.

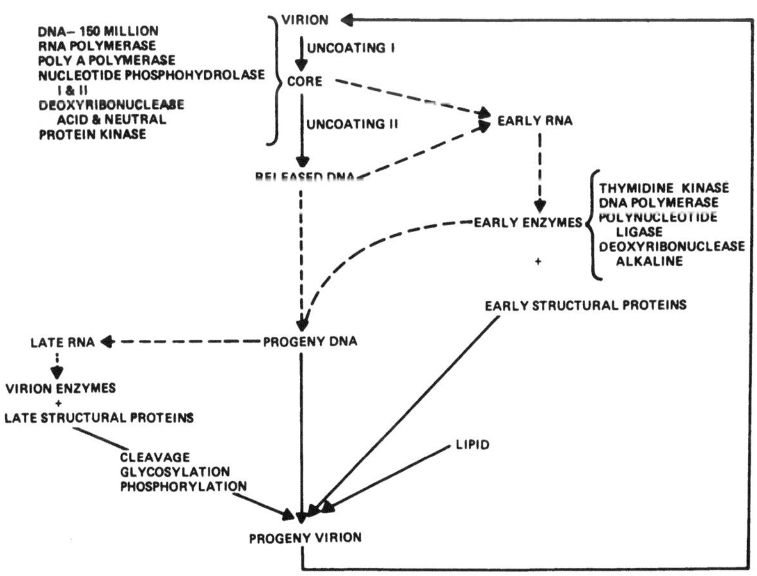

Fig. 18. Reproduction of poxviruses.

3. The viral RNA species are translated on host ribosomes and certain early proteins including thymidine kinase are formed.
4. Host protein synthesis and subsequently RNA and DNA synthesis are inhibited by a process which may not require the synthesis of viral macromolecules.
5. The core is degraded or ruptures and DNA is released.
6. Additional early genes are transcribed and additional early proteins, including DNA polymerase and several structural proteins, are made.
7. DNA replication occurs within discrete areas of the cytoplasm.
8. Synthesis of late RNA is initiated, and late proteins, including the virion-associated enzymes and the majority of structural proteins, are synthesized.
9. Formation of early proteins is depressed.
10. DNA replication terminates.
11. Poxvirus morphogenesis occurs by a complex process that may include the *de novo* formation of viral membranes and the cleavage of higher-molecular-weight precursors of structural proteins.
12. Maturing virions migrate away from the sites of biosynthesis; some pass through the plasma membrane and others may eventually be released by cell lysis.

ACKNOWLEDGMENTS

I thank Drs. P. Geshelin, K. Berns, E. Paoletti, H. Rosemond-Hornbeak, and J. Kleiman for permission to use unpublished material; Drs. S. Dales, E. Medzon, R. Granados, D. Roberts, W. Joklik, and N. Salzman for photographs of published material; Academic Press, Cambridge University Press, American Society for Microbiology, and National Academy of Sciences for permission to use published material; and E. Paoletti for reading the manuscript.

8. REFERENCES

Abdussalam, M., 1957, Elementary bodies of sheeppox, *Am. J. Vet. Res.* **18,** 614.

Abel, P. M., 1963, Reactivation of heated vaccinia virus *in vitro*, *Z. Vererbungsl.* **94,** 249.

Abel, P. M., and Trautner, T. A., 1964, Formation of an animal virus within a bacterium, *Z. Vererbungsl.* **95,** 66.

Allison, A. C., and Burke, D. C., 1962, The nucleic acid content of viruses, *J. Gen. Microbiol.* **27**, 181.

Anderson, C. W., Baum, P. R., and Gesteland, R. F., 1973, Processing of adenovirus 2-induced proteins, *J. Virol.* **12**, 241.

Andrewes, C., and Pereira, H. G., 1972, "Viruses and Vertebrates," Williams & Wilkins, Baltimore.

Andrewes, C., Allison, A. C., Armstrong, J. A., Bearcroft, G., Niven, J. S. F., and Pereira, H. S., 1959, A virus disease of monkeys causing large superficial growths, *Acto Unio Int. Contra Cancrum* **15**, 760.

Appleyard, G., and Way, H. J., 1966, Thiosemicarbazone-resistant rabbitpox virus, *Brit. J. Exp. Pathol.* **47**, 144.

Appleyard, G., and Westwood, J. C. N., 1964, The growth of rabbitpox virus in tissue culture, *J. Gen. Microbiol.* **37**, 391.

Appleyard, G., Westwood, J. C. N., and Zwartouw, H. T., 1962, The toxic effect of rabbitpox virus in tissue culture, *Virology* **18**, 159.

Appleyard, G., Hume, V. B. M., and Westwood, J. C. N., 1965, The effect of thiosemicarbazones on the growth of rabbitpox virus in tissue culture, *Ann. N.Y. Acad. Sci.* **130**, 92.

Appleyard, G., Hapel, A. J., and Boulter, E. A., 1971, An antigenic difference between intracellular and extracellular rabbitpox virus, *J. Gen. Virol.* **13**, 19.

Aubertin, A. M., and McAuslan, B. R., 1972, Virus-associated nucleases: Evidence for endonuclease and exonuclease activity in rabbitpox and vaccinia viruses, *J. Virol.* **9**, 554.

Aubertin, A. M., Guir, J., and Kirn, A., 1970, The inhibition of vaccinia virus DNA synthesis in KB cells infected with frog virus 3, *J. Gen. Virol.* **8**, 105.

Babber, O. P., Chowdhury, B. L., and Rani, G., 1966, Pock formation by vaccinia virus deoxyribonucleic acid after passage through *E. coli* spheroplasts, *Acta Virol.* **10**, 15.

Bablanian, R., 1968, The prevention of early vaccinia-induced cytopathic effects by inhibition of protein synthesis, *J. Gen. Virol.* **3**, 51.

Bablanian, R., 1970, Studies on the mechanism of vaccinia virus cytopathic effects: Effect of inhibitors of RNA and protein synthesis on early virus-induced cell damage, *J. Gen. Virol.* **6**, 221.

Bach, M. K. and Magee, W. E., 1962, Biochemical effects of isatin β-thiosemicarbazones on development of vaccinia virus, *Proc. Soc. Exp. Biol. Med.* **110**, 565.

Bakay, M., and Burke, D. C., 1972, The production of interferon in chick cells infected with DNA viruses: A search for double-stranded RNA, *J. Gen. Virol.* **16**, 399.

Ball, F. R., and Medzon, E. L., 1973, Sedimentation changes of L cells in a density gradient early after infection with vaccinia virus, *J. Virol.* **12**, 588.

Bandlow, G., Koszinowski, U., and Thomssen, R., 1973, Gesteigerte Immunogenität von Plasmamembranen vacciniavirusinfizierter BHK Zellen, *Arch. Ges. Virusforsch.* **40**, 63.

Barbanti-Brodano, G., Portolani, M., Bernardini, A., Stirpe, F., Mannini-Palenzona, A., and Placa, M. L., 1968, Thymidine kinase activity in human amnion cell cultures infected with Shope fibroma virus, *J. Gen. Virol.* **3**, 471.

Bauer, D. J., 1961, The zero effect dose (E_0) as an absolute numerical index of antiviral chemotherapeutic activity in the pox virus group, *Brit. J. Exp. Pathol.* **42**, 201.

Bauer, D., and Sadler, P., 1960, The structure–activity relationships of the anti-viral chemotherapeutic activity of isatin β-thiosemicarbazone, *Brit. J. Pharmacol.* **15,** 101.

Bearcroft, W. C. C., and Jamieson, M. F., 1958, An outbreak of subcutaneous tumors in rhesus monkeys, *Nature* (*Lond.*) **182,** 195.

Beaud, G., Kirn, A., and Gros, F., 1972, *In vitro* protein synthesis directed by RNA transcribed from vaccinia DNA, *Biochem. Biophys. Res. Commun.* **49,** 1459.

Becker, Y., and Joklik, W. K., 1964, Messenger RNA in cells infected with vaccinia virus, *Proc. Natl. Acad. Sci. USA* **51,** 577.

Becker, Y., and Sarov, I., 1968, Electron microscopy of vaccinia virus DNA, *J. Mol. Biol.* **34,** 655.

Bedson, H. S., and Cruickshank, J. G., 1968, Multiple protein functions in the replication of poxviruses, *J. Gen. Virol.* **3,** 147.

Bellett, A. J. D., and Fenner, F., 1968, Studies of base-sequence homology among some cytoplasmic deoxyriboviruses of vertebrate and invertebrate animals, *J. Virol.* **2,** 1374.

Ben-Ishai, Z., Heller, E., Goldblum, N., and Becker, Y., 1969, Rifampicin and poxivirus replication, *Nature* (*Lond.*) **224,** 29.

Bergoin, M., and Dales, S., 1971, Comparative observations on poxviruses of invertebrates and vertebrates, *in* "Comparative Virology" (K. Maramorosch and E. Kurstak, eds.), pp. 171–205, Academic Press, New York.

Bergoin, M., Devauchelle, G., and Vago, C., 1969, Electron microscopy study of the pox-like virus of *Melolontha melolontha* L.: Virus morphogenesis, *Arch. Ges. Virusforsch.* **28,** 285.

Bergoin, M., Devauchelle G., and Vago, C., 1971, Electron microscopy study of melolontha poxvirus: The fine structure of occluded virions *Virology* **43,** 453.

Bernkopf, H., Nishmi, M., and Rosin, A., 1959, Effect of active and inactive vaccinia virus preparations on human amnion cell cultures, *J. Immunol.* **83,** 635.

Berns, K. I., and Silverman, C., 1970, Natural occurrence of cross-linked vaccinia virus deoxyribonucleic acid, *J. Virol.* **5,** 299.

Berns, K. I., Silverman, C., and Weissbach, A., 1969, Separation of a new deoxyribonucleic acid polymerase from vaccinia-infected HeLa cells, *J. Virol.* **4,** 15.

Bialy, J. S., and Colby, C., 1972, Inhibition of early vaccinia virus ribonucleic acid synthesis in interferon-treated chicken embryo fibroblasts, *J. Virol.* **9,** 286.

Bodo, G., and Jungwirth, C., 1967, Effect of interferon on deoxyribonuclease induction in chick fibroblast cultures infected with cowpox virus, *J. Virol.* **1,** 466.

Bodo, G., Scheirer, W., Suh, M., Schultze, B., Horak, I., and Jungwirth, C., 1972, Protein synthesis in pox-infected cells treated with interferon, *Virology* **50,** 140.

Braunwald, J., Guir, J. Engel, M. L., and Kirn, A., 1969, Thermosensibilité des enzymes impliquées dans la réplication de l'ADN d'un mutant froid du virus vaccinal, *C. R. Hebd. Seances Acad. Sci. Ser. D Sci. Nat.* **268,** 1816.

Braunwald, J., Burkard, G., Guir, J., Engel, M. L., and Kirn, A., 1970, Synthesis of DNA polymerase at a supra-optimal temperature in the cytoplasm of KB cells infected with a "cold" mutant of the vaccinia virus, *Ann. Inst. Pasteur* (*Paris*) **118,** 88.

Brown, A., Mayyasi, S. A., and Officer, J. E., 1959, The "toxic" activity of vaccinia virus in tissue culture, *J. Infect. Dis.* **104,** 193.

Brown, M., Dorson, J. W., and Bollum, F. J., 1973, Terminal riboadenylate transferase: a poly A polymerase in purified vaccinia virus, *J. Virol.* **12,** 203.

Cairns, J., 1960, The initiation of vaccinia infection, *Virology* **11,** 603.
Chang, L. M. S., and Hodes, M. E., 1967, induction of DNA nucleotidyl transferase by the Shope fibroma virus, *Virology* **32,** 258.
Change, L. M. S., and Hodes, M. E., 1968, RNA polymerase in cells infected with Shope fibroma virus, *Virology* **36,** 323.
Cheevers, W., O'Callaghan, D. J., and Randall, C. C., 1968, Biosynthesis of host and viral deoxyribonucleic acid during hyperplastic fowlpox infection *in vivo*, *J. Virol.* **2,** 421.
Chernos, V. I., Libshits, B. A., Yakobson, E., and Ghendon, Y. Z., 1972, Mechanism of antiviral action of acetone on rabbitpox virus replication, *J. Virol.* **9,** 251.
Citarella, R. V., Muller, R., Schlabach, A., and Weissbach, A., 1972, Studies on vaccinia virus-directed deoxyribonucleic acid polymerase, *J. Virol.* **10,** 721.
Clarkson, S. G., and Runner, M. N., 1971, Transfer RNA changes in HeLa cells after vaccinia virus infection, *Biochim. Biophys. Acta* **238,** 498.
Cohen, G. H., and Wilcox, W. C., 1966, Soluble antigens of vaccinia infected mammalian cells. I. Separation of virus-induced soluble antigens into two classes on the basis of physical characteristics, *J. Bacteriol.* **92,** 676.
Cohen, G. H., and Wilcox, W. C., 1968, Soluble antigens of vaccinia-infected mammalian cells. III. Relation of "early" and "late" proteins to virus structure, *J. Virol.* **2,** 449.
Colby, C., and Duesberg, P. H., 1969, Double-stranded RNA in vaccinia virus-infected cells, *Nature* (*Lond.*) **222,** 940.
Colby, C., Jurale C., and Kates, J. R., 1971, Mechanisms of Synthesis of vaccinia virus double-stranded ribonucleic acid *in vivo* and *in vitro*, *J. Virol.* **7,** 71.
Craigie, J., 1932, The nature of the vaccinia flocculation reaction, and observations on the elementary bodies of vaccinia, *Brit. J. Exp. Pathol.* **13,** 259.
Craigie, J., and Wishart, F. O., 1936, Studies on the soluble precipitable substances of vaccinia. II. The soluble precipitable substances of dermal vaccine, *J. Exp. Med.* **64,** 819.
Dahl, R., and Kates, J. R., 1970*a*, Intracellular structures containing vaccinia DNA: Isolation and characterization, *Virology* **42,** 453.
Dahl, R., and Kates, J. R., 1970*b*, Synthesis of vaccinia virus "early" and "late" messenger RNA *in vitro* with nucleoprotein structures isolated from infected cells, *Virology* **42,** 463.
Dales, S., 1963, The uptake and development of vaccinia virus in L strain cells followed with labeled viral deoxyribonucleic acid, *J. Cell Biol.* **18,** 51.
Dales, S., 1965*a*, Penetration of animal viruses into cells, *Progr. Med. Virol.* **7,** 1.
Dales, S., 1965*b*, Effects of streptovitacin A on the initial events in the replication of vaccinia and reovirus, *Proc. Natl. Acad. Sci. USA* **54,** 462.
Dales, S., and Kajioka, R., 1964, The cycle of multiplication of vaccinia virus in Earle's strain L cells. 1. Uptake and penetration, *Virology* **24,** 278.
Dales, S., and Mosbach, E. H., 1968, Vaccinia as a model for membrane biogenesis, *Virology* **35,** 584.
Dales, S., and Silverberg, H., 1968, Controlled double infection with unrelated animal viruses, *Virology* **34,** 531,
Dales, S., and Siminovitch, L., 1961, The development of vaccinia virus in Earle's L strain cells as examined by electron microscopy, *J. Biophys. Biochem. Cytol.* **10,** 475.
deHarven, E., and Yohn, D. S., 1966, The fine structure of the Yaba monkey tumor poxvirus, *Cancer Res.* **26,** 995.

Douglas, H. W., Williams, B. L., and Rondle, C. J. M., 1969, Microelectrophoresis of poxviruses in molar sucrose, *J. Gen. Virol.* **5,** 391.

Downer, D. N., Rogers, H. W., and Randall, C. C., 1973, Endogenous protein kinase and phosphate acceptor protein in vaccinia virus, *Virology* **52,** 13.

Downie, A. W., and Dumbell, K. R., 1956, Poxviruses, *Annu. Rev. Microbiol.* **10,** 237.

Dubbs, D. R., and Kit, S., 1964, Isolation and properties of vaccinia mutants deficient in thymidine kinase-inducing activity, *Virology* **22,** 214.

Duesberg, P. H., and Colby, C., 1969, On the biosynthesis and structure of double-stranded RNA in vaccinia virus-infected cells, *Proc. Natl. Acad. Sci. USA* **64,** 396.

Eagles, G. H., and Ledingham, J. C. G., 1932, Vaccinia and the Paschen body: Infection experiments with centrifugalised virus filtrates, *Lancet* **1,** 823.

Easterbrook, K. B., 1962, Interference with the maturation of vaccinia virus by isatin B-thiosemicarbazone, *Virology* **17,** 245.

Easterbrook, K. B., 1966, Controlled degradation of vaccinia virions *in vitro*: An electron microscopic study, *J. Ultrastruct. Res.* **37,** 132.

Easterbrook, K. B., 1967, Morphology of deoxyribonucleic acid extracted from cores of vaccinia virus, *J. Virol.* **1,** 643.

Easterbrook, K. B., 1972, Crystalline aggregates observed in the vicinity of freeze-etched poxvirus inclusions, *Can. J. Microbiol.* **18,** 403.

Easterbrook, K. B., and Rozee, K. R., 1971, The intracellular development of vaccinia virus as observed in freeze-etched preparations, *Can. J. Microbiol.* **17,** 753.

Elford, W. J., and Andrewes, C. H., 1972, Filtration of vaccinia virus through gradocol membranes, *Brit. J. Exp. Pathol.* **13,** 36.

Epstein, M. A., 1958, Structural differentiation in the nucleoid of mature vaccinia virus, *Nature* (*Lond.*) **181,**784.

Eron, L. J., and McAuslan, B. R., 1966, The nature of poxvirus-induced deoxyribonucleases, *Biochem. Biophys. Res. Commun.* **22,** 518.

Esteban, M., and Metz, D. H., 1973*a*, Early virus protein synthesis in vaccinia virus-infected cells, *J. Gen. Virol.* **19,** 201.

Esteban, M., and Metz, D. H., 1973*b*, Inhibition of early vaccinia virus protein synthesis in interferon-treated chicken embryo fibroblasts, *J. Gen. Virol.* **20,** 111.

Ewton, D., and Hodes, M. E., 1967, Nucleic acid synthesis in HeLa cells infected with Shope fibroma virus, *Virology* **33,** 77.

Fenner, F., 1959, Genetic studies with mammalian poxviruses. II. Recombination between two strains of vaccinia virus in single HeLa cells, *Virology* **8,** 499.

Fenner, F., 1968, "The Biology of Animal Viruses," Vol. I, Academic Press, New York.

Fenner, F., 1969, Conditional lethal mutants of animal viruses, *Curr. Top. Microbiol.* **48,** 1.

Fenner, F., and Burnet, F. M., 1957, A short description of the poxvirus group (vaccinia and related viruses), *Virology* **4,** 305.

Fenner, F., and Woodroofe, G. M., 1960, The reactivation of poxviruses. II. The range of reactivating viruses, *Virology* **11,** 185.

Fiel, R. J., Mark, E. H., and Munson, B. R., 1970, Small-angle light scattering of bioparticles. 3. Vaccinia virus, *Arch. Biochem. Biophys.* **141,** 547.

Follett, E. A. C., and Pennington, T. H., 1971, Antiviral effect of constituent parts of the rifampicin molecules, *Nature* (*Lond.*) **230,** 117.

Fournier, F., Tovell, D. R., Esteban, M., Metz, D. H., Ball, L. A., and Kerr, I. M., 1973, The translation of vaccinia virus messenger RNA in animal cell-free systems, *FEBS* (*Fed. Eur. Biochem. Soc.*) *Lett.* **30,** 268.

Freda, C. E., and Buck, C. A., 1971, System of double infection between vaccinia virus and mengovirus, *J. Virol.* **8,** 293.
Gafford, L. G., and Randall, C. C., 1967, The high molecular weight of the fowlpox virus genome, *J. Mol. Biol.* **26,** 303.
Gafford, L. G., and Randall, C. C., 1970, Further studies on high molecular weight fowlpox virus DNA and its hydrodynamic properties, *Virology* **40,** 298.
Gafford, L. G., Sinclair, F., and Randall, C. C., 1969, Growth cycle of fowlpox virus and change in plaque morphology and cytopathology by contaminating mycoplasma, *Virology* **37,** 464.
Gafford, L. G., Sinclair, F., and Randall, C. C., 1972, Alteration of DNA metabolism in fowlpox-infected chick embryo monolayers, *Virology* **48,** 567.
Garon, C. F., and Moss, B., 1971, Glycoprotein synthesis in cells infected with vaccinia virus. II. A glycoprotein component of the virion, *Virology* **46,** 233.
Gemmel, A., and Fenner, F., 1960, Genetic studies with mammalian poxviruses III. White (u) mutants of rabbitpox virus, *Virology* **11,** 219.
Ghendon, Y. Z., 1972, Conditional-lethal mutants of animal viruses, *Progr. Med. Virol.* **14,** 68.
Ghendon, Y. Z., and Chernos, V. I., 1972, Mutants of poxviruses resistant to aceton and N-methyl-isatin-beta-thiosemicarbazone, *Acta Virol.* **16,** 308.
Ghendon, Y., and Samilova, G., 1968, Antiviral effect of acetone, *J. Gen. Virol.* **3,** 271.
Ghosh, S. N., and Gifford, G. E., 1965, Effect of interferon on the dynamics of ^{3}H-thymidine incorporation and thymidine kinase induction in chick fibroblast cultures infected with vaccinia virus, *Virology* **27,** 186.
Gifford, G. E., and Klapper, D. G., 1969, Effect of proteolytic enzymes on vaccinia virus replication, *Arch. Ges. Virusforsch.* **26,** 321.
Giorno, R., and Kates, J. R., 1971, Mechanism of inhibition of vaccinia virus replication in adenovirus-infected HeLa cells. *J. Virol.* **7,** 208.
Gold, P., and Dales, S., 1968, Localization of nucleotide phosphohydrolase within vaccinia virus, *Proc. Natl. Acad. Sci. USA* **60,** 845.
Granados, R. R., 1973, Entry of an insect poxvirus by fusion of the virus envelope with the host cell membrane, *Virology* **52,** 305.
Granados, R. R., and Roberts, D. W., 1970, Electron microscopy of a poxlike virus infecting an invertebrate host, *Virology* **40,** 230.
Green, M. and Piña, M., 1962, DNA polymerase and deoxyribonucleotide kinase activities in cells infected with vaccinia virus, *Virology* **17,** 603.
Grimley, P. M., 1971*a*, High-voltage electron microscopy of poxvirus nucleoid material, *Proc. EMSA 29th Ann. Meet.* pp. 380–381.
Grimley, P. M., 1971*b*, High-voltage electron microscopy of poxvirus bioassembly, *in* "Biomedical Applications of High-Voltage Electron Microscopy," National Institutes of Health, Division of Research Resources, Bethesda, Md.
Grimley, P. M., and Moss, B., 1971, Similar effect of rifampin and other rifamycin derivatives on vaccinia virus morphogenesis, *J. Virol.* **8,** 225.
Grimley, P. M., Rosenblum, E. N., Mims, S. J., and Moss, B., 1970, Interruption by rifampin of an early stage in vaccinia virus morphogenesis: Accumulation of mebranes which are precursors of virus envelopes, *J. Virol.* **6,** 519.
Groyon, R. M., and Kniazeff, A. J., 1967, Vaccinia virus infection of synchronized pig kidney cells, *J. Virol.* **1,** 1255.
Hanafusa, H., 1960, Killing of L cells by heat and U.V.-inactivated vaccinia virus, *Biken J.* **3,** 191.

Hanafusa, T., 1961, Enzymatic synthesis and breakdown of deoxyribonucleic acid by extracts of L cells infected with vaccinia virus, *Biken J.* **4,** 97.

Hanafusa, H., Hanafusa, T., and Kamahora, J., 1959, Transformation phenomena in the pox group of viruses. II. Transformation between several members of pox group, *Biken J.* **2,** 85.

Harford, C. G., Hamlin, A., and Rieders, E., 1966, Electron microscopic autoradiography of DNA synthesis in cells infected with vaccinia virus, *Exp. Cell. Res.* **42,** 50.

Harford, C. G., Rieders, E., and Osborn, R., 1972, Inhibition of arc-like fragments and immature forms of vaccinia virus by methisazone, *Proc. Soc. Exp. Biol. Med.* **139,** 558.

Hayashi, K., Rosenthal, J., and Notkins, A. L., 1972, Iodine-125-labeled antibody to viral antigens: Binding to the surface of virus-infected cells, *Science* (*Wash., D.C.*) **176,** 516.

Heller, E., Argaman, M., Levy, H., and Goldblum, N., 1969, Selective inhibition of vaccinia virus by the antibiotic rifampicin, *Nature* (*Lond.*) **222,** 273.

Hiller, G., Jungwirth, C., Bodo, G., and Schultze, B., 1973, Biological activity of poly rI: poly rC: Effect on poxvirus-specific functions, *Virology,* **52,** 22.

Hinze, H. C., and Walker, D. C., 1964, Response of cultured rabbit cells to infection with the Shope fibroma virus, I. Proliferation and morphological alteration of the infected cells, *J. Bacteriol.* **88,** 1185.

Hinze, H. C., and Walker, D. L., 1971, Comparison of cytocidal and noncytocidal strains of Shope rabbit fibroma virus, *J. Virol.* **7,** 577.

Hoagland, C. L., Smadel, J. E., and Rivers, T. M., 1940, Constituents of elementary bodies of vaccinia. I. Certain basic analyses and observations on lipid components of the virus, *J. Exp. Med.* **71,** 737.

Hoagland, C. L., Ward, S. M., Smadel, J. E., and Rivers, T. M., 1942, Constituents of elementary bodies of vaccinia. VI. Studies on the nature of the enzymes associated with purified virus, *J. Exp. Med.* **76,** 163.

Holowczak, J. A., 1970, Glycopeptides of vaccinia virus. I. Preliminary characterization and hexosamine content, *Virology* **42,** 87.

Holowczak, J. A., 1972, Uncoating of poxviruses. I. Detection and characterization of subviral particles in the uncoating process, *Virology* **50,** 216.

Holowczak, J. A., and Joklik, W. K., 1967*a*, Studies of the structural proteins of vaccinia virus. I. Structural proteins of virions and cores, *Virology* **33,** 717.

Holowczak, J. A., and Joklik, W. K., 1967*b*, Studies of the structural proteins of vaccinia virus. II. Kinetics of the synthesis of individual groups of structural proteins, *Virology* **33,** 726.

Horak, I., Jungwirth, C., and Bodo, G., 1971, Poxvirus-specific cytopathic effect in interferon-treated L cells, *Virology* **45,** 456.

Hyde, J. M., and Peters, D., 1971, The organization of nucleoprotein within fowlpox virus, *J. Ultrastruct. Res.* **35,** 626.

Hyde, J. M., Gafford, L. G., and Randall, C. C., 1967, Molecular weight determination of fowlpox virus DNA by electron microscopy, *Virology* **33,** 112.

Ichihashi, Y., and Matsumoto, S., 1966, Studies on the nature of Marchal bodies (A-type inclusion) during ectromelia virus infection, *Virology* **29,** 264.

Ichihashi, Y., and Matsumoto, S., 1969, Genetic study of pox group viruses, *Virusu* **19,** 155.

Ichihashi, Y., Matsumoto, S., and Dales, S., 1971, Biogenesis of poxviruses: Role of A-type inclusions and host cell membranes in virus dissemination, *Virology* **46,** 507.

Israeli, E., 1966, Mechanism of pock formation by Shope fibroma virus, *J. Bacteriol.* **92,** 727.

Israeli, E., and Sachs, L., 1964, Cell–virus interactions with the Shope firbroma virus on cultures of rabbit and rat cells, *Virology* **23,** 473.

Jacobson, M. F., Asso, J., and Baltimore, D., 1970, Further evidence on the formation of poliovirus proteins, *J. Mol. Biol.* **49,** 657.

Jacquemont, B., Grange, J., Gazzolo, L., and Richard, M. H., 1972, Composition and size of Shope fibroma virus deoxyribonucleic acid, *J. Virol.* **9,** 836.

Jefferts, E. R., and Holowczak, J. A., 1971, RNA synthesis in vaccinia-infected L cells: Inhibition of ribosome formation and maturation, *Virology* **46,** 730.

Joklik, W. K., 1962*a*, The purification of four strains of poxvirus, *Virology* **18,** 9.

Joklik, W. K., 1962*b*, The preparation and characteristics of highly purified radioactively labeled poxvirus. *Biochim. Biophys. Acta* **61,** 290.

Joklik, W. K., 1962*c*, Some properties oi poxvirus deoxyribonucleic acid, *J. Mol. Biol.* **5,** 265.

Joklik, W. K., 1964*a*, The intracellular uncoating of poxvirus DNA. I. The fate of radioactively-labeled rabbitpox virus, *J. Mol. Biol.* **8,** 263.

Joklik, W. K., 1964*b*, The intracellular uncoating of poxvirus DNA. II. The molecular basis of the uncoating process, *J. Mol. Biol.* **8,** 277.

Joklik, W. K., 1964*c*, The intracellular fate of rabbitpox virus rendered noninfectious by various reagents, *Virology* **22,** 620.

Joklik, W. K., 1966, The poxvirus, *Bacteriol. Rev.* **30,** 33.

Joklik, W. K., 1968, The poxviruses, *Annu. Rev. Microbiol.* **22,** 359.

Joklik, W. K., and Becker, Y., 1964, The replication and coating of vaccinia DNA, *J. Mol. Biol.* **10,** 452.

Joklik, W. K., and Merigan, T. C., 1966, Concerning the mechanism of action of interferon, *Proc. Natl. Acad. Sci. USA* **56,** 558.

Jungwirth, C., and Dawid, I. B., 1967, Vaccinia DNA: Separation of viral from host cell DNA, *Arch. Ges. Virusforsch.* **20,** 464.

Jungwirth, G., and Joklik, W. K., 1965, Studies on "early" enzymes in HeLa cells infected with vaccinia virus, *Virology* **27,** 80.

Jungwirth, C., and Launer, J., 1968, Effect of poxvirus infection on host cell deoxyribonucleic acid synthesis, *J. Virol.* **2,** 401.

Jungwirth, C., Launer, J., Dombrowski, G., and Horak, I., 1969, Characterization of deoxyribonculeases induced by poxviruses, *J. Virol.* **4,** 866.

Jungwirth, C., Horak, I., Bodo, G., Lindner, J., and Schultze, B., 1972, The synthesis of poxvirus-specific RNA in interferon-treated cells *Virology* **48,** 59.

Kaku, H., and Kamahora, J., 1964, Giant cell formation in L cells infected with active vaccinia virus, *Biken J.* **6,** 299.

Kates, J., and Beeson, J., 1970*a*, Ribonucleic acid synthesis in vaccinia virus. I. The mechanism of synthesis and release of RNA in vaccinia cores, *J. Mol. Biol.* **50,** 1.

Kates, J., and Beeson, J., 1970*b*, Ribonucleic acid synthesis in vaccinia virus. II. Synthesis of polyriboadenylic acid, *J. Mol. Biol.* **50,** 19.

Kates, J. R., and McAuslan, B. R., 1967*a*, Relationship between protein synthesis and viral deoxyribonucleic acid synthesis, *J. Virol.* **1,** 110.

Kates, J. R., and McAuslan, B. R., 1967*b*, Messenger RNA synthesis by a "coated" viral genome, *Proc. Natl. Acad. Sci. USA* **57,** 314.

Kates, J. R., and McAuslan, B. R., 1967*c*, Poxvirus DNA-dependent RNA polymerase, *Proc. Natl. Acad. Sci. USA* **58,** 134.

Kates, J., Dahl, R., and Mielke, M., 1968, Synthesis and intracellular localization of vaccinia virus deoxyribonucleic acid-dependent ribonucleic acid polymerase, *J. Virol.* **2,** 894.

Kato, S., Kameyama, S., and Kamahora, 1960, Autoradiography with tritium-labeled thymidine of pox virus and human amnion cell system in tissue culture, *Biken J.* **3,** 135.

Kato, S., Hara, M., Miyamoto, H., and Kamahora, J., 1963, Inclusion markers of cowpox and alastrim virus, *Biken J.* **6,** 233.

Kato, S., Ogawa, M., and Miyamoto, H., 1964, Nucleocytoplasmic interaction in poxvirus-infected cells. 1. Relationship between inclusion formation and DNA metabolism of the cells, *Biken J.* **7,** 45.

Kato, S., Keichiro, T., and Miyamoto, H., 1965, Autoradiographic studies on the Yaba monkey tumor virus and host cell interactions, *Biken J.* **8,** 45.

Kato, S., Ono, K., Miyamoto, H., and Mantani, M., 1966, Virus–host cell interaction in rabbit fibrosarcoma produced by Shope fibroma virus. *Biken J.* **9,** 51.

Katz, E., and Margalith, E., 1973, Location of vaccinia virus structural polypeptides on the surface of the virus particle, *J. Gen. Virol.* **18,** 381.

Katz, E., and Moss, B., 1969, Synthesis of vaccinia viral proteins in cytoplasmic extracts. I. Incorporation of radioactively labeled amino acids into polypeptides, *J. Virol.* **4,** 416.

Katz, E., and Moss, B., 1970*a*, Formation of vaccinia virus structural polypeptide from a higher-molecular-weight precursor: Inhibition by rifampicin, *Proc. Natl. Acad. Sci. USA* **66,** 677.

Katz, E. and Moss, B., 1970*b*, Vaccinia virus structural polypeptide derived from a high-molecular-weight precursor: Formation and integration into virus particles, *J. Virol.* **6,** 717.

Katz, E., Grimley, P., and Moss B., 1970, Reversal of anti-viral effects of rifampicin, *Nature* (*Lond.*) **227,** 1050.

Katz, E., Margalith, E., Winer, B., and Goldblum, N., 1973*a*, Synthesis of vaccinia viral polypeptides in the presence of isatin β-thiosemicarbazone, *Antimicrob. Ag. Chemother.* **4,** 44.

Katz, E., Winer, B. Margalith, E., and Goldblum, N., 1973*b*, Isolation and characterization of an IBT-depenent mutant of vaccinia virus, *J. Gen. Virol.* **19,** 161.

Kit, S., and Dubbs, D. R., 1965, Properties of deoxythymidine kinase partially purified from non-infected and virus-infected mouse fibroblast cells, *Virology* **26,** 16.

Kit, S., Dubbs, D. R., and Piekarski, L. J., 1962, Enhanced thymidine phosphorylating activity of mouse fibroblasts (strain LM) following vaccinia infection, *Biochem. Biophys. Res. Commun.* **8,** 72.

Kit, S., Dubbs, D. R., and Hsu, T. C., 1963*a*, Biochemistry of vaccinia-infected mouse fibroblasts (strain LM). III. Radioautographic and biochemical studies of thymine-^{3}H uptake into DNA of L-M cells and rabbit cells in primary culture, *Virology* **19,** 13.

Kit, S., Dubbs, D. R., and Piekarski, L. J., 1963*b*, Inhibitory effects of puromycin and fluorophenylalanine on induction of thymidine kinase by vaccinia infected L-cells, *Biochem. Biophys. Res. Commun.* **11,** 176.

Kit, S., Piekarski, L. J., and Dubbs, D. R., 1963*c*, Induction of thymidine kinase by vaccinia-infected mouse fibroblasts, *J. Mol. Biol.* **6,** 22.

Kit, S., Valladores, Y., and Dubbs, D. R., 1964, Effect of age of culture and vaccinia infection on uridine kinase activity of L-cells, *Exp. Cell Res.* **34,** 257.

Kit, S., Nakajima K., and Dubbs, D. R., 1970, Transfer RNA methylase activities of SV40-transformed cells and cells infected with animal viruses, *Cancer Res.* **30,** 528.
Klagsbrun, M., 1971, Changes in the methylation of transfer RNA in vaccinia infected HeLa cells. *Virology* **44,** 153.
Kleiman, J., and Moss, B., 1973, Protein kinase activity from vaccinia virions: Solubilization and separation into heat-labile and heat-stable components, *J. Virol.* **12,** 684.
Koziorowska, J., Wlodarski, K., and Mazurowa, N., 1971, Transformation of mouse embryo cells by vaccinia virus, *J. Natl. Cancer Inst.* **46,** 225.
Laemmli, U. K., 1970, Cleavage of structural proteins during the assembly of the head of bacteriophage T4, *Nature* (*Lond.*) **227,** 680.
Lancini, G., Gricchio, R., and Thiry, L., 1971, Antiviral activity of rifamycins and N-aminopiperazines, *J. Antibiot.* (*Tokyo*) **24,** 64.
Levine, S., Magee, W. E., Hamilton, R. D., and Miller, O. V., 1967, Effect of interferon on early enzyme and viral DNA synthesis in vaccinia virus infection, *Virology* **32,** 33.
Levinson, W., 1973, Inhibition of viruses, tumors and pathogenic microorganisms by isatin β-thiosemicarbazone and other thiosemicarbazones, *in* "Selective Inhibitors of Viral Functions" (W. A. Carter, ed.), pp. 213–226, CRC Press, Cleveland, Ohio.
Loh, P. C., and Riggs, J. L., 1961, Demonstration of the sequential development of vaccinial antigens and virus in infected cells: Observations with cytochemical and differential fluorescent procedures, *J. Exp. Med.* **114,** 149.
McAuslan, B. R., 1963*a*, Control of induced thymidine kinase activity in the poxvirus-infected cell, *Virology* **20,** 162.
McAuslan, B. R., 1963*b*, The induction and repression of thymidine kinase in the poxvirus-infected HeLa cell, *Virology* **21,** 383.
McAuslan, B. R., 1965, Deoxribonuclease activity of normal and poxvirus infected HeLa cells, *Biochem. Biophys. Res. Commun.* **19,** 15.
McAuslan, B. R., 1967, "Symposium on Enzyme Regulation and Metabolic Control, Mexico City, 1966," National Cancer Institute Monograph 27, National Cancer Institute, Bethesda, Md.
McAuslan, B. R., 1969*a*, The biochemistry of poxvirus replication, *in* "Virus Growth and Cell Metabolism" (H. B. Levy, ed.), pp. 361–413, Marcel Dekker, New York.
McAuslan, B. R., 1969*b*, Rifampicin inhibition of vaccinia replication, *Biochem. Biophys. Res. Commun.* **37,** 289.
McAuslan, B. R., and Joklik, W. K., 1962, Stimulation of the thymidine phosphorylating system in HeLa cells on infection with poxvirus, *Virology* **8,** 486.
McAuslan, B. R., and Kates, J. R., 1967, Poxvirus-induced acid deoxyribonuclease: Regulation of synthesis; control of activity *in vivo*; purification and properties of the enzyme, *Virology* **33,** 709.
McAuslan, B. R., Herde, P., Pett, D., and Ross, J., 1965, Nucleases of virus-infected animal cells, *Biochem. Biophys. Res. Commun.* **20,** 586.
McCrea, J. F., and Lipman, M. B., 1967, Strand length measurements of normal and 5-iodo-2′-deoxyuridine–treated vaccinia virus deoxyribonucleic acid released by the Kleinschmidt method, *J. Virol.* **1,** 1037.
Macfarlane, M. G., and Salaman, M. H., 1938, The enzymatic activity of vaccinial elementary bodies, *Brit. J. Exp. Pathol.* **19,** 184.
McGuire, P. M., and Sharp, D. G., 1972, The composition of a population of vaccinia virions, *Proc. Soc. Exp. Biol. Med.* **139,** 587.

McGuire, P. M. Dunlap, R. C., and Sharp, D. G., 1973, Physical markers of change in a virion population during passage, *J. Infect. Dis.* **127,** 278.

Magee, W. E., 1962, DNA polymerase and deoxyribonucleotide kinase activities in cells infected with vaccinia virus, *Virology* **17,** 604.

Magee, W. E., and Bach, M. K., 1965, Biochemical studies on the antiviral activities of the isatin β-thiosemicarbazones, *Ann. N.Y. Acad. Sci.* **130,** 80.

Magee, W. E., and Miller, O. V., 1967, Immunological evidence for the appearance of a new DNA polymerase in cells infected with vaccinia virus, *Virology* **31,** 64.

Magee, W. E., Levine, S., Miller, O. V., and Hamilton, R. D., 1968, Inhibition by interferon of the uncoating of vaccinia virus, *Virology* **35,** 505.

Mantani, M., Miyamoto, H., and Kato, S., 1968, Effect of viral DNA synthesis of poxvirus upon nuclear DNA synthesis of synchronized FL cells, *Biken J.* **11,** 71.

Marquardt, J., S. E. Holm, and E. Lycke, 1965, Immunoprecipitating factors of vaccinia virus, *Virology* **27,** 170.

Medzon, E. L., and Bauer, H., 1970, Structural features of vaccinia virus revealed by negative staining, *Virology* **40,** 860.

Metz, D. H., and Esteban, M., 1972, Interferon inhibits viral protein synthesis in L cells infected with vaccinia virus, *Nature* (*Lond.*) **238,** 385.

Mitchiner, M. B., 1969, The envelope of vaccinia and orf viruses: An electron-cytochemical investigation, *J. Gen. Virol.* **5,** 211.

Miyamoto, H., and Kato, S., 1971, Cell surface antigens induced by poxviruses. I. Effects of antimetabolites on cell surface antigens, *Biken J.* **14,** 311.

Morgan, C., Ellison, S. A., Rose, H. M., and Moore, D. H., 1954, Structure and development of viruses observed in the electron microscope. II. Vaccinia and fowlpox viruses. *J. Exp. Med.* **100,** 301.

Morgan, C. S., Ellison, A., Rose, H. M., and Moore, D. H., 1955, Serial sections of vaccinia virus examined at one stage of development in the electron microscope, *Exp. Cell Res.* **9,** 572.

Moshkowitz, A., Goldblum, N., and Heller, E., 1971, Studies on the antiviral effect of rifampicin in volunteers, *Nature* (*Lond.*) **229,** 422.

Moss, B., 1968, Inhibition of HeLa cell protein synthesis by the vaccinia virion, *J. Virol.* **2,** 1028.

Moss, B., 1973, Rifamycin SV derivatives, *in* "Selective Inhibitors of Viral Functions" (W. A. Carter, ed.), pp. 313–328, CRC Press, Cleveland, Ohio.

Moss, B., and Filler, R., 1970, Irreversible effects of cycloheximide during the early period of vaccinia virus replication, *J. Virol.* **5,** 99.

Moss, B., and Katz, E., 1969, Synthesis of vaccinia viral proteins in cytoplasmic extracts. II. Identification of early and late viral proteins, *J. Virol.* **4,** 596.

Moss, B., and Rosenblum, E. N., 1972, Hydroxylapatite chromatography of protein–sodium dodecyl sulfate complexes. A new method for the separation of polypeptide subunits, *J. Biol. Chem.* **247,** 5194.

Moss, B., and Rosenblum, E. N., 1973, Protein cleavage and poxviral morphogenesis: Tryptic peptide analysis of core precursors accumulated by blocking assembly with rifampicin, *J. Mol. Biol.* **81,** 267.

Moss, B., and Salzman, N. P., 1968, Sequential protein synthesis following vaccinia virus infection, *J. Virol.* **2,** 1016.

Moss, B., Katz, E., and Rosenblum, E. N., 1969*a*, Vaccinia virus directed RNA and protein synthesis in the presence of rifampicin, *Biochem. Biophys. Res. Commun.* **36,** 858.

Moss, B., Rosenblum, E. N., Katz, E., and Grimley, P. M., 1969*b*, Rifampicin: A specific inhibitor of vaccinia virus assembly, *Nature* (*Lond.*) **224,** 1280.
Moss, B., Rosenblum, E. N., and Garon, C. F., 1971*a*, Glycoprotein synthesis in cells infected with vaccinia virus. I. Non-virion glycoproteins, *Virology* **46,** 221.
Moss, B., Rosenblum, E. N., and Grimley, P. M., 1971*b*, Assembly of vaccinia virus particles from polypeptides made in the presence of rifampcin, *Virology* **45,** 123.
Moss, B., Rosenblum, E. N., and Grimley, P. M., 1971*c*, Assembly of virus particles during mixed infection with wild-type vaccinia and a rifampicin-resistant mutant, *Virology* **45,** 135.
Moss, B., Rosenblum, E. N., Grimley, P. M., and Mims, S. J., 1972, Rifamycins: Modulation of specific anti-poxvirus activity by small substitutions on the piperazinyliminomethyl side chain, *Antimicrob. Ag. Chemother,* **2,** 181.
Moss, B., Rosenblum, E. N., and Garon, C. F., 1973*a*, Glycoprotein synthesis in cells infected with vaccinia virus. III. Purification and biosynthesis of the virion glycoprotein, *Virology,* **55,** 143.
Moss, B., Rosenblum, E. N., and Paoletti, E., 1973*b*, Polyadenylate polymerase from vaccinia virions, *Nature* (*Lond.*) **254,** 59.
Munyon, W., and Kit, S., 1965, Inhibition of thymidine kinase formation in LM (TK^-) cells simultaneously infected with vaccinia and a thymidine kinaseless vaccinia mutants, *Virology,* **26,** 374.
Munyon, W. H., and Kit, S., 1966, Induction of cytoplasmic ribonucleic acid synthesis in vaccinia-infected LM cells during inhibition of protein synthesis, *Virology* **29,** 303.
Munyon, W. E., Paoletti, E., and Grace, J. T., Jr., 1967, RNA polymerase activity in purified infectious vaccinia virus, *Proc. Natl. Acad. Sci. USA* **58,** 2280.
Munyon, W., Paoletti, E., Ospina, J., and Grace, J. T., Jr., 1968, Nucleotide phosphohydrolase in purified vaccinia virus, *J. Virol.* **2,** 167.
Munyon, W., Mann, J., and Grace, J. T., Jr., 1970, Protection of vaccinia from heat inactivation by nucleotide triphosphates, *J. Virol.* **5,** 32.
Murialdo, H., and Siminovitch, L., 1972, The morphogenesis of bacteriophage lambda. IV. Identification of gene products and control of expression of the morphogenetic information, *Virology* **48,** 785.
Nagayama, A., Pogo, B. G. T., and Dales, S., 1970, Biogenesis of vaccinia: Separation of early stages from maturation by means of rifampicin, *Virology* **40,** 1039.
Nagington, J., Newton, A. A., and Horne, R. W., 1964, The structure of orf virus, *Virology* **23,** 461.
Nishmi, M., and Keller, R., 1962, The microepidemiology of vaccinial infection as studied in HeLa cell stationary cultures, *Virology* **18,** 109.
Noyes, W. F., 1962*a*, The surface fine structure of vaccinia virus, *Virology* **17,** 282.
Noyes, W. F., 1962*b*, Further studies on the structure of vaccinia virus, *Virology* **18,** 511.
Obert, G., Tripier, F., and Guir, J., 1970, Arginine requirement for late mRNA transcription of vaccinia virus in KB cells, *Biochem. Biophys. Res. Commun.* **44,** 362.
Obijeski, J. J., Palmer, E. L., Gafford, L. G., Randall, C. G., 1973, Polyacrylamide gel electrophoresis of fowlpox and vaccinia virus proteins, *Virology* **51,** 512.
Oda, K., and Joklik, W. K., 1967, Hybridization and sedimentation studies on "early" and "late" vaccinia messenger RNA, *J. Mol. Biol.* **27,** 395.
Ohno, S., 1967, Studies on the mechanism of action of interferon: Inhibition of thy-

midine kinase induction in vaccinia virus-infected chick embryo cells, *J. Biochem.* (*Tokyo*) **61,** 277.

Oki, T., Fujiwara, Y., and Heidelberger, C., 1971, Utilization of host cell DNA by vaccinia virus replicating in HeLa cells irradiated intranuclearly with tritium, *J. Gen. Virol.* **13,** 401.

Olsen, R. G., and Yohn, D., 1970, Immunodiffusion analysis of Yaba poxvirus structural and associated antigens, *J. Virol.* **5,** 212.

Padgett, B. L., and Tomkins, J. K. N., 1968, Conditional lethal mutants of rabbitpox virus. III. Temperature-sensitive (ts) mutants; physiological properties, complementation and recombination, *Virology* **36,** 161.

Pagett, B. L., and Walker, D. L., 1970, Effect of persistent fibroma virus infection on susceptibility of cells to other viruses, *J. Virol.* **5,** 199.

Paoletti, E., and Moss, B., 1972*a*, Protein kinase and specific phosphate acceptor proteins associated with vaccinia virus cores, *J. Virol.* **10,** 417.

Paoletti, E., and Moss, B., 1972*b*, Deoxyribonucleic acid-dependent nucleotide phosphohydrolase activity in purified vaccinia virus, *J. Virol.* **10,** 866.

Paoletti, E., and Moss, B., 1974, Two nucleic acid-dependent nucleoside triphosphate phosphohydrolases from vaccinia virus: Nucleotide substrate and polynucleotide cofactor specificities, *J. Biol. Chem.* **249,** 3281.

Paoletti, E., Rosemond-Hornbeak, H., and Moss, B., 1974, Two nucleic acid-dependent nucleoside triphosphate phosphohydrolases from vaccinia virus: Purification and characterization, *J. Biol. Chem.* **249,** 3273.

Parker, R. F., 1938, Statistical studies of the nature of the infectious unit of vaccinia virus, *J. Exp. Med.* **67,** 725.

Parker, R. F., and Rivers, T. M., 1936, Immunological and chemical investigations of vaccine virus. IV. Statistical studies of elementary bodies in relation to infection and agglutination, *J. Exp. Med.* **64,** 439.

Pennington, T. H., 1973, A comparison of virus-induced antigens and polypeptides, *J. Gen. Virol.* **19,** 65.

Pennington, T. H., and Follett, E. A., 1971, Inhibition of poxvirus maturation by rifamycin derivatives and related compounds, *J. Virol.* **7,** 821.

Pennington, T. H., Follett, E. A., and Szilagyi, J. F., 1970, Events in vaccinia virus-infected cells following the reversal of the antiviral action of rifampicin, *J. Gen. Virol.* **9,** 225.

Peters, D., 1956, Morphology of resting vaccinia virus, *Nature* (*Lond.*) **178,** 1453.

Peters, D., and Müller, G., 1963, The fine structure of the DNA containing core of vaccinia virus, *Virology,* **21,** 266.

Pfau, C. J., and McCrea, J. F., 1962*a*, Release of deoxyribonucleic acid from vaccinia virus by 2-mercaptoethanol and pronase, *Nature* (*Lond.*) **194,** 894.

Pfau, C. J., and McCrea, J. F., 1962*b*, Some unusual properties of vaccinia virus deoxyribonucleic acid, *Biochim. Biophys. Acta* **55,** 271.

Pfau, C. J., and McCrea, J. F., 1963, studies on the deoxyribonucleic acid of vaccinia virus. II. Characterization of DNA isolated by different methods and its relation to virus structure, *Virology* **21,** 425.

Pickles, E. G., and Smadel, J. E., 1938, Ultracentrifugation studies on the elementary bodies of vaccine virus. I. General methods and determination of particle size, *J. Exp. Med.* **68,** 583.

Pitkanen, A., McAuslan, B., Hedgpeth, J., and Woodson, B., 1968, Induction of poxvirus ribonucleic acid polymerase, *J. Virol.* **2,** 1363.

Planterose, D. N. C., Nishimura, C., and Salzman, N. P., 1962, The purification of vaccinia virus from cell cultures, *Virology* **18,** 294.

Pogo, B. G. T., 1971, Biogenesis of vaccinia: Effect of rifampicin on transcription, *Virology* **4,** 576.

Pogo, B. G. T., and Dales, S., 1969, Two deoxyribonculease activities within purified vaccinia virus, *Proc. Natl. Acad. Sci. USA* **63,** 820.

Pogo, B. G. T., and Dales, S., 1971, Biogenesis of vaccinia: Separation of early stages from maturation by means of hydroxyurea, *Virology* **43,** 144.

Pogo, B. G. T., and Dales, S., 1973, Biogenesis of poxviruses: Inactivation of host DNA polymerase by a component of the invading inoculum particle, *Proc. Natl. Acad. Sci. USA* **70,** 1726.

Pogo, B. G. T., Dales, S., Bergoin, M., and Roberts, D. W., 1971, Enzymes associated with an insect poxvirus, *Virology* **43,** 306.

Polisky, B., and Kates, J., 1972, Vaccinia virus intracellular DNA–protein complex: Biochemical characteristics of associated protein, *Virology* **49,** 168.

Prescott, D. M., Kates, J., and Kirkpatrick, J. B., 1971, Replication of vaccinia virus DNA in enucleated L-cells, *J. Mol. Biol.* **59,** 505.

Prescott, D. M., Myerson, D., and Wallace, J., 1972, Enucleation of mammalian cells with cytochalasin B, *Exp. Cell Res.* **71,** 480.

Quintrell, N. A., and McAuslan, B. R., 1970, Inhibition of poxvirus replication by streptovaricin, *J. Virol.* **6,** 485.

Randall, C. C., Gafford, L. G., and Darlington, R. W., 1962, Bases of the nucleic acid of fowlpox virus and host deoxyribonucleic acid, *J. Bacteriol.* **83,** 1037.

Randall, C. C., Gafford, L. G., Soehner, R. L., and Hyde, J. M., 1966, Physicochemical properties of fowlpox virus deoxyribonucleic acid and its anomalous infectious behavior, *J. Bacteriol.* **91,** 95.

Roberts, R. B., Cowie, D. B., Abelson, P. H., Bolton, E. T., and Britten, R. J., 1955, "Studies of Biosynthesis in *Escherichia coli,*" p. 521, Carnegie Instn., Wash. D.C.

Rodriguez-Burgos, A., Chordi, A., Diaz, R., and Tormo, J., 1966. Immunoelectrophoretic analysis of vaccinia virus, *Virology* **30,** 569.

Rosemond, H., and Moss, B., 1973, Phosphoprotein component of vaccinia virions, *J. Virol.* **11,** 961.

Rosemond-Hornbeak, H., and Moss, 1974, Single-stranded DNA specific nuclease from vaccinia virus: Endonucleolytic and exonucleolytic activities, *J. Biol. Chem.* **249,** 3292.

Rosemond-Hornbeak, H., Paoletti, E., and Moss, B., 1974, Single-stranded DNA-specific nuclease V_1 from vaccinia virus: Purification and characterization, *J. Biol. Chem.* **249,** 3287.

Rosenkranz, H. S., Rose, H. M., Morgan, C., and Hsu, K. C. 1966, Effect of hydroxyurea on virus development. II. Vaccinia virus, *Virology* **28,** 510.

Salzman, N. P., 1960, The rate of formation of vaccinia deoxyribonucleic acid and vaccinia virus, *Virology* **10,** 150.

Salzman, N. P., and Sebring, E. D., 1967, Sequential formation of vaccinia virus proteins and viral deoxyribonucleic acid, *J. Virol.* **1,** 16.

Salzman, N. P., Shatkin, A. J., and Sebring, E. D., 1964, The synthesis of a DNA-like RNA in the cytoplasm of HeLa cells infected with vaccinia virus, *J. Mol. Biol.* **8,** 405.

Sambrook, J., and Shatkin, A. J., 1969, Polynucleotide ligase activity in cells infected with simian virus 40, polyoma virus or vaccinia virus, *J. Virol.* **4,** 719.

Sambrook, J. F., McClain, M. E., Easterbrook, K. B., and McAuslan, B. R., 1965, A mutant of rabbitpox virus defective at different stages of its multiplication in three cell types, *Virology* **26,** 738.

Sarov, I., and Becker, Y., 1967, Studies on vaccinia virus DNA, *Virology* **33,** 369.

Sarov, I., and Joklik, W. K., 1972*a*, Studies on the nature and location of the capsid polypeptides of vaccinia virions, *Virology* **50,** 579.

Sarov, I., and Joklik, W. K., 1972*b*, Characterization of intermediates in the uncoating of vaccinia virus DNA, *Virology* **50,** 593.

Sarov, I., and Joklik, W. K., 1973, Isolation and characterization of intermediates in vaccinia virus morphogenesis, *Virology* **52,** 223.

Scherrer, R., 1968, Viral and host deoxyribonucleic acid synthesis in Shope fibroma virus-infected cells as studied by means of high-resolution autoradiography, *J. Virol.* **2,** 1418.

Schwartz, J., and Dales, S., 1971, Biogenesis of poxviruses: Identification of four enzyme activities within purified Yaba tumor virus, *Virology* **45,** 797.

Sebring, E. D., and Salzman, N. P., 1967, Metabolic properties of early and late vaccinia virus messenger ribonucleic acid, *J. Virol.* **1,** 550.

Sharp, D. G., and McGuire, P. M., 1970, Spectrum of physical properties among the virions of a whole population of vaccinia virus particles, *J. Virol.* **5,** 275.

Sharp, D. G., Sadhukhan, G. P., and Galasso, G. J., 1964, Quality changes in vaccinia virus during adaptation to growth in cultures of Earle's L cells, *J. Bacteriol.* **88,** 309.

Shatkin, A. J., 1963*a*, The formation of vaccinia virus protein in the presence of 5-fluorodeoxyuridine, *Virology* **20,** 292.

Shatkin, A. J., 1963*b*, Actinomycin D and vaccinia virus infection of HeLa cells, *Nature* (*Lond.*) **199,** 357.

Shatkin, A. J., and Salzman, N. P., 1963, Deoxyribonucleic acid synthesis in vaccinia virus-infected HeLa cells, *Virology* **19,** 551.

Shatkin, A. J., Sebring, E. D., and Salzman, N. P., 1965, Vaccinia virus-directed RNA: Its fate in the presence of actinomycin, *Science* (*Wash., D.C.*) **148,** 87.

Sheek, M. R., and Magee, W. E., 1961, An autoradiographic study of the intracellular development of vaccinia virus, *Virology* **15,** 146.

Sheffield, F. W., Bauer, D. J., and Stephenson, S., 1960, The protection of tissue culture cells by isatin β-thiosemicarbazone from the cytopathic effects of certain poxviruses, *Brit. J. Exp. Pathol.* **41,** 638.

Sheldon, R., Jurale, C., and Kates, J., 1972*a*, Detection of polyadenylic acid sequences in viral and eukaryotic RNA, *Proc. Natl. Acad. Sci. USA* **69,** 417.

Sheldon, R., Kates, J., Kelley, D. E., and Perry, R. P., 1972*b*, Polyadenylic acid sequences on 3′ termini of vaccinia messenger ribonucleic acid and mammalian nuclear and messenger ribonucleic acid, *Biochemistry* **11,** 3829.

Shope, R. E., 1932*a*, A transmissible tumor-like condition in rabbits, *J. Exp. Med.* **56,** 793.

Shope, R. E., 1932*b*, A filtrable virus causing a tumor-like condition in rabbits and its relationship to virus myxomatosum, *J. Exp. Med.* **56,** 803.

Sime, E. H., and Bedson, H. S., 1973, A comparison of ultraviolet action spectra for vaccinia virus and T2 bacteriophage, *J. Gen. Virol.* **18,** 55.

Smadel, J. E., and Hoagland, C. L., 1942, Elementary bodies of vaccinia, *Bacteriol. Rev.* **6,** 79.

Smadel, J. E., Pickels, E. G., and Shedlovsky, T., 1938, Ultracentrifugation studies on the elementary bodies of vaccine virus. II. The influence of sucrose, glycerol, and urea solutions on the physical nature of vaccine virus, *J. Exp. Med.* **68,** 607.

Smadel, J. E., Rivers, T. M., and Pickels, E. G., 1939, Estimation of the purity of preparation of elementary bodies of vaccinia, *J. Exp. Med.* **70,** 379.

Smadel, J. E., Rivers, T. M., and Hoagland, C. L., 1972, Nucleoprotein antigen of vaccinia virus. I. A new antigen obtained from elementary bodies of vaccinia, *Arch. Pathol.* **34,** 275.

Stevenin, J., Peter, R., and Kirn, A., 1970, Action des temperatures supraoptemales sur la transcription du genome du virus vaccinal, *Biochim. Biophys. Acta* **199,** 363.

Subak-Sharpe, J. H., Burk, R. R. Crawford, L. V., Morrison, J. M., Hay, J., and Keir, H. M., 1966, An approach to the evolutionary relationships of mammalian DNA viruses through analysis of the pattern of nearest neighbor base sequences, *Cold Spring Harb. Symp. Quant. Biol.* **31,** 737.

Subak-Sharpe, J. H., Timbury, M. C., and Williams, J. F., 1969, Rifampicin inhibits the growth of some mammalian viruses, *Nature* (*Lond.*) **222,** 341.

Summers, D. F., Maizel, J. V., Jr., and Darnell, J. E., Jr., 1965, Evidence for virus-specific noncapsid proteins in poliovirus-infected HeLa cells, *Proc. Natl. Acad. Sci. USA* **54,** 505.

Sutton, W. D., 1971, A crude nuclease preparatio suitable for use in DNA reassociation experiments, *Biochim. Biophys. Acta* **240,** 522.

Szilagyi, J. F., and Pennington, T. H., 1971, Effect of rifamycins and related antibiotics on the deoxyribonucleic acid-dependent ribonucleic acid polymerase of vaccinia virus particles, *J. Virol.* **8,** 133.

Szybalski, W., Erikson, R. L., Gentry, G. A., Gafford, L. G., and Randall, C. C., 1963, Unusual properties of fowlpox virus DNA, *Virology* **19,** 586.

Takehara, M., 1970, Nucleic acid synthesis during the focus formation by shope fibroma virus on green monkey kidney cells, *Arch. Ges. Virusforsch.* **31,** 303.

Takehara, M., and Schwerdt, C. E., 1967, Infective subviral particles from cell cultures infected with myxoma and fibroma viruses, *Virology* **31,** 163.

Takahashi, M., Kameyama, S., Kato, S. and Kamahora, J., 1959, The immunological relationship of the poxvirus group, *Biken J.*, **2,** 27.

Tan, K. B., and McAuslan, B. R., 1970, Effect of rifampicin on poxvirus protein synthesis, *J. Virol.* **6,** 326.

Tan, K. B., and McAuslan, B. R., 1972, Binding of deoxyribonucleic acid-dependent deoxyribonucleic acid polymerase to poxvirus, *J. Virol.* **9,** 70.

Thiry, L., and Lancini, G., 1970, Inhibition of vaccinia virus by 1-methyl-4-aminopiperazine. *Nature* (*Lond.*) **227,** 1048.

Thompson, R. L., Davis, J., Russell, P. B., and Hitchings, G. H., 1953*a*, Effect of aliphatic oxime and isatin thiosemicarbazones on vaccinia infection in the mouse and in the rabbit, *Proc. Soc. Exp. Biol. Med.* **84,** 496.

Thompson, R. L., Minton, S. A., Officer, J. E., and Hitchings, G. H., 1953*b*, Effect of heterocyclic and other thiosemicarbazones on vaccinia infection in the mouse, *J. Immunol.* **70,** 229.

Tsuchiya, Y., and Tagaya, I., 1972*a*, General characteristics of enhanced plaque formation by poliovirus in poxvirus-infected cells, *J. Gen. Virol.* **14,** 229.

Tsuchiya, Y., and Tagaya, I., 1972*b*, Mechanism of enhanced plaque formation by poliovirus in poxvirus-infected cells, *J. Gen. Virol.* **14,** 237.

Tsuruhara, T., 1971, Immature particle formation of Yaba poxvirus studied by electron microscopy, *J. Natl. Cancer Inst.* **47,** 549.

Turner, G. S., and Squires, E. J., 1971, Inactivated smallpox vaccine: Immunogenicity of inactivated intracellular and extracellular virus, *J. Gen. Virol.* **13,** 19.

Ueda, Y., Ito, M., and Tagaya, I., 1969, A specific surface antigen induced by poxvirus, *Virology* **38,** 180.

Verna, J. F., 1965, Cell culture response to fibroma virus, *J. Bacteriol.* **89,** 524.
Vilagines, R., and McAuslan, B. R., 1970, Interference with viral messenger RNA and DNA synthesis by superinfection with a heterologous deoxyvirus, *Virology* **42,** 1043.
Walen, K. H., 1971, Nuclear involvement poxvirus infection, *Proc. Natl. Acad. Sci. USA* **68,** 165.
Wehrli, W., and Staehelin, M., 1971, Actions of the rifamycins, *Bacteriol. Rev.* **35,** 290.
Westwood, J. C. N., Harris, W. J. Zwartouw, H. T., Titmuss, D. H. J., and Appleyard, G., 1964, *J. Gen. Microbiol.* **34,** 67.
Westwood, J. C. N., Zwartouw, H. T., and Appleyard, G., 1965, Comparison of the soluble antigens and virus particle antigens of vaccinia virus, *J. Gen. Microbiol.* **38,** 47.
White, H. B., Jr., Powell, S. S., Gafford, L. G., and Randall, C. C., 1968, The occurrence of squalene in lipid of fowlpox virus, *J. Biol. Chem.* **243,** 4517.
Wilcox, W. C., and Cohen, G. H., 1967, Soluble antigens of vaccinia-infected mammalian cells. II. Time course of synthesis of soluble antigens and virus structural proteins, *J. Virol.* **1,** 500.
Wilcox, W. C., and Cohen, G. H., 1969, the poxvirus antigens, *Curr. Top. Microbiol.* **47,** 1.
Woodroofe, G. M., and Fenner, F., 1960, Genetic studies with mammalian poxviruses. IV. Hybridization between several different poxviruses, *Virology* **12,** 272.
Woodroofe, G. M., and Fenner, F., 1962, Serological relationships within the poxvirus group: An antigen common to all members of the group, *Virology* **16,** 334.
Woodruff, C. E., and Goodpasture, E. W., 1929, The infectivity of isolated inclusion bodies of fowlpox, *Am. J. Pathol.* **5,** 1.
Woodson, B., 1967, Vaccinia mRNA synthesis under conditions which prevent uncoating, *Biochem. Biophys. Res. Commun.* **27,** 169.
Woodson, B., 1968, Recent progress in poxvirus research, *Bacteriol. Rev.* **32,** 127.
Woodson, B., and Joklik, W. K., 1965, The inhibition of vaccinia virus multiplication by isatin β-thiosemicarbazone, *Proc. Natl. Acad. Sci. USA* **54,** 946.
Wyatt, G. R., and Cohen, S. S., 1953, The bases of the nucleic acids of some bacterial and animal viruses: The occurrence of 5-hydroxymethyl cytosine, *Biochem. J.* **55,** 774.
Yohn, D. S., and Gallagher, J. F., 1969, Some physiochemical properties of Yaba poxvirus deoxyribonucleic acid, *J. Virol.* **3,** 114.
Yohn, D. S., Marmol, F. R., and Olsen, R. G., 1970, Growth kinetics of Yaba tumor poxvirus after *in-vitro* adaptation to cercopithecus, *J. Virol.* **5,** 205.
Zakay-Rones, Z., and Becker, Y., 1970, Anti-poxvirus activity of rifampicin associated with hydrazone side chain, *Nature* (*Lond.*) **226,** 1162.
Zarling, J. M., and Tevethia, S. S., 1971, Expression of concanavalin A binding sites in rabbit kidney cells infected with vaccina virus, *Virology* **45,** 313.
Zwartouw, H. T., 1964, The chemical composition of vaccinia virus, *J. Gen. Microbiol.* **34,** 115.
Zwartouw, H. T., Westwood, J. C. N., and Appleyard, G., 1962, Purification of poxviruses by density gradient centrifugation, *J. Gen. Microbiol.* **29,** 523.
Zwartouw, H. T., Westwood, J. C. N., and Harris, W. J., 1965, Antigens from vaccinia virus particles, *J. Gen. Microbiol.* **38,** 39.
Zweig, M., and Cummings, D. J., 1973, Cleavage of head and tail proteins during bacteriophage T5 assembly: Selective host involvement in the cleavage of a tail protein, *J. Mol. Biol.* **80,** 505.

Addendum to Chapter 2

Reproduction of Papovaviruses

Norman P. Salzman and George Khoury

[The following material was added in proof.]

THE GENETIC APPROACH TO SV40 AND POLYOMA

It would appear from the discussion in previous sections that the interaction between papovaviruses and their host cells is quite complex. Yet the SV40 and polyoma virus genomes are small, with a coding capacity for three to seven proteins. By isolating and studying the properties of conditionally lethal mutants, it was hoped that particular biological properties could be directly related to the function of specific viral proteins, and that these functions, in turn, could be mapped on the viral genome. This approach, has been particularly successful in the last three years, largely through the efforts of several investigators who have isolated and characterized large numbers of temperature-sensitive (*ts*) mutants.

SV40 *ts* MUTANTS

At this time, more than 150 SV40 *ts* mutants have been isolated and partially characterized (Tegtmeyer *et al.*, 1970; Kit *et al.*, 1968; Tegtmeyer and Ozer, 1971; Robb and Martin, 1972; Kimura and Dulbecco, 1972, 1973; Chou and Martin, 1974, Dubbs *et al.*, 1974).* These mutants can be organized into five classes, on the basis of

* Some of the references in this Addendum will be found in the list at the end of Chapter 2 (pages 126–141), others in a supplementary list at the end of the Addendum.

physiological function and/or their behavior in complementation assays.

Group A

The *ts* mutants of complementation group A appear to be defective in an early viral function. At the nonpermissive temperature they are able to infect cells and synthesize early virus-specific RNA (Cowan *et al.*, 1973; R. Saral, G. Khoury, J. Chou, and R. Martin, unpublished results), but T-antigen appears to be abnormal (Tegtmeyer; Osborn and Weber, personal communications).

Furthermore, the group A mutants are neither able to induce host cell (Chou and Martin, personal communication) nor SV40 (Tegtmeyer, 1972) DNA synthesis. If, however, a cycle of viral DNA synthesis is initiated at the permissive temperature and the cells are then shifted to the nonpermissive temperature, that cycle of DNA synthesis is completed (Tegtmeyer, 1972) and late virus-specific RNA synthesis continues (Cowan *et al.*, 1973). Group A mutants are able to transform a variety of cells at the permissive temperature but do not transform these cells at the nonpermissive temperature. Of particular interest are the recent finding with cells that have been transformed by group A *ts* mutants at the permissive temperature. When these transformed cells are shifted to the restrictive temperature, certain transformed phenotypic properties may revert (R. Tegtmeyer; R. Martin, J. Chou, J. Avila, and R. Saral; J. Butel, J. S. Brugge, and C. A. Noonan, personal communications).

The mutants of group A are not thermolabile and infection under nonpermissive conditions does not result in the production of detectable structural proteins (Tegtmeyer and Ozer, 1971; Chou and Martin, personal communication). Finally, in elegant marker rescue experiments Lai and Nathans (*Virology*, in press) showed that these group A mutants map in *Hin* fragments H and I, which are located in the middle of the early SV40 region (see Fig. 12). Whether there are one or more subgroups within group A is still a matter of speculation. However, the recent finding of Kimura (1974) suggest there may be at least two subgroups. His results show that a particular group A mutant could not assist the replication of adenovirus in monkey cells (see Sect. 4.6). These results are in contrast to those of Jerkofsky and Rapp (1973), who found an early SV40 mutant which did provide the helper function at the nonpermissive temperature. One possible explanation for these results would be the presence of two distinct early subgroups.

Whether the difference in these early mutants relates to their ability to induce host cell DNA synthesis is presently under investigation.

In summary, then, it seems likely that group A mutants represent a defect in the early SV40 gene region which codes for an early viral protein(s). The early region now appears to represent 45–50% of the genome (see Sect. 3.7) and therefore could code for a protein as large as T-antigen. At any rate, the early function(s) appears to directly or indirectly induce the synthesis of host cell DNA, viral DNA, and in turn, late viral mRNA. A candidate protein would most likely be a DNA-binding protein, and such properties have recently been ascribed to T-antigen (R. Carrol; P. Tegtmeyer, personal communications). This protein might be a polymerase, an unwinding protein and/or an endonuclease, and considerable effort is presently being devoted to its purification and characterization.

Groups B, C, and BC

SV40 *ts* mutants of groups B and C (complementing) and BC (noncomplementing) are thermolabile and induce the synthesis of an altered V-antigen or capsid protein (Tegtmeyer *et al.*, 1970; Tegtmeyer and Ozer, 1971; Kimura and Dulbecco, 1972; Dubbs *et al.*, 1974; Chou and Martin, 1974). Since they have been shown by marker rescue experiments (Lai and Nathans, *Virology*, in press) to map in a relatively circumscribed late region of the genome corresponding to *Hin* fragments F, J, and G (see Fig. 12), it is thought that they might all contain a defect in the major capsid protein, VP1. As might be expected, the early viral functions appear normally in cells infected with these mutants and cells can be transformed at the nonpermissive temperature by them.

Group D

Mutants of this group proved difficult to study because of their "leakiness" at the nonpermissive temperature. Nevertheless, extensive studies with these mutants, and in particular *ts* 101, have shown that this group is essentially noncomplementing. While infection with viral DNA leads to a normal lytic cycle and the production of temperature-sensitive progeny, viral infection is itself blocked at a very early stage (Robb and Martin, 1972; Chou *et al.*, 1974; Chou and Martin, 1974). The virus is adsorbed, but there is no subsequent synthesis of early

SV40 mRNA (R. Saral, G. Khoury, J. Chou, and R. Martin; unpublished results), early viral proteins, or viral DNA. In addition the "D function" is necessary only very early in the lytic cycle, since infection of cells at the permissive temperature followed by a shift to the restrictive temperature after 10 to 20 hours results in no reduction in the rate of viral DNA synthesis. From a consideration of all these properties, it seems most likely that the D mutants are blocked at the restrictive temperature in some stage of uncoating. Thus one might conclude that the D-protein is structural (either a capsid-, or perhaps more likely, a core-associated protein), and the removal of this protein is necessary for expression of the viral genome. Nathans and Lai have mapped a D mutant in *Hin* fragment E (Fig. 12) which, as might be expected, is located in the late region of the viral genome.

POLYOMA *ts* MUTANTS

As was the case for SV40, a number of temperature-sensitive mutants of polyoma have been isolated from viral stocks treated with chemical mutanges (Fried, 1965; DiMayorca *et al.*, 1969; Eckhart, 1969).

These mutants can be classified into four or five groups on the basis of physiological or complementation experiments. In most respects the groups of polyoma mutants are analogous to those of SV40 and will be discussed only briefly.

Groups I and IV

The polyoma mutants of these groups are late mutants, analogous to the group B and C SV40 *ts* mutants. At the nonpermissive temperature early functions appear to be normal but there are detectable defects in the synthesis of structural proteins as demonstrated either by the absence of V-antigen in infected cells or by heat lability of the mutant virion capsids. Group I and IV mutants will complement each other at restrictive temperatures.

Groups II and III

These early polyoma mutants are in many respects analogous to the SV40 group A mutants. They are unable to synthesize viral DNA at nonpermissive temperatures, and at least group II mutants are defective in the function necessary for establishment of transformation. It is not yet certain that group III mutants can transform at the restrictive temperature (the only characteristic which separates them from group II mutants) or whether this phenomenon is simply a result of multiplicity-dependent leakiness (Eckhart, 1969; Oxman *et al.*, 1972).

At the present time, it appears that the group II function is not necessary for maintenance of the transformed state; BHK-21 cells transformed at the permissive temperature by a group II mutant (*ts-a*) do not appear to lose their phenotypic characteristics of transformation when shifted to the restrictive temperature (Fried, 1965; Eckhart, 1969; DiMayorca *et al.*, 1969). This property is extremely important; it suggests that once transformation has been initiated, the continued synthesis of the viral function which was necessary for initiation of transformation is no longer required. In the light of recent contrasting data which suggest that the early SV40 function may be necessary for maintenance of the transformed state, these early polyoma mutants deserve further examination.

Group V

Mutant *ts*-3 is representative of polyoma group V mutants and has certain properties which resemble the SV40 D mutants. Virions are not able to induce host cell DNA synthesis or early viral functions at the nonpermissive temperature (Dulbecco and Eckhart, 1970), yet viral DNA is fully infectious under these same conditions (Eckhart, unpublished results). On the other hand, BHK cells transformed by *ts*-3 at the permissive temperature lose certain of their transformed characteristics (morphology, wheat germ agglutinin sites, and topoinhibition) (Eckhart *et al.*, 1971) when shifted to the nonpermissive temperature, while other properties of the transformed phenotype (e.g., ability to grow in soft agar) are retained. These temperature-sensitive characteristics of transformation are somewhat analogous to the properties of SV40 group A mutants. It is possible, therefore, that *ts*-3 is a double mutant, and with recent advances in mapping of the polyoma genome and the application of marker rescue studies to animal virus systems this question should be answered in the near future.

In summary, temperature-sensitive mutants of SV40 and polyoma have been of tremendous value in the investigation of the viral contribution to virus–cell interactions. There is no doubt that they will continue to be useful in approaching many of the still unanswered questions related to the lytic cycle and cell transformation.

REFERENCES

Chou, J. Y., Avila, J., and Martin, R. G., 1974, Viral DNA synthesis in cells infected by temperature sensitive mutants of SV40, *J. Virol.* **14,** 116.

Chou, J. Y., and Martin, R. G., 1974, Complementation analysis of SV40 mutants, *J. Virol.* **13,** 1101.

DiMayorca, G., Callender, J., Marin, G., and Giordano, R., 1969, Temperature-sensitive mutants of polyoma virus, *Virology* **38,** 126.

Dubbs, D. R., Rachmeler, M., and Kit, S., 1974, Recombination between temperature sensitive mutants of SV40, *Virology* **57,** 161.

Dulbecco, R., and Eckhart, W., 1970, Temperature-dependent properties of cells transformed by a thermosensitive mutant of polyoma virus, *Proc. Natl. Acad. Sci., USA* **67,** 1775.

Eckhart, W., 1969, Complementation and transformation by temperature-sensitive mutants of polyoma virus, *Virology* **38,** 120.

Eckhart, W., Dulbecco, R., and Burger, M., 1971, Temperature-dependent surface changes in cells infected or transformed by a thermosensitive mutant of polyoma virus, *Proc. Natl. Acad. Sci. USA* **68,** 283.

Fried, M., 1965, Isolation of temperature-sensitive mutants of polyoma virus, *Virology* **25,** 669.

Kimura, G. and Dulbecco, R., 1972, Isolation and characterization of temperature-sensitive mutants of SV40, *Virology* **49,** 394.

Kimura, G., and Dulbecco, R., 1973, A temperature-sensitive mutant of SV40 affecting transforming ability, *Virology* **52,** 529.

Kimura, G., 1974, Genetic evidence for SV40 gene function in enhancement of human adenovirus in similar cells, *Nature* **248,** 590.

Tegtmeyer, P., Dohan, C. Jr., and Reynikoff, C., 1970, Inactivating and mutagenic effects of nitrosoguanidine on SV40, *Proc. Natl. Acad. Sci. USA* **66,** 745.

Tegtmeyer, P., and Ozer, H. L., 1971, Temperature sensitive mutants of SV40: Infection of permissive cells, *J. Virol.* **8,** 516.

Index

MIX
Papier aus verantwortungsvollen Quellen
Paper from responsible sources
FSC® C105338

If you have any concerns about our products,
you can contact us on
ProductSafety@springernature.com

In case Publisher is established outside the EU,
the EU authorized representative is:
Springer Nature Customer Service Center GmbH
Europaplatz 3, 69115 Heidelberg, Germany

Printed by Libri Plureos GmbH
in Hamburg, Germany